Petroleum Engineering

Editor-in-Chief

Gbenga Oluyemi, Robert Gordon University, Aberdeen, Aberdeenshire, UK

Series Editors

Amirmasoud Kalantari-Dahaghi, Department of Petroleum Engineering, West Virginia University, Morgantown, WV, USA

Alireza Shahkarami, Department of Engineering, Saint Francis University, Loretto, PA, USA

Martin Fernø, Department of Physics and Technology, University of Bergen, Bergen, Norway

The Springer series in Petroleum Engineering promotes and expedites the dissemination of new research results and tutorial views in the field of exploration and production. The series contains monographs, lecture notes, and edited volumes. The subject focus is on upstream petroleum engineering, and coverage extends to all theoretical and applied aspects of the field. Material on traditional drilling and more modern methods such as fracking is of interest, as are topics including but not limited to:

- Exploration
- Formation evaluation (well logging)
- Drilling
- Economics
- Reservoir simulation
- Reservoir engineering
- Well engineering
- Artificial lift systems
- Facilities engineering

Contributions to the series can be made by submitting a proposal to the responsible publisher, Anthony Doyle at anthony.doyle@springer.com or the Academic Series Editor, Dr. Gbenga Oluyemi g.f.oluyemi@rgu.ac.uk.

More information about this series at http://www.springer.com/series/15095

Theis Solling · Muhammad Shahzad Kamal ·
Syed M. Shakil Hussain
Editors

Surfactants in Upstream E&P

Springer

Editors
Theis Solling
Center for Integrative Petroleum Research (CIPR)
King Fahd University of Petroleum and Minerals
Dhahran, Saudi Arabia

Muhammad Shahzad Kamal
Center for Integrative Petroleum Research (CIPR)
King Fahd University of Petroleum and Minerals
Dhahran, Saudi Arabia

Syed M. Shakil Hussain
Center for Integrative Petroleum Research (CIPR)
King Fahd University of Petroleum and Minerals
Dhahran, Saudi Arabia

ISSN 2366-2646 ISSN 2366-2654 (electronic)
Petroleum Engineering
ISBN 978-3-030-70025-6 ISBN 978-3-030-70026-3 (eBook)
https://doi.org/10.1007/978-3-030-70026-3

This Springer imprint is published by the registered company Springer Nature Switzerland AG
The registered company address is: Gewerbestrasse 11, 6330 Cham, Switzerland

Preface

In principle the history of surfactants can be traced all the way back to the Babylonians approximate 2800 BC. At the time there was of course no preexisting knowledge of molecular structure but yet the empirical observation that certain substances, usually what resulted from boiling animal fat with ash, are able to release grease from for example kitchen utensils. The molecules that are active in such soaps are the fatty acids released as potassium carboxylates then lipids undergo saponification with the basic potassium oxides that are present in most ashes. They have a non-polar hydrocarbon end that mixes well with grease and a carboxylate group ($-COO^-$) that mixes well with water and it is this bipolarity that results in a "grease wash of". The bipolar property of a molecule is, indeed, what is characterizing the class of compounds that is generally referred to as "surfactants". The first discovery of a compound that if is working effectively in reducing surface tension was most like done by Neergard in 1929. He did not classify his mixture as a surfactant, that term arose in the 60s as a contraction of surface active agent. The class of compounds since then gained traction for application in most branches of science and in most industries. Here the focus is on the application of surfactants in the upstream, and some might say midstream, sector(s) of the oil and gas industry. This is admittedly not the first edited book deals with the topic; there have been contributions that focus on enhanced oil recovery where surfactants are instrumental in changing the wetting state of the reservoir and in reducing the interfacial tension between oil and water so that (in both cases) the oil flows more easily in the reservoir. Here we have taken a broader approach in the sense that we have attempted to address a range of the most important challenges in key upstream thrust areas. We attempt to cover an individual thrust area from several different angles in each of the 10 chapters that comprise the book. The chapters that make up this compilation are delivered by a broad spectrum of academicians and industrial contributors alike to capture an interdisciplinary vibe across the field of application of surfactant and surfactant research in upstream.

Dhahran, Saudi Arabia

Theis Solling
Muhammad Shahzad Kamal
Syed M. Shakil Hussain

Contents

Synthesis

Design and Synthesis of Low Molecular Weight and Polymeric Surfactants for Enhanced Oil Recovery

Patrizio Raffa

Abstract Surfactants are defined as molecules able to lower the surface (or interfacial) tension at the gas/liquid, liquid/liquid, and liquid/solid interfaces. Due to their properties, they are typically employed as detergents, emulsifiers, dispersants, wetting and foaming agents. In chemical enhanced oil recovery (cEOR), surfactants are used as flooding agents, alone or in combination with polymers, alkali, and more recently nanoparticles, to increase the microscopic displacement efficiency. From a chemical point of view, surfactants are amphiphiles, meaning that they bear in their structure both hydrophilic and hydrophobic moieties. Some naturally occurring surfactants exists, but the majority are synthetic. The availability of synthetic surfactants, allows a big variety of structures and properties. In this chapter, the main classes of surfactants will be reviewed, with focus on those used or proposed for use for chemical enhanced oil recovery. After a general introduction about surfactants and their main structural and physico-chemical properties, specific aspects of design and synthesis will be discussed. Particular emphasis will be given to the most recent developments, which includes zwitterionic, gemini and polymeric surfactants. Own work of the author of this chapter in the field of polymeric surfactants will be highlighted.

1 Introduction

An Italian legend is believed by many to be at the origin of the name "soap". In ancient Rome (0–300 AD) people used to bring animals carcasses on a mount Sapo, to be burned as sacrifice to the gods. At the first rain, the remains were washed down to a river nearby. The water of that river resulted to be particularly effective in washing clothes. The animal's fats (triglycerides) were supposedly reacting with the (alkaline) wood ashes, to form salts of fatty acids. These molecules were excellent cleaning agents when dissolved in water, and we now know them as soaps, from the

P. Raffa (✉)
University of Groningen, Groningen, The Netherlands
e-mail: p.raffa@rug.nl

T. Solling et al. (eds.), *Surfactants in Upstream E&P*, Petroleum Engineering,
https://doi.org/10.1007/978-3-030-70026-3_1

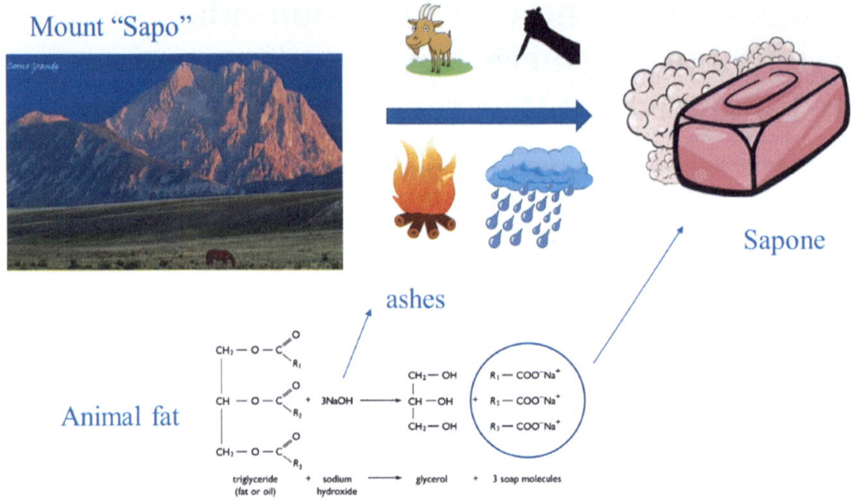

Fig. 1 Representation of Mount Sapo legend

name of the mountain (in latin: *sapone*). A schematic representation of the legend and the corresponding chemical reaction is reported in Fig. 1.

The story is almost certainly false (there is no record of a mountain with such a name, for example), but it tells us that ancient romans already knew how to make soap. A more probable origin of the name comes from the word *sebum*, which means tallow. Indeed, similarly to what told by the legend, in ancient times, soap was made by reacting tallow (beef or mutton fat) with ashes. Salts of fatty acids are good detergents because they contain in their structure a hydrophilic moiety and a hydrophobic one. In other words, they are amphiphilic. Because of their ability to lower the surface tension of water, they are also commonly known as surfactants.

The ongoing COVID-19 pandemic is harshly reminding all humanity about the importance of surfactants in the modern society. One of the most significant measures adopted worldwide to prevent the diffusion of the infection, is the frequent use of surfactants to wash hands and sanitize surfaces and objects.

Surfactants are not only used as detergents, but they are included in the formulation of many industrial products, such as emulsifiers, paints, wetting agent, dispersants, foaming agents [1]. In this chapter we will look more closely at surfactants used or designed for use as flooding agent for chemical enhanced oil recovery (cEOR) [2, 3].

In the context of cEOR, surfactants are used alone or in combination with polymers (SP flooding), alkali (AS and ASP flooding) [4–7], and more recently nanoparticles [8–11], to increase the microscopic displacement efficiency. This is accomplished mainly via reduction of capillary forces, responsible of trapping the oil in the small pores of the reservoir rocks, with subsequent reduction of residual oil saturation. In some surfactant flooding processes, emulsification of the oil occurs, with the formation of a viscous emulsion phase, which increases mobilization of the oil [12].

Additionally, surfactants alter the wetting properties of the reservoir and emulsify the oil, both phenomena contributing positively to oil recovery. In the design and synthesis of surfactants for application in cEOR, several important factors need to be taken into account, such as reservoir temperature, salinity and pH, rock permeability, formation type, adsorption, costs and efficiency. For these reasons, the optimal choice of a surfactant will be very dependent on the particular reservoir where it should be employed. However, from the vast amount of literature available, some general criteria can be established for their design [13, 14].

2 Main Properties of Surfactants

Numerous books and reviews have been published about general surfactants properties. In this section, content from selected sources will be very briefly summarized [15–19].

2.1 *Adsorption at Interfaces and Surface Excess*

A surfactant molecule is commonly constituted by a polar (hydrophilic) head attached to an elongated apolar (hydrophobic) tail (Fig. 2a). When such molecules are dissolved in water, they will preferentially migrate at the air/water interface, where they can "stick" their tails out of direct contact with water (Fig. 2b). In a two-phase water/oil system, they will have analogous behavior (Fig. 2c), with the difference that molecules will be also dissolved in the oil phase.

At a microscopical level interfaces are not plane, but they occupy a finite volume, where properties of one phase change gradually into properties of the other. However, they can mathematically be treated as geometric planes (Gibbs dividing surface) [15, 18], therefore an interface will have no volume. If we dissolve n_s moles of surfactant in the water/oil system, they will be distributed in the three phases: water (w), oil (o) and interface (σ) as in Eq. 1:

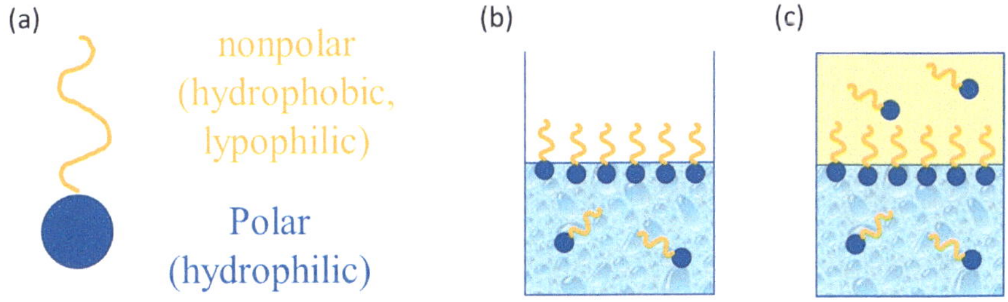

Fig. 2 Representation of surfactants and their adsorption at interfaces

$$n_s = n_s^w + n_s^o + n_s^\sigma \quad (1)$$

Rearranging for n_s^σ, we have:

$$n_s^\sigma = n_s - (c_s^w V_w + c_s^o V_o) \quad (2)$$

We have used here the assumption that the interface has no volume ($V = V_w + V_o$). Equation 2 shows that if a substance adsorb at the interface ($n_s^\sigma > 0$), then the concentration of surfactant in the water and oil phase (c_s^w and c_s^o respectively) would be lower than expected, considering the total number of moles introduced. For surfactants, this value is non negligible. When normalized to the interfacial area, it takes the name of surface excess $\Gamma_s = n_s^\sigma / A$, by definition expressed as mol/m^2. Adsorption at the interface of immiscible phases has several consequences on the system properties, such as lowering of surface (interfacial) tension and altering surface elasticity and rheology. Analogously, adsorption on solid surfaces alters wetting and adhesion properties. This is very relevant for cEOR applications, as adsorption on rocks plays a major role in the selection of the proper surfactant.

2.2 *Surface/Interfacial Tension*

Surface tension (or energy) of a liquid, can be defined as the energy required to increase the surface of that liquid per unit surface, against air [18]. It is generally expressed by the symbol γ and measured in mN/m (equivalent to mJ/m^2). Existence of surface tension comes from the unbalance between interaction forces (dipolar and van der Waals) experienced by molecules from their "neighbors", which number is different at the surface and in the bulk (Fig. 3). This difference in energy tends to minimize the surface, and it is at the origin of γ. Analogous reasoning can be made for interfaces between immiscible liquids, but interactions between molecules from both liquids need to be considered. In this case we talk about interfacial tension (IFT). The parameter considered of importance for EOR is the IFT between water phase and crude oil.

Another important parameter used to define surfactants is the surface excess. As surfactants molecules are dissolved in water, they will adsorb at the interface and their surface excess (Γ_s) will increase. This causes lowering of the surface tension, according to the Gibbs isotherm, which can be derived by the Gibbs–Duhem equation (Eq. 3) applied at the interface [15]:

$$dG^\sigma = V^\sigma dP - S^\sigma dT + Ad\gamma + \sum n_i^\sigma d\mu_i \quad (3)$$

If we consider equilibrium condition and constant temperature (isotherm), the terms dG^σ (the Gibbs free energy at the interface) and SdT are = 0, as well as the term VdP (the interface has no volume by definition). Therefore Eq. 3 can be simplified to:

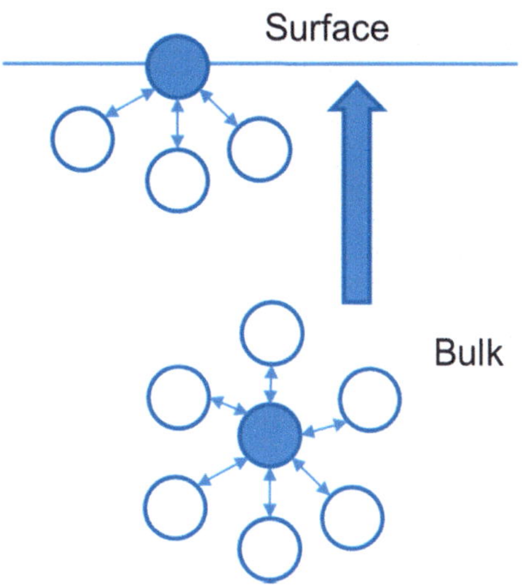

Fig. 3 Representation of interaction forces between molecules at the surface and in the bulk

$$-Ad\gamma = \sum n_i^{\sigma} d\mu_i \tag{4}$$

$$d\gamma = -\sum \frac{n_i^{\sigma}}{A} d\mu_i = -\sum \Gamma_i d\mu_i \tag{5}$$

In Eq. 5 the definition of surface excess Γ given above is used. In a binary system water + surfactant, if we chose the Gibbs dividing surface in a way that $\Gamma_w = 0$, we have:

$$d\gamma = -\Gamma_s d\mu_s = -\Gamma_s d(RTln[s]) \tag{6}$$

$$\Gamma_s = -\frac{1}{RT}\left(\frac{d\gamma}{d(\ln[s])}\right) = -\frac{[s]}{RT}\left(\frac{d\gamma}{d[s]}\right) \tag{7}$$

Equation 7 is known as Gibbs isotherm, and indicates that adsorption at the interface results in lowering of the surface tension. Most water soluble substances adsorb at the water/air interface; however, surfactants have very high values of Γ_s compare to regular solutes, therefore the decrease in surface tension will be much more pronounced.

2.3 Self-assembly of Surfactant and Critical Micellar Concentration

Curves for Γ_s and γ as a function of concentration of a generic surfactant in water solution looks like the ones in Fig. 4.

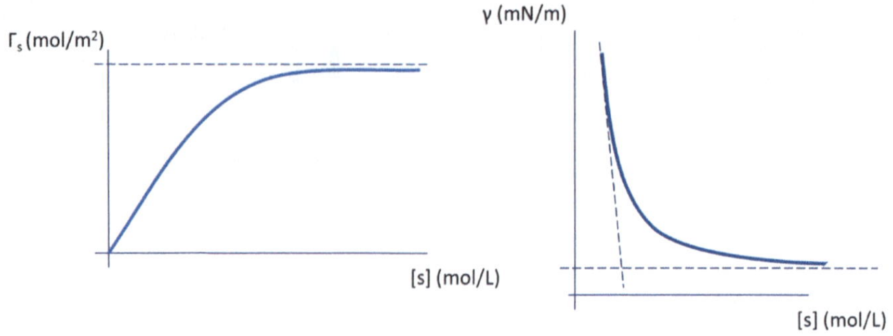

Fig. 4 Surface excess (left) and surface tension (right) of water solutions of a surfactants

From both curves it is possible to observe an asymptotic behavior. As surfactant is added in solution, it will adsorb at the water surface, increasing its surface excess (Fig. 4, left side), and lowering the surface tension of water (Fig. 4, right side). As the concentration increases, we can observe that the surface gets "crowded" with surfactants molecules and the adsorption is reduced, until it reaches a saturation value and additional molecules of surfactants added will not change values of Γ and γ anymore. The excess molecules will self-assemble into aggregates, constituted by several molecules arranged in order to exclude solvation water form the hydrophobic tails. These aggregates can have complex shapes, depending on several factors, such as the critical packing parameter (discussed later), solubility and concentration. The most simple aggregates are spherical and they are generally called micelles [16]; other aggregates include vesicles, reverse micelles and membranes. According to a thermodynamic model of micellization, the phenomena occurs spontaneously when a certain concentration is reached in solution. This goes under the name of critical micellar concentration (CMC), and it will depend on surfactant structure but also to parameters such as salinity of water and temperature. This concentration corresponds approximately to the intersection of dotted lines in the graph in Fig. 4, right side. It is important to keep in mind that micellization is a dynamic equilibrium, therefore there is continuous exchange of molecules between micelles and solution. The same happens to molecules at the surface/interface.

2.4 Critical Packing Parameter

One important quantity to take into account in the design of surfactants for many applications, including cEOR, is the critical packing parameter (CPP). This depends on the structure of the surfactants and it can be measured as the ratio between the cross sections of the polar and apolar moieties of the molecules, respectively, as illustrated in Fig. 5 [19]. Which kind of aggregates will be formed in solution by a certain surfactant will be largely determined by the CPP. The most common aggregate

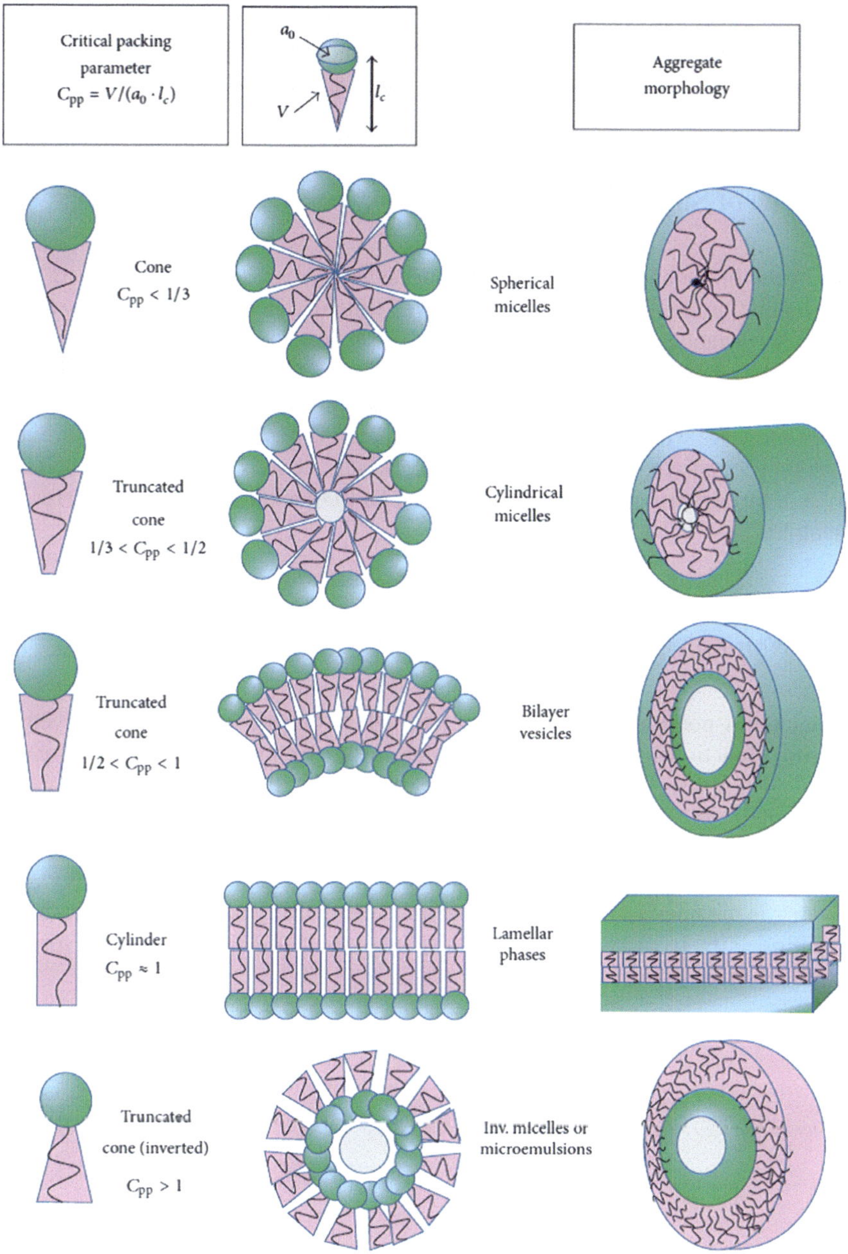

Fig. 5 Dependence of self-assembled micellar structures from critical packing parameter. Reproduced with permission from [19]

is spherical micelles, but worm-like micelles, vesicles, membranes and so on, are all possible (Fig. 5). Also the adsorption on surfaces and emulsion stabilization are governed to some extent form the CPP.

3 Type of Surfactants

The most important classification of surfactant is based on the nature of the hydrophilic group. Therefore, we can have nonionic, cationic, anionic or zwitterionic surfactants. The hydrophobic group is almost always constituted by long alkyl chains, sometimes containing aromatic rings. In the context of cEOR, other important classes of surfactants are gemini surfactants, biosurfactants and polymeric surfactants. We will now briefly discuss all these classes, in a general context and mention general synthetic methods [15, 20]. In later sections, we will illustrate more in detail surfactants actually used or proposed for EOR applications, including recent examples.

3.1 Nonionic Surfactants

Polar groups not bearing a charge are most likely containing hydroxyl, polyether or amino functionalities. The most common classes of nonionic surfactants are so-called ethoxylates. These molecules generally possess a hydrophobic chain attached to few ethoxy (CH_2CH_2O) units and terminating with a hydroxyl group. Depending on the hydrophobic chain, they can be alcohol ethoxylates, alkyl phenol ethoxylates or fatty acid ethoxylates (Fig. 6). Alkyl ethoxylates are sometimes labeled as C_xE_y, where x is the number of carbons in the alkyl chain and y is the number of ethoxy repeating units. Nonionic surfactants are usually less effective in reducing surface

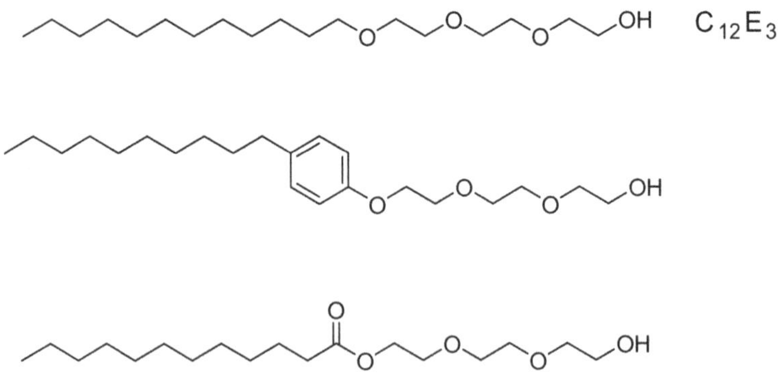

Fig. 6 Examples of nonionic surfactants

tension of charged ones and their CMC are 1 or 2 order of magnitude lower than a corresponding ionic one. A common feature of nonionic surfactants is the presence of a cloud point of their water solutions as the temperature increases. This is due to the lowered solubility of ethoxy chains with temperature, due to weakening of hydrogen bonds with water. Surfactant properties such CMC, surface tension, aggregation number and micellar size can be tuned by varying the length of the hydrophobic and hydrophilic groups.

These kind of ethoxylated nonionic surfactants can easily be produced by reaction of fatty alcohols or acid with ethylene oxide, via ring opening reaction at 150–180 °C and 2–5 atm of pressure, initiated by a base catalyst, such as NaOH or KOH [21]. As it is not easy to control the number of ethoxylated units attached, the products are generally constituted by mixtures [15].

Nonionic surfactants are generally non-toxic and less irritating for the skin compared to charged ones, therefore their main application is in detergents and skin care products. They are often used in combination with charged surfactants, often showing synergistic effects.

As discussed later, they can also be employed in cEOR, usually in combination with other chemicals. A special class of nonionic surfactants is constituted by poloxamers, that are block copolymers of ethylene oxide (PEO) and propylene oxide (PPO). As the latter are polymeric surfactants, these will be discussed later in more detail.

3.2 *Cationic Surfactants*

Cationic surfactants are typically quaternary ammonium salts, with one or more long alkyl chains (typically a C8–C18) as substituents (Fig. 7). The number of long chains can be varied to switch solubility from water (one chain) to oil (two or more chains). Cationic surfactants adsorb on negatively charged surfaces and often display antibacterial activity. Unlike anionic surfactants (next section), they can tolerate salinity, and are not negatively affected by water hardness. Cationic surfactants are used to a minor extent for EOR application, mostly in mixture with anionic ones [22]

The most common way to prepares quaternary ammonium surfactants is by alkylation of tertiary amines. The procedure is very simple, requiring just the mixing of the two reagent is an opportune refluxing solvent. A relatively recent application of this reaction in the synthesis of various cationic surfactants [23] is shown in Fig. 8.

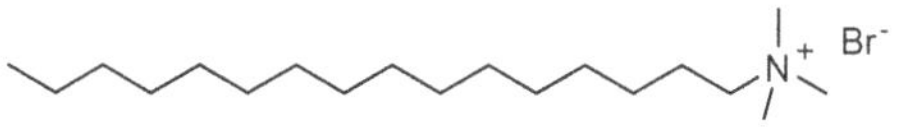

Fig. 7 CTAB (cetyl trimethylammonium bromide), a typical cationic surfactant

Fig. 8 Example of synthesis of quaternary ammonium cationic surfactants

3.3 Anionic Surfactants

These are the most common kind of surfactants, due to their relatively low costs and easy synthetic methods. The hydrophilic group is generally constituted by a carboxylate, sulfate, sulfonate or phosphate functionality (Fig. 9).

Alkyl carboxylate, also known as fatty acids (or salts), are the earliest known surfactants (see introduction) and can be directly produced from animals or plants fats by a process called saponification, which is a base-catalyzed hydrolysis (see Fig. 1). Plant based fatty acids also present unsaturations in the alkyl chain, which allows for further chemical transformations. The main drawback of carboxylate surfactants, is their low tolerance to hardness. In fact, in presence of divalent cations such as Ca^{2+} and Mg^{2+}, their form insoluble salts.

Sulfates are prepared by reaction of alcohols with sulfuric acid, or more commonly chlorosulfonic acid (Fig. 10a). Sodium dodecyl sulfate (SDS) belongs to this class and it is possibly the most studied and used industrial surfactant. Sulfonates have the sulfur atom directly attached to a carbon atom, which makes them more resistant to hydrolysis. These can be easily prepared by direct sulfonation of aromatic compounds, such as alkylbenzenes or naphthalene derivatives (Fig. 10b), or by sulfonation of terminal or internal alkenes.

Either organic sulfates or sulfonates can present some ethoxylated units, to make them more soluble in high salinity conditions. ENORDET, a series of surfactants

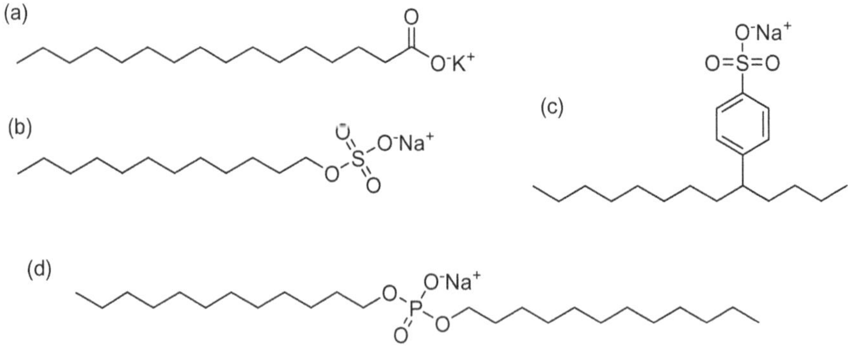

Fig. 9 Examples of anionic surfactants: carboxylate (**a**), sulfonate (**b**), sulfate (**c**) and phosphate (**d**)

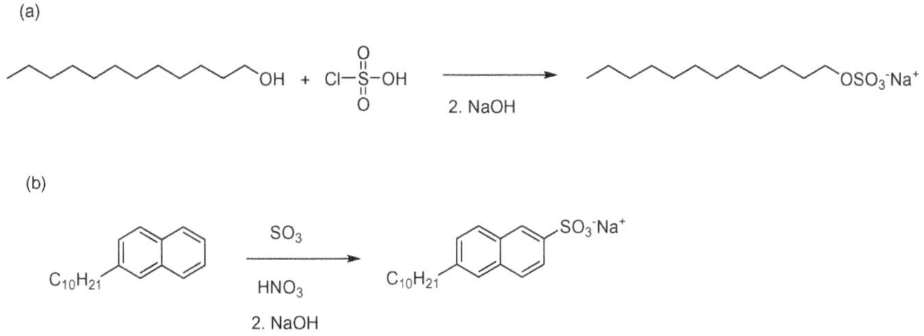

Fig. 10 Preparation of sulfates (**a**) and arylsulfonates (**b**) surfactants

used by Shell for cEOR, belong to this class of surfactants. These surfatctants will be discussed later.

Alkyl or arylalkyl phosphates (Fig. 10d) are analogous to sulfates, prepared by reaction of phosphorous oxides with fatty alcohols. Notable examples of these surfactant are phospholipids, natural surfactants that are the main constituents of cell membranes.

3.4 *Zwitterionic Surfactants*

Zwitterionic, or amphoteric surfactants, possess both a positive and a negative charge [17]. The positive group is generally a quaternary ammonium, while the anionic one can be a carboxylate, a sulfate, or a phosphate. The most common zwitterionic surfactants are N-alkyl (carboxy)betaines and sulfobetaines (Fig. 11, top and bottom respectively). The behavior of zwitterionic surfactants is usually pH dependent, with acidic pH favoring the cationic form and basic pH the anionic one. They may possess an isoelectric point, where their solubility in water reaches a minimum, and they resemble nonionic surfactants.

Fig. 11 Examples of zwitterionic surfactants; (carboxy)betaine (top) and sulfobetaine (bottom)

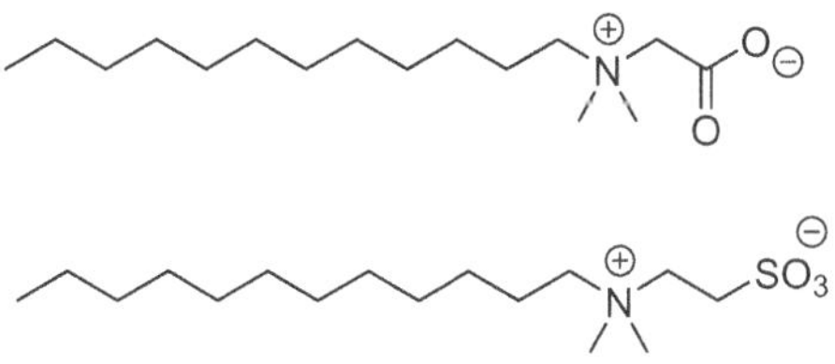

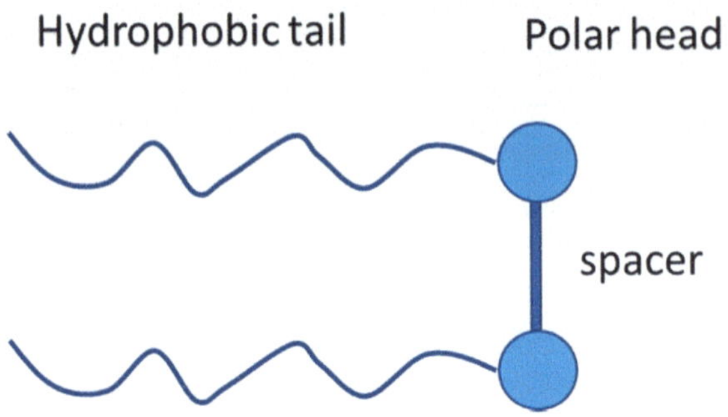

Fig. 12 Schematic representation of a gemini surfactant

3.5 *Gemini Surfactants*

Gemini surfactants became popular for cEOR applications in recent times [7, 24, 25]. These are generally constituted by nonionic, cationic or anionic surfactants, with multiple hydrophilic head and hydrophobic tail (generally two, hence the name) with the hydrophilic heads connected by a spacer. A schematic representation is given in Fig. 12. All combinations are possible: double cationic, double anionic, cationic/anionic, etc. Compared to traditional surfactants, they usually possess a much lower CMC and lower surface tension values, as well as unique self-assembly and rheological properties [19, 25].

3.6 *Biosurfactants*

The term biosurfactant is used to describe surfactants produced via "in vivo" processes by microorganism [26]. This definition does not generally include surfactants prepared by chemical transformation of bio-based sources, of which there are also numerous examples, including sorbitan esters, ethoxylated fats and oils, fatty esters of mono and polysaccharides, that will be discussed more in detail later.

Biosurfactants are produced from microbial cultures grown on hydrophobic substrates. Their composition is not exactly defined, but they are constituted mainly of glycolipids and lipopeptides. Biosurfactants received interest in the context of oil recovery [27, 28].

3.7 *Polymeric Surfactants*

Polymeric surfactants are largely used in industry as emulsion stabilizers and dispersants, but they find application in many diverse areas [29–32]. The most common types are block copolymers of ethylene oxide and propylene oxide, known under the commercial names of Pluronics (BASF) or Symperionics (Croda), or the collective name of poloxamers. As nonionic surfactants, these are non-toxic and can be used in cosmetics, detergents, healthcare products and pharmaceuticals. Their molecular weights are not very high, generally in the order of magnitude of 10.000 Da. Polymeric surfactants usually display lower surface activity than low molecular weight surfactants, and slower (or absent) dynamics.

The structure and composition of polymeric surfactants can be very different. Block copolymers (such as Pluronics) can be referred to as macrosurfactants, while polymers with statistic distribution of hydrophobic and hydrophilic monomers, or polymerized amphiphiles, are called polysoaps. More complex structures, such as star, graft, comb and multiblock are also known [29].

Block structures can form intermolecular micellar aggregates, analogously to their low molecular weight counterparts, while polysoaps can form unimeric micelles by intramolecular aggregation (Fig. 13) [29]. This generally result in quite different solution properties.

Polymeric surfactants can also have responsive behavior (to pH, salinity, temperature, etc.), which is of interest for biomedical applications such as drug delivery. For example, poloxamers show interesting thermally induced aggregation, due to the presence of a PPO block, which hydrophobicity increases with temperature. NIPAM based block copolymers also display analogous behavior.

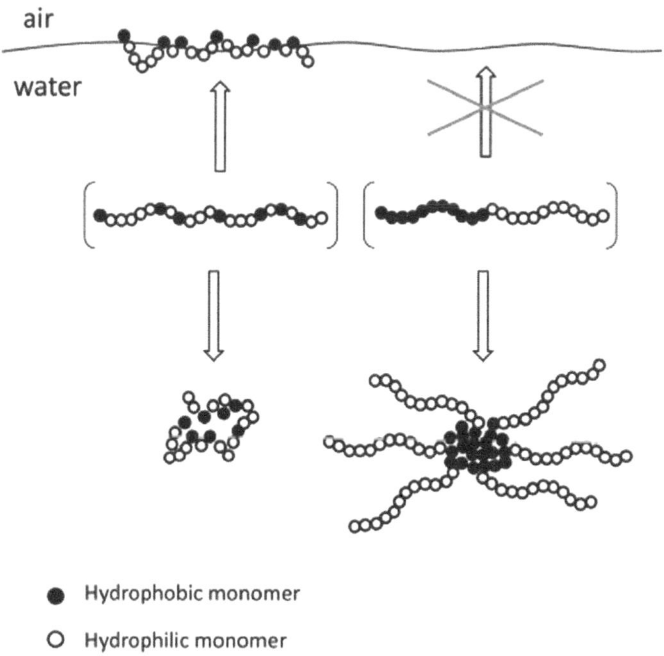

Fig. 13 Different adsorption and aggregation behavior of polymeric surfactants. Adapted with permission from [29]

The presence of charged blocks, as in amphiphilic block polyelectrolytes, usually results in remarkable rheological behavior, of interest for cEOR, as well as responsiveness to pH and salinity [33]. Polymeric surfactants will be discussed more in detail in the context of cEOR in a following section.

4 Surfactants in Enhanced Oil Recovery

Chemical enhanced oil recovery (cEOR) is a collection of methods used to recover oil after primary and secondary techniques have exhausted their potential. Many books and reviews have been published about this topic [2, 3, 7, 34]. The idea of using surfactants in oil recovery was introduced in the beginning of the twentieth century, but it was implemented successfully only from the 1960s [35]. Since then, many other chemicals have been proposed and used. The most typical chemicals used in cEOR are nowadays polymers, surfactants, alkali, and nanoparticles, used alone or in combination [4–7, 9, 11, 12].

The mechanism by which the oil is recovered, varies depending on the nature of the chemicals used. Polymer flooding aims at increasing macroscopic sweep efficiency, by adjusting the mobility ratio between displacing fluid (water) and mobilized oil [34, 36–38]. Water soluble polymers act as viscosifiers of the water phase, allowing to optimize the viscosity ratio between water and oil, and therefore the mobility ratio which depends on it. In this condition, phenomena such as viscous fingering are minimized and the reservoir is swept more efficiently, increasing the oil recovered at the production wells. Polymer flooding is treated in detail in other sources, cited here for the interested reader [34, 38–40].

Surfactant flooding also has been the subject of various recent reviews [13, 35, 41–43]. In surfactant flooding, the main recovery mechanisms are different, and they are mostly based on reducing the residual oil saturation [13, 30]. This can be accomplished in two ways: (1) improving the microscopic efficiency, by reducing the interfacial tension (γ or IFT) and therefore mobilizing the oil trapped in small pores due to capillary forces, and (2) altering the rock wettability from oil-wet to water-wet. We will briefly discuss these two aspects. As a consequence of these mechanisms, emulsification of the oil can occur, with the formation of a new phase. This will also be briefly discussed. Nowadays, surfactant flooding is mostly used in combination with other chemicals, because of the proven synergy between the systems; therefore we have surfactant-polymer flooding (SP), alkali-surfactant flooding (AS) and alkali-surfactant-polymer flooding (ASP) [4–6], as well as, more recently, surfactant-nanoparticle flooding [8]. Also CO_2 flooding can make use of surfactants (CO_2 foams) [42].

4.1 Interfacial Tension Reduction

The main recovery mechanism that governs surfactant flooding is based on mobilization of oil trapped in rock pores and throats by capillary forces. In order to mobilize the oil, the displacing fluid needs to work against the Laplace pressure, causing a change in the radius of curvature of the droplet. This work is done by viscous forces. The balance between viscous and interfacial force is expressed by the adimensional capillary number *Ca* (Eq. 8) [5, 30]:

$$Ca = \frac{G\eta_d r}{\gamma} \tag{8}$$

where G and η_d are gradient velocity and the viscosity of the displacing fluid, r is the curvature radius of the droplet, and γ is the interfacial tension (often also indicated as IFT) between the two fluids. It has been demonstrated that the residual oil saturation correlates inversely with the capillary number. In particular, it is usually shown in capillary desaturation curves [2, 44], that in a typical EOR process, the residual oil saturation drops of several order of magnitudes by increasing the capillary number above values of 10^{-2}. This can be accomplished by decreasing the interfacial tension between oil and water (see Eq. 8). As typical values of *Ca* for waterflooding are in the order of magnitude of 10^{-6}, ultralow values of γ (10^{-3}–10^{-4} mN/m) are required in order to have an efficient surfactant flooding [2, 3]. Ionic surfactants are able to provide ultralow IFT values, therefore they find use in cEOR, as it will be discussed later.

Some authors claim that ultralow values of interfacial tension might not be necessary to improve oil recovery, and actually in SP flooding intermediate values might be preferred [45]. Other authors concludes from their EOR studies that pressure drop has more effect than IFT reduction on recovery efficiency [46].

An increase in viscosity and a decrease in interfacial tension, still resulting in high Ca values, can be accomplished simultaneously by the use of polymeric surfactants [30].

4.2 Wettability Alteration

Surfactants are not only able to adsorb at the water/oil interface, decreasing IFT, but they can also easily adsorb on the rock surface, altering their wetting properties. Although surfactant adsorption is actually a problem, because it causes loss of material in the reservoir, with subsequent environmental issues and increase of costs, it can also have beneficial effects on oil recovery.

As surfactant molecules adsorb on the rock surface, the wettability of the rocks can switch from oil-wet to water-wet as a consequence of the increased hydrophilicity of the rock. This happens especially if the surfactants are charged. As a result, water can displace oil from the rock surface more easily and the recovery increases [43].

4.3 Emulsification and Emulsion Phase

As the IFT between water and oil is reduced by action of the surfactants, the mobilized oil droplet can become small enough, by action of shear forces, to form an emulsion phase [47]. The formed emulsion is stabilized by the presence of a surfactant layer around the oil droplets, according to the DLVO theory. The formation of an emulsion is considered to be necessary in surfactant (or SP) flooding, in order to guarantee a favorable mobility ratio during the process. In fact, surfactant solutions normally have very low viscosities, which determines poor sweep efficiency of the displacing fluid, while emulsions possess good rheological behavior in this respect. Emulsion stability and droplet size play a relevant role in oil recovery [47], therefore a proper choice of surfactant needs to be made.

5 Design and Synthesis of Low Molecular Weight Surfactants for EOR

The general criteria adopted in the design of surfactants for cEOR, as discussed in Chapter "IFT Role on Oil Recovery During Surfactant Based EOR Methods", are based on reducing IFT and increasing water wetting of the reservoirs. Other important aspects to take into account are tolerance to temperature and salinity, especially divalent cations such as Ca^{2+} and Mg^{2+}. It is known that the most challenging reservoirs are the ones with high salinity and high temperature, so the majority of surfactants more recently synthesized and tested for EOR are designed for these kinds of reservoirs. Another very relevant aspect is connected to costs: the surfactant has to be cheap and produced easily from readily available starting materials. Environmental aspects are unfortunately not much considered in the design of surfactants for EOR, although they should. In this respects, biocompatible surfactants should be employed.

Based on the various criteria, several surfactants have been proposed and used for cEOR in the years. The most common ones have been recently reviewed [5, 14, 35, 41, 48]. They all belong to the classes discussed in Sect. 3. Here we will discuss surfactants specifically designed and tested for cEOR application. The main classes of surfactants investigated for cEOR are without a doubt sulfates and sulfonates (anionic). Other systems include various polyethers (nonionic), bio-derived surfactants such as lignin sulfonate, gemini surfactants, and polymeric surfactants such as Pluronics (BASF) [14].

5.1 Nonionic Surfactants

Nonionic surfactants are usually less surface active than ionic ones, but they are also less sensitive to salinity, in particular divalent cations, and have lower tendency to adsorb on the rocks, therefore they find use in surfactant flooding, mostly in combination with ionic surfactants [49].

5.1.1 Alcohol Alkoxylates

Among nonionic surfactants, the most common are based on ethylene/propylene glycols. Examples are given in Sect. 3.1. The common way to synthesize these surfactants is by reaction of a fatty alcohol or acid with EO and/or PO via ring opening reaction initiated by a base (Fig. 14). By this method, it is usually not easy to control the number of units incorporated, therefore the products generally consist of mixtures of molecules with slightly different molecular weights [15].

A recent study of alkoxylated nonionic surfactants designed for EOR application, specifically for SP flooding, has been reported by Yan et al. [50]. These systems are characterized by a double hydrophobic octyl chain and variable amount of ethoxylated groups, and are obtained as illustrated in Fig. 15. The influence of the number of ethoxy group on the solubility, surface activity and IFT with crude oil was evaluated.

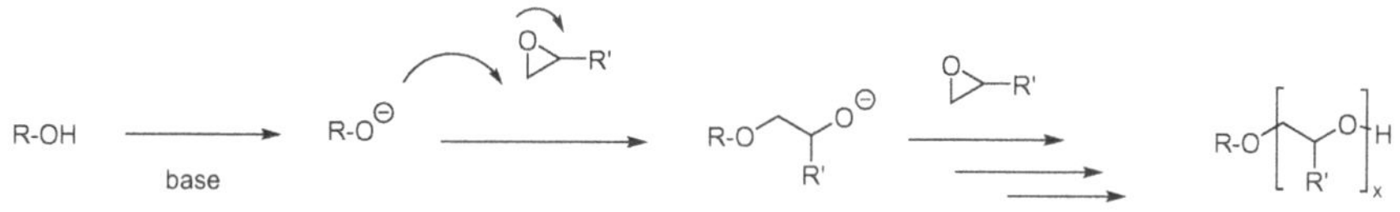

Fig. 14 Mechanism of formation of polyalkoxylated surfactants

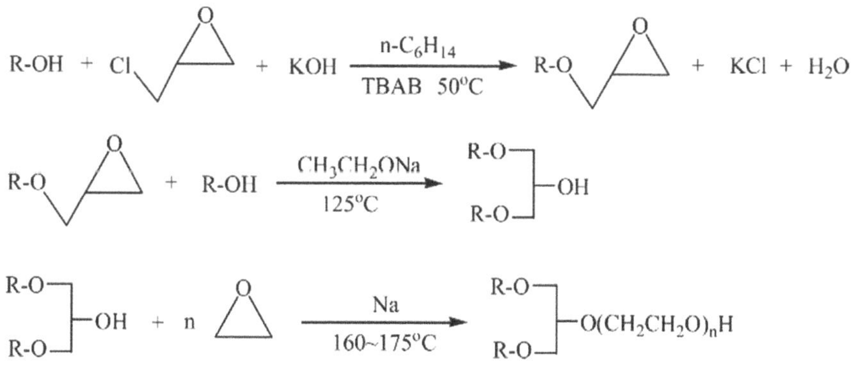

Fig. 15 Synthesis of dioctyl glyceryl ether ethoxylates. Reproduced with permission from [50]

The number of ethoxy units in nonionic surfactants designed for EOR, has been found to be also important for determining the adsorption of surfactants on rock pores and surfaces [51].

5.1.2 Alkoxylated Amines

Analogously to alcohols, reaction of amines with EO, affords ethoxylated nonionic surfactants. These surfactants are so-called "switchable", because they become cationic at acidic pH (Fig. 16). This characteristic can be exploited in CO_2 flooding, due to the intrinsic acidity of CO_2 [52–54]. The nonionic character is more pronounced as longer ethoxy chains are attached to the amine.

5.2 *Anionic Surfactants*

As anticipated in paragraph 3, anionic surfactants are the most studied and used in many applications, including EOR, due to low costs and generally easy synthesis. Among these, sulfates, sulfonates, and carboxylates are the most often found in the context of EOR.

5.2.1 Sulfates and Sulfonates

Most of the surfactants proposed and used for cEOR contain a sulfur-derived group as the hydrophilic moiety. They have been briefly described in Sect. 3.3.

Initially, in the 1970s, oil company were producing petroleum sulfonates directly from crude oil, by reaction with sulfuric acid, which caused problems of sludge disposal, and later with SO_3 [55]. The product were obviously not well defined mixtures of alkyl and aryl sulfonates. This approach is still used in some recent work [56], but it is rare in academic research.

In later years, synthetic sulfonates with better defined structures, displaying more efficiency, were developed. The most popular ones are alkyl aryl sulfonates, alpha olefines sulfonates (AOS), internal olefines sulfonates (IOS), alcohol alkoxy sulfates/sulfonates.

Alkyl aryl sulfonates synthesis is shown in Sect. 3.3. As these compounds tend to be harmful for aquatic systems, and they are not biodegradable, they are not preferred

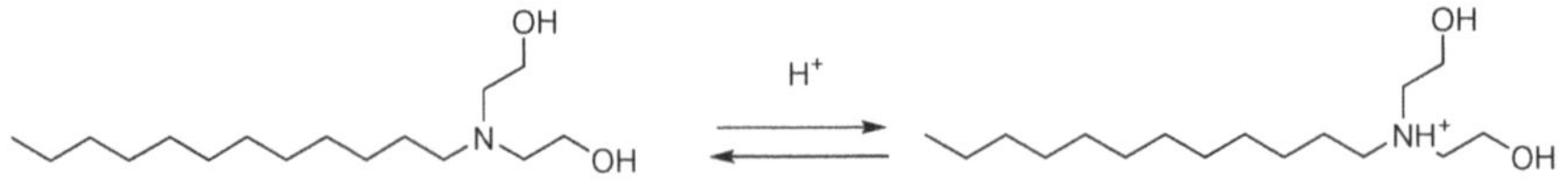

Fig. 16 Alkoxylated amine "switchable" surfactants

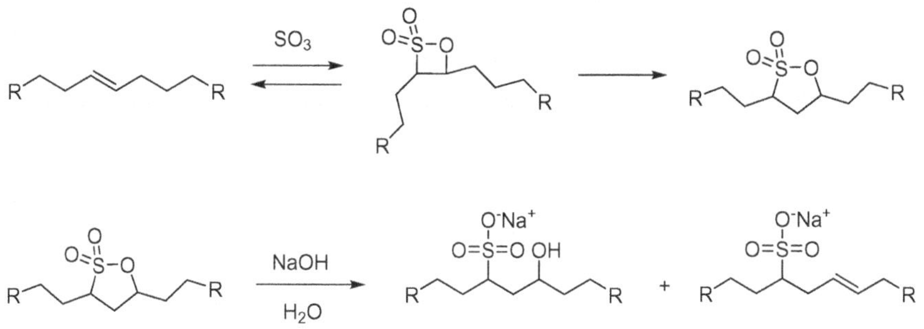

Fig. 17 Mechanism of sulfonation of IOS according to Barnes et al. [57]

for cEOR applications. However, due to the easy preparation, many studies appeared in literature about their employ in surfactant flooding [35].

IOS were developed and extensively studied in Shell [57]. Their synthesis involves three steps, summarized in Fig. 17: sulfonation, neutralization and hydrolysis. AOS are analogous to these ones, but terminal olefines are used instead of internal ones. AOS possess a more linear structure, are biodegradable, and have good resistance to divalent cations [35].

Research developed in Shell led to the design of the ENORDET series of surfactants, optimized for EOR in difficult reservoir conditions (high temperature and salinity). These includes the just mentioned IOS and branched C16,17 alcohol alkoxy sulfonates, prepared as shown in Fig. 18 [58]. The first step is analogous to what seen for nonionic alkoxylated surfactants. These classes of surfactants are suitable because they have reduced tendency to form ordered liquid crystal structures in the reservoir. Moreover, the presence of alkoxy units increases the solubility of the surfactants in high salinity brines [15, 35]. By varying the relative amount of EO and PO groups, the hydrophobic/hydrophilic balance can be tuned to meet the desired requirements in terms of IFT, optimal salinity, etc. When highly branched alcohols are used, the performances are known to improve significantly. Guerbet's method is used to conveniently dimerize alcohols, that can be later turned by the same procedure in the so-called Guerbet alkoxy sulfate (GAS) surfactants [59].

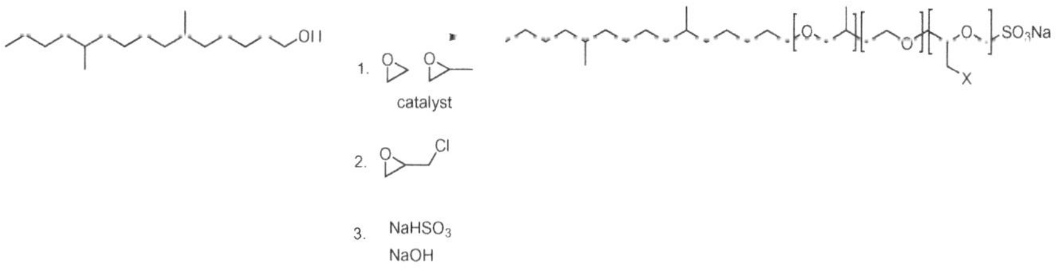

Fig. 18 Synthesis of Shell alcohol alkoxy sulfonates

Structurally very similar, are surfactants of the series ALFOTERRA [13], owned by Sasol. These typically contain a C12-13 hydrophobic tail, few PO units, and a terminal sulfate group.

Sodium methyl ester sulfonates (SMES) were prepared and tested for EOR from crude (non-edible) Jatropha oil [60]. This surfactant is of interest because it can be produced from a bio-based source, that does not interfere with the food chain. The crude oil is first treated with methanol in acidic conditions, to convert free fatty caids into the corresponding methyl esters. In a second step, transesterification with Methanol/KOH is performed, to complete the conversion to methyl esters from the triglyceride portion of the oil. The final step involves the reaction with Chlorosulfonic acid in pyridine, followed by neutralization with Na_2CO_3 to obtain the sodium salt form. The structure of the final product is not clearly specified.

5.2.2 Alcohol Alkoxy Carboxylates

This is another class of anionic surfactants largely used for cEOR [14, 35, 61]. These are based on the reaction of alkoxylated alcohols with Chloroacetic acid (Fig. 19). Also for these surfactants, variable quantities of EO and PO units can be added (R' being H and CH_3 respectively), to tune the properties of the surfactant to reservoir conditions. Compared to analogous sulfates and sulfonates, these surfactants are more stable at thigh temperatures, in a larger range of pH [61].

5.3 Zwitterionic Surfactants

Zwitterionic surfactants are becoming very popular in recent times for cEOR applications, due to their superior salt tolerance and temperature resistance [62, 63]. They can be used alone or in combination with other surfactants [64]. These are usually N-alkyl betaines or sulfobetaines and their synthesis can require multi-step processes. Simple N-alkyl sulfobetaines prepared as shown in Fig. 20 have been synthesized and tested for SP flooding in core flood experiments with good results [62].

A zwitterionic surfactant for EOR was derived from coconut oil alcohols polyoxethylene, as illustrated in Fig. 21 [65]. The presence of EO chains increases the water solubility, compared to other sulfobetaines. This surfactant proved to be effective in reducing IFT between connate water and crude oil of the Daqing field.

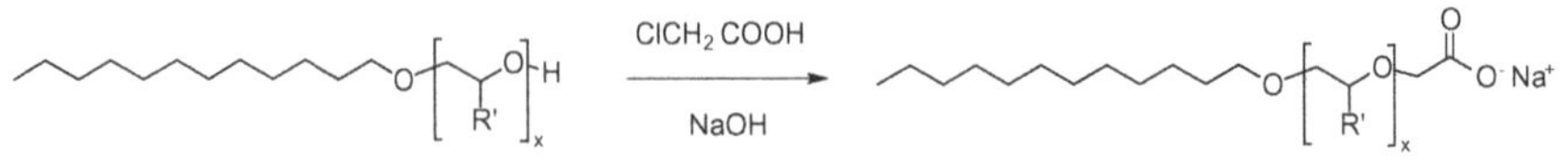

Fig. 19 Example of synthesis of an alcohol alkoxy carboxylate

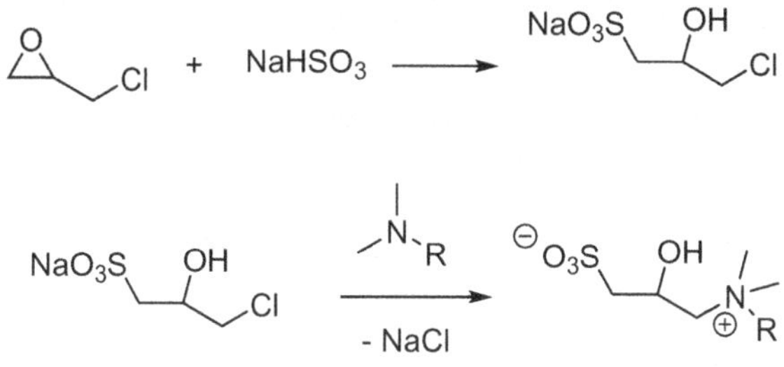

Fig. 20 Synthesis of zwitterionic sulfobetaine surfactants

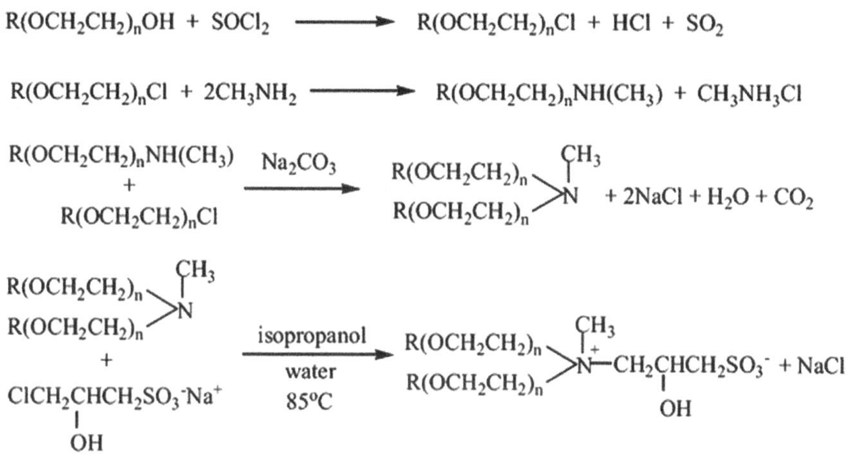

Fig. 21 Synthesis of sulfobetaine from coconut alcohol alkoxylate. Reproduced from [65]

Kumar and Mandal prepared N-alkyl carboxybetaines for use in EOR [66, 67], as illustrated in Fig. 22. The procedure is very similar to the one used for quaternary ammonium cationic surfactants and sulfobetaines, but in this case the tertiary amine is reacted with sodium chlorocetate.

Kamal, Hussain and Fogang prepared various betaines and sulfobetaines, and performed an extensive study of rheological, interfacial and thermal properties, as well as performance in oil recovery in combination with poly(AM-AMPS) [68, 69].

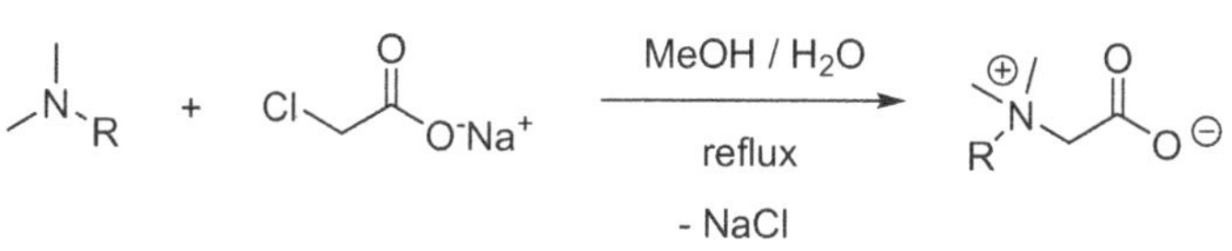

Fig. 22 Synthesis of carboxybetaine surfactants

Interestingly, they found out that the presence of unsaturations on the alkyl chain can improve water solubility.

Some examples of zwitterionic sulfobetaine surfactants for EOR, obtained by synthetic transformation of naturally derived molecules, such as castor oil [70] and lignin [71], have been reported.

5.4 Gemini Surfactants

Also gemini surfactants attracted great interest recently for cEOR, due to their very low IFT values and interesting rheological behavior [7, 24, 25]. A very comprehensive review on gemini surfactants for EOR application has been presented by Kamal in relatively recent times [25]. Structures and synthetic methods vary significantly across the board, and possibilities are numerous. Here we will show representative examples of gemini surfactants that have actually been evaluated or proposed for EOR applications.

Dicationic gemini surfactants that have been studied for EOR applications [72–74], can be easily synthesized by reaction of N,N,N,N tetraalkyl diamines with alkyl bromides (Fig. 23) or, conversely, by reacting tertiary amines with dibromides. The choice will depend on the availability of starting materials.

An example of the latter method was very recently used as the final step in the synthesis of novel ethoxylated quaternary ammonium gemini surfactants for EOR (Fig. 24) [75]. These surfactants are stable in water at high temperature for months.

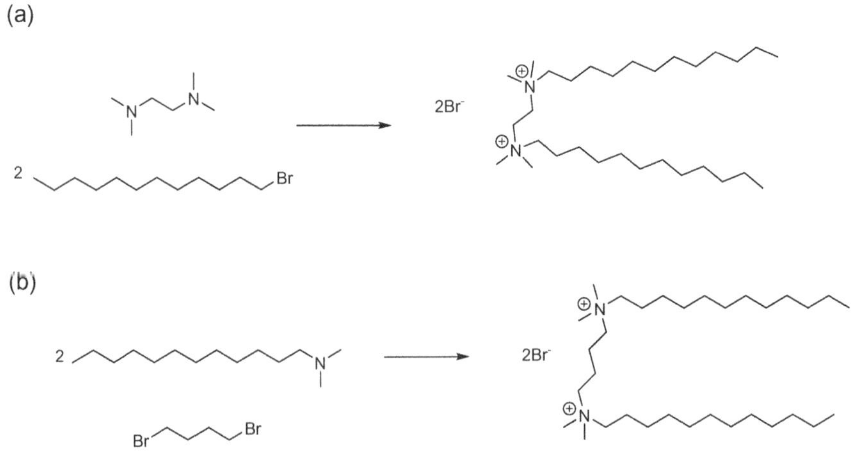

Fig. 23 Synthesis of gemini cationic surfactants from diamine and alkylbromide (**a**) or amines and dibromide (**b**)

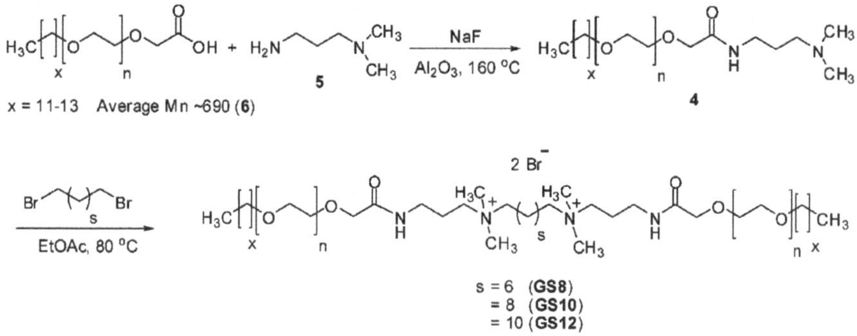

Fig. 24 Synthesis of ethoxylated quaternary ammonium gemini surfactants studied for EOR. Reproduced with permission from [75]

Another similar approach, involves the reaction of tertiary amines with epichlorohydrin (Fig. 25) [76]. In this case, the spacer contains an extra hydroxyl group, which increases water solubility.

Diamines have also been used to prepare dianionic gemini surfactants for EOR, as illustrated in Fig. 26 [77]

A different approach to dianionic gemini surfactants, not involving the use of amines, was previously reported (Fig. 27) [78], and later used to prepare systems for EOR applications [79, 80]. The obtained surfactants can also be hydrolyzed to remove the sulfonate group, affording nonionic gemini surfactants.

Gemini aromatic sulfonate surfactants, particularly stable to high temperature and high salinity, were also recently synthesized as shown in Fig. 28 [81]. Once the diaromatic system is synthesized, sulfonation with chlorosulfonic acid affords the final surfactant product.

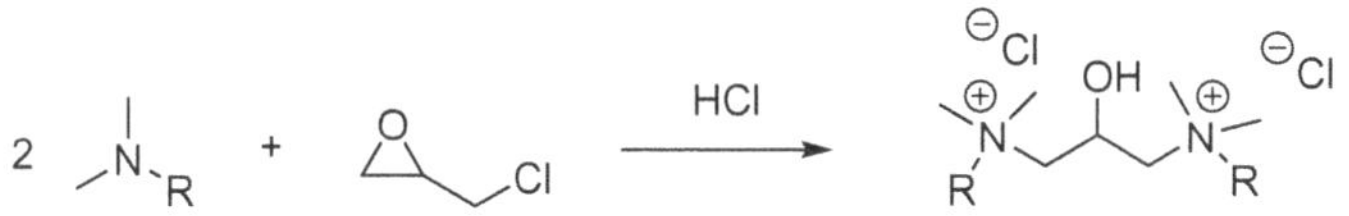

Fig. 25 Dicationic gemini surfactant synthesized from epichlorohydrin

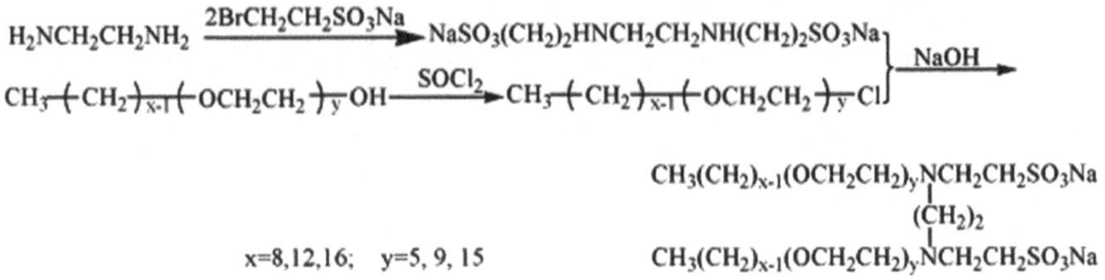

Fig. 26 Dianionic gemini surfactant prepared from diamines. Reproduced with permission from [77]

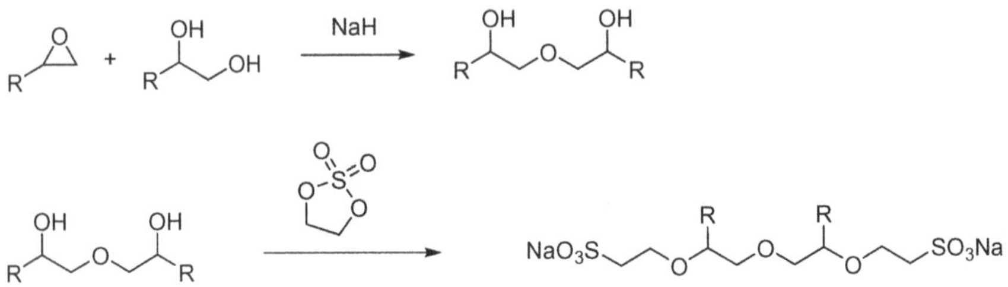

Fig. 27 Example of dianionic gemini surfactant prepared with a different approach. R is a long alkyl chain

Zwitterionic gemini surfactants are also very interesting systems [24]. The synthetic approach varies, but in some case it is similar to the previously observed ones. An example of such a surfactant actually tested for EOR is illustrated in Fig. 29 [82].

Ionic liquid-type gemini imidazolium have also been proposed as a new class of gemini surfactants for EOR [83] The synthesis of imidazolium based ionic liquid surfactants is shown in Fig. 30 [84]. It requires protection-deprotection steps, that makes the procedure not straightforward.

5.5 *Other Bio-based and Miscellaneous Surfactants*

Especially in recent times, as the society became more and more aware about environmental issues, the search for new bio-based and bio-compatible surfactants has been particularly active, and this has been true also for systems designed for application in oil recovery. Those are distinguished from biosurfactants, which are ill-defined mixture of surface active compounds produced by microorganisms [27, 28].

The structures and characteristics of the surfactant will be a consequence of the biomolecules used as starting materials. We have already mentioned surfactants obtained from castor oil [70, 85, 86] and lignin [71]. These are natural hydrophobic molecules containing functional groups, such as alkoxy, that can be for example

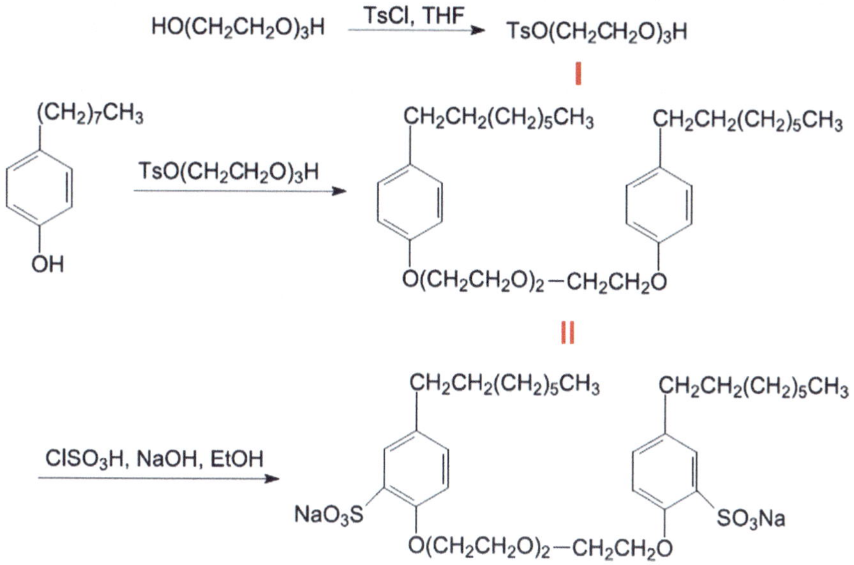

Fig. 28 Gemini aromatic sulfonate surfactants. Reproduced with permission from [81]

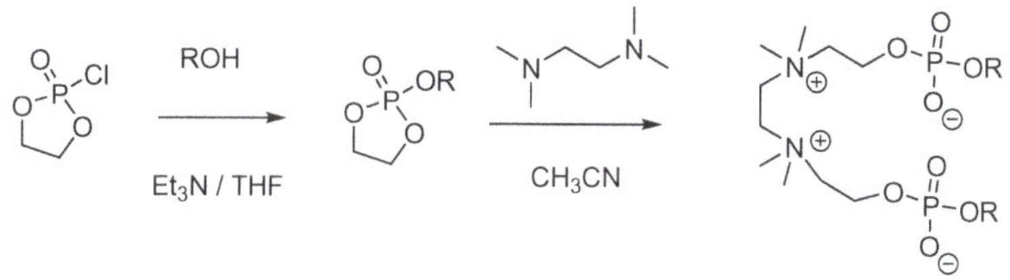

Fig. 29 Example of synthesis of zwitterionic gemini surfactant synthesized in reference [82]. R can be a long alkyl or alkoxy alkyl group

sulfonates, to afford anionic surfactants. Further transformations can be used to obtain zwitterionic surfactants.

An alternative route is transforming an alcohol biobased derivative into a chloride by reaction with $SOCl_2$, then converting into an amine, and subsequently a betaine, either a sulfobetaine or a carboxybetaine. The process is shown in Figs. 20 and 21, and has been used to produce surfactants for EOR from coconut alcohol polyoxyethylene ether [65, 87].

Sugars are also typical substrates for the synthesis of surfactants [88]. As they are hydrophilic, they can be combined with fatty esters or alcohols to make totally biobased non-ionic surfactants. One system that has actually been tested for EOR is a sugar amine sulfonate, with the structure shown in Fig. 31 [89]. This surfactants has been prepared by reaction of anhydrous glucose with an alkylamine, followed by reaction with 2-chloroethylsulfonate.

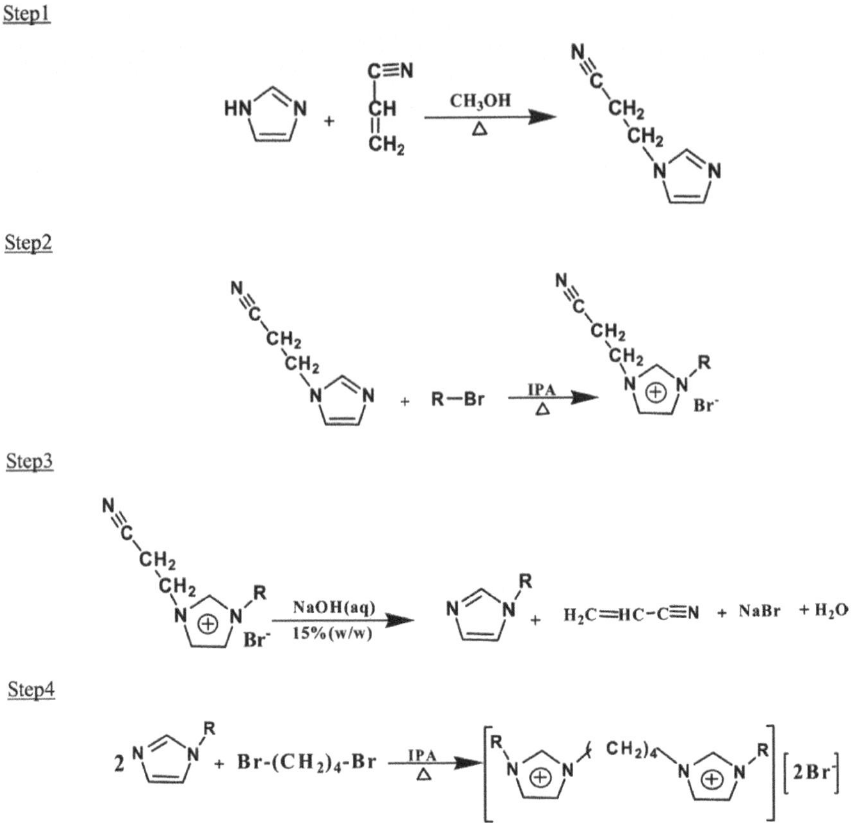

Fig. 30 Ionic liquid-type gemini imidazolium surfactants. Reproduced with permission from [84]

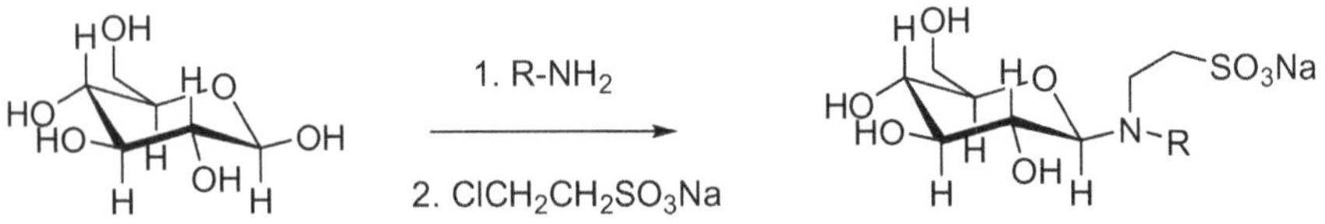

Fig. 31 Example of the synthesis of a sugar based surfactant

The carbohydrate materials can also be a polysaccharide, in which case we have a polymeric surfactant. These will be discussed in the next paragraph.

Among miscellaneous systems, fluorinated surfactants have been proposed for EOR, due to their high thermal stability, and tested [90]. The synthetic procedures can be analogous to what already seen for the preparation of surfactants from alcohols and amines, where fluoroalcohols and fluoroamines are used instead as starting materials.

6 Design and Synthesis of Polymeric Surfactants for EOR

6.1 Polymeric Surfactants in Enhanced Oil Recovery

The use of polymeric surfactants for enhanced oil recovery finds its appeal in the possibility to obtain SP flooding with just one molecule, which has at the same time the characteristics of a water-soluble polymer and a surfactant [29, 30].

The oil recovery mechanism can be either based on IFT reduction as for traditional surfactants (Sect. 4.1) or on mobility control, as is polymer flooding [34]. In fact, although polymeric surfactants usually present surface activities not as good as traditional surfactants, they possess very interesting rheological properties in water. It can be argued that, as polymeric surfactants are not usually able to lower IFT values to ultralow levels, they are not able to increase the capillary number (Sect. 4.1) to an extent to determine a significant oil saturation reduction. Therefore, they should not afford any improved oil recovery. However, some studies suggests that in SP flooding, ultralow IFT values might not be necessary, and rather intermediate values are preferred [45, 46].

6.2 Poloxames

The most common polymeric surfactants are diblock and triblock copolymers of ethylene oxide and propylene oxide (or butylene oxide), usually known under the name of poloxamers or Pluronics (a BASF trademark) [91, 92] and their derivatives. Their use for EOR was deposited in a patent by BASF [93]. Their synthesis is performed by sequential ring opening polymerization, as already seen for analogous low molecular weight surfactants containing EO and PO units (Sect. 5).

Gong et al. [94] prepared poloxamer-like triblock copolymers with an aromatic core, using bisphenol A as initiator, obtaining the structures shown in Fig. 32. They studied their effect on rheological properties of HPAM for cEOR purposes.

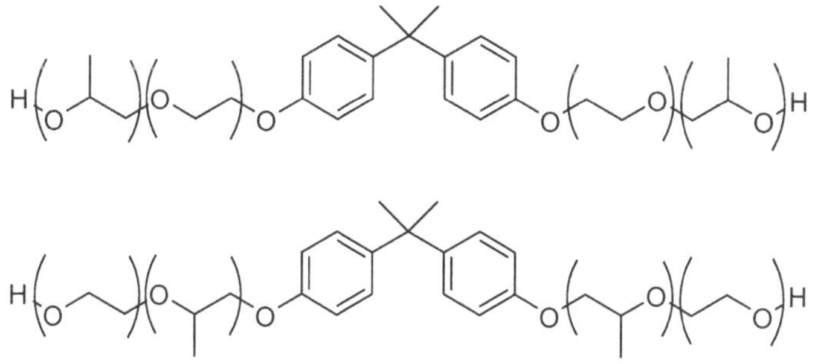

Fig. 32 Poloxamer-like polymeric surfactants proposed for EOR

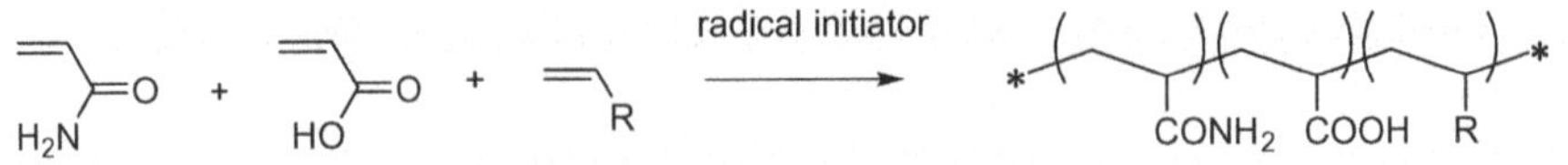

Fig. 33 Synthesis of hydrophobically associative polymers via free radical polymerization

6.3 Hydrophobically Associative Polymers

It has been long recognized that adding hydrophobic monomers in the structure of water soluble polymers used for EOR, such as HPAM, improves the rheological characteristics in solution, via hydrophobic association [30, 36, 38]. These hydrophobically associative polymers can be considered as polymeric surfactants, as the often also display IFT reduction and oil emulsification ability [95–97].

It is not clear if IFT reduction and oil emulsification can have an effect in the oil recovery, or it is rather the viscoelasticity of the system to determine the effectiveness of these polymers. Nonetheless, this class of polymeric surfactants is largely used and studied for application in cEOR.

As many of these polymeric surfactants are derivatives of HPAM, their synthesis is usually performed in the same way as the mentioned polymer, that is via free radical polymerization of a mixture of the opportune polymers (Fig. 33). The polymerization can be usually performed directly in water (or better, in emulsion), therefore typical initiators are water soluble ones, such as $K_2S_2O_8$ [34]. The two hydrophilic components are acrylamide and acrylic acid, to which a third hydrophobic monomer is added, usually in low %, to ensure that the polymer is still water soluble. More than one hydrophobic monomer can be used [95]. A comprehensive list up to 2015 is given in a review by Wever et al. [38]. Also alternating copolymer of the kind styrene-maleic anhydride have been studied and even patented for EOR [30].

Other hydrophobically associative polymers, also proposed and tested for EOR, can be obtained from polysaccharides [98–100], by hydrophobic substitution of the reactive hydroxyl groups present along the whole chain. The hydrophobic groups can be attached by esterification or transesterification reactions, as well as reaction with epoxides. These polymers present the advantage of being obtained from sustainable sources. However, this same advantage constitutes also a problem, as microorganisms can easily degrade biopolymers during their use in the application [30].

6.4 Amphiphilic Block Polyelectrolytes

This class of polymeric surfactants has been proposed and studied for EOR application mostly by the author of this chapter [33, 101–106]. From the point of view of the chemical composition, they resemble hydrophobic associative polymers, but they have a block structure instead of a random distribution of hydrophilic and hydrophobic monomers. Because of this, from the synthetic point of view they offer

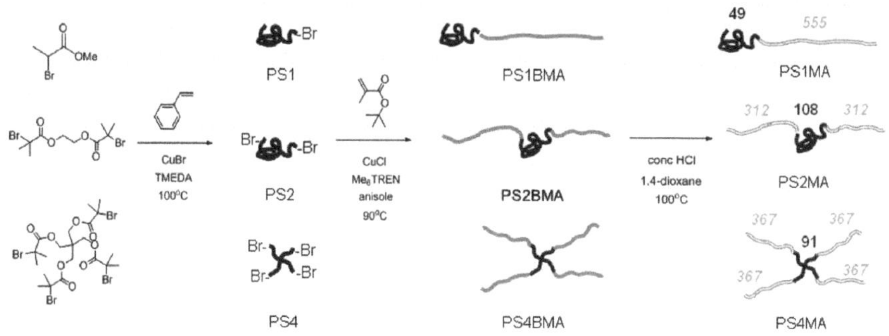

Fig. 34 Synthesis via ATRP and structure of amphiphilic block copolymers proposed for EOR. Reproduced with permission from [33]

a bigger challenge, as controlled radical polymerization methods need to be used. In our work, we used mostly ATRP (atom transfer radical polymerization), to prepare block copolymers constituted mainly by polystyrene for the hydrophobic block and sodium poly (meth)acrylate for the hydrophilic one (Fig. 34). In the hydrophilic block we have in some cases inserted other hydrophilic neutral monomer [104, 106], with the purpose of reducing salt sensitivity.

7 Conclusions and Future Perspective

Surfactants are very interesting molecules, with well-established behavior and characteristics, and a large number of applications. In the context of enhanced oil recovery, there is no reason to think that the use of surfactant will stop any time soon, despite some known environmental issues. On the contrary, the research is still very active in developing new surfactant to improve the performances of surfactant flooding, SP and ASP flooding, and even nanoparticles and CO_2 flooding. Oil extraction form the most challenging high salinity, high temperature, low permeability reservoirs remains a problem in oil industry, therefore new and more efficient systems are still being proposed and developed.

From the first surfactants for EOR developed in the 60s, obtained by simple sulfonation of crude oil, 50 years of research brought many new alternatives, with increasing degree of molecular complexity. Oil companies, as well as research groups in universities all over the world, have produced in time more and more sophisticated molecules, able to meet many of the desired requirement in terms of thermal stability, salinity resistance, high Krafft temperature, low adsorption on rocks, and ultralow IFT in reservoir conditions. This was achieved by carefully tuning the molecular properties of the surfactants. The widespread research resulted in various structural modification to the most simple surfactants, such incorporation of ethoxylated and propoxylated units in ionic surfactants, till the development of zwitterionic and gemini systems. Polymeric surfactants also soon appeared in the panorama of enhanced oil recovery. To obtain all these new structures, usually pure starting materials and several synthetic steps are required, with subsequent increase in the production costs. This investment can still be justified, as the obtained surfactants, when tested in oil recovery and stability experiments, show good promise in harsh condition such as high temperature, high salinity and low permeability. Many studied have been produced, and more are still to come, as every different reservoir requires optimization of the surfactants used, with the synthetic challenges that comes with it.

Polymeric surfactants can represent an interesting class of surfactants for EOR, as they can combine advantages of surfactants and polymers, usually exploited with SP flooding. However, their costs are significantly higher than low-molecular weight surfactants, as the synthetic and purification methods used to prepare them are more sophisticated. Amphiphilic block polyelectrolytes can be considered more similar to nanoparticles, as they form nanometric stable aggregates, therefore they can be associated more to nanoparticle flooding rather than polymer or surfactant flooding.

One important aspect that has become one of the most relevant in modern society, and therefore invests also research and development in enhanced oil recovery processes, is certainly sustainability. For this reason, many examples of bio-based low molecular weight and polymeric surfactants, as well as biosurfactant, appear in the relevant literature. This will most likely constitute a trend in future research.

References

1. Tadros T (2009) Polymeric surfactants in disperse systems. Adv Colloid Interface Sci 147–48:281–299
2. Lake LW (1989) Enhanced oil recovery. Prentice Hall
3. Sheng JJ (2010) Modern Chemical Enhanced Oil recovery: Theory and Practice. Elsevier
4. Chang, H. L. ASP Process and Field Results. in *Enhanced Oil Recovery Field Case Studies* (2013). doi:https://doi.org/10.1016/B978-0-12-386545-8.00010-5
5. Olajire AA (2014) Review of ASP EOR (alkaline surfactant polymer enhanced oil recovery) technology in the petroleum industry: prospects and challenges. Energy 77:963–982
6. Sheng JJ (2014) A comprehensive review of alkaline-surfactant-polymer (ASP) flooding. Asia-Pac J Chem Eng. https://doi.org/10.1002/apj.1824
7. Druetta P, Raffa P, Picchioni F (2019) Chemical enhanced oil recovery and the role of chemical product design. Appl Energy 252:113480
8. Zargartalebi M, Kharrat R, Barati N (2015) Enhancement of surfactant flooding performance by the use of silica nanoparticles. Fuel. https://doi.org/10.1016/j.fuel.2014.11.040
9. Negin C, Ali S, Xie Q (2016) Application of nanotechnology for enhancing oil recovery—a review. Petroleum 2:324–333
10. Pei HH et al (2015) Investigation of nanoparticle and surfactant stabilized emulsion to enhance oil recovery in waterflooded heavy oil reservoirs. SPE Heavy Oil Conf 2015
11. Druetta P, Raffa P, Picchioni F (2018) Plenty of room at the bottom: nanotechnology as solution to an old issue in enhanced oil recovery. Appl Sci 8:2596
12. Nazar MF, Shah SS, Khosa MA (2011) Microemulsions in enhanced oil recovery: a review. Pet Sci Technol 29:1353–1365
13. Kamal MS, Hussein IA, Sultan AS (2017) Review on surfactant flooding: phase behavior, retention, IFT, and field applications. 31 7701–7720 (American Chemical Society)

14. Sheng JJ (2015) Status of surfactant EOR technology. Petroleum 1:97–105
15. Tadros TF (2005) Applied Surfactants: principles and applications. Wiley
16. Tadros TF (2017) Basic principles of interface science and colloid stability (De Gruyter, 2017). https://doi.org/10.1515/9783110540895
17. Nagarajan R (2011) Amphiphilic surfactants and amphiphilic polymers: principles of molecular assembly. In: ACS symposium series, vol 1070, pp 1–22. American Chemical Society
18. Kontogeorgis GM, Kiil S (2016) Introduction to applied colloid and surface chemistry. Introduction to applied colloid and surface chemistry. Wiley. https://doi.org/10.1002/9781118881194
19. Lombardo D, Kiselev MA, Magazù S, Calandra P (2015) Amphiphiles self-assembly: basic concepts and future perspectives of supramolecular approaches. Adv Condens Matter Phys. https://doi.org/10.1155/2015/151683
20. Holmberg K, Jönsson B, Kronberg B, Lindman, B (2002) Surfactants and polymers in aqueous solution. Surfactants and polymers in aqueous solution. Wiley. https://doi.org/10.1002/0470856424
21. Serio M, Di Tesser R, Felippone F, Santacesaria E (1995) Ethylene oxide solubility and ethoxylation kinetics in the synthesis of nonionic surfactants. Ind Eng Chem Res 34
22. Li Y et al (2016) Mixtures of anionic/cationic surfactants: a new approach for enhanced oil recovery in low-salinity high-temperature sandstone reservoir. SPE J 21:1164–1177
23. Zhao Z, Guo X, Jia L, Liu Y (2014) Synthesis and properties of quaternary ammonium surfactants containing a methoxy benzyl substituent†. https://doi.org/10.1039/c4ra07363k
24. Cheng Y et al (2019) Progress in synthesis and application of zwitterionic Gemini surfactants. Front Mater Sci 13:242–257
25. Kamal MS. A review of Gemini surfactants: potential application in enhanced oil recovery. https://doi.org/10.1007/s11743-015-1776-5
26. Kosaric N (1993) Biosurfactants: production, properties, applications. In: Surfactant science series, vol 48. CRC Press
27. Sen R (2008) Biotechnology in petroleum recovery: the microbial EOR. Prog Energy Combust Sci 34:714–724
28. De Almeida DG et al (2016) Biosurfactants: promising molecules for petroleum biotechnology advances. Front Microbiol 7:1718
29. Raffa P, Wever DAZ, Picchioni F, Broekhuis AA (2015) Polymeric surfactants: synthesis, properties, and links to applications. Chem Rev 115
30. Raffa P, Broekhuis AAA, Picchioni F (2016) Polymeric surfactants for enhanced oil recovery: a review. J Pet Sci Eng 145:723–733
31. Riess G, Labbe C (2004) Block copolymers in emulsion and dispersion polymerization. Macromol Rapid Commun 25:401–435
32. Garnier S, Laschewsky A, Storsberg J (2006) Polymeric surfactants: novel agents with exceptional properties. Tenside Surfactants Deterg 43:88–102
33. Raffa P, Brandenburg P, Wever DAZ, Broekhuis AA, Picchioni F (2013) Polystyrene-poly(sodium methacrylate) amphiphilic block copolymers by ATRP: effect of structure, pH, and ionic strength on rheology of aqueous solutions. Macromolecules 46:7106–7111
34. Raffa P, Druetta P (2019) Chemical enhanced oil recovery : advances in polymer flooding and nanotechnology (De Gruyter, 2019)
35. Negin C, Ali S, Xie Q (2017) Most common surfactants employed in chemical enhanced oil recovery. Petroleum. https://doi.org/10.1016/j.petlm.2016.11.007
36. Afolabi RO, Oluyemi GF, Officer S, Ugwu JO (2019) Hydrophobically associating polymers for enhanced oil recovery—part A: a review on the effects of some key reservoir conditions. J Pet Sci Eng 180:681–698
37. Wei B, Romero-Zerón L, Rodrigue D (2014) Oil displacement mechanisms of viscoelastic polymers in enhanced oil recovery (EOR): a review. J Pet Explor Prod Technol 4
38. Wever DAZ, Picchioni F, Broekhuis AA (2011) Polymers for enhanced oil recovery: a paradigm for structure-property relationship in aqueous solution. Prog Polym Sci 36:1558

39. Standnes DC, Skjevrak I (2014) Literature review of implemented polymer field projects. J Pet Sci Eng 122:761–775
40. Wei B (2016) Advances in polymer flooding. In: Viscoelastic and viscoplastic materials (Chapter 1). El-Amin M. (ed). InTech Open, Rijeka. https://doi.org/10.5772/64069
41. Sheng JJ, Leonhardt B, Azri N (2015) Status of polymer-flooding technology. J Can Pet Technol 54:116–126
42. Sagir M, Mushtaq M, Tahir MS, Tahir MB, Shaik AR (2020) Surfactants for enhanced oil recovery applications. Surf Enhanced Oil Recovery Appl. https://doi.org/10.1007/978-3-030-18785-9
43. Hirasaki GJ, Miller CA, Puerto M (2011) Recent advances in surfactant EOR. Spe J 16:889–907
44. Druetta P, Picchioni F (2019) Surfactant flooding: the influence of the physical properties on the recovery efficiency. Petroleum. https://doi.org/10.1016/j.petlm.2019.07.001
45. Yu Q, Jiang H, Zhao C (2010) Study of interfacial tension between oil and surfactant polymer flooding. Pet Sci Technol 28:1846–1854
46. Zhang H, Dong M, Zhao S (2010) Which one is more important in chemical flooding for enhanced court heavy oil recovery, lowering interfacial tension or reducing water mobility? Energy Fuels 24:1829–1836
47. Goodarzi F, Zendehboudi S (2019) A comprehensive review on emulsions and emulsion stability in chemical and energy industries. Can J Chem Eng 97:281–309
48. Pal S, Mushtaq M, Banat F, Al Sumaiti AM (2018) Review of surfactant-assisted chemical enhanced oil recovery for carbonate reservoirs: challenges and future perspectives. Petroleum Sci 15
49. Gupta R, Mohanty KK (2010) Temperature effects on surfactant-aided imbibition into fractured carbonates. SPE J 15:587–597
50. Yan L, Cui Z, Song B, Pei X, Jiang J (2016) Dioctyl glyceryl ether ethoxylates as surfactants for surfactant-polymer flooding. Energy Fuels 30:5425–5431
51. Curbelo FDS et al (2007) Adsorption of nonionic surfactants in sandstones. Colloids Surf A Physicochem Eng Asp 293:1–4
52. Chen Y et al (2014) Switchable nonionic to cationic ethoxylated amine surfactants for CO_2 enhanced oil recovery in high-temperature, high-salinity carbonate reservoirs. SPE J 19:249–259
53. Elhag AS et al (2014) Switchable amine surfactants for stable CO_2/brine foams in high temperature, high salinity reservoirs. In: SPE—DOE improved oil recovery symposium proceedings, vol 1, pp 179–189. Society of Petroleum Engineers (SPE)
54. Chen Y et al (2012) Ethoxylated cationic surfactants for CO_2 EOR in high temperature, high salinity reservoirs. In: SPE—DOE improved oil recovery symposium proceedings, vol 2, pp 1224–1238. Society of Petroleum Engineers
55. Sharma MK, Shah DO (1989) Use of surfactants in oil recovery. Dev Pet Sci 17:255–315
56. Yu F et al (2008) Study on synthesis and applied performance of petroleum sulfonate for enhanced oil recovery. J Dispers Sci Technol 29:975–984
57. Barnes JR et al (2010) Application of internal olefin sulfonates and other surfactants to EOR. Part 1: structure—performance relationships for selection at different reservoir conditions. In: SPE improved oil recovery symposium. Society of Petroleum Engineers.https://doi.org/10.2118/129766-MS
58. Barnes JR et al (2008) Development of surfactants for chemical flooding at difficult reservoir conditions. In: Proceedings—SPE symposium on improved oil recovery, vol 1, pp 435–452. Society of Petroleum Engineers
59. Lu J et al. New surfactant developments for chemical enhanced oil recovery. https://doi.org/10.1016/j.petrol.2014.05.021
60. Elraies KA, Tan IM, Awang M, Saaid I (2010) The synthesis and performance of sodium methyl ester sulfonate for enhanced oil recovery. Pet Sci Technol 28:1799–1806
61. Lu J et al (2014) Novel large-hydrophobe alkoxy carboxylate surfactants for enhanced oil recovery. SPE J. https://doi.org/10.2118/154261-PA

62. Guo S, Wang H, Shi J, Pan B, Cheng Y (2015) Synthesis and properties of a novel alkyl-hydroxyl-sulfobetaine zwitterionic surfactant for enhanced oil recovery. J Pet Explor Prod Technol 5:321–326
63. Zhao J et al (2015) The structure effect on the surface and interfacial properties of zwitterionic sulfobetaine surfactants for enhanced oil recovery. https://doi.org/10.1039/c4ra16235h
64. Kurnia I, Zhang G, Han X, Yu J (2020) Zwitterionic-anionic surfactant mixture for chemical enhanced oil recovery without alkali. Fuel 259:116236
65. Li P et al (2016) A new type of sulfobetaine surfactant with double alkyl polyoxyethylene ether chains for enhanced oil recovery. J Surfactants Deterg 19:967–977
66. Kumar A, Mandal A (2017) Synthesis and physiochemical characterization of zwitterionic surfactant for application in enhanced oil recovery. https://doi.org/10.1016/j.molliq.2017.08.032
67. Kumar A, Mandal A (2018) Characterization of rock-fluid and fluid-fluid interactions in presence of a family of synthesized zwitterionic surfactants for application in enhanced oil recovery. https://doi.org/10.1016/j.colsurfa.2018.04.001
68. Kamal MS, Shakil Hussain SM, Fogang LT (2018) A zwitterionic surfactant bearing unsaturated tail for enhanced oil recovery in high-temperature high-salinity reservoirs. J Surfactants Deterg 21:165–174
69. Shakil Hussain SM et al (2018) Effect of internal olefin on the properties of betaine-type zwitterionic surfactants for enhanced oil recovery. J Mol Liq 266:43–50
70. Zhang QQ et al (2015) Novel zwitterionic surfactant derived from castor oil and its performance evaluation for oil recovery. Colloids Surfaces A Physicochem Eng Asp. https://doi.org/10.1016/j.colsurfa.2015.05.060
71. Chen S et al (2017) Synthesis and physiochemical performance evaluation of novel sulphobetaine zwitterionic surfactants from lignin for enhanced oil recovery. https://doi.org/10.1016/j.molliq.2017.11.021
72. Bi ZC, Qi LY, Liao WS (2005) Dynamic surface properties, wettability and mimic oil recovery of ethanediyl-α, β-bis(cetyldimethylammonium bromide) on dodecane modified silica powder. J Mater Sci 40:2783–2788
73. Chen H, Han L, Luo P, Ye Z (2004) The interfacial tension between oil and gemini surfactant solution. Surf Sci 552:L53–L57
74. Pal N, Saxena N, Mandal A (2018) Studies on the physicochemical properties of synthesized tailor-made gemini surfactants for application in enhanced oil recovery. J Mol Liq 258:211–224
75. Hussain SMS, Kamal MS, Murtaza M (2019) Synthesis of novel ethoxylated quaternary ammonium gemini surfactants for enhanced oil recovery application. Energies 12:1731
76. Yuan T et al (2018) Enhanced oil recovery from high-salinity reservoirs by cationic gemini surfactants. J Appl Polym Sci 135:46086
77. Ma T et al (2019) Property evaluation of synthesized anionic-nonionic gemini surfactants for chemical enhanced oil recovery. Colloids Surfaces A Physicochem Eng Asp 581:123800
78. Rist Ø, Carlsen PHJ (1999) Synthesis of new geminal surfactants. Synth Commun 29:749–754
79. Gao B, Sharma MM (2013) A family of alkyl sulfate gemini surfactants. 1. Characterization of surface properties. J Colloid Interface Sci 404:80–84
80. Gao B, Sharma MM (2013) A family of alkyl sulfate gemini surfactants. 2. Water-oil interfacial tension reduction. J Colloid Interface Sci 407:375–381
81. Hou B et al (2019) A novel high temperature tolerant and high salinity resistant gemini surfactant for enhanced oil recovery. J Mol Liq 296:112114
82. Dong Z, Zheng Y, Zhao J (2014) Synthesis, physico-chemical properties and enhanced oil recovery flooding evaluation of novel zwitterionic gemini surfactants. J Surfactants Deterg 17:1213–1222
83. Zhou H et al (2018) Systematic investigation of ionic liquid-type gemini surfactants and their abnormal salt effects on the interfacial tension of a water/model oil system. J Mol Liq 249:33–39

84. Ao MQ, Xu GY, Zhu YY, Bai Y (2008) Synthesis and properties of ionic liquid-type Gemini imidazolium surfactants. J Colloid Interface Sci 326:490–495
85. Babu K, Pal N, Bera A, Saxena VK, Mandal A (2015) Studies on interfacial tension and contact angle of synthesized surfactant and polymeric from castor oil for enhanced oil recovery. Appl Surf Sci. https://doi.org/10.1016/j.apsusc.2015.06.196
86. Zhang Q-Q et al (2015) Novel zwitterionic surfactant derived from castor oil and its performance evaluation for oil recovery. Physicochem Eng Asp 483:87–95
87. Song B, Hu X, Shui X, Cui Z, Wang Z (2016) A new type of renewable surfactants for enhanced oil recovery: Dialkylpolyoxyethylene ether methyl carboxyl betaines. Colloids Surfaces A Physicochem Eng Asp 489:433–440
88. Queneau Y, Chambert S, Besset C, Cheaib R (2008) Recent progress in the synthesis of carbohydrate-based amphiphilic materials: the examples of sucrose and isomaltulose. Carbohydr Res 343:1999–2009
89. Pu W et al (2017) Laboratory study on the oil displacement properties of sugar amine sulfonate surfactant. J Surfactants Deterg 20:1037–1049
90. Al-Amodi AO, Al-Mubaiyedh UA, Sultan AS, Kamal MS, Hussein IA (2016) Novel fluorinated surfactants for enhanced oil recovery in carbonate reservoirs. Can J Chem Eng 94:454–460
91. Batrakova EV, Kabanov AV (2008) Pluronic block copolymers: evolution of drug delivery concept from inert nanocarriers to biological response modifiers. J ControlRelease 130:98–106
92. Kozlov MY, Melik-Nubarov NS, Batrakova EV, Kabanov AV (2000) Relationship between pluronic block copolymer structure, critical micellization concentration and partitioning coefficients of low molecular mass solutes. Macromolecules 33:3305–3313
93. Huff J et al (2013) Surfactant mixtures for tertiary oil recovery, US8524643
94. Gong H, Xu L, Xu G, Dong M, Li Y (2014) Effect of PEO-PPO-ph-PPO-PEO and PPO-PEO-ph-PEO-PPO on the rheological and EOR properties of polymer solutions. https://doi.org/10.1021/ie404236r
95. Zhao Y et al (2009) Synthesis and characterization of a series of modified polyacrylamide. Colloid Polym Sci 287:237–241
96. Rouzes C, Durand A, Leonard M, Dellacherie E (2002) Surface activity and emulsification properties of hydrophobically modified dextrans. J Colloid Interface Sci 253:217
97. Akiyama E et al (2005) Thickening properties and emulsification mechanisms of new derivatives of polysaccharides in aqueous solution. J Colloid Interface Sci 282
98. Henni W et al (2005) Aggregation of hydrophobically modified polysaccharides in solution and at the air-water interface. J Colloid Interface Sci 281:316–324
99. Cao Y, Li HL (2002) Interfacial activity of a novel family of polymeric surfactants. Eur. Polym. J. 38:1457–1463
100. Desbrieres J, Babak V (2010) Interfacial properties of chitin and chitosan based systems. Soft Matter 6:2358–2363
101. Raffa P, Stuart MCA, Broekhuis AA, Picchioni F (2014) The effect of hydrophilic and hydrophobic block length on the rheology of amphiphilic diblock Polystyrene-b-Poly(sodium methacrylate) copolymers prepared by ATRP. J Colloid Interface Sci 428
102. Raffa P, Broekhuis AAAA, Picchioni F (2016) Amphiphilic copolymers based on PEG-acrylate as surface active water viscosifiers: Towards new potential systems for enhanced oil recovery. J Appl Polym Sci 133
103. Meijerink M et al (2017) Triblock copolymers of styrene and sodium methacrylate as smart materials: synthesis and rheological characterization. Pure Appl Chem 89
104. Boulif N, Sebhaky K, Joosten H, Raffa P (2020) Design and synthesis of novel di- and triblock amphiphilic polyelectrolytes: improving salt-induced viscosity reduction of water solutions for potential application in enhanced oil recovery. Submitt Publ
105. Raffa P, Picchioni F (2020) Preliminary evaluation of amphiphilic block polyelectrolytes as potential flooding agents for chemical enhanced oil recovery. Submitt. Publ

106. Di Sacco F, Pucci A, Raffa P (2019) Versatile multi-functional block copolymers made by atom transfer radical polymerization and post-synthetic modification: switching from volatile organic compound sensors to polymeric surfactants for water rheology control via hydrolysis. Nanomaterials 9:458

Drilling

Application of Surfactants in the Drilling Fluids to Mitigate the Shale Hydration and Swelling

Hafiz Mudaser Ahmad, Tanveer Iqbal, and Saima Yasin

Abstract This chapter is divided into eight different sections. The first three sections describes the introduction of drilling fluid in general and drilling fluid types along with the major drilling fluid additives, major problems of drilling fluids (related to the shale swelling and hydration), Solution of shale swelling and hydration (emphasis on the use of surfactants). It also discusses about the classification and synthesis of surfactants that are used in the formulation of drilling fluids and potential application for shale inhibitions. The next three of this chapter discusses about the impact of surfactants on rheology and filtration properties, and the evaluation of shale inhibition characteristics with surfactant which has explained all the techniques involved in characterizing drilling fluids for shale inhibition. The last two sections of this chapter discusses about the field applications, recommendations and challenges of surfactants for shale inhibition.

Keywords Surfactants · Shale inhibition · Drilling fluids · Wellbore

1 Introduction

Drilling fluids are considered as blood for the drilling operations in the oil and gas industry. Drilling fluids are versatile in their actions in the drilling of oil and gas wells. Drilling fluids broadly classified into three classes such as oil based drilling fluids, synthetic drilling fluids and water based drilling fluids. The use of specific drilling fluid for the drilling of wellbore bore mainly depends on the characteristics of wellbore such as temperature and pressure of wellbore, sensitivity towards swelling upon interacting with drilling fluid component, ease of drilling operations and recycling and disposal of fluid after the drilling operation. The selection of drilling fluid type

H. M. Ahmad (✉) · T. Iqbal
Department of Chemical, Polymer and Composite Materials Engineering,
University of Engineering and Technology Lahore (New Campus), Lahore, Pakistan
e-mail: h.m.ahmad@uet.edu.pk

S. Yasin
Department of Chemical Engineering, University of Engineering and Technology Lahore, Lahore, Pakistan

T. Solling et al. (eds.), *Surfactants in Upstream E&P*, Petroleum Engineering,
https://doi.org/10.1007/978-3-030-70026-3_2

for the drilling operation also considers the sustainability, cost and environmental concerns. The schematics of drilling fluid circulation is shown in Fig. 1 [1, 2].

Drilling fluids are broadly classified in three different categories such as water-based drilling fluids (WBDF), oil-based drilling fluids (OBDF), and synthetic drilling fluids (SDF). The synthetic drilling fluids include the mist, forms and stiff foams which have very specific and limited applications in the oil and gas industry for drilling applications. Oil-based drilling fluids are considered as superior in terms of performance in the drilling operations having enhanced rheological, filtration and shale inhibition properties. The oil based drilling fluids have severe effects on environment with relatively high operational and disposal cost compared to other drilling fluids. The desire of oil-based drilling fluids for marine drilling application has been decreased from last few years due to the strict regulation of environmental protection agency. The water-based drilling fluids on the other hand are widely employed for the offshore and onshore drilling operations [3–5]. The classes of additives used in the formulation of water-based drilling fluids is shown in Fig. 2. The desire of using water-based drilling fluids in the drilling operations is due to the easy preparation, environmental friendliness, recyclable and easy disposal [6–10].

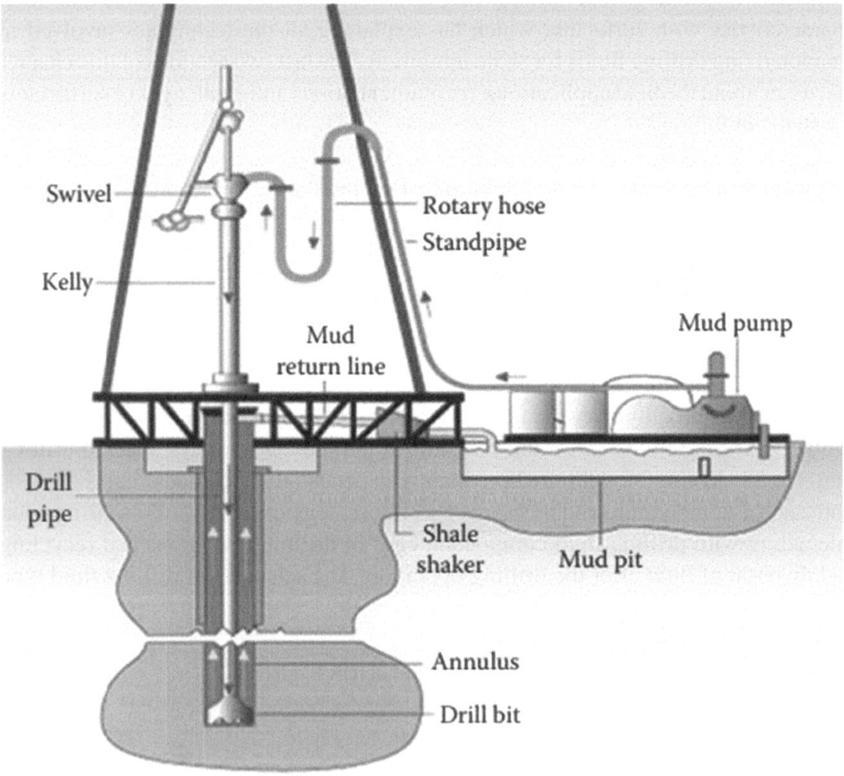

Fig. 1 Schematics of drilling fluid circulation in the rig [2]

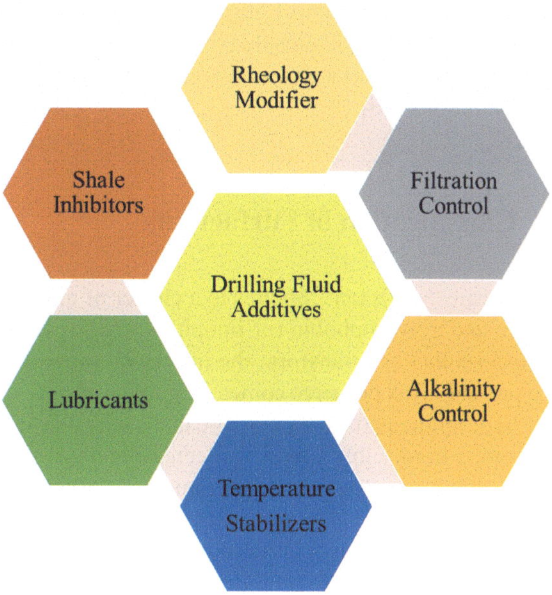

Fig. 2 Drilling fluid additives

The wellbore formation often encounter stability issues during the drilling process of oil and gas wells. The maximum number of complications occur during the horizontal drilling process of wellbore formations. The wellbore instability issues include the pipe sticking, tight hole, wellbore collapse due to the swelling and hydration which leads to the higher drilling cost and prolong wellbore completion process. The wellbore instability problems frequently occur in the formations which have shale rock which often interact with the water content of drilling fluid and leads to the swelling and hydration of wellbore formations [11–14]. Approximately 90% wellbore instability problems occur in the wellbore formations which are composed of shale rock. The shale rock contains the clay minerals which interacts with the water in water-based drilling fluids and decrease the shale rock strength and results in the wellbore instability issues. These wellbore instability issues related to the shale swelling and hydration can be avoided using such additives in the drilling fluids which are less prone to the swelling of wellbore formations [15–18].

To minimize the shale swelling and hydration during the oil and gas well drilling process, various shale inhibitors are employed in the drilling fluid formulations. There are several conventional shale inhibitors are used to reduce the wellbore instabilities such as polymers, nanoparticles and their derivatives [19–23], inorganic salts [24–26], alcohols [27, 28], surfactants [29–31] and ionic liquids [32–35]. Among all the available shale inhibitors, surfactants are considered as most versatile class for shale inhibition due to the unique and extraordinary inhibition characteristics. Cationic surfactants chemically interacts with clay minerals in the shale rock through electrostatic forces which leads to the attachment of positive part of surfactant molecule with the negative charges surface while tail end long alkyl chain act as barrier to minimize the interaction of clay with water. Huang et. al has explored the impact of cationic alkyl polyglycoside on shale inhibition performance and it was observed that this cationic surfactant is favorable for the inhibition of reactive shale at lower concentrations [36].

2 Classification of Surfactants

The phrase "surfactant" applies to a cluster of molecules having both a hydrophilic nature and a hydrophobic (or lipophilic) nature. Surface Active Agents is a set of molecules that can transform the interfacial properties of the liquids in which they are present. This property comes from their affinity as they have both hydrophobic and hydrophilic parts in their nature that's why they can concentrate on the interfaces separating immiscible phases by decreasing their surface tension. Depending on the hydrophobic moiety of the surfactants most of the surfactants have the same 'tails' of hydrocarbons chain in their structure and this chain could be linear, branched, or aromatic like surfactants of siloxane and fluorocarbon. But in some rare cases, these tails could also be consisting of non-hydrocarbon chains such as per-fluorocarbon and poly-dimethyl-siloxane. Although a small number of surface-active agents are naturally formed such as saponins or lecithin, there is a fashion to fabricate more and more surfactants comes from natural resources [37–41].

The hydrophilic nature of the surfactants which ensure the water affinity of molecules is a base to classify the four classes of the surfactants into nonionic, anionic, cationic, and amphoteric. The generic classification of shale inhibitors is shown in Fig. 3. This categorization is based on the composition of the polarity of the head group. The choice of surfactants for any application is a cautious task in which frequent points must be considered for competent outputs. Among others, one should judge those directly associated with functions to be satisfied (detergency, emulsification, foam excellence, instability, etc.), and those interrelated to expenditure, toxicity, and environmental concerns.

2.1 Nonionic Surfactants

As from their name, non-ionic surfactants have a charge less head. Non-ionic surfactants composed up of a hydrophilic "head" and a hydrophobic" tail". This type of surfactant is very valuable because of its low sensitivity to water hardness and pH. Nonionic surfactants are superior to anionic surfactants at removing organic soils. The two are frequently used together to produce double-action, flexible cleaners that can not only pick up and hang up particulate soils but also soften oily soils. They can be blended with other ionic surfactants to get aimed results because they don't have any charge on them. Cocamide, ethoxylates, and alkoxylates are some common

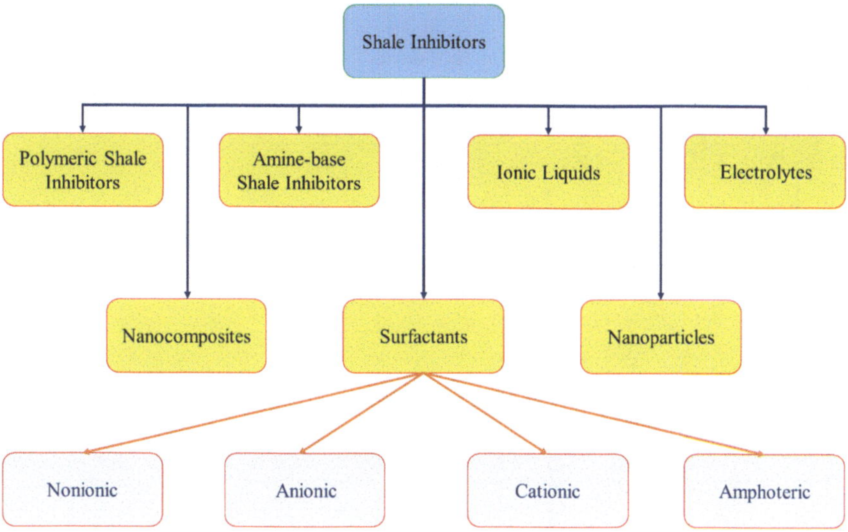

Fig. 3 Classification of shale inhibitors

non-ionic surfactants. The examples of nonionic surfactants used in drilling fluid formulations is shown in Table 1.

2.2 *Anionic Surfactants*

Anionic surfactants hold an anionic charge at their head such as carboxylates, sulfate, sulfonate, or phosphate. Anionic surfactants are used, as secondary surfactants, due to their soft profile and their less-foaming attributes. Anionic surfactants generate a lot of foam when assorted. While anionic surfactants are brilliant for stimulating and suspending particulate soils, they are not as good at emulsifying oily soils Anionic surfactants are generally employed in collaboration with other surfactants (non-ionic or **zwitterionic**), which bring up gradation in their performance. The examples of anionic surfactants used in drilling fluid formulation is shown in Table 2.

2.3 *Cationic Surfactants*

Cationic surfactants are involved in a positively charged group at their head. Most of the cationic surface-active agents are anti-microbial, anti-fungal, etc. from the practical point of view they are usually characterized by having an incredibly high substantiate on various substrates. A straight significance of this fact is that a cationic modify the surface behavior and make a hydrophilic to act as a hydrophobic and

Table 1 Nonionic surfactants used in the formulation of drilling fluids

S. No.	Name	Structure	References
1	2-Hexa-decyloxyethnol		[42]
2	Active ethoxylated amine (Ultramine 20)	$H_3C-(CH_2)_{17}-N(CH_2CH_2OH)(CH_2CH_2OH)$	[43]
3	Active ethoxylated amine (Ultramine 50)	$H_3C-(CH_2)_{17}-N(CH_2CH_2O-CH_2CH_2OH)((CH_2CH_2O)_2-CH_2CH_2OH)$	[43]
4	Polyoxyethylene		[44]
5	Oleyl ether	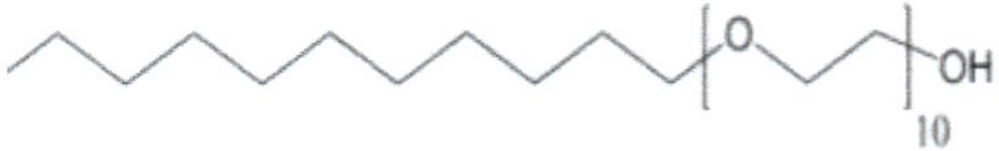	[44]

(continued)

Table 1 (continued)

S. No.	Name	Structure	References
6	Polyoxyethylene	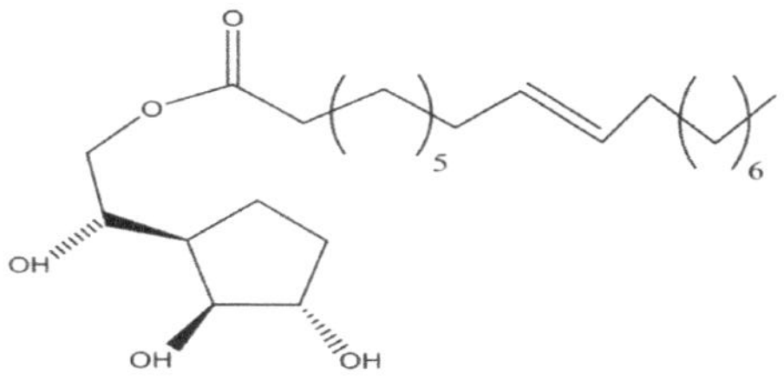	[44]
7	Tallow amine polyethylene glycol ether	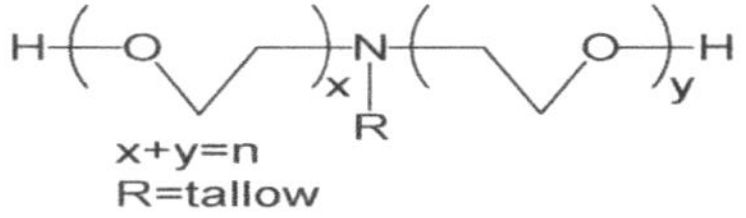	[45]

Table 2 Anionic surfactants used in the formulation of drilling fluids

S. No.	Name	Structure	References
1	Alkyl benzene sulfonate	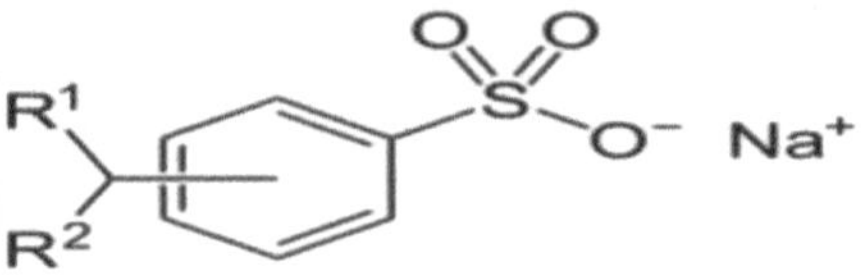	[42]
2	Sodium dodecyl sulfate	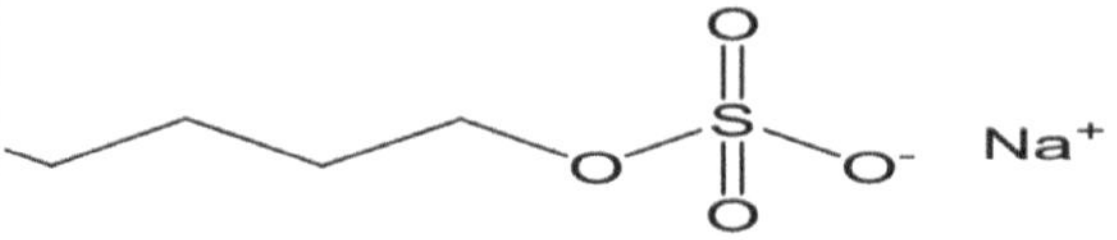	[44]
3	Sodium dodecyl benzene sulfonate	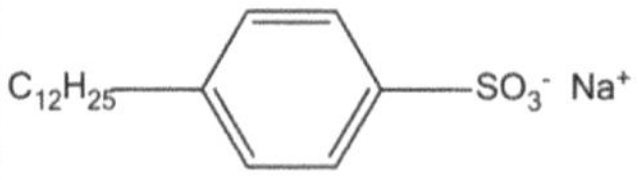	[46]

vice versa. Cationic surfactants cannot be employed with anionic surfactants. If positively functional head cationic surfactants are mixed with negatively functional head anionic surfactants, they will drop out of solution and no longer be effectual. Cationic and nonionic surface-active agents, however, are well-matched. Besides a cationic makes water-insoluble complexes with anionic and these complexes remain soluble in organic solvents which have their application in surface protections. The examples of cationic surfactants used in the formulation of drilling fluids is shown in Table 3.

2.4 *Amphoteric Surfactants*

Amphoteric surfactants are categorized by the reality that these can hold both a positive charge on a cationic site and a negative charge on an anionic site. The dual charges cancel each other and create a net charge of zero, mentioned to as zwitterionic Amphoteric surfactants are habitually sensitive to pH and will act as anionic or cationic depending on the pH. In an alkaline situation, the anionic appearance is major, and in acidic circumstances, the cationic structure succeeds. amino oxide and betaines are commonly used zwitterionic surfactants. There is a table below that represents the Name, Type, and the Chemical structure of the surfactants that are employed in the drilling fluids as an additive to get the ambient outputs in the petroleum industries. The examples of amphoteric surfactants used in the formulation of drilling fluids is shown in Table 4.

Table 3 Cationic surfactants used in the formulation of drilling fluids

S. No.	Name	Structure	References
1	Cetrimonium bromide	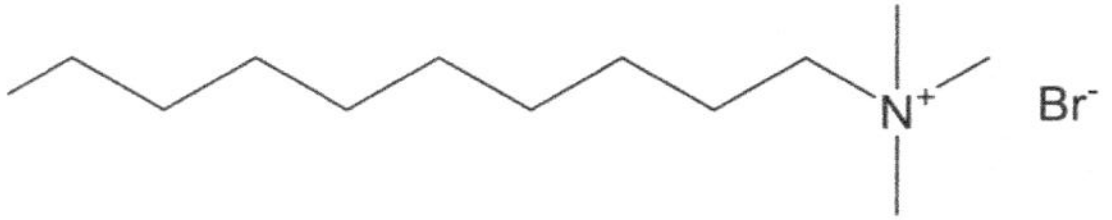	[44]
2	Tetra methyl ammonium chloride	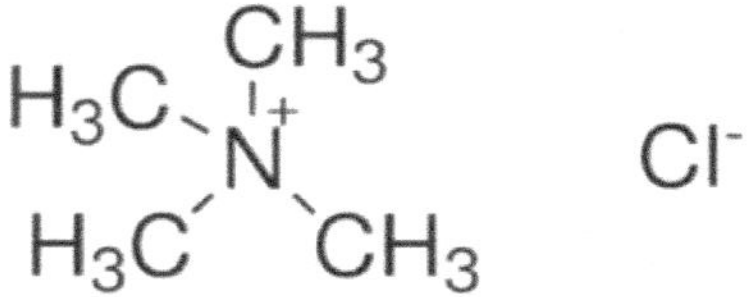	[47]
3	1-Ethyl-3 methylimidazolium chloride	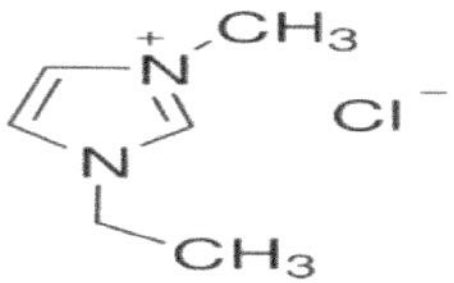	[47]
4	Tetra methyl ammonium bromide	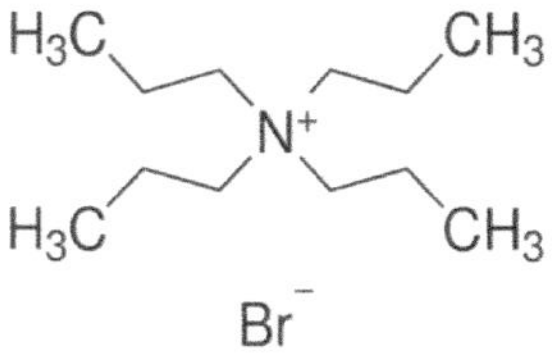	[47]

(continued)

Table 3 (continued)

S. No.	Name	Structure	References
5	Cetyl tri methyl ammonium bromide	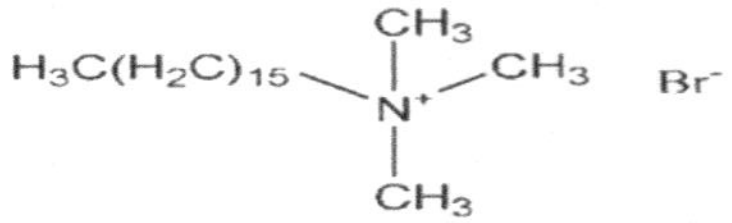	[46]
6	1-Octyl-3-methyl midazolium tetra fluoro borate	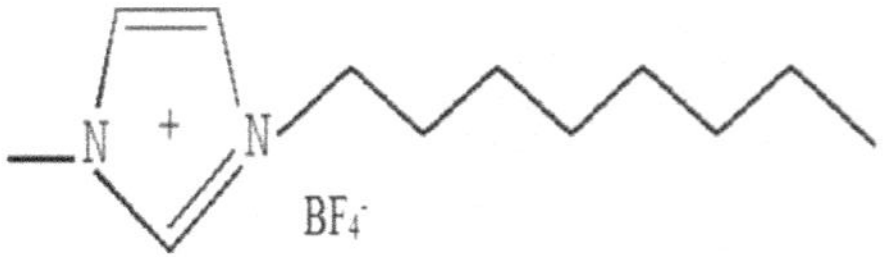	[48]
7	Alkyl dimethyl benzyl ammonium chloride	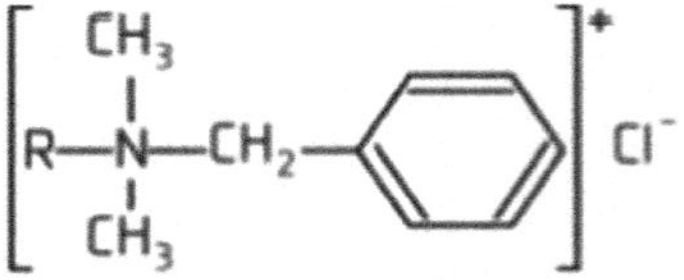	[49]

Table 4 Amphoteric surfactants used in the formulation of drilling fluids

S. No.	Name	Structure	References
1	Polyacrylamide	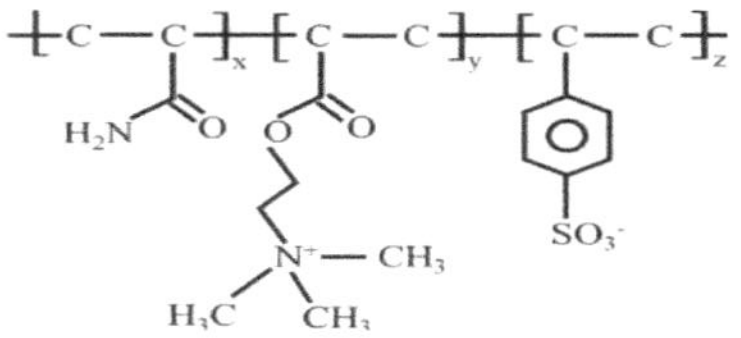	[50]
2	Oleylamidopropyl betaine	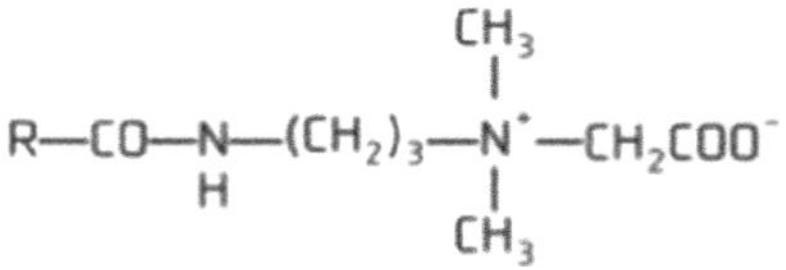	[49]
3	Cocamidopropyl betaine	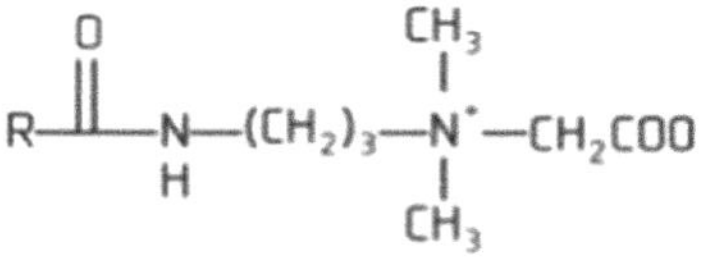	[51]
4	Lauryl myristyl betaine	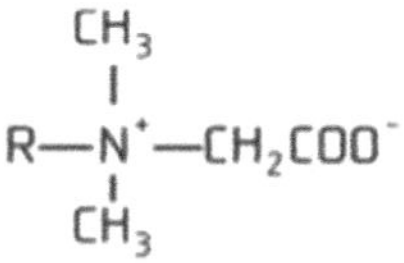	[49]

(continued)

Table 4 (continued)

S. No.	Name	Structure	References
5	Cross-Linked Polyethylene (PEX)	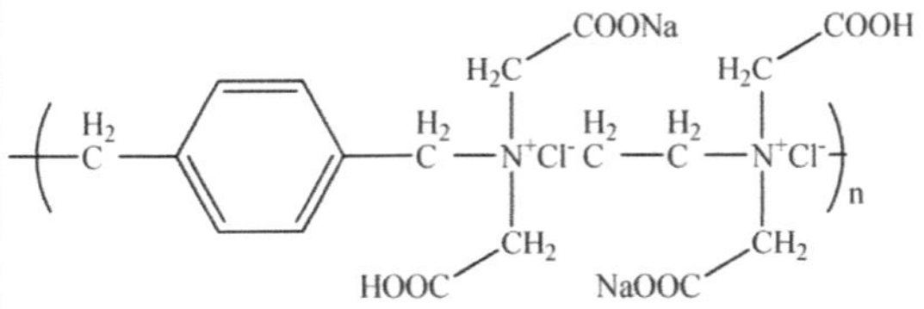	[52]
6	Cocamidopropyl betaine	$C_{12}H_{25}-C(=O)-NH-(CH_2)_3-N^+(CH_3)_2-CH_2-COO^-$	[49]

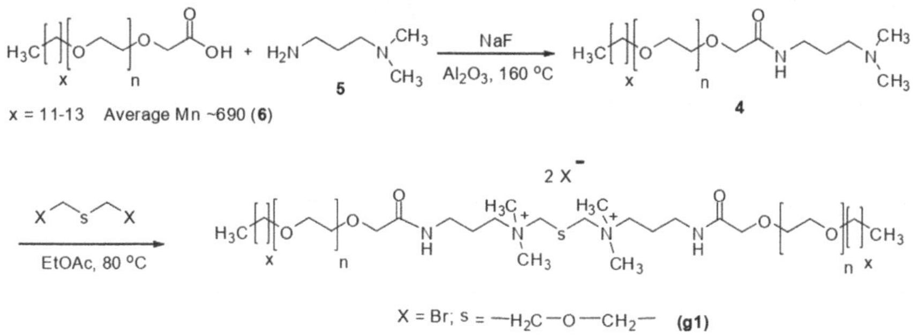

Fig. 4 Synthesis routine of cationic gemini surfactant

3 Synthesis of Surfactants

3.1 Synthesis of Cationic (Gemini) Surfactants

The synthesis route of gemini surfactant is mentioned below. The gemini cationic surfactants have two head groups attached through an organic spacer. The gemini cationic surfactant was synthesized having a diethyl ether spacing by the amidation reaction of glycolic acid ethoxylate lauryl ether 3-(dimethylamino)-1-propylamine and NaF in a 500 mL flask in the inert gas (argon) environment for 8 h at 160 °C. The water produced in the chemical reaction was absorbed by alumina (A12O3) absorbent. The addition amount of compound 3-(dimethylamino)-1-propylamine was added to the reaction mixture after 8 h followed by continuous stirring for 6 h. After the chemical reaction, the unreacted compound 3-(dimethylamino)-1-propylamine was vaporized by reducing the pressure of the reaction vessel and sodium fluoride NaF was separated by filtration process. The product was then treated with bis(2-bromoethyl) ether in the presence of ethyl alcohol up to 8 h at 80 °C. After the extraction of solvent at the end of the reaction, the column chromatography was carried out using ethanol as mobile to get the gemini surfactant [53]. The synthesis route of cationic gemini surfactant is shown in Fig. 4.

3.2 Synthesis of Nonionic Surfactants

The synthesis route for nonionic surfactant is mentioned below. The nitrogen atom is the central atom while benzene rings are attached with three branches of surfactant molecule. The chemical reaction was carried out with 0.02 mol of Tris(4-dodecylphenethyl) amine and 0.08 mol of chlorosulphonic acid dissolved in 1,2-dichloroethane using ice bath with temperature below 15 °C up to 3 min under the constant pressure environment and constant stirring. The reaction mixture was then heated to 25 °C for the next 3 h. In the next phase 50 ml of pure water and

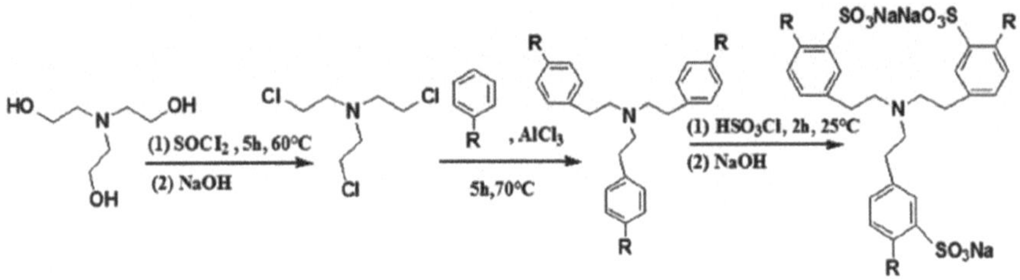

Fig. 5 Synthesis route of nonionic surfactant [54]

30 ml of petroleum ether was added to the reaction mixture dropwise and stirring was applied. The reaction mixture left undisturbed and upper layer was separated and mixed with 20 wt.% solution of sodium hydroxide to adjust the pH to 8.0. Eventually, the product mixture was mixed with the ethanol solution at 60 °C and brown colored product tris(2-dodecyl benzene sulfonic acid sodium) ethyl amine was obtained through filtration process [54]. The synthesis route of nonionic surfactant is shown in Fig. 5.

4 Impact of Surfactants on Rheology and Filtration Properties of Drilling Fluids

Surfactants used as additives in drilling mud to withstand desired rheological properties for potential drilling applications [55]. In literature many rheological tests were performed to analyze rheological properties like viscosity, gel strength, plastic viscosity thermal aging and Yield point and to check stability at temperature and pressure counters in reservoir. Addition of surfactants in base mud increases the viscosity because long chain molecules interact with mud platelets and form a compact structure which causes the increase in viscosity and molecules having small chains at equal concentration did not have considerable effect on viscosity. Simple base mud cannot withstand because at high temperature and aging time decreases the Plastic viscosity(PV) of drilling fluid so addition of surfactant is much more necessary [56, 57]. In drilling applications there is need of a particular value of PV to transport cuttings of formation as PV is majorly due to the linkage of base mud and additive (surfactant) and might be decreases at elevated temperature due to lesser interaction and clashing at higher temperature [38]. Gel strength (GS) in other factors to predict the rheological properties for drilling fluid having surfactant as additive. GS is the ability of fluid to hold formations cuttings when drilling operations are paused and when operations is stopped cutting of formations tends to settle down at bottom hole so high gel strength would require high pressure of pumps to restart the process. From literature it is reported that by increasing temperature it causes reduction in gel strength so a specific gel strength would be required to carry out formation cuttings and high strength would not be preferable in drilling and may cause pipe sticking and formation damage [58]. Yield point is measure of bonding (Chemical and electrical) between platelets of mud and additive and higher the yield point, higher the viscosity and addition of ionic surfactant increases the yield point of drilling mud. Lower yield point of base mud is due to force of repulsion between particles that push molecule away and by adding surfactant it will causes force of attraction between platelets of base mud and surfactant. Threshold value of yield point is required for smooth running and cleaning and decrease in yield point may cause cuttings to transport at bottom which might cause bit balling and potential pipe sticking [59].

Filtration loss decreases the efficiency of overall process as drilling fluids (DFs) penetrates the formation so therefore to inhibit the fluid loss into formation surfactant added to clay to control the filtration properties of DFs. Basic clay as DFs gives less filtrate and high fluid loss unwanted in drilling operation and it may cause the formation damage and unreliability of wellbore [6, 60]. As explained earlier surfactant causes the formation of cake due to compact structure which inhibits the fluid loss into formations and promising results were obtained as addition of surfactant increases the filtrate volume [23].

Swelling of clay is detrimental phenomenon encountered in reservoir due to absorbing water and so therefore addition of surfactant reduces the swelling ability of clay. Increasing concentration of surfactant decreases swelling ability from 99 to 91% and this is due to the adsorption of surfactant molecules between the platelets of clay. There is a threshold value of concentration up to and also known as which swelling ability decreases but after it starts increasing and may affect the process directly [58]. Gemini surfactants are most widely used and showed promising results in field applications. Adsorption of surfactant on clay surface is dependent on concentration and Gemini surfactant and traditional surfactant showed similar results by using sodium based clay to check the swelling of clay. Both showed similar behavior as only hydrophilic part of Gemini surfactant adsorb on clay surface while the second is located towards water phase.

5 Evaluation of Shale Inhibition Characteristics with Surfactants

The inhibition characteristics of drilling fluids can be determined by performing different tests such as linear swelling, hot rolling dispersion test, pressure transmission test, uniaxial compression test. The presence of shale inhibitor in the drilling fluids can alter the inhibition characteristics altogether by changing the chemical interactions of shale with water. The effectiveness of shale inhibitors can be determined by XRD analysis of shale powder, zeta potential analysis, particle size measurement, sedimentation test. The interactions of shale inhibitors with the shale surface can be analyzed with the help of scanning electron microscope. The explanation of these inhibition tests is mentioned below.

5.1 Shale Swelling Properties

The linear swelling test of shale usually carried out in the presence of drilling fluids having shale inhibitors and performance of shale inhibitors was determined over a specific period. The linear swelling of shale experiments performed by using shale pellet made by compacting shale powder in the compactor at 6000 psi pressure for half hour. The linear swelling of shale was determined by placing the shale pellet in the cup of linear swelling cell followed by the filling of drilling fluid have specific shale inhibitor. The linear swelling of shale usually analyzed for 24 h experiments of by immersing the shale pellet in the drilling fluid. Murtaza et. al. has performed linear shale swelling experiment by using three different gemini surfactants such as 12-EO-12 (GS-A), 12-NH-12 (GS-B), and 12-OH-12 (GS-C) at room temperature and linear swelling results were compared with the KCl and dodecyltrimethylammonium bromide (DTAB) based drilling fluid as shown in Fig. 6.

The inhibition performance of three gemini surfactants was evaluated with various concentrations and it was observed that optimum concentration of all three gemini surfactants ranges from 0.05 to 0.1%. Increasing the surfactants concentration results in the decrease in shale swelling due to adsorption of surfactant in the interlayers of clay minerals. After the critical concentration (0.1%) of surfactants in the drilling formulation, the linear swelling of shale increases due to the formation of bilayer of alkyl chains in the interlayers and also shows the tilting arrangement of surfactant molecules which leads to the increase int the d-spacing of clay minerals [58].

5.2 Cutting Dispersion Test

The drilling of wellbore formation results in the huge amount of shale cuttings that needs to be transported to the wellbore surface. The purpose of drilling fluid formulations in the wellbore is to keep the drilled cuttings suspended and to minimize the dispersion of cuttings in the wellbore. The dispersion of drilled cuttings into the wellbore mar results in severe issues such as bit balling, high density of drilling fluid and ultimately high power would be required to rotate the drilling bit. Ahmad et. al has studied the effect of cationic gemini surfactant in the formulation of drilling fluid to minimize the dispersion of drilled cuttings. The dispersion of drilling cutting was observed minimum in the presence of deionized water, whereas the maximum dispersion of drilled cuttings was observed approximately 88% in the presence of modified drilling fluid which has gemini surfactant as one of the drilling fluid additives as shown in Fig. 7. The dispersion of drilled cuttings can be minimized with the addition of such additives to the drilling fluids which protects and forms a layer around cutting and minimizes the dispersion of shale cuttings [16].

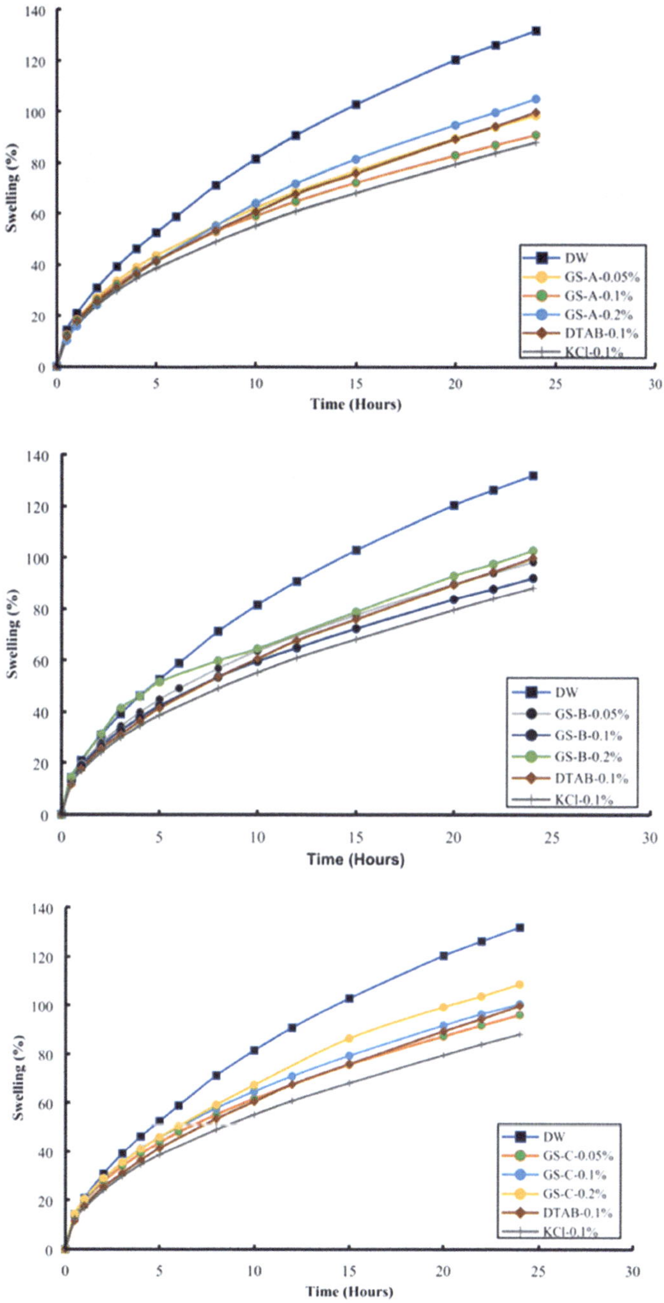

Fig. 6 Linear swelling test in the presence of 12-EO-12 (GS-A), 12-NH-12 (GS-B), and 12-OH-12 (GS-C) [58]

Fig. 7 Shale recovery with different drilling formulations [16]

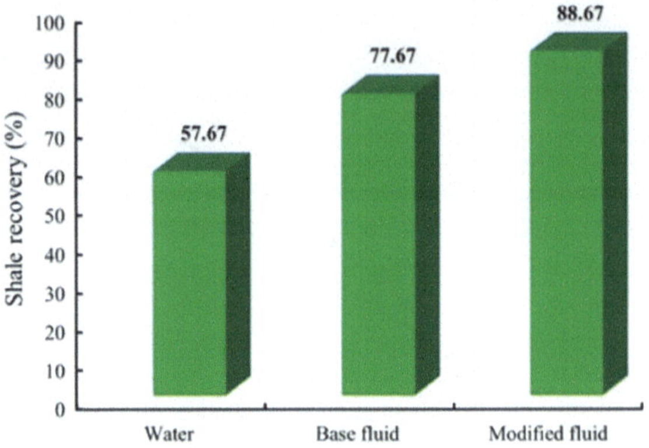

5.3 Zeta Potential Analysis

Colloidal dispersion stability affected by the charge density on clay particles and chemical interaction of additives (inhibitors) with clay particles. Colloidal dispersion stability is mainly measured by zeta potential. If the clay particles have high charge density, then the colloidal dispersion will have high value of zeta potential. The clay particles in the colloidal dispersion tend to coagulate if the dispersion has less value of zeta potential. The water-based drilling fluids usually accompanied with shale inhibitors (surfactants) and these additives interact with the clay contents of drilling fluids which affect the stability of drilling fluids. Barati et al. has studied the effect of KETALO-15 concentration on zeta potential of clay dispersed in deionized water as shown in Fig. 8. The zeta potential of clay dispersed in deionized water in the absence of KETALO-15 was observed 39.6 mV, while the addition of KETALO-15 has reduced the zeta potential to 21.2 mV at 3 wt.% of KETALO-15 in the colloidal dispersion. The zeta potential results showed that the addition of surfactant (KETALO-15) into the colloidal dispersion decreased the interaction of clay with water, whereas the clay chemical interacts with surfactant molecules. These results indicate that the surfactant can be use as shale inhibitors because they can limit the interaction of water with clay [45].

Fig. 8 Effect of KETALO-15 concentration of zeta potential on bentonite dispersion [45]

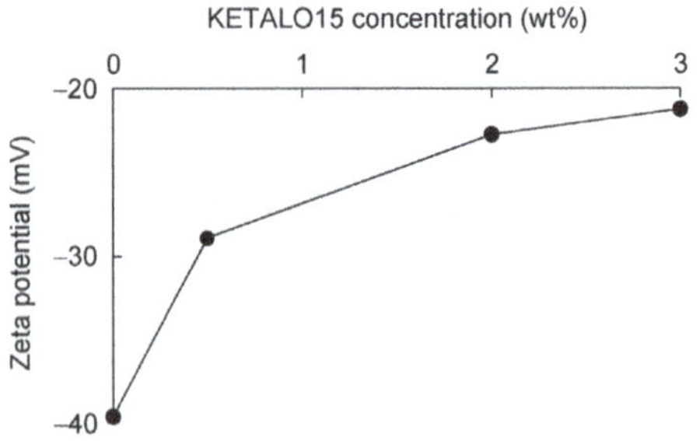

5.4 Uniaxial Compression Test

The wellbore formations having shale contents in the wellbore rock mainly affected by the drilling fluids in the drilling process. The strength of shale in the wellbore formation mainly governed by the type of clay minerals and other impurities in the structure. The greater the strength of shale rock is less likely to deform or become unstable during the drilling process. On the other hand, the shale rock having less strength is more likely to destabilize the wellbore formations. The effect of shale inhibitors on the compression strength of shale was determined by treated the shale core sample with various shale inhibitors (surfactants) at different concentrations. Liu et al. has used the combination of two different surfactants such as Tween 60 and Span 20 to study the inhibition characteristics of shale with various concentrations. The collective 1:1 of these two surfactants were referred as SP in the experimental procedure. He has used the Taw-2000 rock triaxial tester to the study the uniaxial compression strength of shale at the deformation rate of 0.00125 mm/s. The pristine shale sample was soaked in the deionized water and compression strength of shale was observed 31.39 MPa. However, the addition of 0.5% of SP in the deionized water has increased the compressive strength of shale up to 46.37 MPa and further increase the concentration of SP to 4% in the deionized water resulted in the increase of compressive strength of shale up to 77.46 MPa as shown in Fig. 9. The results showed that the addition of surfactants in the drilling fluid formulation can increase the compressive strength of shale which reduced the ability of the shale swelling and hydration [61].

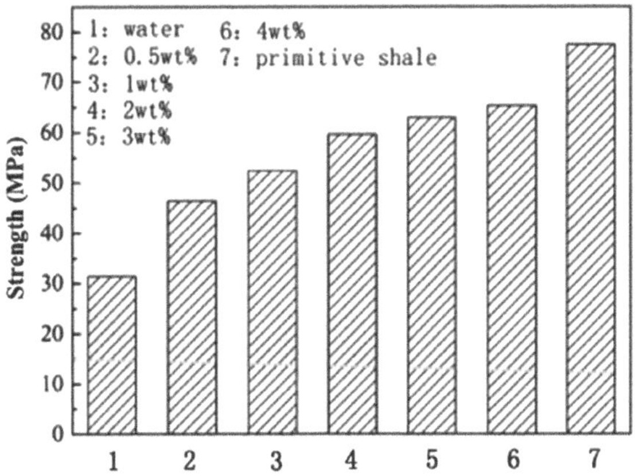

Fig. 9 Compressive strength of shale in different concentrations of SP inhibitors [61]

5.5 *Particle Size Measurement*

The determination of particles size in the presence of water and in the presence of inhibitor solution gives the indication about swelling and dispersion of clay minerals. The montmorillonite particles has strong affinity towards water which leads to delamination and dispersion of clay particles in the water and decreased the montmorillonite particles along with the increase in the number of small platelets of montmorillonite. The delamination is due to migration of water molecules to the interlayers of clay surface which weakens the interlayer interactions and leads to dispersion of clay particles in the water. The inhibitors (surfactant) in the water-based drilling fluid react chemical with the montmorillonite clay and prevent the delamination and dispersion of clay particles. Aghdam et al. has used different concentrations (mass %) of Seidlitzia Rosmarinus leaf and stem extract (SRLSE) to determine the particles size of montmorillonite in the deionized water as shown in Fig. 10. Montmorillonite particle size measured in deionized water was approximately 526 nm, indicating the extreme delamination and swelling of clay particles. Conversely, the addition of SRLSE (1.5 and 3%) concentration in the deionized water having dispersed clay particles resulted in the increase of particles size to 1636 nm and 3050 nm, respectively. This increase in the particle size was due to strong affinity of SRLSE towards the clay particles which leads to the reduction delamination, dispersion and swelling of clay particles in the deionized water. The increase in particle size also indicates the superior inhibition characteristics of SRLSE [62].

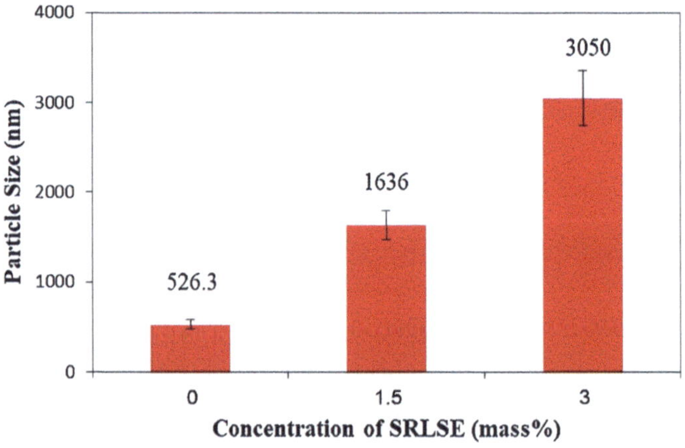

Fig. 10 Effect of mass% on particle size of clay [62]

5.6 *Wettability Alteration Test*

The interactions of drilling fluids with shale rock mainly affected by the presence of drilling fluid additives and water content. The additives of drilling fluids reacts with the shale formations and results in the swelling and hydration which leads to other drilling issues such as wellbore collapse, and pipe sticking. The shale inhibitors (surfactants) in the drilling fluids reacts with the shale surface and alter the hydrophilicity of wellbore formations. A schematics shows the modification of clay surface with surfactant molecules as shown in Fig. 11. It was observed that the surfactants with cationic head reacts with negatively charged clay minerals in the shale formations and the tail end of surfactant molecules which are mainly alkyl chains make the outer surface hydrophobic to the water contents in the drilling fluid. The hydrophobic surface of shale formations due to the adsorption of surfactants minimizes the interactions of water with shale resultantly reduces the swelling and hydration of shale formations. The schematic diagram showing the adsorption of surfactant on the shale surface is given below.

Shi et al. has studied the effect of several cationic surfactants to understand the wettability of shale surface after treating with shale inhibitors (surfactants). After performing several experiments of contact angle, three cationic surfactants YS-1, FC-1 and FS-1 were chosen to study the impact of these materials on shale wettability. The combination of these surfactants with the concentrations 0.001%, 0.01%, 0.05% respectively can increase the contact angle of water as on shale surface as shown in the figure. The hydrophobic tails (alkyl chains) repel the water from the surface of shale and resulting in the increase in the contact angle of water droplet on the shale surface. The results of increasing contact angle shows that surfactants have ability to change the wettability of shale surface as shown in Fig. 12 [17].

In another study, Aghasemi et al. has studied the effect of biosurfactant on wettability of clay surface. The hydrophilicity of clay surface has changed altogether

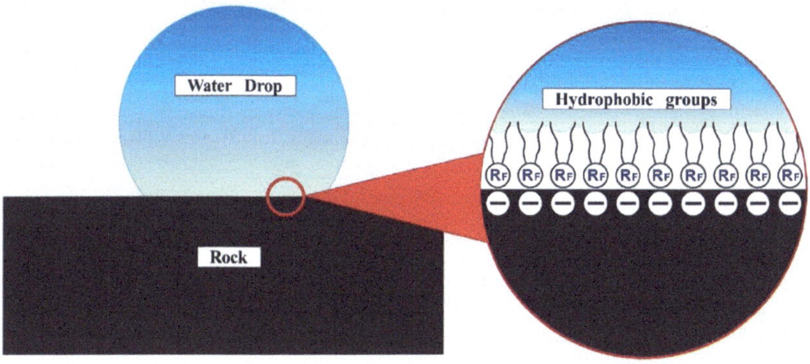

Fig. 11 Schematic diagram showing the adsorption of surfactant on shale surface [17]

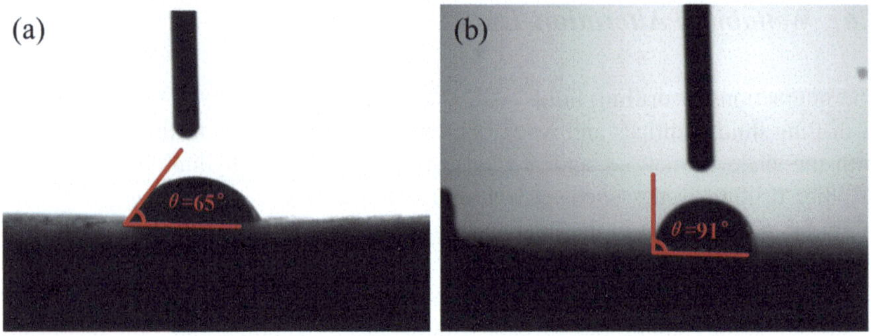

Fig. 12 Contact angle measurement **a** before and **b** after treating with surfactants [17]

after treating the clay samples with biosurfactants at different concentrations. It was observed by measuring the contact angle of small water droplet on the clay surface treated with biosurfactant. The contact angle of water droplet increased by increasing the concentration of biosurfactant in the treatment of clay sample. This increase in the contact angle reflect the lesser affinity of clay sample towards water after treating with biosurfactant. It shows that the wettability of wellbore formation can be tuned to minimized the clay swelling and hydration by using biosurfactant in the drilling fluid formulations [37]. A schematics showing the modification of clay in the wellbore formation during the drilling process in shown in Fig. 13.

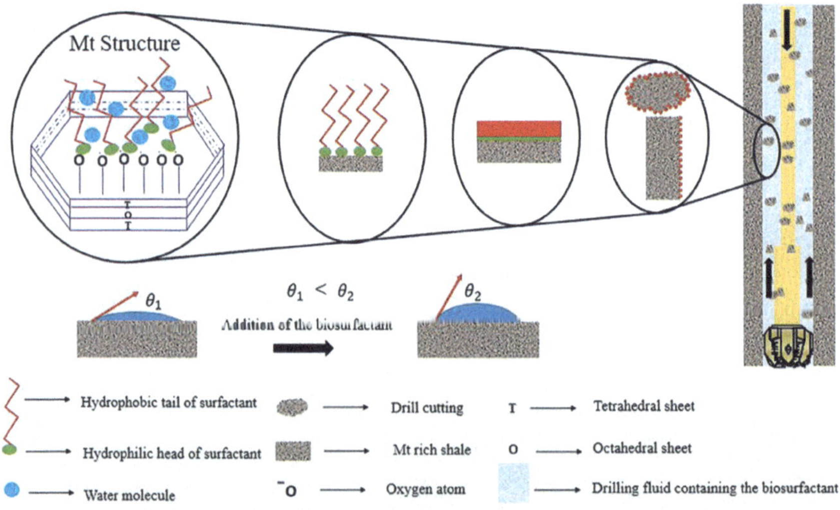

Fig. 13 Schematic diagram showing the interaction of biosurfactant with clay particles [37]

Fig. 14 Shale surface **a** before **b** after treating with QS natural surfactant [40]

5.7 SEM Analysis of Shale

Scanning electron microscopic analysis of shale before and after treating with shale inhibitor is performed to investigate the surface morphology. Aghdam et al. has studied the effect of natural surfactant quillaja saponin (QS) on shale surface morphology and dispersion properties. The SEM analysis of shale before treating with QS natural surfactant showed that particles have very small in size and are extremely dispersion, whereas the shale after treating with QS natural surfactant has bigger size agglomerates with minimum dispersion of particles. The SEM analysis confirms that this QS natural surfactant has ability to minimize the disintegration and dispersion of shale in the drilling fluid as shown in Fig. 14 [40].

5.8 XRD Analysis of Shale

The inhibition of shale swelling and hydration in the presence of shale inhibitors also determined by the measuring the d-spacing of clay minerals found in shale sample [63]. The shale inhibitors have ability to chemically interact with negatively charges clay platelets and forms an inhibitive monolayer, bilayer or multilayer on the clay platelets which minimizes the interaction of water and clay minerals. The adsorption of inhibitors (surfactants) on clay results in the increment of interlayer d-spacing. Huang et al. has studied the effect of Biosurfactant cationic alkyl polyglycoside (CAPG) on d-spacing of bentonite clay as shown in Fig. 15. It was observed that the bentonite d-spacing without treating with biosurfactant 1.239 nm, whereas by treating the bentonite with CAPG, the d-spacing of bentonite increased to 1.387 nm at 2 wt.% concentration of CAPG. The increase in the d-spacing with biosurfactant depicts the swelling and hydration inhibition capacity of clay [36]. The schematics showing the interaction of CAPG surfactant with clay particles is shown in Fig. 16.

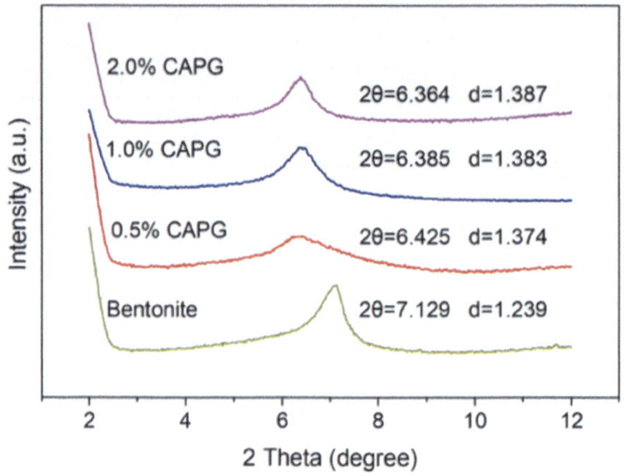

Fig. 15 Effect of biosurfactant CAPG on d-spacing of bentonite [36]

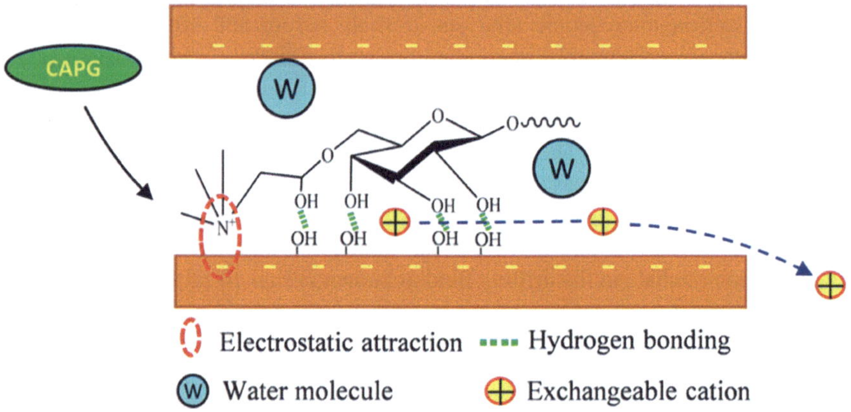

Fig. 16 Schematic diagram explaining the adsorption mechanism of CAPG on interlayers of clay [36]

6 Effect of Different Parameters on Shale Inhibition Properties

6.1 Effect of Temperature on Shale Inhibition Characteristics

Drilling operations carried out at deep bottom hole may face the high temperature and high pressure conditions up to 150 °C. The high pressure and temperatures affects the drilling fluid rheological, filtration and shale inhibition properties. These severe conditions at bottom hole such as high pressure up to 10,000 psi and 150 °C temperature may result in several mechanical issues and other difficult challenges. Therefore, the selection of drilling fluid additives for high temperature drilling application is very important for smooth drilling operations. Shettigar et al. has observed the effect of Hexadecyl-trimethyl-ammonium bromide or cetyltrimethyl ammonium bromide (CTAB) surfactant along with the combination of anionic surfactant on rheology, filtration and shale inhibition properties especially at higher temperature (120 °C). The impact of individual additive and combined effect of anionic polymer CTAB was analyzed at different temperature and it was observed that with the increase in the CTAB concentration from 0.1 to 0.3% (keeping the anionic polymer concentration constant) the fluid loss was reduced to 10.4–5.4 cc and fluid loss was slightly increased to 12.8 ml by increasing temperature up to 120 °C. Similarly, the thermal stability test of drilling fluid formulation was carried out by hot rolling the formulation at 120 °C followed by the measurement of rheological properties. It was observed that polymer was not stable at higher temperature due to hydrolytic and oxidative process. While the addition of CTAB surfactant to the drilling fluid formulation, the thermal stability of formulation increase. The increase in the thermal stability was due to the CTAB surfactant which is stable at higher temperature 120 °C. it was believed the higher thermal stability of formulation due to the formation of complex between long alkyl chains of CTAB molecules and alkyl chains of anionic polymer chains which results in better rheological properties after hot rolling test of drilling fluid [64].

Another research study carried out by Aggrey et al. to investigate the interactions of nonionic surfactant with shale and shale inhibition characteristics at different temperatures. The shale inhibition is directly related to the uniaxial compression strength of shale. The higher the value of uniaxial compression strength of shale, then less will be the swelling and hydration of shale. The shale was treated with nonionic surfactant at 25 and 120 °C followed by the determination of uniaxial compression strength. It was observed that by increasing the temperature to 120 °C, the uniaxial compression strength was higher than the strength of shale treated with nonionic surfactant at 25 °C. The higher value of compressive strength directly related to the less swelling and hydration of shale. The inhibition characteristics of shale at higher temperature was attributed to the chemical interactions of nonionic surfactant with shale rock [65].

6.2 *Effect of Electrolyte on Surfactant Based Drilling Fluids Properties*

Shale swelling in the wellbore formation leads to the instability of reservoir which is the result of physio-chemical interactions of drilling fluid with clay contents in the wellbore and mechanical forces of drilling operations. Swelling of shale in the wellbore formation due to the presence of electrolytes often explained by two kinds of swelling mechanisms such as osmotic swelling and crystalline swelling. In osmotic swelling, the clay minerals in the wellbore having high concentration of metal cations compared to the surrounding fluid (drilling fluid) which results in the migration of water molecules to the interlayers of clay minerals. The water molecules have strong affinity towards the changes clay minerals and results in the hydration and swelling of wellbore formations. On the other hand, crystalline swelling in the wellbore occurs due to the migration of metal cations from surrounding solution (drilling fluid) to the interlayers of clay minerals which leads to the formation monolayer, bilayer and multilayer of hydrated cations in between the layers of clay minerals.

The disruption of ionic interaction of drilling fluid components with shale content results in the severe hydration and swelling of wellbore formations. These interactions can be controlled using different fluid additives such as electrolytes and surfactants with appropriate type and concentration which favors the interaction of these additives with shale compared to the interaction of shale with water. The drilling fluids containing surfactants as drilling fluid additive often face dissolution issue due to the type and concentration of surfactant. When certain surfactant solution were heated, the phase separation phenomenon occur which results in the cloudy solution of surfactants. It was believed that phase separated surfactant aggregates can be useful for shale inhibition which minimizes the migration of water to the shale formations. The nonionic surfactants also named as glycols because of the presence of -OH group in their structures. The presence of nonionic surfactants in the drilling fluid formulations considered as shale inhibitors due to the superior interactions (hydrogen bonding) of -OH group of nonionic surfactant with oxygen atoms of silicates and aluminates in the shale formations. The presence of aluminates and silicates in the shale formations also have great affinity towards the water which makes hydrogen bonding and results in the swelling of shale. The nonionic surfactant competes between the interaction of water molecules with oxygen atoms of aluminates and silicates of shale. Due to the strong hydrogen bonding, nonionic surfactant's hydrogen bonding prevails and forms the protective layer across the shale surface which limits the shale hydration and swelling.

The presence of electrolyte such as potassium chloride along with the nonionic surfactant synergistically inhibit the shale swelling and hydrations. The potassium chloride have metal cation that can be exchanges with the metal cations present in the interlayers of clay of shale formations. The potassium ion first exchange the free ions available in the clay interlayers which results in the compact structure of clay interlayers and secondly, the nonionic surfactant, which are glycols, have tendency to make strong hydrogen bonding with oxygen atoms of aluminates and silicates of clay in the shale formations. The molecular size of nonionic surfactant is comparatively bigger than the molecules of water and creates hindrance for the affinity of water with shale. The strong interactions of surfactant with shale and substitution of potassium ions in the clay contents resulted a complex structure which prevents the interaction of water with shale and prevent the shale swelling and hydration.

7 Field Applications

Surfactants have been widely used for several oilfield application due to the unique and diversified properties. Surfactants are employed as wetting agents and emulsifiers in the drilling wells where oil-based drilling fluids are preferred. However, surfactants have been widely used in the formulations of water-based drilling fluids and the use of surfactants in the formulations have been continuously increasing due to numerous applications in the oilfield drilling operation. The main attributes of surfactant based drilling fluids includes the prevention of shale swelling and wellbore instabilities, to prevent the sticking of shale cuttings to the drill bit and differential pipe sticking. The surfactants are also employed as foaming agents in the drilling fluids where high gas to water ratio forms are required for the drilling of hard rock wellbore and low pressure reservoirs. Surfactants are also employed as defoaming agents where less foaming characteristics of drilling fluids are required. In the formulation of water-based drilling fluids, the surfactants are also coupled with water soluble polymers to make complex formulations to enhance the shale inhibition characteristics for low pressure reservoirs. In the recent years, the major applications of surfactants are reported in the enhanced oil recovery. The studies in the literature reported that surfactants used in the EOR applications results in the greater recovery of oil compared to the conventional recovery techniques. The surfactant flooding in the reservoirs increases the production of oil by decreasing the interfacial tension, improving the oil mobility in the reservoir. Surfactant flooding in the reservoir alters the wettability of reservoir rock and allow the water to pass through them which results of enhanced oil recovery. The polymer-surfactant flooding also proves fruitful results in EOR applications with greater recovery of oil.

8 Recommendation and Challenges of Surfactants for Shale Inhibition

Surfactants have been employed in the formulation of drilling fluid for the inhibition of shale swelling and hydration during the drilling of oil and gas reservoirs. The selection of surfactant type (Cationic, anionic, non-ionic) for the formulation of drilling fluid, their concentration and compatibility with other drilling fluid additives (polymers, electrolytes and clay contents) is still a main challenge before starting the drilling operations for a reservoir. The reservoir conditions vary with increasing depth and at the deep bottom hole high temperature and high pressure condition are being observed for the drilling fluids. The drilling fluid formulations or the additives in the drilling fluid may face deterioration or degradation at high temperature and high pressure conditions. The properties of surfactants such as the solubility in the drilling fluids and the interactions of surfactants with other drilling fluid additives mainly affected which ultimately leads to the change in drilling fluid properties. The desired inhibition results might not be achieved with surfactant based drilling fluids for high temperate and high pressure reservoirs. The possible future work related to the formulation of surfactant based drilling fluid formulation include the use of surfactants that can bear high temperatures such as 150 °C at the bottom hole conditions of reservoir with environmentally friendly nature. The other recommendations include the characterization of surfactant based drilling fluids and hydration inhibition especially for marine drilling applications which will also considers the electrolytes impact on the performance of drilling fluids. The recent research focuses on the extraction of surfactants from biological sources and the potential use for the inhibition characteristics in water-based drilling fluids.

9 Conclusion

Drilling fluids play a key role in the extraction of fossil fuels from reservoirs. Drilling fluids have several important applications of in the oil and gas industry during drilling operations such as lubricating the drill bit, minimize the filtration loss, suspension of drilled cuttings, transport the cuttings to the surface, conformation control and prevent the shale swelling and hydration. The selection of drilling fluid and additives are equally important for the easy and uninterrupted drilling operations. Drilling fluid additives such as rheology modifiers, fluid loss control agents, and shale inhibitors are the key components which control the properties of drilling fluids and the proper selection with acceptable concentration of each component along with their mutual compatibility results in smooth drilling operation in the reservoir. The water-based drilling fluids are widely accepted in the drilling operation of reservoirs due to the ease the preparation, superior rheology and filtration properties, enhanced shale inhibition characteristics and environmentally friendly nature. Shale inhibition properties of all type of surfactant are reported in the literature but the selection of surfactant for inhibition of swelling and hydration mainly depends on reservoirs properties, types of swelling clay content in the reservoir formation, and the interaction of surfactant molecules with the formation rock. The recent studies reported that the use of cationic gemini surfactant in the formulation of water-based drilling fluid resulted in the improved shale inhibition properties.

Acknowledgements The authors would like to say thanks to the Center for Integrative and Petroleum Research of King Fahd University of Petroleum and Minerals, Department of Chemical, Polymer, and Composite Materials Engineering of University of Engineering and Technology (New Campus) for providing the technical support in writing this chapter. The authors would like to acknowledge the support provided by the students, Zubair Khalid and Nauman Zahid for compiling and editing this chapter.

References

1. Moslemizadeh A, Aghdam SK, Shahbazi K, Aghdam HK, Alboghobeish F (2016) Assessment of swelling inhibitive effect of CTAB adsorption on montmorillonite in aqueous phase. Appl Clay Sci 127:111–122
2. Bourgoyne Jr AT, Millheim KK, Chenevert ME, Young Jr FS (1991) Applied drilling engineering
3. Attia M, Elsorafy W, D'Angelo S (2010) New engineered approach to replace oil-based drilling fluids with high performance water-based drilling fluids in Mediterranean Sea. In: North Africa technical conference and exhibition
4. Zhuang G, Zhang H, Wu H, Zhang Z, Liao L (2017) Influence of the surfactants' nature on the structure and rheology of organo-montmorillonite in oil-based drilling fluids. Appl Clay Sci 135:244–252. https://doi.org/10.1016/j.clay.2016.09.033
5. Abduo MI, Dahab AS, Abuseda H, AbdulAziz AM, Elhossieny MS (2016) Comparative study of using water-based mud containing multiwall carbon nanotubes versus oil-based mud in HPHT fields. Egypt J Pet 25(4):459–464. https://doi.org/10.1016/j.ejpe.2015.10.008
6. Dias FTG, Souza RR, Lucas EF (2015) Influence of modified starches composition on their performance as fluid loss additives in invert-emulsion drilling fluids. Fuel 140:711–716
7. Jain R, Mahto V, Sharma VP (2015) Evaluation of polyacrylamide-grafted-polyethylene glycol/silica nanocomposite as potential additive in water based drilling mud for reactive shale formation. J Nat Gas Sci Eng 26:526–537. https://doi.org/10.1016/j.jngse.2015.06.051
8. Liu F, Jiang G, Peng S, He Y, Wang J (2016) Amphoteric polymer as an anti-calcium contamination fluid-loss additive in water-based drilling fluids. Energy Fuels 30(9):7221–7228. https://doi.org/10.1021/acs.energyfuels.6b01567
9. Yang L, Jiang G, Shi Y, Lin X, Yang X (2017) Application of ionic liquid to a high-performance calcium-resistant additive for filtration control of bentonite/water-based drilling fluids. J Mater Sci 52(11):6362–6375. https://doi.org/10.1007/s10853-017-0870-7
10. Hossain ME, Wajheeuddin M (2016) The use of grass as an environmentally friendly additive in water-based drilling fluids. Pet Sci 13(2):292–303. https://doi.org/10.1007/s12182-016-0083-8
11. Fattah KA, Lashin A (2016) Investigation of mud density and weighting materials effect on drilling fluid filter cake properties and formation damage. J African Earth Sci 117:345–357
12. Zhao X, Qiu Z, Sun B, Liu S, Xing X, Wang M (2019) Formation damage mechanisms associated with drilling and completion fluids for deepwater reservoirs. J Pet Sci Eng 173:112–121
13. Roosta H, Dashti A, Mazloumi SH, Varaminian F (2016) Inhibition properties of new amino acids for prevention of hydrate formation in carbon dioxide–water system: experimental and modeling investigations. J Mol Liq 215:656–663
14. Sivasankar P, Kumar GS (2019) Influence of bio-clogging induced formation damage on performance of microbial enhanced oil recovery processes. Fuel 236:100–109
15. Sun Y et al (2019) Large-scale experiments of the borehole instability on shale formation influenced by drill pipe rotation. Energy Sci Eng 7(6):2895–2920
16. Ahmad HM, Kamal MS, Mahmoud M, Shakil Hussain SM, Abouelresh M, Al Harthi MA (2019) Organophilic clay-based drilling fluids for mitigation of unconventional shale reservoirs instability and formation damage. J Energy Resour Technol Trans ASME 141(9). https://doi.org/10.1115/1.4043248
17. Shi Y, Chen S, Yang X, Yu L, Cai J (2019) Enhancing wellbore stability of coal measure strata by electrical inhibition and wettability control. J Pet Sci Eng 174:544–552. https://doi.org/10.1016/j.petrol.2018.11.052
18. Li MC et al (2015) Soy protein isolate as fluid loss additive in bentonite-water-based drilling fluids. ACS Appl Mater Interfaces 7(44):24799–24809. https://doi.org/10.1021/acsami.5b07883
19. Ahmad HM, Kamal MS, Hussain SMS, Al-Harthi M (2020) Synthesis of novel polymer nanocomposite for water-based drilling fluids. AIP Conf Proc 2205(1):20057

20. Ahmad HM, Kamal MS, Hussain SMS, Al-Harthi M (2020) Synthesis of novel copolymer based on N-Vinyl caprolactam and acrylamide monomers for high temperature drilling applications. AIP Conf Proc 2205(1):20049
21. Li M-C, Wu Q, Song K, Qing Y, Wu Y (2015) Cellulose nanoparticles as modifiers for rheology and fluid loss in bentonite water-based fluids. ACS Appl Mater Interfaces 7(8):5006–5016
22. Kamal MS, Adewunmi AA, Sultan AS, Al-Hamad MF, Mehmood U (2017) Recent advances in nanoparticles enhanced oil recovery: rheology, interfacial tension, oil recovery, and wettability alteration. J Nanomater 2017
23. Sunintaboon P, Duangphet S, Tangboriboonrat P (2009) Polyethyleneimine-functionalized poly (methyl methacrylate) colloidal nanoparticles for directly coating natural rubber sheet. Colloids Surf A Physicochem Eng Asp 350(1–3):114–120
24. de Carvalho Balaban R, Vidal ELF, Borges MR (2015) Design of experiments to evaluate clay swelling inhibition by different combinations of organic compounds and inorganic salts for application in water base drilling fluids. Appl Clay Sci 105:124–130
25. Zhang R et al (2020) Preparation and performance of ammonium-malic salts as shale swelling inhibitor and a mechanism study. Inorg Nano-Metal Chem 1–5
26. Al-Homadhi ES (2009) Improving Local bentonite performance for drilling fluids applications. J King Saud Univ Eng Sci 21(1):45–52. https://doi.org/10.1016/S1018-3639(18)30522-1
27. Clegg F, Breen C, Khairuddin (2014) Synergistic and competitive aspects of the adsorption of poly(ethylene glycol) and poly(vinyl alcohol) onto Na-bentonite. J Phys Chem B 118(46):13268–13278. https://doi.org/10.1021/jp507772t.
28. Mondal D et al (2013) Effect of poly (vinyl pyrrolidone) on the morphology and physical properties of poly (vinyl alcohol)/sodium montmorillonite nanocomposite films. Prog Nat Sci Mater Int 23(6):579–587
29. Murtaza M, Ahmad HM, Kamal MS, Hussain SMS, Mahmoud M, Patil S (2020) Evaluation of clay hydration and swelling inhibition using quaternary ammonium dicationic surfactant with phenyl linker. Molecules 25(18):4333
30. Kamal MS, Al-harthi MA, Ahmad HM, Hussain SSM, Mahmoud M (2020) Organophilic clay based drilling fluids containing gemini surfactant and polyacrylamide. US Patent Application. 16/578,786, 15-Oct-2020
31. Ahmed HM, Kamal MS, Al-Harthi M (2019) Polymeric and low molecular weight shale inhibitors: a review. Fuel 251:187–217. https://doi.org/10.1016/j.fuel.2019.04.038
32. Ahmad HM, Kamal MS, Murtaza M, Khan S, Al-Harthi M (2019) Alteration of wettability and hydration properties of shale using ionic liquids in water-based drilling fluids. In: Abu Dhabi international petroleum exhibition & conference
33. Bera A, Agarwal J, Shah M, Shah S, Vij RK (2020) Recent advances in ionic liquids as alternative to surfactants/chemicals for application in upstream oil industry. J Ind Eng Chem 82:17–30. https://doi.org/10.1016/j.jiec.2019.10.033
34. Jia H et al (2020) Study of a gemini surface active ionic liquid 1, 2-bis (3-hexylimidazolium-1-yl) ethane bromide as a high performance shale inhibitor and inhibition mechanism. J Mol Liq 301:112401
35. Yang L et al (2019) Effect of alkyl chain length on shale hydration inhibitive performance of vinylimidazolium-based ionic liquids. Ind Eng Chem Res 58(20):8565–8577
36. Lv K, Huang X, Li H, Sun J, Du W, Li M (2020) Modified biosurfactant cationic alkyl polyglycoside as an effective additive for inhibition of highly reactive shale. Energy Fuels 34(2):1680–1687
37. Ghasemi M, Moslemizadeh A, Shahbazi K, Mohammadzadeh O, Zendehboudi S, Jafari S (2019) Primary evaluation of a natural surfactant for inhibiting clay swelling. J Pet Sci Eng 178:878–891
38. Ratkievicius LA, Da Cunha Filho FJV, De Barros Neto EL, Santanna VC (2017) Modification of bentonite clay by a cationic surfactant to be used as a viscosity enhancer in vegetable-oil-based drilling fluid. Appl Clay Sci 135:307–312. https://doi.org/10.1016/j.clay.2016.10.011
39. Zhang Y, Zhao Y, Zhu Y, Wu H, Wang H, Lu W (2012) Adsorption of mixed cationic-nonionic surfactant and its effect on bentonite structure. J Environ Sci (China) 24(8):1525–1532. https://doi.org/10.1016/S1001-0742(11)60950-9

40. Aghdam SK, Kazemi A, Ahmadi M (2020) A laboratory study of a novel bio-based nonionic surfactant to mitigate clay swelling. Petroleum
41. Qasemi M, Moslemizadeh A, Shahbazi K, Mohammadzadeh O, Zendehboudi S, Jafari S (2019) Primary evaluation of a natural surfactant for inhibiting clay swelling. J Pet Sci Eng
42. Yunita P, Irawan S, Kania D (2015) Evaluation of non-ionic and anionic surfactants as additives for water-based mud. Am J Chem 5:52–55
43. Silva IA, Sousa FKA, Menezes RR, Neves GA, Santana LNL, Ferreira HC (2014) Modification of bentonites with nonionic surfactants for use in organic-based drilling fluids. Appl Clay Sci 95:371–377
44. Numkam GL, Akbari B (2019) Effect of surfactant chemistry on drilling mud performance. J Pet Sci Eng 174:1309–1320
45. Barati P, Shahbazi K, Kamari M, Aghajafari A (2017) Shale hydration inhibition characteristics and mechanism of a new amine-based additive in water-based drilling fluids. Petroleum. https://doi.org/10.1016/j.petlm.2017.05.003
46. Yue Y et al (2018) Improving wellbore stability of shale by adjusting its wettability. J Pet Sci Eng 161:692–702. https://doi.org/10.1016/j.petrol.2017.12.023
47. Bavoh CB, Yuha YBM, Tay WH, Ofei TN, Lal B, Mukhtar H (2019) Experimental and modelling of the impact of quaternary ammonium salts/ionic liquid on the rheological and hydrate inhibition properties of xanthan gum water-based muds for drilling gas hydrate-bearing rocks. J Pet Sci Eng 183:106468
48. Luo Z, Wang L, Yu P, Chen Z (2017) Experimental study on the application of an ionic liquid as a shale inhibitor and inhibitive mechanism. Appl Clay Sci 150:267–274. https://doi.org/10.1016/j.clay.2017.09.038
49. Lazorenko G, Kasprzhitskii A, Yavna V (2020) Comparative study of the hydrophobicity of organo-montmorillonite modified with cationic, amphoteric and nonionic surfactants. Minerals 10(9):732
50. Zhao M et al (2020) Dynamic cross-linking mechanism of acid gel fracturing fluid. Colloids Surf A Physicochem Eng Asp 607:125471
51. Keshavarzi B, Javadi A, Bahramian A, Miller R (2019) Formation and stability of colloidal gas aphron based drilling fluid considering dynamic surface properties. J Pet Sci Eng 174:468–475
52. Hamad BA et al (2020) A novel amphoteric polymer as a rheology enhancer and fluid-loss control agent for water-based drilling muds at elevated temperatures. ACS Omega 5(15):8483–8495
53. Hussain SM, Mahboob A, Kamal MS (2020) Poly (Oxyethylene)-amidoamine based gemini cationic surfactants for oilfield applications: effect of hydrophilicity of spacer group. Materials (Basel) 13(5):1046
54. Zhang R, Cai Y, Wang J, Zhao Z, Mao X, Peng Z (2017) A star-shaped anionic surfactant: synthesis, alkalinity resistance, interfacial tension and emulsification properties at the crude oil-water interface. J Surfactants Deterg 20(5):1027–1035
55. Mahmoud SA, Dardir MM (2011) Synthesis and evaluation of a new cationic surfactant for oil-well drilling fluid. J Surfactants Deterg 14(1):123–130. https://doi.org/10.1007/s11743-010-1214-7
56. Yunita P, Irawan S, Kania D (2016) Optimization of water-based drilling fluid using non-ionic and anionic surfactant additives. Proc Eng 148:1184–1190. https://doi.org/10.1016/j.proeng.2016.06.628
57. Negm NA, Tawfik SM, Abdou MI, Badr EA, Ghuiba FM (2015) Evaluation of some nonionic surfactants derived from tannic acid as additives for water-based mud. J Surfactants Deterg 18(2):309–319. https://doi.org/10.1007/s11743-014-1627-9
58. Murtaza M, Khan RA, Kamal MS, Hussain SMS, Mahmoud M (2020) Poly (Oxyethylene)-amidoamine based gemini cationic surfactants with hydrophilic spacers as clay stabilizers. https://doi.org/10.1021/acs.energyfuels.0c01010
59. El-Sukkary MMA et al (2014) Evaluation of some vanillin-modified polyoxyethylene surfactants as additives for water based mud. Egypt J Pet 23(1):7–14. https://doi.org/10.1016/j.ejpe.2014.02.002

60. Barry MM, Jung Y, Lee JK, Phuoc TX, Chyu MK (2015) Fluid filtration and rheological properties of nanoparticle additive and intercalated clay hybrid bentonite drilling fluids. J Pet Sci Eng 127:338–346. https://doi.org/10.1016/j.petrol.2015.01.012
61. Liu J et al (2019) Inhibition of the hydration expansion of sichuan gas shale by adsorption of compounded surfactants. Energy Fuels 33(7):6020–6026
62. Aghdam SK, Moslemizadeh A, Madani M, Ghasemi M, Shahbazi K, Moraveji MK (2019) Mechanistic assessment of Seidlitzia Rosmarinus-derived surfactant for restraining shale hydration: a comprehensive experimental investigation. Chem Eng Res Des 147:570–578
63. Yao R, Jiang G, Li W, Deng T, Zhang H (2014) Effect of water-based drilling fluid components on filter cake structure. Powder Technol 262:51–61. https://doi.org/10.1016/j.powtec.2014.04.060
64. Shettigar RR, Misra NM, Patel K Cationic surfactant (CTAB) a multipurpose additive in polymer-based drilling fluids. J Pet Explor Prod Technol 1–10
65. Aggrey WN, Asiedu NY, Adenutsi CD, Anumah P (2019) A novel non-ionic surfactant extract derived from Chromolaena odarata as shale inhibitor in water based drilling mud. Heliyon 5(5):e01697. https://doi.org/10.1016/j.heliyon.2019.e01697

Effect of Surfactants on the Performance of Water-Based Drilling Fluids

Basim Abu-Jdayil and Mamdouh Ghannam

Abstract In the formulation of drilling fluids, different additives are used to optimize their rheological behavior and control the fluid loss throughout the drilling process. Surfactants, as one of these additives, play a vital role in sealing off the lost circulation zones and in controlling the rheological properties of the dispersions to meet the specification of the desired applications. This chapter is divided into three sections. The first section reviews the definition, functions, and properties of drilling fluids, bentonites and surfactants. The second section provides an overview on the main and recent research on utilization of surfactants in drilling fluid formulations. The last section describes an experimental work on the effect of cationic surfactant CTAB and anionic surfactant SDS on the performance of water based drilling fluid. Adding of CTAB surfactant to water-based drilling fluid reduced significantly its viscosity and shifted its rheological behavior from shear thinning fluid with a yield stress towards Newtonian behavior. On the other hand, the SDS surfactant was effective in modifying the rheological properties of water-based drilling fluid in the concentration range that corresponds to critical micelle concentration (CMC) and critical coagulation concentrations (CCC) of SDS.

Keywords Water-based drilling fluid · Bentonite · Surfactants · CTAB · SDS · Rheology

1 Introduction

Human survival and economic activities all depend on the supply of energy. Nowadays, the worldwide industrial growth led to increase the global demands of energy while the production of conventional crude oil throughout the world has almost reached to its peak. Recent reports claim that there is a decrease in light oil production, while the composition of produced oil is getting heavier [1]. Therefore, the utilization of the unconventional hydrocarbon resources such as heavy oil and tight

B. Abu-Jdayil (✉) · M. Ghannam
Chemical & Petroleum Engineering Department, UAE University, Al Ain, UAE
e-mail: babujdayil@uaeu.ac.ae

T. Solling et al. (eds.), *Surfactants in Upstream E&P*, Petroleum Engineering,
https://doi.org/10.1007/978-3-030-70026-3_3

reservoirs would play an important role in the future of the ever-growing world's energy consumption [2]. The unconventional hydrocarbons are known as one of the major energy source in the twenty first century [3], where their reserves account for more than 80% of the total quantity of petroleum in the world [4]. It is expected for oil and gas industry to transform exploration development domain from conventional to unconventional hydrocarbon accumulations that are different in types, geological features and genesis [5]. Drilling in unconventional reservoirs can help to meet the world's energy demand. Although the concept of drilling fluids remained the same, their technology became much more complex. Drilling fluids are necessary to drilling success, both maximizing recovery and reducing the time it takes to achieve first oil [6].

1.1 Drilling Fluid

A drilling fluid, also known as drilling mud, is defined as the fluid that is used in a drilling operations, where the fluid is circulated from the surface, down the drill string, through the bit, and pumped back to the surface via the annulus [7]. Drilling fluids are suspensions of clays in liquid phase with chemical additives used to modify their properties. There are many types of drilling fluids including water-based, oil-based, synthetic-based, and air drilling fluids such as mist, foams, and stiff foams [7]. In water-based drilling fluid (WBDF), the carrier fluid is fresh water or brine, while those with oil as a continuous phase are called oil-based fluids. In most drilling processes, water-based fluids are used, owing to the simplicity of their preparation, low cost, environmental friendliness, cutting based removal and fast formation/breaking-down rate [8].

Clay minerals are the basic elements of drilling fluids. Clays provide and maintain a suitable rheological behavior for drilling fluid and control the fluid loss throughout the drilling process. Selecting the type of clay and its quantity are critical factors for successful drilling process [9]. In most of drilling fluids, the used clay is sodium bentonite which is naturally occurring clay mineral contains mainly sodium montmorillonite. In some cases, the palygorskite-sepiolite group clay minerals are used in drilling fluid formulations [8]. Owing to their absorptive properties, sepiolite and palygorskite are used in different applications as carriers, fillers, clarifying agents, and lubricant reclamation.

Additives including electrolytes, polymers, surfactants and nanomaterials were used to optimize the performance of the drilling fluid, most likely to modify its rheological properties [9–12].

Drilling fluid plays an important role in oil and gas exploring process. In addition to carry out effective drilling operations, drilling fluids improve the productivity of wells. The main functions of drilling fluids are to carry the rock cuttings to the surface (needs a minimum yield stress), to maintain a sufficient pressure against the rock formation (needs a viscous fluid), to lubricate and cool the bit (need enough fluidity that comes from the shear thinning behavior) [13].

In addition to the characteristics and chemical composition of the well, the environmental impact should be taken into consideration, when the drilling mud engineer is prescribing the type of drilling fluid necessary in a well. While water-based drilling fluids have generally less impact on the environment during offshore drilling, oil-based drilling fluids may work better with a saltier rock.. Disposal of the used drilling fluids can also be another challenge. Recently, new methods have been established to recycle drilling fluids [14].

The properties of drilling fluids, namely the rheological properties (plastic viscosity, apparent viscosity and yield stress), mud density, fluid loss, pH and gel strength, have been investigated extensively in the literature. Most of the studies, focused on the rheological behavior of drilling fluids, as they are often used in a variety of changing environmental conditions, such as changing pressure and temperature, and inflow through various geometrical shapes of a conduit (e.g. in pipes and strange annuli for oil and gas wells and in rivers and hills) [15].

The temperature, type and concentration of clay and additives, and pH are important factors influencing the rheological properties of the drilling fluids.

1.2 Bentonite

Drilling fluids are mainly based on dispersed sodium bentonites. Bentonite is a clay generated often from the alteration of volcanic ash and consisting mainly of montmorillonite that consist of individual crystallites with dimensions are less than 2 μm. Montmorillonite crystallites themselves are three-layer clay minerals: two tetrahedral layers and one octahedral layer. In montmorillonite tetrahedral layers consisting of [SiO4]—tetrahedrons enclose the [M(O5, OH)]-octahedron layer, where M is mainly Al, Mg, but Fe is also found, see Fig. 1 [16]. The small negative charge of the silicate layers is usually compensated by exchangeable ions in the intercrystallite region. The charge is so weak that the cations like $Ca2^{+}$, $Mg2^{+}$ or Na^{+} can be adsorbed in this region with their hydrate shell. The intercrystalline swelling results from the extent of hydration [17]. The contact between montmorillonite and water leads to swell montmorillonite particles several times by its original volume. However, the wide range of industrial applications is related to the cation exchange in the interlayer space of montmorillonite with two compositional endmembers (Na-bentonite and Ca-bentonite). While Na-bentonites exhibit great swelling capacity, Ca-bentonites can be considered as non-swelling clays [18]. Moreover, other impurities can be found bentonite such as feldspar, calcite, quartz, cristobalite,, mica, pyrite and ferrous carbonate [19].

With exceptional rheological behavior, bentonites are utilized in different industries including drilling fluids, paper, dyes, pharmaceuticals, cement, composites, and ceramics (e.g. [20–26]). In addition and owing to the fact that bentonite is a porous material with high capability of adsorption, it has an application in environmental engineering as a sorbent for different heavy metals and organic pollutants (e.g. [27–29]).

Fig. 1 Schema of bentonite structure [14]

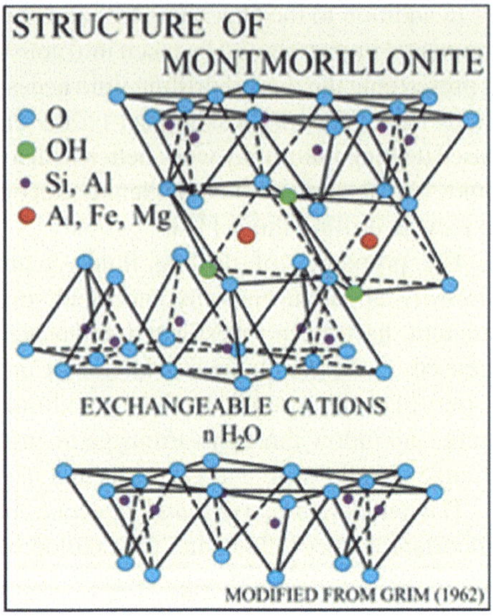

One of the most important and efficient applications of sodium-bentonite suspensions Fig. 2 is their use as drilling fluid in the oil and gas industry [30–32]. Sodium bentonite's main feature in drilling comes from its sexcellent colloidal properties and its unique rheological behaviour, where small solid concentration (2.0–9.0 wt%) sin water can form a viscous, shear thinning material with significant yield stress [11], Abdou and El Sayed Ahmed [33, 34]. High viscosity with yield stress behaviour exhibited at low shear rates is representing solids lifting capacity away from the drill bit, and relatively low viscosity at high shear rates (shear thinning behaviour) in the vicinity of the drill bit is necessary to reduce torque requirements. In addition, dispersions of bentonites in water exhibit a thixotropic behaviour, which is often observed at high enough concentrations [11, 35–37].

1.3 Surfactants

On daily basis, surfactants are one of the most chemicals that are consumed in large quantities. Surfactants are substances that create self-assembled molecular clusters called micelles in a water or oil solution and adsorb to the liquid–gas/solid interface. To exhibit these two behaviors, a surfactant molecule should have a chemical structure with two different functional groups that have different chemical affinity.Usually the surfactant molecules have both an alkyl chain with 8–22 carbons, which is called a hydrophobic group (does not show affinity to water). The surfactant molecules also have a functional group, which has affinity to water and it is

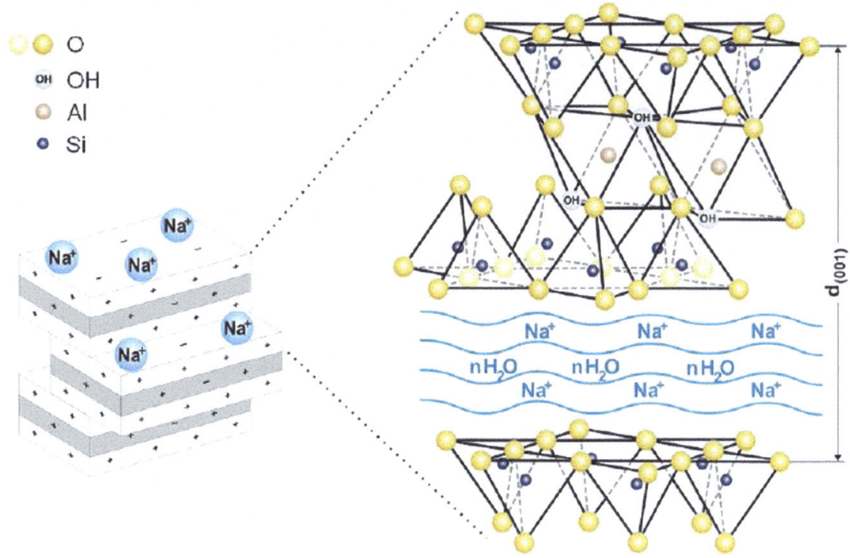

Fig. 2 Schema of sodium montmorillonite structure [38]

called the hydrophilic group [39]. Based on this behavior, surfactants are classified as amphiphilic molecules. As shown in Fig. 3, and due to the different chain lengths in many surfactants, reference is often made to a hydrophilic "head" and a hydrophobic "tail".

Surfactants are soluble in both aqueous and organic solvents. The surfactant reduces the surface tension of saqueous solution by adsorbing at the interface of liquid–gas system. By adsorbing at the liquid–liquid interface, surfactant can also reduce the interfacial tension in oil–water system [40]. When there are a sufficient concentration of surfactant in a solution, the surfactant molecules combine together to form certain structures called micelles, see Fig. 4. When the micelle forms, the surfactant heads position themselves so they are exposed to aqueous phase, while the tails are grouped together in the center of the micelle structure far away from the aqueous sphase [41]. Many factors play significant role in the determination of the critical micelle concentration (CMC) where micelle formation becomes significant in liquid phase. The most important factors that are known to have an influence on the CMC in aqueous solution are (i) the chemical structure of the surfactant, (ii) nature of counter ions in the case of ionic surfactants the presence of electrolyte in the solution, especially in the case of ionic surfactants, (iii) the presence of electrolyte in the solution, especially in the case of ionic surfactants, (iv) the presence of different organic compounds in solution, and (v) solution temperature [42]. Moreover, it was found that different types of aggregates can be formed in surfactant solution that depend on the surfactant geometry. Different possible aggregate morphologies in surfactant solutions were identified by Kunitake et al. [43], which are including vesicles, globules, rods, lamellae, tubes, and disks. Varying the length of the tail and/or

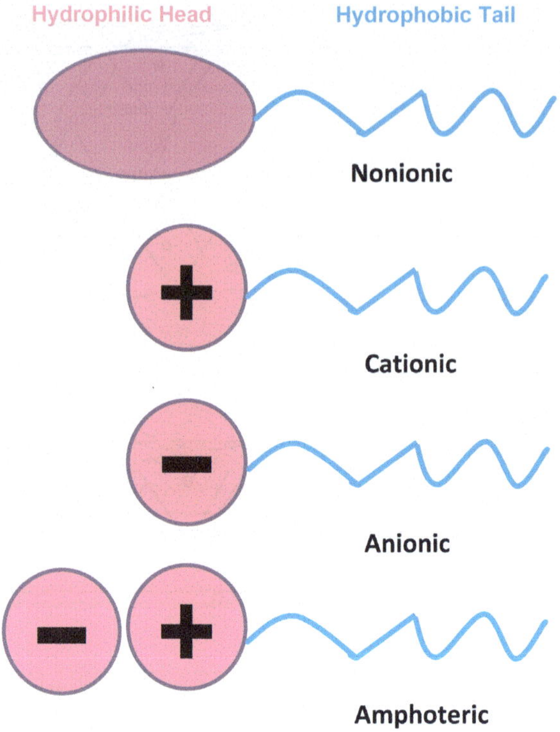

Fig. 3 Types and schematic structure of surfactant

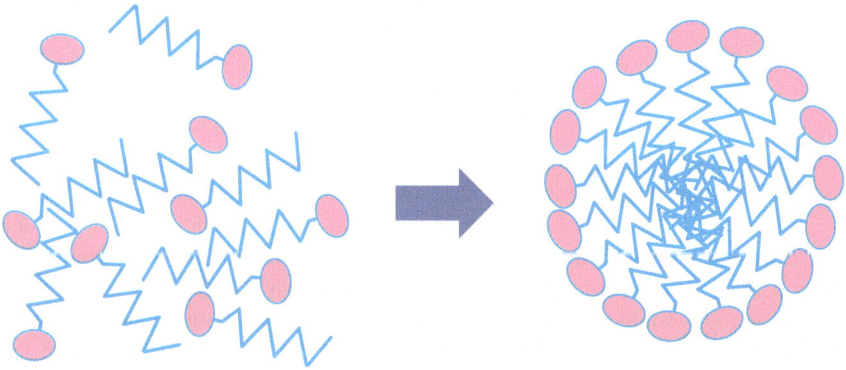

Fig. 4 With high enough concentration, individual molecules of surfactant aggregate to for form a micell (right)

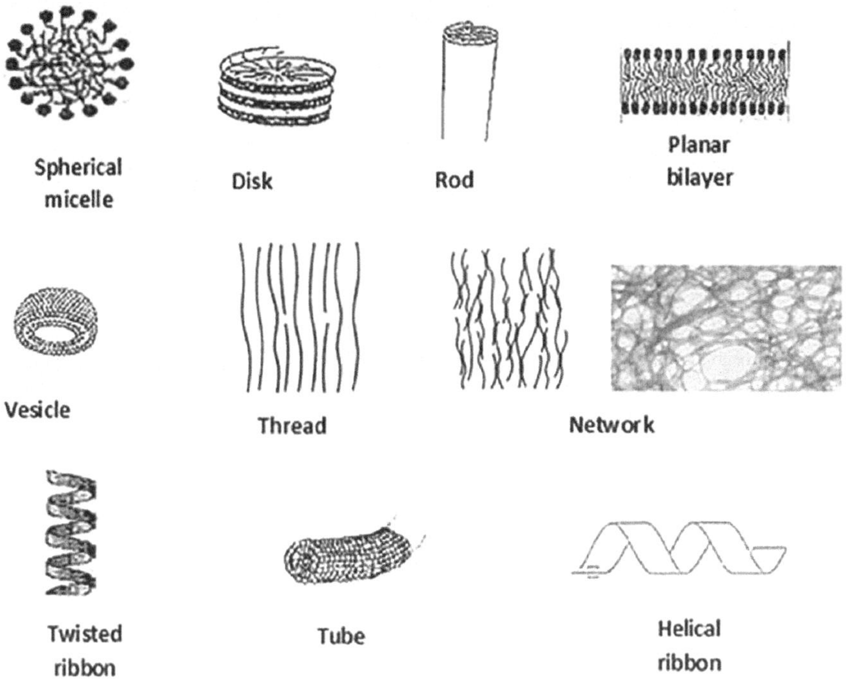

Fig. 5 Shapes of aggregates formed by surfactants in water [42]

the flexible linker can lead to change the structure of the aggregate. Some possible types of aggregates formed in aqueous solutions are shown in Fig. 5.

According to the nature of the hydrophilic group, surfactants are classified into ionic and nonionic surfactants, see Fig. 3. Ionic surfactants are sub-classified into anionic, cationic and amphoteric surfactants. In anionic surfactants, the hydrophilic group dissociates into anions in aqueous solutions (e.g. carboxylates (COO^-) or sulphonates (SO_3^-)), while in cationic surfactants the hydrophilic group dissociates into cations (e.g. quaternary amines ($-R_4N^+$)).The hydrophilic group in amphoteric surfactants dissociates into anions and cations often depending on the solution pH (frequently carboxylate and quaternary amine group). On the other hand, nonionic surfactants do not dissociate into ions in aqueous solutions, where they bond with polar, non-ionic groups such as ether, alcohol, or ethoxylate. (see Fig. 6 [39]).

Other types of surfactants called Zwitterionic, double chain, gemini and bolaforms can also be found. The Zwitterionic surfactant molecules contain both anionic and cationic centers at the head group. In double chain surfactants, the polar head is attached with two hydrocarbon chains. On the other hand, "gemini" surfactants contain two hydrophobic and two hydrophilic groups. The gemini surfactants can be thought of "twin" surfactants as being made up of two typical surfactant molecules chemically linked at or near the head group. "Bolaforms" surfactants have two head groups (both cationic, both anionic, or one anionic and the other cationic) joined by

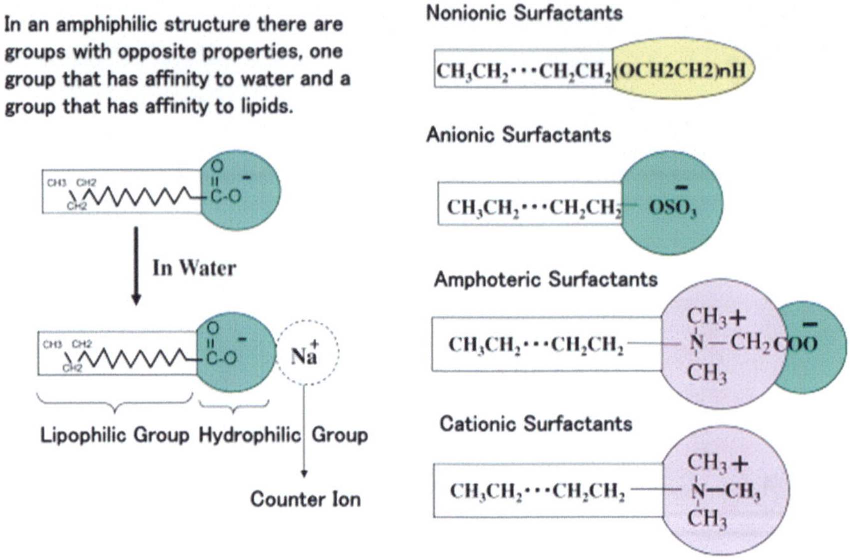

Fig. 6 Classification of surfactants [39]

hydrophobic spacer. The nature of head group(s) and the hydrocarbon chain length control the surface activity of "bolaforms". Surfactants with longer hydrocarbon chains are found to be more surface-active than those with shorter hydrocarbon tail [42].

Most of the surfactants have similar tails that consist of a hydrocarbon chain, which can be linear, branched or aromatic. For example, Fluoro-surfactants have fluorocarbon chains and Siloxane surfactants have siloxane chains. However, surfactants with mixed chains and complex structures have been recently developed [44]. In Table 1, most famous surfactants with their applications are summarized.

2 Effect of Surfactants on Drilling Fluid Performance

Many difficulties face the drilling fluid engineers such as developing a drilling fluid system that yields acceptable well stability and prevents fluid invasion. One of the main challenges faces the drilling fluid engineers is the loss of drilling fluid, due to the formation openings. The loos of drilling fluid results in loss of hydrostatic pressure at the bottom of the well and allows influx of formation fluids and finally could lead to loss of well control. It is essential that circulation of drilling fluid be regained for drilling process to continue. Formation of filter cake or formation of mud solids layer on the wall of the hole in the formation can cause stuck drill pipe problem, which can occur after drilling has been halted due to a rig breakdown or when conducting other nondrilling operations [7]. Therefore, different additives, mainly polymers and

Table 1 Examples of surfactants [44]

Surfactant type	Structure/name	Applications
Nonionic surfactant	Polyoxyethylene glycol octylphenol ethers: C8H17–(C6H4)–(O-C2H4)1–25–OH	Wetting agent and coatings
	Polyoxyethylene glycol alkylphenol ethers: C9H19–(C6H4)–(O-C2H4)1–25–OH	Spermacide
	Polyoxyethylene glycol sorbitan alkyl esters	Food ingredient
	Sorbitan alkyl esters	Polishes, cleaners, and fragrance carriers
	Block copolymers of polyethylene glycol and polypropylene glycol	Various
Anionic surfactant	Dioctyl sodium sulfosuccinate (DOSS)	Wetting agent, coatings, and toothpaste
	Perfluorooctanesulfonate (PFOS)	Scotchguard™, and Skydrol™
	Linear alkylbenzene sulfonates	Laundry detergents, and dishwasher detergents
	Sodium lauryl ether sulfate	Shampoos and bath products
	Lignosulfonate	Concrete plasticizer, plasterboard, and DMSO
	Sodium stearate	Handsoap and HI&I products
Cationic surfactant	Quaternary ammonium cations include: Cetyl trimethylammonium bromide (CTAB) Cetyl trimethylammonium chloride (CTAC)	Softeners in textile, and anti-static additives
	Benzalkonium chloride (BAC), Cetylpyridinium chloride (CPC), Benzethonium chloride (BZT)	Anti-microbials, and anti-fungals
Zwitterionic surfactants	Sultaines; CHAPS (3-[(3-Cholamidopropyl)dimethylammonio]-1-propanesulfonate)	Cosmetics, and in special use
	Bataines: cocamidopropyl betaine	

surfactants, have been used in drilling fluids to seal off the lost circulation zones and to control the rheological behaviour of the suspensions to fit with the specification of the desired applications, such as poly vinyl pyrolidone [45], poly vinyl alcohol [46, 47], xanthan [48], polyanionic scellulose [49], polyethyleneimine [50], poly (ethylene glycol) [51], polypropylene glycol [52] and carboxymethyl cellulose sodium salt

[12, 53], polyacrylamide [54], polyethyleneimine [55], and sodium dodecyl sulphate (SDS) [56]. In addition, the reduction of bentonite concentration in the mud system is another target of the mud engineers to avoid the undesirable effects of high bentonite concentrations. High soild concentarion in drilling fluid leads to a reduction in the drilling rate of a wellbore a given depth. Moreover, the high bentonite concentration in drilling fluids has two other main disadvantages that the high cost of transportation and storage, which can be critical for offshore drilling sites and sites located in distant and hostile environments [57].

Owing to the fact that bentonites are porous materials with high capabilities for physical and chemical adsorption, they were used extensively as a sorbent for different heavy metals and organic pollutants. However, the efficiency of bentonite as a sorbent was improved by using different kinds of surfactants to switch the surface properties of the bentonite from hydrophilic to hydrophobic [28, 58]. This process allows the pollutants to be adsorbed on the hydrophobic bentonite surface, where the pollutants are captured by hydrophobic interaction between alkyl chains. Many investigators have used the cationic surfactant cetyltrimethyl ammonium bromide (CTAB) and the anionic surfactant sodium dodecyl sulphate (SDS) to modify the surface properties of bentonites and to improve their sorption properties (e.g. [28, 59–61]).

On the other hand, the addition of surfactants to bentonite suspensions can influence their stability. It is expected that the surfactants can induce flocculation of bentonite particles by bridging, charge neutralization, surfactant-particle surface complex formation and depletion flocculation, or by a combination of two or more of these mechanisms.

The first important flocculation mechanism, which is the bridging mechanism needs that the surfactant chains be adsorbed on the bentonite surface with only a few points of attachment, while the bulk of the surfactant chains projecting into the solution for contact and adherence to other bentonite particles. It should be mentioned that the strong adsorption of surfactants on bentonite surface can cause surface saturation, preventing effective bridging and destabilizing bentonite suspension [62]. This means that the strong adsorption of surfactants on bentonite particles does not support the flocculation process. Cationic and anionic surfactants can have clear effect on the flow properties of bentonite suspensions, which is more complex than the surfactant effects on other clays. Vali and Bachmann [63] found that delamination of bentonite produces very thin particles, where the flow behavior of these particles can changed by modest changes in experimental conditions, such as pretreatment reactions, degree of dispersion, and pH. In the presence of anionic surfactants, it has been proposed that the adsorption of surfactant molecules on the bentonite particles and the subsequent flocculation process occur via hydrogen bonding between the solid surfaces and the hydroxyl groups on the surfactant. While in the case of cationic surfactants, adsorption on negatively charged bentonite particles and consequent flocculation occur by simple electrostatic attraction [64].

Modification of the surface properties of bentonite by adding surfactants permits them to be used as rheological control agents in different engineering applications such as drilling fluid, paints, oils, composites and pharmaceuticals.

2.1 *Surfactants in Water-Based Drilling Fluids (WBDF)*

2.1.1 Ionic Surfactants

Modification of drilling fluid performance with anionic surfactants received little attention comparing to cationic surfactants. The effect of surfactants on the flow behavior of sodium bentonite dispersions (2 wt%) was studied for a cationic (cetylpyridinium chloride, CPC1) and an anionic surfactant (sodium dodecylsulfate, SDS) by Permien and Lagaly [10]. Cationic and anionic surfactants affected the rheological behavior of sodium bentonite suspensions by several mechanisms. In strongly acidic media, where the pH < 4, the surfactants have clear effect on the stability of the card-house structure. The cationic surfactant promoted disintegration of the card-houses. The same also took place at higher concentrations of the anionic surfactant (>10^{-3} M), whereas lower SDS concentrations stiffened the network. The electroviscous effect was responsible for the reduction in the shear stress and yield point at pH > 4, where the surfactants acted as 1: 1 electrolytes. However, at high surfactant concentrations, the alkyl chains of opposed particles interacted and flocs were formed. Still higher concentrations caused repeptization by recharging the particles.

An early study on the effect of the anionic surfactant (linear alkyl benzenesulfonate, LABS) and the cationic surfactant (distearly dimethyl ammonium chloride, DDAC) on the rheological behavior of Na-bentonite was carried out by Gungor [64]. The effect of the surfactants have been tested at different concentrations. At low values of pH (pH = 2.4), it was found that the distearly dimethyl ammonium positive ion was adsorbed on the negative faces of bentonite leading to decrease the negative charge there. This reduction in bentonite negative charge resulted in lowering the face-edge interactions and so decreasing the yield stress. At high values of pH (pH = 9.5), the flow behavior of bentonite changed with the DDAC concentration, as it was governed by the electroviscous effect. At DDAC concentrations > 10^{-4} M, the exchange of the counter ions on bentonite particles by DDA + cations reduced its hydrophilicity, which cause the particles settling as flocs. On the other hand, LABS anions at low concentration (< 10^{-5} M) and low pH (= 2.4) were not adsorbed by the bentonite particles and did not affect the flow properties of the dispersions.). At these conditions, the yield stresses were high due to formation of the card-house structure. At LABS concentrations > 10^{-3} M, the yield stress value decreased because the surfactant anions broke up the edge-face contacts. At high pH (= 9.5), the values of yield stress were constant up to about 5×10^{-4} M and then increased.

The flow behavior of Egyptian bentonite dispersions was investigated in the presence of cationic polyethylene imine followed by anionic polyacrylamide at different bentonite concentrations and temperatures (20–80 °C) [65]. All tested samples were aged for 24 h. An inverse relation between temperature and dispersions plastic viscosity was detected. Contrary to that, increasing the temperature led to increase the apparent viscosity, yield point, gel strength, and consistency index of the dispersions.. Moreover, the results showed that there was a strong dependence of the zeta potential on dispersions temperature. The constructed potential energy profile at different bentonite concentrations gave a high-repulsion potential energy between bentonite surfaces by increasing temperature, which means that the dispersions stability improved.

In the study of Yunita et al. [66], a non-ionic and anionic surfactants were used with the aim to optimize the flow and filtration properties of water-based drilling fluid. Alkyl benzene sulphonate was used as anionic surfactant, while 2-hexadecyloxyethnol was utilized as non-ionic surfactant. The results of this investigation showed that both surfactants can improve the rheological and filtration properties of water-based drilling fluid. The filtration loss was improved by 41.3% in the presence of surfactants. It was expected that the long chain of both surfactants increased the drilling fluid viscosity, which has a positive effect in blocking the pores and forming a tight filter cake. Moreover, the addition of these surfactants to the water-based drilling fluids enhanced their resistance at the higher temperature.

Data presented in the work of Elochukwu et al. [67] focused on the using of Methyl Ester Sulphonate (MES) surfactant and nano-polystyrene in water-based drilling mud. The researchers utilized Bingham and Power law models to study the rheology of the drilling fluid, which showed that its formulation was improved in the presence of MES and nano-polystyrene. Filtration experiments under low pressure low temperature (LPLT) and high pressure high temperature (HPHT) conditions revealed that MES and nano-polystyrene reduced the filtration loss of drilling fluid by 50.7% at LPLT and 61.1% at HPHT conditions.

Two series of cationic Gemini surfactants, alkanediyl-α,ω-bis[N,N-dimethyl alkyl (octyl or dodecyl)ammonium] dibromide (R-s-R; s = 6, 10, 12 and R = 8 and 12) were prepared by Ahmed et al. [68] to be used as additives for water-based drilling muds. FTIR and mass spectroscopy were used to confirm the chemical structures of the prepared surfactants, which demonstrated noteworthy surface activity that was more pronounced with those of longer hydrophobic chain length. Moreover, the effect of the prepared Gemini surfactants on the rheology and fluid loss properties of water-based drilling fluid formulated from local sodium bentonite was tested. The results revealed that the Gemini surfactants have improved the rheological and filtration properties of the sodium bentonite according to API specifications.

Three cationic bolaform surfactants with different spacer lengths were prepared by reaction two moles of triisopropanolamine with one mole of each of the following 1,4-dibromobutane, 1,5-dibromopentane and 1,6-dibromohexane [69]. Then, the effect of the prepared surfactants on the properties of water-based drilling fluids was investigated. Weak surface activity of the prepared cationic surfactants was detected from the surface tension measurements. The prepared water-based drilling fluids with the presence of the cationic bolaform surfactants demonstrated an acceptable yield stress, mud density, swelling ratio, rheological and filtration properties. The prepared cationic surfactants qualified a clay sample to be used as drilling mud as the rheological behavior satisfied the API standard and Oil Company Materials Association (OCMA) specification.

The effects of alkali metal, calcium and lanthanum chlorides, alkali metal sulphates, sodium dodecyl sulphate, cetyltrimethylammonium bromide and pH on the zeta potential and electrical conductivity of sodium bentonite dispersions were studied by Mészáros et al. [70]. The behavior of the dispersion in an external electric field can be summarized as follows: zeta potential of particles displayed relatively high values both in water and in supporting electrolytes solution (from −35 mV to −52 mV) with low sensitivity to pH,the zeta potential dependencies for RbCl and CsCl concentration were characterized by a constant value at concentration of $10^{-5} - 10^{-3}$ mol/L, and for LiCl, NaCl and KCl by a maximum at concentration of $10^{-3} - 10^{-2}$ mol/L,zeta potential in alkali metal sulphates solutions was measured to be higher compared to that for chlorides at equivalent concentration of salts; addition of electrolytes with bi-and trivalent counter-ions or cationic surfactant caused a clear reduction in zeta potential values and surface charge reversal, respectively; while moving from LiCl to CsCl in the lyotropic series, a regular reduction in the surface conductivity of particles was measured. The researchers explained the observed regularities by accounting for two opposite trends related to the polarization of the electrical double layer (EDL) in the external electric field (increase of zeta potential at low concentration) and the effect of the EDL compression (decrease of zeta potential at high concentration) as well as the increase in filling up the Stern-layer with alkali metal ions as their radius increases. It was linked the bigger zeta potential values for sulphates compared to that for chlorides with dehydration effect of sulphate co-ions which results in the thinning the surface hydrate layer and shift of the shear plane toward the surface. The rheological and other drilling fluid properties were not investigated.

Pickering emulsions showed high potential for different applications in the oil and gas industry, especially those with pH-responsive behavior that allows them to be easily implemented. However, the main concern on the pH-responsive emulsions is the narrow pH window in the stabilization of emulsions, which is not preferred for some complicated reservoir conditions. Liu et al. [71] reported recently a Pickering water-in-oil (W/O) emulsion system with a wide pH window, which can be produced from the interactions between fumed silica nanoparticles (NPs) and amine surfactant. FTIR spectroscopy, turbidity, surface tension and other measurements were used to confirm the pH-dependent interaction between NPs and amine surfactant. Liu et al. [71] found that the stable W/O emulsions generated by this system were formed within the pH window that showed strong flocculation of NPs due to the adsorption of surfactant onto the particle surfaces, which imparts a certain hydrophobicity to the NPs. Outside of this pH range, the similar charge of the surfactant and the nanoparticles generates strong repulsion between them, so the fumed silica NPs cannot be supplied with such hydrophobicity. The pH-responsive Pickering emulsion-based drilling fluid exhibited adjustable rheological behavior and was reversible by a cyclic emulsification-demulsification process. Liu et al. [71] anticipated that this system can be switched "off" in follow up procedures and waste disposal at the surface., while it can be switched "on" over a wide pH range with high stability when running underground.

The swelling of shale formation when it is exposed to water-based drilling fluids leads to many operational problems. Many common shale inhibitors are used in the industry to overcome this swelling problem, however, they suffer from many drawbacks, which adversely affect the performance of drilling fluids. The study of Murtaza et al. [72] focused on the development and application of new quaternary ammonium Gemini cationic surfactants as shale swelling inhibitors. The developed cationic Gemini surfactants have similar structure with different spacer lengths. It was found that the new cationic surfactants have the ability to reduce the clay swelling. The increased hydrophobicity of the GS12 surfactant (containing flexible large spacer C12) allowed it to be a beneficial swelling inhibitor as compared to GS8 and GS10 with lower hydrophobicity. In addition, GS12 showed comparable performance with the common shale inhibitor used in the oil industry. SEM images revealed the surfactant-treated clays have fine, dense, and smooth surface. Moreover, it was found that the developed surfactants adsorbed and intercalated between layers of clay through hydrogen bonding and attractive electrostatic forces. Therefore, the orientation of lipophilic tails inhibited the water adsorption in the clay which resulted in decreased clay swelling. It was also reported that the developed cationic surfactants decreased the clay swelling without changing the other drilling fluid properties such as stability, rheology, and filtration properties.

To develop environmentally friendly high performance drilling fluid, effective and green shale hydration inhibitors are needed. For this purpose, Jiang et al. [73] have prepared composites of gelatin and inorganic salt (KCl) or organic salt (2, 3-epoxypropyl-trimethylammonium chloride, EPTAC) as green shale inhibitors. The inhibitive properties of the prepared composites were assessed by comparison with the properties of common inhibitors. Results revealed that when 2.0 wt% gelatin and 1.0 wt% salt were mixed, the composites reduced the swelling height of bentonite to less than 1.70 mm, improved shale recovery to 80% at 150 °C and inhibited 16.0 wt.% bentonite mud-making, suggesting great inhibition performance. Moreover, it was found that the gelatin composite with EPTAC has better synergistic inhibition than that with KCl. The inhibitive mechanisms of prepared composites were also investigated by observing aggregation in the drilling fluid system using electron microscopy, measuring interlayer space through XRD observing hydrophobic modification degree by measuring the water contact angle, and measuring the zeta potential and water activity. Both composites showed lower water activity and effectively prevented water from invading bentonite. In these composites, gelatin mainly adsorbed on the surface of bentonite particles, encapsulated them, and improved their hydrophobic capacity to a certain degree, whereas the salts entered the interlayer, expelled water, and decreased the interlayer space. The main conclusion was that the combinations of gelatin and KCl or EPTAC had significant effect on inhibiting shale hydration.

2.1.2 Nonionic Surfactants

The geological characteristics and the potential formation damage of the Permian formation of the reservoir was investigated by Zhang et al. [74], who identified that the water-blocking due to invasion of drilling or completion fluids is one of the most severe causes of damage to gas well deliverability. Zhang et al. [74] have developed a low-damage drilling fluid containing efficient water-blocking preventing surfactants, optimized temporary bridging agents (TBAs), and film-forming agents. The development of such drilling fluid was based on the phase trap prevention method, ideal packing theory, and film-forming technology. The results obtained from the evaluation of the new drilling fluid showed that the fluid has acceptable performance in terms of rheological properties, shale-swelling inhibition, temporary plugging effect, ultra-low filtration, and lubricity. Moreover, the developed drilling fluid can efficiently reduce water-blocking. Drilling fluid comprising nonionic surfactants to reduce the interfacial tension between the water-based filtrate and reservoir gas, contributed to mitigation of water-blocking in low-permeability gas reservoirs.

Four vanillin-based nonionic surfactant derivatives were synthesized by Negm et al. [7] to show good surface-active property in solution. These surfactants showed enhancement of the properties of water-based fluid. The results revealed that the performance of water-based drilling fluid formulated with the new prepared surfactants varied within an acceptable range comparable to the performance of field water-based drilling fluid.

In another work, Negm et al. [75] used tannic acid to synthesize a new type of nonionic surfactant. It was found that the ratio of the attached hydrophobic chains controls the surface activity of the produced surfactants. Moreover, the experimental works on the new produced nonionic surfactants proved that those surfactants exhibited good performance when added as viscosifiers to the water-based drilling fluid formulation—Comparing with a reference viscosifier, the rheological and filtration properties of the most synthesized surfactants were very promising.

Using of cost effective and environmentally friendly shale inhibitor in drilling industry steadily increased. Shadizadeh et al. [76] used a newly developed nonionic surfactant (*Zizyphus spina-christi* extract, ZSCE for inhibiting shale hydration. The conductivity technique was used to evaluate the adsorption behavior of ZSCE onto shale cuttings. In addition, different inhibition evaluation methods were used to assess the inhibitive properties of ZSCE. All experimental results showed that ZSCE surfactant can work as a potential shale inhibitor,however, its inhibitive capacity had no clear improvement at concentrations above the CMC. Comparing with polyamine and potassium chloride, ZSCE surfactant showed a better performance. Moreover, it was found that the ZSCE surfactant was compatible with other conventional additives can be found in water-based drilling fluids. The stability of Na-bentonite particles in ZSCE solution was indicated from the SEM micrographs. It was expected that the main inhibition mechanism for ZSCE surfactant is due the hydrogen bonding between oxygen atoms available on silica surface of bentonite and hydrophilic tail of ZSCE molecules, which can produce f a hydrophobic shell on the bentonite surface. The results of this investigation opened the door to utilize the plant-based surfactants as shale inhibitors in water-based drilling fluids.

In their recent work, Aggrey et al. [77] reported the utilization of the nonionic surfactant saponins, which was extracted from *Chromolaena odorata* (CO) leaf, as shale hydration inhibitor. The inhibitive performance of the plant-based nonionic surfactant was evaluated through measurements of surface-active, rheological, strength and inhibition properties. It was found that the saponins surfactant has high compatibility with the water-based drilling fluids and can produce a stable highly effective shale inhibitor. Concentration range of 2.5–3.5 wt% CO could be used as inhibitor component in water-based drilling fluid with a CMC of 3.5 wt% as an optimum concentration to improve the inhibitive properties.

2.1.3 Mixed Surfactants

In the work of González et al. [78], the effects of a surfactant additive (SA) and its dissolution in diesel (SB), on the rheological and tribological properties of WBDFs formulated with hematite and calcium carbonate were investigated. The ionic/anionic mixture (80/20) surfactant was used in this work. The non-ionic surfactant was a lauryl alcohol ethoxilate with an ethoxilation degree of four units, while the anionic surfactant was a Trilaureth-4 Phosphate. SB surfactant was prepared by mixing the surfactant SA with diesel in a ratio of 40:60 (w/v).The tribological properties were determined by measuring the coefficient of friction (CF) in conjunction with optical surface profilometry used to evaluate the wear behavior. In addition, the dispersion stability of solid particles in the aqueous surfactant solutions was identified by light scattering techniques. The findings of this work verified that the used surfactant can reduce significantly the CF, and that SA formulation was much better than SB in CF reduction. Moreover, it was found that the viscosity increases in the polymeric drilling fluids formulated with hematite and SA, reflecting strong interactions in the polymer–surfactant–solid system. However, the kind of these interactions was not discussed. In the case of the drilling fluid containing calcium carbonate the addition of surfactant had no effects on the rheological behavior.

Zhang et al. [79] have mixed the nonionic surfactant Triton X-100 (TX100) with the cationic surfactant hexadecylpyridinium bromide (HDPB) to study their effects on bentonite structure. The findings showed that the surfactant intercalation led to decrease the bentonite specific surface area, pore volume, and surface roughness and irregularities. While the co-adsorption of both surfactants increased the ordering conformation of the adsorbed surfactants on bentonite, it decreased the thermal stability of the clay system. The effect of mixed surfactants on the rheological behavior of bentonite was not one of the objectives of this study.

The effects of anionic sodium tetradecyl sulfonate (NaTDS), cationic tetradecytrimethylammonium bromide (TDTMABr), and nonionic Brij 30 surfactants on the electro-kinetic and rheological behaviors of sodium bentonite and sepiolite dispersions were studied by Tunc et al. [80]. It has been found that the surface charges of sodium bentonite and sepiolite were negative in water, which converted to positive after the addition of TDTMABr at a certain concentration. As the concentration of NaTDS is increased in the dispersion, the zeta potential value of sepiolite decreased gradually, but that of bentonite did not change significantly. All clay dispersions exhibited Bingham flow behavior in the presence of surfactants. Depending on the surfactant concentration, sepiolite–surfactant suspensions displayed rheopectic or thixotropic behavior, while sodium bentonite–surfactant dispersions showed a thixotropic behavior. On the contrary of sodium bentonite–surfactant system, the surfactant concentration led to significant changes in the Bingham viscosity of sepiolite dispersions. An evidence about the interactions between clay particles and surfactant molecules had been observed in the FTIR spectra of dispersions. This study concluded that the clay dispersions having the desired rheological properties and zeta potential can be obtained by using different clay–surfactant mixtures.

Colloidal *aphrons* are multi-layered stable bubbles or droplets, surrounded by a thin surfactant film. The ability of two natural surfactants extracted from Henna and *Seidlitzia Rosmarinus* to produce aphronized fluid was investigated by Ahmadi et al. [81]. Both surfactants showed comparable performance in generation Colloidal gas aphron (CGA) fluids with conventional used surfactants in drilling fluids. Increasing surfactant or polymer concentration decreased the average aphron diameters. In addition, more aphron bubbles were generated by increasing the surfactant concentration. On the contrary, increasing the polymer concentration increased the fluid density (less aphron bubbles were generated). While the viscosity of CGA fluid was controlled by the concentration of both surfactant and polymer, the gel strength of CGA fluid depended mainly on the polymer concentration. Comparing to the base fluid, the aphronized fluids generated by both natural surfactants demonstrated their high ability to reduce filtration loss. Stability tests showed that CGA fluid prepared by henna extract has less stability than the fluid prepared by *Seidlitzia Rosmarinus* extract.

The sequential intercalation of three surfactants into calcium bentonite (Ca-Bent) was investigated by Monteiro et al. [82]. These surfactants were the cationic surfactant cetyltrimethylammonium bromide (CTAB), the anionic surfactant sodium dodecyl sulfate (SDS), and the nonionic surfactant lauryl alcohol ethoxylated (ALEO). Three calcium bentonite-surfactant aqueous solutions were prepared: OBent-I with CTAB, OBent-II with CTAB + SDS, and OBent-III with CTAB + SDS + ALEO. The findings of this investigation indicated that OBent-III has the highest values of basal spacing (13 nm), thermal stability (400 °C), superficial lipophilicity index (79.76°) and swelling index in polar or apolar medium (6–10 mL/g).

In a recent study, the effect of temperature (0 to 90 °C) on the drilling fluids phase inversion, and resulting changes in their rheology was investigated [83]. Nonionic (Brij 93: polyoxyethylene (2) oleyl ether, Brij O10: polyoxyethylene (10) oleyl ether, Span 80: sorbitan monooleate) and ionic (SDS and CTAB) surfactants were used to prepare water- (WBM) and oil-based muds (OBM). In addition, the effect of surfactant nature and dispersed phase volume fraction on fluid loss properties was studied. For the range of investigated conditions, phase inversion of WBM samples stabilized by nonionic polyoxyethylene oleyl ether surfactants was noticed. Differences in rheological behavior of the sample, as a result of inversion from one mud type to another, was equally observed. Generally, the findings of this study underlined the importance of evaluating potential application of nonionic surfactants to induce phase inversion in drilling fluids used for offshore operations.

It is expected that the drilling engineers will face more and more and more complex technical challenges with the rapid development of deepwater drilling operations, where hard conditions encountered. One of main challenges is that the drilling fluid used should exhibit the desired rheological properties at low temperatures with high ability to inhibit hydrate formation. Synthetic biodegradable drilling fluids have been widely used in deepwater drilling operations as they possess acceptable rheological properties, have high penetration rate with good ability to prevent hydrate formation. Calcium phytate and an amide surfactant (CEMU) were used by Hu et al. [84] to prepare a synthetic drilling fluid. It was found that no gas hydrate were formed in the developed synthetic drilling fluid at 20 MPa and 0 °C. In addition, the developed fluid exhibited excellent rheological behavior at low temperatures. The apparent viscosity and the yield stress of the drilling fluid changed significantly with temperature when the content of CEMU was relatively high (~5%). When the CEMU content was 3%, the apparent viscosity and the yield stress of the drilling fluid did not change with temperature. Moreover, the results of this study revealed that increasing the CEMU concentration leads to decrease the average particle size of the dispersed droplets, which can strengthen the interaction between the interfacial area and dispersed droplets and between the dispersed and continuous phases. The authors claimed that this optimized synthetic drilling fluid could be used in deepwater drilling operations.

2.2 *Surfactants in Organic-Based Drilling Fluids*

Different types of surfactants were also utilized extensively in modifications of organic-based drilling fluids. A cationic surfactant was synthesized by alkylation of coal tar phenol with tetradecyl alcohol to be used as an additive for drilling fluids [85]. The produced alkyl phenol was then reacted with formaldehyde and amine to produce the Mannich base product, which was further reacted sodium chloroacetate to produce a cationic surfactant. The prepared surfactant, which was charctized by FTIR analysis, surface tension, emulsification power, and wetting power, displayed promising results when utilized as the primary emulsifier in the formulation of both oil-based drilling fluid and synthetic-based drilling fluid.

The purpose of the study of [86] was to modify a procedure for clay dispersing in the organic phase of a vegetable-oil-based drilling fluid. The drilling fluid formulation consists of bentonite (modified by cationic surfactant CTABr), barite, soybean oil, lauryl alcohol triethoxylate and brine. The bentonite was modified by three different concentrations of CTABr. The measurements of surface tension measurements and contact angles on soybean oil were used to evaluate the surface modification of the bentonite. Increasing amounts of CTABr cations were adsorbed on bentonite surface, led to decrease of the contact angle, suggesting an increasing affinity for soybean oil. The results of the study revealed that temperature is the most controlling factor in the rheological behavior of the developed drilling fluid. Moreover, the results showed that CTABr surfactant modifications enhance the interaction between bentonite and the organic phase of the fluid, which has increased the ability of the drilling fluid to maintain cuttings in suspension, therefore favoring separation.

Zhuang et al. [87] aimed in their work to select appropriate organo-montmorillonites (OMt) for the formulation of oil-based drilling fluids. They studied also the effects of surfactants' nature on the structure and rheological behavior of oil-based drilling fluids prepared with OMt. The following cationic surfactants were used to modify the surface of OMt: lauryl trimethyl ammoniumchloride (C12), octadecyl trimethyl ammonium chloride (C18-A), Benzenemethanaminium, N,N-dimethyl-Noctadecyl chloride (C18-B) and Dimethyl dioctadecyl ammonium chloride (DC18). From the XRD analysis, it was concluded that bigger surfactants produced higher basal spacing and more ordered internal arrangement. On the other hand, the SEM images indicated a trend that surfactants with bigger size and low Hydrophile-Lipophile Balance (HLB) value resulted in more OMt lamellaes (thickness of ca. 50 nm) scattered. XRD analysis of OMt in oil-based drilling fluids testified a tendency that modifiers with bigger size and low HLB values led to easier exfoliation in oil. OMt modified with surfactants of small size and high HLB values obviously shrank while OMt modified with surfcatnts of bigger size and low HLB values kept exfoliation above 180 °C. The surfactants' size and polarity have clear effect on the rheological behavior of oil-based drilling fluids prepared with the modified OMt. Exfoliation of OMt in oil improved the viscosity and thixotropy of oil-based drilling fluids.

Recently, more attention was paid to the utilization of organo-palygorskite (OPal) in oil-based drilling fluids that is derived from palygorskite (Pal), which is a hydrated magnesium aluminum silicate clay mineral. It is expected that the high thermal stability and salt resistance of Pal may make OPal be suitable for offshore and deep drilling operations. The work of Zhuang et al. [88] aimed to reveal the effects of surfactants on the properties of oil-based drilling fluids containing OPal. Three quaternary ammonium salts surfactants with different lipophilicity were used to modify the surface of palygorskite. The same cationic surfactants were used in a previous study [87]. The disparity of OPal in oil increased with increasing the lipophilicity of surfactants. However, high disparity can't provide oil-based drilling fluid with high viscosity and gel strength. In addition, a little polarity can improve the rheological behavior of drilling fluids. Although high temperature facilitates the disaggregation of OPal in oil, it may result in thermal decomposition of surfactants on the surface of palygorskite.

Weng et al. [89] have prepared a group of cationic-anionic organo-sepiolite (CA-OSep) to be used in oil-based drilling fluids. The following surfactants were used in this investigation: dodecyltrimethylammonium bromide (DTAB), n-octyltrimethylammonium bromide (OTAB), cetyltrimethylammonium bromide (CTAB), tetradecyltrimethylammonium bromide (TTAB), octadecytrimethylammonium bromide (STAB), and sodium dodecyl sulfate (SDS). The effect of surfactant chain length,concentration of surfactants, and the ratio of cationic to anionic surfactants on OSep were investigated. Cationic surfactant with long chains showed better adhesion to the sepiolite surface than those with short chains. Moreover, CA-OSep showed higher thermal stability comparing with cationic organosepiolite (C-OSep). The effects of OSep addition on the rheological properties of oil-based drilling fluids were investigated at different temperatures. Results of this work revealed that organic surfactants had been adsorbed on the surface of sepiolite, and the CA-OSep showed higher surface polarity and thermal stability in diesel oil than C-OSep. This work concluded that CA-OSep can be used as rheological control additive for oil-based drilling fluids with very promising thermal stability.

Bentonites modified with nonionic surfactants Ultramine 20 (TA20) and Ultramine 50 (TA50) were used to prepare organic-based drilling fluids [90]. An effective intercalation of the nonionic surfactants in organoclays was observed, and the content of incorporated surfactants was also quantified. Chemical compatibility between diesel and kerosene organic media and the organoclays produced was indicated from the swelling measurements. In terms of rheological properties, some of the tested organophilic clay samples showed a potential for use as additives in organic-based drilling fluids. The investigators claimed that the nonionic surfactants have advantages over ionic surfactants as they have higher ability in improving the thermal and chemical properties of organoclays.

To optimize the new wells performance, the drilling engineers need to choose between drilling with water-based drilling fluid and risk possible drilling instability or drill with oil-based drilling fluid to achieve stability in difficult trajectories and risk permeability impairment. To overcome this problem, reversible emulsion based drilling fluids were suggested in recent studies. By using acid–base chemical switch, reversible emulsion can be converted from water-in-oil (W/O) to oil-in-water (O/W) emulsion.. The reaction this acid–base chemical switch with drilling system was studied by Okoro et al. [91]. Silicon Ethoxide was used in this study as a reversible surfactant to formulate a reversible drilling fluid that will not react with the formation nor the drilling equipment. Different volumes of Silicon Ethoxide ranging from 5 to 45 ml were investigated, where the reversibility was achieved from 30 ml. Three different base oils, namely,EDC-99, Palm Kernel Oil and Polytriethanolamine were also investigated. With the reversed emulsion drilling fluid, reduction in permeability within the range of 16–19 mD was achieved, which facilitates its using close to the reservoir during drilling operation without facing significant permeability impairment.

The work of [92] investigated the effects of carboxymethyl cellulose and tragacanth gum on the filtration, rheological, and interfacial properties of oil-in-water emulsion-based drilling fluids. Tragacanth gum was used as a surfactant in the emulsification process of the drilling fluid. The rheological properties and filtrate loss characteristics of the emulsion based drilling fluid were controlled by adding carboxymethyl cellulose and tragacanth gum. Moreover, the presence and concentration of these polymers affected clearly the stability of emulsion systems. The drilling fluids exhibited a shear thinning behavior in the presence of oil and additives. In addition to its role as an emulsifier for the emulsion-based drilling fluids containing diesel oil, tragacanth gum acted as a perfect viscosifier and filtrate loss control agent. The study concluded that optimizing the concentrations of carboxymethyl cellulose and tragacanth gum in the emulsion system, main properties such as fluid rheology, filtrate loss, and pH can be improved.

Shettigar et al. [93] studied the impact of cationic surfactant cetyltrimethylammoniumbromide (CTMAB) on rheological, filtrate loss and biocide properties of a polymer-based drilling fluid. Inhibitive polymer drilling fluid was prepared using shale inhibitors (KCl, K_2SO_4 and $CaCO_3$) and XC polymer. Addition of CTMAB enhanced fluid retention capacity of the drilling fluid even at very low concentration 0.1% (w/v) without altering the other properties. In addition, the cationic surfactant showed good compatibility with conventional fluid loss preventers such as pre-gelatinized starch when used in combination. The dissolved oxygen test results revealed that no need for any other biocide to prevent degradation of bio-polymer-based drilling fluid. Moreover, the thermal stability studies showed that the CTMAB is for applications in high salinity conditions and at high temperatures up to 120 °C without effecting the rheology behavior of the fluid.

In another utilization of CTAB, [94] formulated a reversible emulsion system induced by bentonite and the cationic CTAB. In this work, the contact angle and zeta potential measurements were used to evaluate surface wettability of bentonite particles in the process of phase transmission behavior, which was characterized by the electrical conductivity and microscopic analysis. The adsorption behavior and adsorption layer structure of CTAB at bentonite particles were studied at different surfactant concentrations. The results showed that changing the cationic surfactant concentration can reverse the wettability of bentonite particles, and then the two phase transition behavior of bentonite based emulsion can be induced. Additionally, the reversible emulsion oil-based drilling fluid system demonstrated good thermal stability and a small amount of fluid loss, and it reduced successfully the damage produced from the traditional oil-based drilling fluids on oil well completion.

Using of oil-based drilling fluid forms a thin layer of solids adsorbed on the well surface, called filter cake. To obtain acceptable well cementation this filter cake should be removed. Recently, microemulsion systems containing SDS, kerosene as oil phase, n-butanol as co-surfactant and distilled water were developed by da Silva et al. [95] to remove the filter cake from the well surface. The results revealed that the developed microemulsion systems composed of inverse and bicontinuous micelles removed 100% of the filter cake regardless of the content of the constituents in the formulation. It was also found that the co-surfactant/surfactant ratio is a controlling parameter for direct micelles to remove all of the of the filter cake. The compatibility of the developed reversible emulsion with cement, providing desirable thermal stability and enhancing the wettability of the rock in water.

2.3 General Remarks

From the overview presented in the previous section on the main and recent works in the modifications of the WBDF performance by different types of surfactants, we can draw the following remarks:

1. The above studies show that the drilling fluids having the desired rheological properties and stability that satisfies the API standard and OCMA specification can be achieved using different surfactant-clay systems.
2. The using of cationic surfactants in drilling fluid formulations is dominant over other surfactant types.

3. Mixtures of non-ionic and anionic surfactants can be presented as a new kind of potential rheological behavior control additive for drilling fluids, with very promising thermal stability.
4. Environmentally friendly surfactants with additive compatibility, ease of handling and cost for effective placement in water-based drilling fluids are in high demand. Most surfactants currently utilized in drilling systems are inorganic based and very few studies on the bio-based materials were carried out.
5. Surfactants can be also used in drilling fluid formulations to minimize water-blocking problems and to prevent hydrate formation in the deepwater drilling operation, where the temperature is low.
6. The chemical structures of the surfactants used in drilling fluid modifications are confirmed using FTIR and mass spectroscopy. The evaluation of these surfactants includes studying the surface activity of these compounds and their surface properties including surface tension, critical micelle concentration, emulsification power, effectiveness, minimum surface area and maximum surface excess.
7. The modified drilling fluids are evaluated by: XRD analysis to determine the interactions between the surfactants and the inter layers of the clay structure. Rheological properties, thixotropy, gel strength, filtration loss properties and temperature effect on rheological properties (apparent viscosity, plastic viscosity, yield stress, thixotropy and gel strength) of the water-based fluid are usually studied. Zeta-potential measurements are used to evaluate the stability of the drilling fluids in presence of the surfactants. In some cases, contact angle tests are used to characterize surface wettability of clay solid particles. Thermogravimetric Analysis (TGA) can also be used to evaluate the thermal stability of the developed mud.
8. Most of the studies focus on measuring the local rheological properties of drilling fluid like the single point apparent viscosity, plastic viscosity and yield stress. Little number of studies provide full rheological characterization of the drilling fluid, which gives better chance for understanding the fluid behavior and subsequent developments.
9. It is noted the absence of viscoelastic properties of the developed drilling fluid from the investigations, although these properties are very important in drilling fluids used in deepwater operations experience large variations in temperature and pressure.

3 Cationic Surfactant CTAB and Anionic Surfactant SDS in Water-Based Drilling Fluid

Although the effects of CTAB and SDS surfactants on the rheological and physical properties of clay dispersions were presented in many reports in the literature [56, 82, 83, 86, 89, 94, 96, 97], low solid concentrations were used in these investigations. Moreover, most of the conducted rheological studies did not include the time-dependent behavior of clay suspensions, which is an a main property for many industrial applications including the drilling fluids. Therefore, this work aims at investigating the impacts of the anionic surfactant SDS and the cationic surfactant CTAB as additives on the rheological behavior including thixotropy of water-based drilling fluids.

3.1 *Experimental*

3.1.1 Materials

Sodium-bentonite used in this investigation was a laboratory grade clay, which was supplied by Sigma-Aldrich CHEMIE GmbH (Germany). The bentonite sample has the following chemical composition (wt.%): SiO_2 63.020; Al_2O_3 21.080; H_2O5.64; Fe_2O_3 3.250; MgO 2.670; Na_2O 2.425; CaO 0.650; FeO 0.350; and K_2O 0.245. The average particle size of bentonite was 5.34 μm, where all particles were less than 74 μm. Cetyltrimethyl ammonium bromide (CTAB) and sodium dodecyl sulfate (SDS) surfactants were supplied by Sigma-Aldrich CHEMIE GmbH (Germany). The chemical structure of SDS and CTAB are shown in Fig. 7.

3.1.2 Drilling Fluid Preparation

Two water-based drilling fluids with bentonite concentration of 4.0 and 8.0 wt% were prepared in this work. To avoid the formation of aggregates and ensure homogeneous suspension, surfactant free drilling fluids were prepared by adding bentonite particles gradually to deionized water under continuous mixing conditions. For drilling fluid-surfactant suspensions, the following surfactant solutions were prepared in advanced: 1.0×10^{-3}, 2.0×10^{-3}, 5.0×10^{-3}, 1.0×10^{-2}, 2.0×10^{-2}, 5.0×10^{-2} and 1.0×10^{-1} M. Then, the clay particles were added gradually to the surfactant solution under mild continuous mixing conditions. After receiving a homogenous suspension, the sample was kept in a closed container for 24 h at room temperature before conducting the rheological tests. Before start recording the rheological properties, the sample was stirred in the rheometer at shear rate of 5 s^{-1} for 1 min, followed by a rest time for two minutes to remove the stress history from the sample.

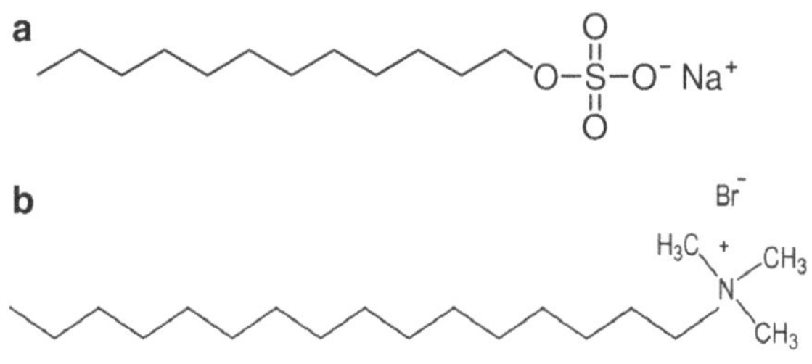

Fig. 7 Chemical structure of **a** SDS, **b** CTAB

3.1.3 Measurements of Rheological Properties

Rheolab QC viscometer (Anton Paar, Germany) equipped with coaxial cylinder measuring system was used to measure the rheological properties of the prepared drilling fluids. The Coaxial cylinder measuring system has the follow dimensions: the radius of the measuring cup = 14.460 mm; and the radius of the measuring bob = , 13.329 mm. The gab width available with this geometry a was 1.132 mm. The flow curves (shear stress (τ) versus shear rate ($\dot{\gamma}$)) of drilling fluid samples were measured at a constant temperature of 25 °C $\mp$ 0.1 in the shear rate range of 500–1500 s^{-1}. Many industrial applications are covered in the tested shear rate range. Among of these applications is the drilling fluid which is usually tested at shear rate of 511 and 1022 s^{-1} [98]. All flow curves were measured by increasing (forward measurements) and decreasing (backward measurements) shear rates. The hysteresis area formed between the forward and backward flow curves was determined as a measure of thixotropy using data analysis option of RHEOPLUS/32 V3.31 software. All flow curves measurements were performed in duplicate and the average values were reported. The measured shear stresses exhibited less than 3% difference. Eliminating of slip conditions in the used concentric cylinder measuring system was conducted by following the prehearing approach prior to measurements and using of coated measuring system. However, the absence of slip conditions from the used system was further confirmed from comparative measurements on the same suspension using different measuring system, which showed no significant difference between the reported stresses.

3.1.4 Measurement of Zeta Potential

Zeta/nano particle analyzer (NanoPlus, Otsuka Electronics CO LTD, Japan) was used to measure the zeta potential values of the prepared drilling fluids. The zeta potential values were used to evaluate the effect of surfactant type and concentration on the stability of the drilling fluid. In these measurements, the bentonite concentration was kept at 4.0 wt%. The zeta potential of fluid samples was measured at 24 h after preparation.

3.2 Investigation of Surfactant-WBDF Properties

3.2.1 Rheology of Surfactant Free Drilling Fluids

The rheological properties of surfactant free drilling fluid (SFDF) were discussed in details for bentonite concentration of 0.5–10.0 wt% in former studies [10, 11]. Figure 8a shows the flow curves of the SFDF with 4.0 and 8.0 wt% bentonite concentration. SFDFs tested in this work exhibited non-Newtonian rheological behavior at both concentrations. The flow curves of the prepared SFDFs were fitted well with

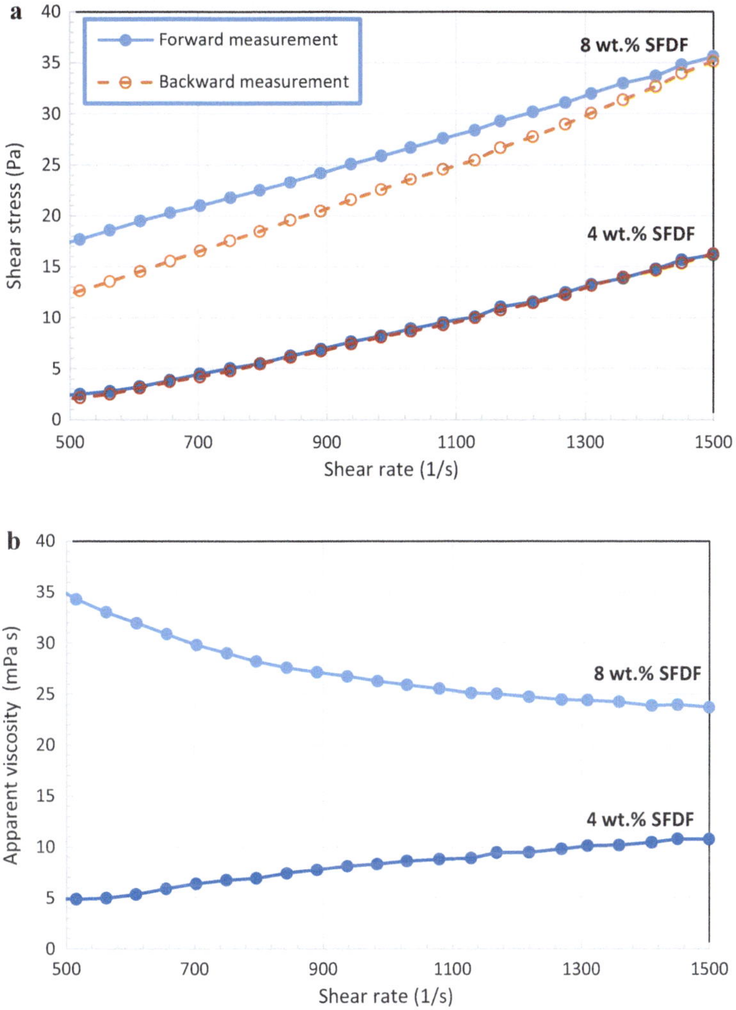

Fig. 8 **a** Flow curves of SsFDF at different solid concentration, **b** Apparent viscosity of SFDF

the Herschel-Bulkley model:

$$\tau = \tau_0 + m\dot{\gamma}^n \tag{1}$$

where τ is the shear stress, $\dot{\gamma}$ is the shear rate, τ_o is the yield stress, m is the consistency coefficient and n is the flow behavior index. As can be seen in Table 2, where the regressed parameters of Herschel-Bulkley model are reported, that the 8 wt% SFDF demonstrated pseudo-plastic behavior (shear thinning) with a noticeable yield stress. On the other side, the 4 wt% drilling fluid behaved like a shear thickening material

Table 2 Herschel-Bulkley parameters of WBDF for different surfactant concentration

Bentonite Conc. (wt%)	Surfactant Conc (wt%)	SDS			CTAB		
		τ_0 (Pa)	m × 10^5 (Pa · s^n)	n	τ_0 (Pa)	m × 10^5 (Pa · s^n)	n
4.0	0.0	0.0	10.0	1.65	0.0	10.0	1.65
	1.0×10^{-3}	0.0	12.5	1.60	0.0	5.7	1.71
	2.0×10^{-3}	0.0	13.1	1.58	0.0	2.3	1.82
	5.0×10^{-3}	0.0	6.6	1.66	0.0	1.6	1.85
	1.0×10^{-2}	0.0	9.0	1.63	0.0	1.9	1.84
	2.0×10^{-2}	0.0	8.3	1.64	0.0	2.1	1.78
	5.0×10^{-2}	0.0	7.7	1.65	0.0	1.3	1.85
	1.0×10^{-1}	0.0	6.3	1.67	0.0	1.2	1.84
8.0	0.0	8.65	2618.9	0.95	8.65	2618.9	0.95
	1.0×10^{-3}	6.90	2333.6	0.98	0.91	1508.0	1.0
	2.0×10^{-3}	6.40	1875.0	1.0	0.0	1400.0	1.0
	5.0×10^{-3}	2.74	1690.3	1.0	0.0	1380.0	1.0
	1.0×10^{-2}	4.37	2070.0	1.0	0.0	1220.0	1.0
	2.0×10^{-2}	11.5	2149.0	1.0	0.0	1190.0	1.0
	5.0×10^{-2}	7.31	1994.0	1.0	0.0	1010.0	1.0
	1.0×10^{-1}	6.67	1660.0	1.0	0.0	975.3	1.0

in the tested shear rate range without remarkable yield stress. These behaviors can also be seen in Fig. 8b, where the apparent viscosity of 8 wt% SFDF decreased with shear rate (shear thinning behavior), while the apparent viscosity of 4 wt% SFDF increased slightly with shear rate presenting a shear thickening behavior.

It is obvious that increasing the bentonite solid concentration will lead to increase all rheological parameters like the yield stress τ_o, the apparent viscosity and the consistency coefficient (m). It is expected that in dispersions with high sloid concentration (like 8.0 wt% SFDF), a continuous networked structure will be formed as a result of solid particles flocculation instead of individual flocs that usually observed in dispersions with lower solid concentrations. In dispersions with high solid concentration, the particles orient themselves towards positions of minimum free energy under the effect of Brownian motion to build up slowly the network structure [31].

To test the thixotropic behavior of the SFDF, the flow curves were measured by increasing shear rate values in the forward measurement followed by decreasing the shear rate values in the backward measurement), see Fig. 8a. The form of hysteresis loops between the flow curves of the forward and backward measurements, indicates that the SFDF exhibited a thixotropic behavior. It should be recalled that if the apparent viscosity of a material decreases and immediately recovers after shearing, the material has a shear thinning not a thixotropic behavior. As can be seen in Fig. 8a, the thixotropic behavior of the 8.0 wt% SFDF is very clear, while that for 4.0 wt% SFDF is not detectable. In addition, the calculated area of hysteresis loops for both fluids shows that the extent of thixotropy of 8.0 wt% sample (3189 Pa/*s*) is orders of magnitude higher than that of 4.0 wt% sample (577 Pa/*s*). The thixotropic behavior has been reported for many clay dispersions [99], Besq et al. [51, 100, 101], which was solid concentration dependent. Dispersions with higher solid concentration showed pronounced thixotropic behavior [11]. This behavior can be attributed to the fact that the pieces of the network structure which arebroken under shearing effect, need time to be rebuild again the network structure [31]. The thixotropic behavior of a material means that the rate of bonds breakdown in the network structure of this material is higher than the rate of restructuring process.

3.2.2 Rheology of Surfactant-WBDF

The effect of different SDS concentrations on the flow curves of WBDF is shown in Fig. 9. Addition of SDS surfactant to 4.0 wt% bentonite dispersion led initially to decrease the viscosity/shear stress values to reach a minimum at 0.005 M of SDS, then these values moved up to show a maximum at 0.01 M SDS concentration and decreased again at higher SDS concentration, see Fig. 9a. The same trend was detected in 8.0 wt% bentonite dispersion with one difference that the maximum shear stress, which was higher than the pure bentonite shear stress, shifted to 0.02 M SDS concentration, see Fig. 9b. The effects of SDS concentration on the apparent viscosity of WBDF at shear rates of 511 s^{-1} are shown in Fig. 10. This value of shear rate is usually used to evaluate the apparent viscosity of water-based drilling fluids [98]. The apparent viscosity of drilling fluid showed a minimum at 0.005 M SDS concentration and a maximum at 0.01 and 0.02 for 4.0 wt% and 8.0 wt% bentonite concentration, respectively. The apparent viscosity of 8.0 wt% WBDF in the presence of 0.02 M SDS is 43 mPa s, which is greater than that of SFDF (35.8 mPa s).

It was reported in the literature that the surfactant concentration has pronounced effect on the behavior surfactant based formulations [102]. As the concentration of a surfactant in aqueous phase increased, the individual surfactant molecules began to aggregate, with their hydrophilic heads being housed by the aqueous phase and their hydrophobic tails extending out of the aqueous phase to minimize the system free energy. These aggregates are called micelles and occur at the critical micelle concentration (CMC). Micelles have, in general, spherical shape with a dimeter about 4–10 nm [103]. These micelles can act as burier between bentonite-bentonite solid particles and can also reduce the bridging mechanism between the surfactant and bentonite. Increasing the surfactant concentration above the CMC, other strong and elastic micelle forms (cylindrical, hexagonal-packed, and lamellar structures) may produce, which would improve the bridging mechanism between the solid particles and the surfactant leading to increase the system viscosity.

It can be seen that the SDS concentration range tested here 0.001–0.1 M covers both the CMC of SDS in water which is around 0.0095 M according to [104], and the critical coagulation concentration of SDS, CCC, which is the minimum surfactant concentration needed to origin coagulation of a colloidal dispersion like bentonite dispersion [97]. Three kinds of interactions between anionic surfactant SDS and

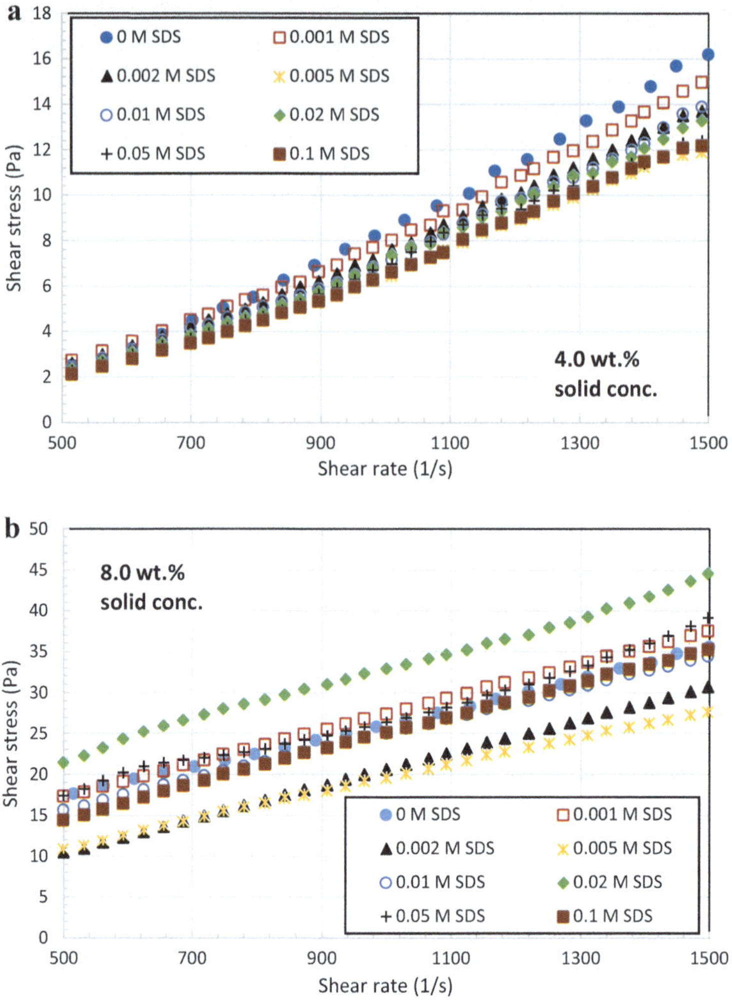

Fig. 9 Flow curves of SDS-drilling fluid for **a** 4.0 wt% bentonite concentration, **b** 8.0 wt% bentonite concentration

negative charged bentonite particles can take place [97]. The first kind is that ion exchange can take place between the anionic part of surfactant $CH_3(CH_2)_{11}OSO_3^-$ and the OH^- ions on the surface of bentonite particles. The second type of interactions is the formation of hydrogen bonding between bentonite particles and SDS molecules. In addition, it is possible that Na^+ cation on the bentonite surface can establish electrostatic bridges between the surface of bentonite and the anionic part of SDS. All these three kinds for the interactions between the surfactant SDS and bentonite particles can occur in the current system. It is clear from the rheological

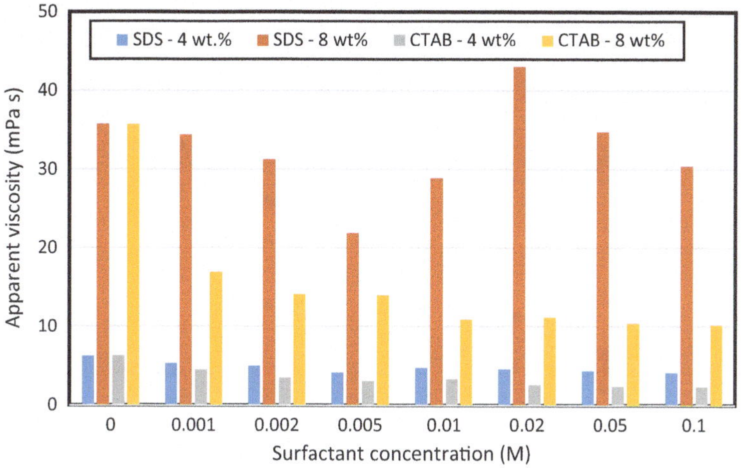

Fig. 10 Apparent viscosity of SDS-drilling fluids at 511 s^{-1}

results that the SDS surfactant is adsorbed by bentonite particles in the concentration range of 0.01 and 0.02 M producing aggregates as a result of the interactions between the hydrophobic tails of the anionic surfactant and the positive edge of the bentonite particles, which resulted in the formation of a structure with higher resistance against shearing process. Increasing the SDS concentration from 0.05 M to 0.1 M, decreased the drilling fluid viscosity. The adsorption of SDS surfactant on the bentonite surface will be increased by increasing the surfactant concentration. However, this strong adsorption can cause surface saturation, preventing later the effective bridging between the bentonite particles and causing reduction in the dispersion viscosity. Tunc et al. [80] have observed similar behavior.

Adding the cationic surfactant CTAB to the bentonite dispersions resulted in a clear decrease in the shear stress values of the WBDF, see Fig. 11. The effect of CTAB concentration on the apparent viscosity of WBDF at 511 s^{-1} is presented in Fig. 10. The apparent viscosity of CTAB-drilling fluid system decreased clearly with surfactant concentration until the addition of 0.02 M of CTAB. Beyond this concentration, the effect of cationic surfactant CTAB on the apparent viscosity of bentonite dispersions was insignificant. Adding of surfactant CTAB reduced the sdegree of flocculation in the bentonite dispersion which was resulted in viscosity decrease.

It has been shown that the replacing of exchange ions on the surface of clay by a cationic surfactant will change the clay surface properties from hydrophilic to hydrophobic [28]. This should resulted in an increase in the viscosity of clay dispersion due mainly to the electrostatic interactions that will produce from binding of positively charged surfactant molecules onto the negatively charged clay particles. In addition, the hydrophobic tails of CTAB molecules adsorbed onto the clay particles surface interact with each other by hydrophobic interaction mechanism to produce large flocs [58, 80]. In the current study, it is expected that the previous discussed

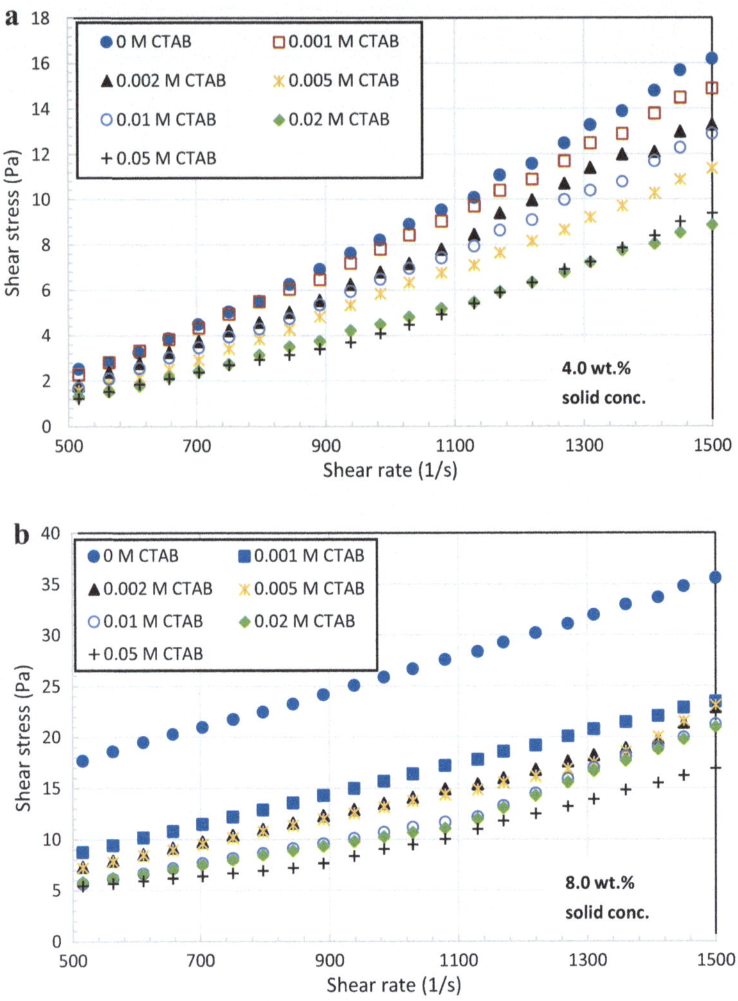

Fig. 11 Flow curves of CTAB-drilling fluid for **a** 4.0 wt% bentonite concentration, **b** 8.0 wt% bentonite concentration

mechanisms were not dominant in the tested CTAB concentration (0.001 – 0.10 M). The same tend of viscosity reduction upon addition of cationic surfactant to clay dispersion was also observed by Isci et al. [58] when they tested the effect dodecyltrimethylammonium bromide (DTABr) on the rheological behavior of sodium bentonite. This decrease in viscosity was attributed to the formations of second layers onto the bentonite particles with positive excess charges that produced dense aggregates and settled to sediment. In our study, the initial increase in the viscosity of bentonite-CTAB system was not detected as it occurs usually at very low surfactant concentration, which was out of the concentration range investigated here.

It was reported by Tunc et al. [80] that the rheological properties of bentonite dispersions have not been affected by the addition of cationic surfactant tetradecytrimethylammonium bromide. However, adding low concentration of the same surfactant (0.0001 M) to sepiolite increased slightly the dispersion viscosity. Further addition of surfactant led to pronounced reduction in sepiolite viscosity. On the other hand, [89] found that the addition of CTAB to sepiolite had no effect on clay structure. When the concentration of CTAB reached to 20 wt% of sepiolite, the CTAB-Sepiolite system had the biggest surface contact angle, lowest surface energy and highest decomposition temperature. In addition, the 20 wt% CTAB-Sepiolite-diesel oil system exhibited optimum rheological properties, with an apparent viscosity of 50 mPa s, a plastic viscosity of 45 mPa s, and a yield stress of 5 Pa.

The flow curves of CTAB- and SDS-WBDF dispersions were fitted well with Herschel-Bulkley model (Eq. 1), and the regressed parameters are reported in Table 2. For 4.0 wt% bentonite, the shear thickening behavior is still dominated in the presence of both surfactants. While the addition of SDS did not change significantly the flow behavior index, n. On the other hand, the CTAB surfactant has increased the n values, which means that the addition of CTAB to the bentonite suspension has improved the shear thickening properties.

More pronounced effect of surfactants was observed with the 8.0 wt% bentonite suspension. Initial addition of SDS surfactant led to diminish the yield stress of 8.0 wt% bentonite dispersion, which then increased suddenly at SDS concentration of 0.02 M demonstrating a maximum value of 11.5 Pa. This behavior goes along with the trend of the apparent viscosity. Moreover, the rheological of 8.0 wt% bentonite dispersion changed gradually from shear thinning behavior ($n < 1$) with a yield stress to Bingham behavior ($n = 1$) upon the addition of SDS surfactant. On the other hand, the addition of CTAB surfactant to 8.0 wt% bentonite suspensions resulted in removing completely the yield point and changing the rheological behavior from shear thinning to Newtonian ($n = 1$). This behavior along with the apparent viscosity results suggested that the interactions between the bentonite particles in the presence of CTAB surfactant were reduced to the minimum.

The evaluation and analysis of the time-dependent rheological behavior of surfactant-WBDF is another aspect of interest in this discussion. The thixotropic behavior, which is defined as an increase in the system viscosity with time of shearing at constant shear rate, was more noticeable in SDS-WBDF than that in CTAB-WBDF, as most of the CTAB-WBDF samples showed Newtonian behavior. The areas of hysteresis loops presented in Fig. 12 for both bentonite samples are in accord with the trend of the steady rheological properties (apparent viscosity vs. shear rate). In general, the thixotropic behavior of 4.0 wt% surfactant-WBDF was unnoticeable and the addition of either anionic or cationic surfactant did not affect the time-dependent behavior of bentonite dispersions.

On the other hand, the 8.0 wt% drilling fluid demonstrated clear thixotropic behavior in the presence of the SDS surfactant. The formation presence of flow curves hysteresis loops is an indication of the effects of shearing process on the surfactant-bentonite molecular structure. This means that the rate of structure damaging due to the shearing process is greater than the rate of rebuilding. Figure 12 shows that the

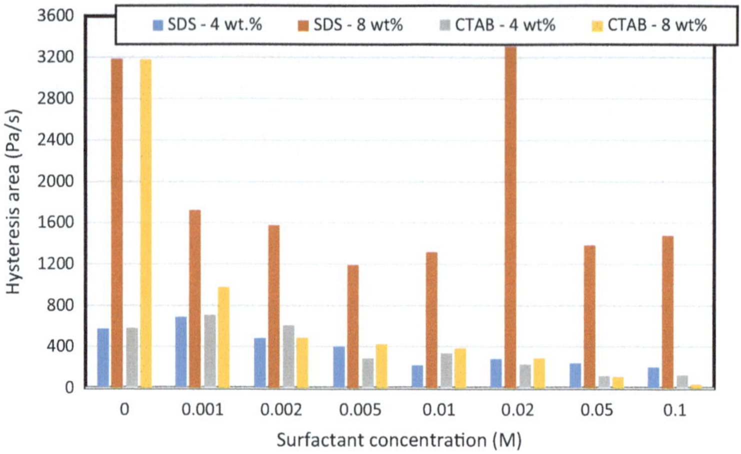

Fig. 12 Hysteresis Area of surfactant-drilling fluids

maximum degree of thixotropy occurred at 0.02 M SDS concentration, which is very close to the thixotropy extent of SFDF, see Fig. 13. The area of hysteresis loops For 8.0 wt% CTAB-bentonite dispersions decreased gradually with CTAB concentration to reach negligible value at high surfactant concentration. It should be mentioned that most of the calculated area of hysteresis loops for CTAB-bentonite dispersions are within the error range of the experimental results and can be neglected.

It is clear from the above results that using SDS surfactant will be very useful in WBDF where it needs to increase the yield stress of the drilling fluid and keep

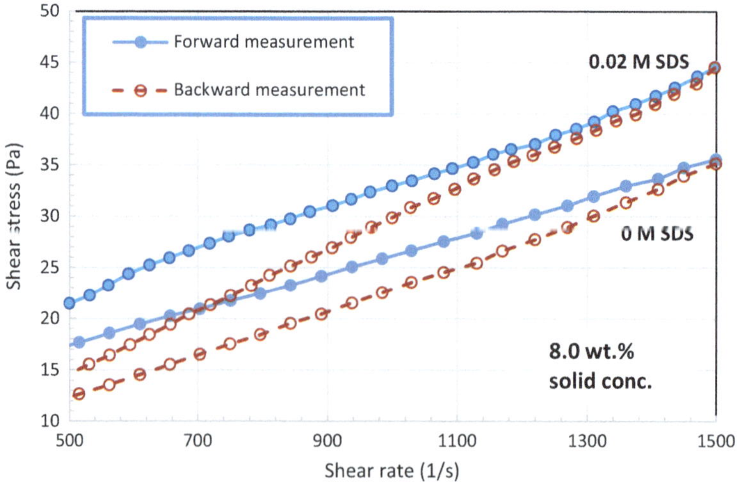

Fig. 13 Hysteresis loops of SFDF and drilling fluid containing 0.02 M SDS

good thixotropic behavior. On the other hand, adding cationic surfactant like CTAB into bentonite-based dispersion will be advantageous in other applications such as in slurry transportation, where minimum resistance to flow is required to facilitate the mixing and filtering processes.

3.2.3 Zeta Potential

The zeta potential is used as an indicator of the stability of clay suspensions system. The zeta potential value reflects the degree of electrostatic repulsion between similarly charged, adjacent particles in suspension. Figure 14 shows the measured zeta potential of surfactant-drilling fluid dispersions as a function of surfactant concentration. Increasing the SDS concentration led to increase the absolute value of zeta potential. The WBDF becomes more stable by adding 0.02 M of SDS surfactant as the zeta potential values increased from −7.4 to −36.4 mV. This increase in zeta potential value was accompanied with general decrease in the bentonite dispersion viscosity. Moreover, switching of the rheological behavior of drilling fluid from shear thinning with a yield stress to Bingham behavior upon the addition of SDS surfactants indicates that the dispersion system became less flocculated.

On the other hand, the zeta potential values of CTAB-bentonite dispersion decreased with CTAB concentration, indicating that the system became less stable and more flocculated. Formation of flocculated particles led to increase the sedimentation of solid particles, which can explain the observed rheological behavior of bentonite dispersions in the presence of CTAB surfactant, as it changed from non-Newtonian to Newtonian. CTAB surfactant forms second layers onto the bentonite particles with positive excess charges that lead to decrease the dispersion viscosity [58].

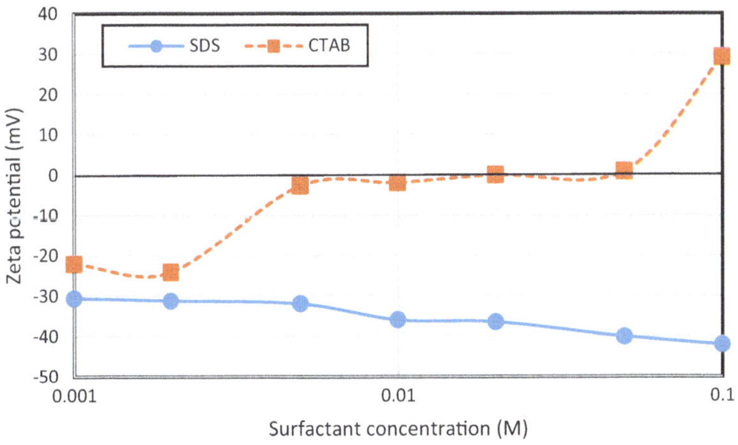

Fig. 14 Hysteresis loops of SFDF and drilling fluid containing 0.02 M SDS

It is well known that if a dispersion has a large positive or negative zeta potential, the particles within it tend to successfully repel each other and there will be no tendency for the particles to come together. On the other hand, smaller negative or positive zeta potential values increase the likelihood of flocculation. The dividing line between stable and unstable suspensions is generally taken as $\pm$ 30 mV. In the current study, the zeta potential of SDS-bentonite system was in the range of −30 to −42 mV providing stable dispersions. In the case of CTAB, the zeta potential was very close to zero for most of surfactant concentrations indication aggregates formation in the dispersion.

3.3 Conclusions

Addition of the anionic SDS surfactant in the concentration range of 0.01–0.02 M to the water-based drilling fluid which contains sodium bentonite particles led to increase the dispersion viscosity and improve its degree of thixotropy. This range of surfactant concentration corresponds to the CMC and CCC values of SDS surfactant, respectively. Moreover, the presence of SDS surfactant in WBDF system switched its modified the rheological behavior from shear thinning to Bingham plastic. The zeta potential measurements indicated that SDS-drilling fluid is stable in the SDS concentration range of 0.01 − 0.10 M. One the other hand, adding the cationic surfactant CTAB with concentration range of 0.001 − 0.1 M to the WBDF decreased its viscosity and diminished its yield stress and thixotropic behavior. The CTAB-WBDF dispersions demonstrated a Newtonian rheological behavior in the most of the tested surfactant concentration. The addition of cationic surfactant CTAB to the bentonite particles formed dense aggregates in the dispersion with higher tendency for sedimentation, which led to reduce the dispersion viscosity and reflected on the low value of zeta potential.

References

1. Shah AA, Fishwick RP, Leeke GA, Wood J, Rigby SP, Greaves M (2010) Experimental optimization of catalytic process in-situ for heavy oil and bitumen upgrading. In: Canadian unconventional resources and international petroleum conference. society of petroleum engineers
2. Mohsenzadeh A, Al-Wahaibi Y, Jibril A, Al-Hajri R, Shuwa S (2015) The novel use of deep eutectic solvents for enhancing heavy oil recovery. J Petrol Sci Eng 130:6–15
3. Shuwa S, Jibril B, Al-Wahaibi Y, Al-Hajri R (2014) Heavy-oil-recovery enhancement with choline chloride/ethylene glycol-based deep eutectic solvent. SPE J 20(01):79–87
4. Hein FJ (2006) Heavy oil and oil (tar) sands in North America: an overview & summary of contributions. Nat Resour Res 15(2):67–84
5. Zou C, Zhu RK, Wu ST, Yang Z, Tao SZ, Yuan XJ, Bai B (2012) Types, characteristics, genesis and prospects of conventional and unconventional hydrocarbon accumulations: taking tight oil and tight gas in China as an instance. Acta Petrolei Sinica 33(2):173–187

6. Pryazhnikov MI, Mikhienkova EI, Minakov AV, Rudyak VY (2019) Investigation of temperature dependence of drilling mud viscosity with silica nanoparticles addition. J Phys Conf Series 1385(1):012038. IOP Publishing
7. Negm NA, Tawfik SM, Abdou MI (2014) Evaluation of nonionic 13 surfactants in drilling muds. Surfactants Tribol 4(4):313
8. Zhang JR, Xu MD, Christidis GE, Zhou CH (2020) Clay minerals in drilling fluids: functions and challenges. Clay Miner 55(1):1–11
9. Al-Zubaidi NS, Alwasiti AA, Mahmood D (2017) A comparison of nano bentonite and some nano chemical additives to improve drilling fluid using local clay and commercial bentonites. Egypt J Pet 26(3):811–818
10. Permien T, Lagaly G (1995) The rheological and colloidal properties of bentonite dispersions in the presence of organic compounds V. Bentonite and sodium montmorillonite and surfactants. Clays Clay Mineral 43(2):229–236
11. Abu-Jdayil B (2011) Rheology of sodium and calcium bentonite–water dispersions: Effect of electrolytes and aging time. Int J Miner Process 98(3–4):208–213
12. Abu-Jdayil B, Ghannam M (2014) The modification of rheological properties of sodium bentonite-water dispersions with low viscosity CMC polymer effect. Energy Sour Part A: Recovery, Utilization, Environ Eff 36(10):1037–1048
13. Ramsey MS (2019) Practical wellbore hydraulics and hole cleaning: unlock faster, more efficient, and trouble-free drilling operations. Gulf Professional Publishing
14. RIGZONE (2020) https://www.rigzone.com/training/insight.asp?insight_id=291&c_id=. Accessed on 18 July 2020
15. Afolabi RO, Orodu OD, Efeovbokhan VE (2017) Properties and application of Nigerian bentonite clay deposits for drilling mud formulation: recent advances and future prospects. Appl Clay Sci 143:39–49
16. *la* BENTONITE (2020) https://www.bentonite.it/bentonite-structure.php. Accessed 20 July, 2020.
17. Industrial Minerals Association—North America IMA-NA (2020) https://www.ima-na.org/page/what_is_bentonite. Accessed 14 July 2020
18. Salduondo J, Comerio M, Cravero F, Etcheverry R (2020) Mineralogical and geochemical analysis of sodium bentonites in continental settings: The Uspallata Group (Triassic) of the Cuyana Basin, Mendoza province, Argentina. J South Am Earth Sci 102548
19. Kumar A, Lingfa P (2020) Sodium bentonite and kaolin clays: comparative study on their FT-IR, XRF, and XRD. Mater Today: Proc 22:737–742
20. Abu-Jdayil B, Al-Malah K, Sawalha R (2002) Study on bentonite-unsaturated polyester composite materials. J Reinf Plast Compos 21(17):1597–1607
21. Al-Malah K, Abu-Jdayil B (2007) Clay-based heat insulator composites: thermal and water retention properties. Appl Clay Sci 37(1–2):90–96
22. Cara S, Carcangiu G, Padalino G, Palomba M, Tamanini M (2000) The bentonites in pelotherapy: chemical, mineralogical and technological properties of materials from Sardinia deposits (Italy). Appl Clay Sci 16(1–2):117–124
23. Darvishi Z, Morsali A (2011) Synthesis and characterization of Nano-bentonite by sonochemical method. Ultrason Sonochem 18(1):238–242
24. Kaci A, Chaouche M, Andréani PA (2011) Influence of bentonite clay on the rheological behaviour of fresh mortars. Cem Concr Res 41(4):373–379
25. Menezes RR, Marques LN, Campos LA, Ferreira HS, Santana LNL, Neves GA (2010) Use of statistical design to study the influence of CMC on the rheological properties of bentonite dispersions for water-based drilling fluids. Appl Clay Sci 49(1–2):13–20
26. Soy U, Demir A, Caliskan F (2011) Effect of bentonite addition on fabrication of reticulated porous SiC ceramics for liquid metal infiltration. Ceram Int 37(1):15–19
27. Banat F, Al-Asheh S, Abu-Aitah L (2002) Competitive adsorption of phenol, copper ions and nickel ions on to heat-treated bentonite. Adsorpt Sci Technol 20(2):107–117
28. Guo J, Chen S, Liu L, Li B, Yang P, Zhang L, Feng Y (2012) Adsorption of dye from wastewater using chitosan–CTAB modified bentonites. J Colloid Interface Sci 382(1):61–66

29. Shirsath SR, Hage AP, Zhou M, Sonawane SH, Ashokkumar M (2011) Ultrasound assisted preparation of nanoclay Bentonite-FeCo nanocomposite hybrid hydrogel: a potential responsive sorbent for removal of organic pollutant from water. Desalination 281:429–437
30. Coussot P, Bertrand F, Herzhaft B (2004) Rheological behavior of drilling muds, characterization using MRI visualization. Oil Gas Sci Technol 59(1):23–29
31. Luckham PF, Rossi S (1999) The colloidal and rheological properties of bentonite suspensions. Adv Coll Interface Sci 82(1–3):43–92
32. Mohammed AS (2018) Electrical resistivity and rheological properties of sensing bentonite drilling muds modified with lightweight polymer. Egypt J Pet 27(1):55–63
33. Abdou MI, El-Sayed Ahmed H (2011) Effect of particle size of bentonite on rheological behavior of the drilling mud. Pet Sci Technol 29(21):2220–2233
34. Choo KY, Bai K (2015) Effects of bentonite concentration and solution pH on the rheological properties and long-term stabilities of bentonite suspensions. Appl Clay Sci 108:182–190
35. Abdou MI, Ahmed HES (2013) A study on the thixotropy of Egyptian bentonite suspensions. Pet Sci Technol 31(19):1980–1991
36. Hammadi L, Boudjenane N, Houdjedje R, Reffis R, Belhadri M (2015) Modeling the time-dependent rheological behavior of clays used in fabrication of ceramic. Int J Mech Aerosp, Ind Mechatron Manuf Eng 9(7):1373–1376
37. Maxey J (2007) Thixotropy and yield stress behavior in drilling fluids. In: AADE 2007 drilling fluids conference (AADE-07-NTCE-37)
38. Ubowska A (2010) Montmorillonite as a polyurethane foams flame retardant. Arch Combust 30(4):459–462
39. Nakama Y (2017) Surfactants. Cosmet Sci Technol 231–244
40. Assadi Y, Farajzadeh MA, Bidari A (2012) Dispersive liquid–liquid microextraction 181–212
41. Shapiro J (2018) An easy guide to understanding surfactants. https://www.ipcol.com/blog/an-easy-guide-to-understanding-surfactants/. Accessed 13 July 2020
42. Ghosh S, Ray A, Pramanik N (2020) Self-assembly of surfactants: An overview on general aspects of amphiphiles. Biophys Chem 106429
43. Kunitake T, Okahata Y, Shimomura M, Yasunami S, Takarabe K (1981) Formation of stable bilayer assemblies in water from single-chain amphiphiles. Relationship between the amphiphile structure and the aggregate morphology. J Am Chem Soc 103(18):5401–5413
44. Hirsch M (2015) Surface active agents (Surfactants). https://knowledge.ulprospector.com/3106/pc-surface-active-agents-surfactants/. Accessed 12 July 2020
45. Güngör N, Alemdar A (1999) Influences of electrolytes, polymers and a surfactant on rheological properties of bentonite–water systems. J Incl Phenom Macrocycl Chem 33(2):155–168
46. İşci S, Günister E, Ece Öl, Güngör N (2004) The modification of rheological properties of clays with PVA effect. Mater Lett 58(12–13):1975–1978
47. Saito M, Asai M (2016) U.S. Patent No. 9,475,900. U.S. Patent and Trademark Office, Washington, DC
48. M'bodj O, Ariguib NK, Ayadi MT, Magnin A (2004) Plastic and elastic properties of the systems interstratified clay–water–electrolyte–xanthan. J Colloid Interface Sci 273(2):675–684
49. Mahto V, Sharma VP (2004) Rheological study of a water based oil well drilling fluid. J Petrol Sci Eng 45(1–2):123–128
50. Alemdar A, Öztekin N, Güngör N, Ece Öl, Erim FB (2005) Effects of polyethyleneimine adsorption on the rheological properties of purified bentonite suspensions. Colloids Surf A: Physicochemical Eng Aspects, 252(2-3):95–98
51. Tunc S, Duman O (2008) The effect of different molecular weight of poly (ethylene glycol) on the electrokinetic and rheological properties of Na-bentonite suspensions. Colloids Surf A: Physicochemical Eng Aspects 317(1-3):93-99
52. Xiao H, Liu S, Chen Y, Han D, Wang D (2017) Impacts of polypropylene glycol (PPG) additive and pH on tribological properties of water-based drilling mud for steel-steel contact. Tribol Int 110:318–325

53. Benyounes K, Mellak A, Benchabane A (2010) The effect of carboxymethylcellulose and xanthan on the rheology of bentonite suspensions. Energy Sour Part A: Recovery, Utilization, Environ Eff 32(17):1634–1643
54. Shaikh SM, Nasser MS, Hussein IA, Benamor A (2017) Investigation of the effect of polyelectrolyte structure and type on the electrokinetics and flocculation behavior of bentonite dispersions. Chem Eng J 311:265–276
55. Xie G, Luo P, Deng M, Su J, Wang Z, Gong R, Duan Q (2017) Intercalation behavior of branched polyethyleneimine into sodium bentonite and its effect on rheological properties. Appl Clay Sci 141:95–103
56. Günister E, Alemdar SA, Güngör N (2004) Effect of sodium dodecyl sulfate on flow and electrokinetic properties of Na-activated bentonite dispersions. Bull Mater Sci 27(3):317–322
57. Glass Jr JE (1985) U.S. Patent No. 4,561,985. U.S. Patent and Trademark Office, Washington, DC
58. İşçi S, Günister E, Alemdar A, Ece OI, Güngör N (2008) The influence of DTABr surfactant on the electrokinetic and rheological properties of soda-activated bentonite dispersions. Mater Lett 62:81–84
59. Lin S-H, Juang R-H (2002) Heavy metal removal from water by sorption using surfactant-modified montmorillonite. J Hazard Mater B92:315–326
60. Bouberka Z, Khenifi A, Sekrane F, Bettahar N, Derriche Z (2008) Adsorption of direct red 2 on bentonite modified by cetyltrimethylammonium bromide. Chem Eng J 136:295–305
61. Al-Asheh S, Banat F, Abu-Aitah L (2003) Adsorption of phenol using different types of activated bentonites. Sep Purif Technol 33:1–10
62. Nasser MS, James AE (2007) Effect of polyacrylamide polymers on floc size and rheological behaviour of kaolinite suspensions. Colloids Surf A: Physicochemical Eng Aspects, 301:311–322
63. Vali H, Bachmann L (1988) Ultrastructure and flow behavior of colloidal smectite dispersions. J Colloid Interface Sci 126(1):278–291
64. Güngör N (2000) Effect of the adsorption of surfactants on the rheology of Na-bentonite slurries. J Appl Polym Sci 75(1):107–110
65. Mostafa BA, Assaad FF, Rashad MA, El-Niklawy AS (2010) The role of temperature on the Egyptian drilling mud performance in the presence of cationic polyethylene imine and anionic polyacrylamide. J Appl Polym Sci 117(5):2958–2963
66. Yunita P, Irawan S, Kania D (2016) Optimization of water-based drilling fluid using non-ionic and anionic surfactant additives. Procedia Eng 148:1184–1190
67. Elochukwu H, Sia LKS, Gholami R, Hamid MA (2018) Data on experimental investigation of Methyl Ester Sulphonate and nanopolystyrene for rheology improvement and filtration loss control of water-based drilling fluid. Data in Brief 21:972–979
68. Ahmed SM, Dardir MM, Mohamed DE, Awad AI (2016) Synthesis and Evaluation of Bis-quaternary-Based Surfactants as Additives for Water-Based Mud. J Surfactants Deterg 19(6):1263–1272
69. Dardir MM, Mohamed DE, Farag AB, Ramdan AM, Fayad MM (2017) Preparation and evaluation of cationic bolaform surfactants for water-based drilling fluids. Egypt J Pet 26(1):67–77
70. Mészáros R, Jobbik A, Varga G, Bárány S (2019) Electrosurface properties of Na-bentonite particles in electrolytes and surfactants solution. Appl Clay Sci 178:105127
71. Liu L, Pu X, Zhou Y, Zhou J, Luo D, Ren Z (2020) Smart pickering water-in-oil emulsion by manipulating interactions between nanoparticles and surfactant as potential oil-based drilling fluid. Colloids Surf Physicochemical Eng Aspects 586:124246
72. Murtaza M, Kamal MS, Hussain SS, Mahmoud M, Syed NA (2020) Quaternary ammonium gemini surfactants having different spacer length as clay swelling inhibitors: Mechanism and performance evaluation. J Mol Liq 113054
73. Jiang G, Li X, Zhu H, Yang L, Li Y, Wang T, Wu X (2019) Improved shale hydration inhibition with combination of gelatin and KCl or EPTAC, an environmentally friendly inhibitor for water-based drilling fluids. J Appl Polym Sci 136(22):47585

74. Zhang H, Yan J, Lu Y, Shu Y, Zhao S (2009) Experimental study of low-damage drilling fluid to minimize waterblocking of low-permeability gas reservoirs. Petr Sci 6(3):271–276
75. Negm NA, Tawfik SM, Abdou MI, Badr EA, Ghuiba FM (2015) Evaluation of some nonionic surfactants derived from tannic acid as additives for water-based mud. J Surfactants Deterg 18(2):309–319
76. Shadizadeh SR, Moslemizadeh A, Dezaki AS (2015) A novel nonionic surfactant for inhibiting shale hydration. Appl Clay Sci 118:74–86
77. Aggrey WN, Asiedu NY, Adenutsi CD, Anumah P (2019) A novel non-ionic surfactant extract derived from Chromolaena odarata as shale inhibitor in water based drilling mud. Heliyon 5(5):e01697
78. González JM, Quintero F, Arellano JE, Márquez RL, Sánchez C, Pernía D (2011) Effects of interactions between solids and surfactants on the tribological properties of water-based drilling fluids. Colloids Surf A: Physicochemical Eng Aspects 391(1–3):216–223
79. Zhang Y, Zhao Y, Zhu Y, Wu H, Wang H, Lu W (2012) Adsorption of mixed cationic-nonionic surfactant and its effect on bentonite structure. J Environ Sci 24(8):1525–1532
80. Tunç S, Duman O, Kancı B (2012) Rheological measurements of Na-bentonite and sepiolite particles in the presence of tetradecyltrimethylammonium bromide, sodium tetradecyl sulfonate and Brij 30 surfactants. Colloids Surf A: Physicochemical Eng Aspects 398:37-47
81. Ahmadi MA, Galedarzadeh M, Shadizadeh SR (2015) Colloidal gas aphron drilling fluid properties generated by natural surfactants: experimental investigation. J Nat Gas Sci Eng 27:1109–1117
82. Monteiro MKS, de Oliveira VRL, dos Santos FKG, de Barros Neto EL, de Lima Leite RH, Aroucha EMM, de Oliveira Silva KN (2018) Influence of the ionic and nonionic surfactants mixture in the structure and properties of the modified bentonite clay. J Mol Liq 272:990–998
83. Numkam GL, Akbari B (2019) Effect of surfactant chemistry on drilling mud performance. J Petrol Sci Eng 174:1309–1320
84. Hu Y, Yue Q, Liu S, Fu Z, Liang S (2011) Research on deepwater synthetic drilling fluid and its low temperature rheological properties. Petrol Sci 8(4):485–489
85. Mahmoud SA, Dardir MM (2011) Synthesis and evaluation of a new cationic surfactant for oil-well drilling fluid. J Surfactants Deterg 14(1):123–130
86. Ratkievicius LA, Da Cunha Filho FJV, Neto ELDB, Santanna VC (2017) Modification of bentonite clay by a cationic surfactant to be used as a viscosity enhancer in vegetable-oil-based drilling fluid. Appl Clay Sci 135:307–312
87. Zhuang G, Zhang H, Wu H, Zhang Z, Liao L (2017) Influence of the surfactants' nature on the structure and rheology of organo-montmorillonite in oil-based drilling fluids. Appl Clay Sci 135:244–252
88. Zhuang G, Zhang Z, Gao J, Zhang X, Liao L (2017) Influences of surfactants on the structures and properties of organo-palygorskite in oil-based drilling fluids. Microporous Mesoporous Mater 244:37–46
89. Weng J, Gong Z, Liao L, Lv G, Tan J (2018) Comparison of organo-sepiolite modified by different surfactants and their rheological behavior in oil-based drilling fluids. Appl Clay Sci 159:94–101
90. Silva IA, Sousa FKA, Menezes RR, Neves GA, Santana LNL, Ferreira HC (2014) Modification of bentonites with nonionic surfactants for use in organic-based drilling fluids. Appl Clay Sci 95:371–377
91. Okoro EE, Dosunmu A, Iyuke S (2020) Silicon ethoxide as reversible surfactant in reversible drilling mud and the mud's effect on permeability. J King Saud Univ –Eng Sci
92. Jha PK, Mahto V, Saxena VK (2015) Effects of carboxymethyl cellulose and tragacanth gum on the properties of emulsion-based drilling fluids. Can J Chem Eng 93(9):1577–1587
93. Shettigar RR, Misra NM, Patel K (2018) Cationic surfactant (CTAB) a multipurpose additive in polymer-based drilling fluids. J Petrol Explor Production Technol 8(2):597–606
94. Zhang R, Zhu X, Cai Y (2019) The phase transformation mechanism of bentonite-stabilized and cetyltrimethylammonium bromide-stabilized emulsions and application in reversible emulsification oil-based drilling fluids. J Surfactants Deterg 22(3):525–534

95. da Silva VL, Ribeiro LS, de Oliveira Freitas JC, da Silva DNN, de Carvalho LS, Rodrigues MAF, Neto ADOW (2020) Application of SDS surfactant microemulsion for removal of filter cake of oil-based drilling fluid: influence of cosurfactant. J Pet Explor Prod Technol 1–12
96. Güngör N, Alemdar A, Atici O, Ece IO (2001) The effect of SDS surfactant on the flow and zeta potential of bentonite suspensions. Mater Lett 51:250–254
97. Yalcın T, Alemdar A, Ece OI, Gungor N (2002) The viscosity and zeta potential of bentonite dispersions in presence of anionic surfactants. Mater Lett 57:420–424
98. Kelessidis VC, Papanicolaou C, Foscolos A (2009) Application of Greek lignite as an additive for controlling rheological and filtration properties of water–bentonite suspensions at high temperatures: a review. Int J Coal Geol 77:394–400
99. Yildiz N, Sarikaya Y, Calimli A (1999) The effect of the electrolyte concentration and pH on the rheological properties of the original and Na_2 CO_3—activated Kutahya bentonite. Appl Clay Sci 14:319–327
100. Besqa A, Malfoyb C, Pantetb A, Monneta P, Righi D (2003) Physicochemical characterisation and flow properties of some bentonite muds. Appl Clay Sci 23:275–286
101. Tombacz E, Szekeres M (2004) Colloidal behavior of aqueous montmorillonite suspensions: the specific role of pH in the presence of different electrolytes. Appl Clay Sci 27:75–94
102. Hoffmann H, Rauscher A, Gradzielski M, Schulz SF (1992) influence of ionic surfactants on the viscoelastic properties of zwitterionic surfactant solutions. Langmuir 8:2140–2151
103. Croce V, Cosgrove T (2003) Rheology, cryogenic transmission electron spectroscopy and small-angle neutron scattering of highly viscoelastic wormlike micellar solutions. Langmuir 19:8536–8541
104. Khan AM, Shah SS (2008) Determination of critical micelle concentration (CMC) of sodium dodecyl sulfate (SDS) and the effect of low concentration of pyrene on its CMC using ORIGIN software. J Chem Soc Pak 30:186–191

Chemical EOR

IFT Role on Oil Recovery During Surfactant Based EOR Methods

Madhar Sahib Azad

Abstract In general, it is believed that the ultra-low IFT provided by surfactant is a requirement for the higher microscopic recovery efficiency during enhanced oil recovery (EOR). In tight oil shale and shale reservoirs, capillary imbibition become a dominant recovery mechanism where ultra-low IFT becomes less significant or even a retarding force in certain scenarios. Recent researches have emphasized that the microscopic efficiency of CO_2 flooding could be improved by adding low IFT surfactants. Surfactants are also used for conformance/mobility control applications in the form of foam. During foam flooding application in naturally fractured reservoirs, the ultra low-IFT conditions is advantageous for oil recovery in dolomite but not in limestone rocks. Although low-IFT conditions positively influences the microscopic recovery during alkali steam-foam flooding, ultra-low IFT is not required. This chapter compiles these cases and sheds insight using fundamental reservoir engineering concepts to understand why the ultra-low IFT conditions, conventionally considered to be a prerequisite for the higher residual oil recovery, are not always beneficial or required or enough during many of the EOR applications.

1 Fundamental Concepts of Enhanced Oil Recovery

1.1 Microscopic Displacement

Total recovery factor during enhanced oil recovery (EOR) is a product of microscopic displacement and macroscopic sweep efficiency (Eq. 1).

$$E = E_d * E_V \tag{1}$$

where

M. S. Azad (✉)
CPG, KFUPM, Dhahran, Saudi Arabia
e-mail: mazad@ualberta.ca

T. Solling et al. (eds.), *Surfactants in Upstream E&P*, Petroleum Engineering,
https://doi.org/10.1007/978-3-030-70026-3_4

E = total recovery factor; E_d = microscopic displacement efficiency; E_V = macroscopic sweep efficiency.

Generally, microscopic displacement efficiency gives a measure of how much trapped oil that is swept by the injection fluids can be mobilized to the total swept oil. It is a pore-scale phenomenon quantifying the amount of well-swept residual oil can that be recovered during flooding (Eq. 2).

$$E_d = \frac{S_{oi} - S_{or}}{S_{oi}} \tag{2}$$

where

S_{oi} = initial oil saturation; S_{or} = residual oil saturation.

For the residual oil to get recovered, it is a must that the injection fluid should contact it. In other words, sweeping effect is a precursor to micro-displacement effect. Not all the swept oil will be mobilized during water flooding because of the higher capillary pressure. A sample calculation shown in the next section illustrates the fact that capillary pressure that traps the oil within the pores would be on order to 1000 psi/ft. Capillary pressure could also be a driving force during spontaneous imbibition. Capillary pressure is directly proportional to interfacial tension (IFT), wettability and radius of the pore (Eq. 3).

$$P_c = \frac{2 * \sigma * \cos\theta}{r} \tag{3}$$

where

P_c = capillary pressure, psi; σ = Interfacial tension between the displacing and displaced fluids, mN/m; r = pore radius, microns

As per the snap-off concepts, the likelihood of the trapping capillary pressure will be higher when the difference between pore body radius and pore throat radius (aspect ratio) is higher (Eq. 4). The readers are suggested to refer Fig. 2.14 of Green and Willhite's [1] for more information about the different capillary pressure gradient caused due to variations in contact angle, IFT etc.

$$\frac{P_c}{L} - \frac{2 * \sigma * \cos\theta}{L}\left(\frac{1}{r1} - \frac{1}{r2}\right) * \left(\frac{14.696 * 30.48}{1.0133 \times 10^6}\right) \tag{4}$$

where

$\frac{P_c}{L}$ = capillary pressure gradient, psi/ft.; $r1$ = radius of pore throat or smaller pores, centimeter; $r2$ = radius of pore body or larger pores, centimeter; L = length of an oil blob, centimeter

For the oil to get mobilized from the pore, driving viscous force should exceed the capillary force (Eq. 4). The driving viscous force can be calculated through the Darcy law (Eq. 5).

$$\frac{dP}{L} = \frac{v * \mu}{K * 0.001127 * 5.615} \tag{5}$$

where

$\frac{dP}{L}$ = pressure gradient, psi/ft; v = flux rate, ft/day; μ = apparent viscosity, cP; k = permeability, mD

The balance between the driving viscous and trapping force can also be explained in terms of the capillary number (N_c). Many versions of capillary number exist in the literature [2]; Al-Quaimi and Rosen [3–5] and the most common number expressed as a ratio of viscous force to the capillary force is represented through Eqs. (6) and (7) respectively.

$$N_c = \frac{v * \mu}{\sigma * \cos\theta} \tag{6}$$

$$N_c = \frac{k * \left(\frac{dp}{L}\right)}{\sigma * \cos\theta} \tag{7}$$

where

N_c = capillary number, dimensionless; v = flux rate of the displacing fluid, ms^{-1}; μ = apparent viscosity of the displacing slug, cP; σ = interfacial tension between the displacing and displaced slugs, mN/m; θ = contact angle; k = permeability, cm^2; $\frac{dp}{L}$ = pressure gradient, psi/cm.

Higher fluxes, higher displacing slug's viscosity, higher pressure gradient and ultra-low IFT leads to higher value of capillary number (Eqs. 6 and 7). Capillary number must be 10^{-3} or 10^{-2} in order to have a significant reduction in S_{or}. Viscous force generated during water flooding at the normal flux rate of 1ft/day would be on the order of 1 psi/ft and corresponding capillary number would be around 10^{-7} [6]. Practical means of enhancing the capillary number is by reducing the IFT between the displacing solutions and displaced oil (Eqs. 6 and 7). By reducing the IFT, trapping capillary pressure will be reduced (Eq. 3), and oil will be mobilized. Low IFT is provided by adding surfactant to the displacing solutions. Surfactant flooding is the common chemical EOR method which aims to improve the microscopic displacement efficiency. The interplay between the gravity forces and capillary forces also influences the microscopic recovery if there is good vertical support. Bond number has been used to characterize the relative importance of gravity forces over capillary forces (Eq. 8).

$$N_b = \frac{\Delta\rho * g * k}{\sigma * \cos\theta} \tag{8}$$

where

N_b = Bond number, dimensionless; $\Delta\rho$ = density difference between the displacing and displaced slugs, Kg/m^3; k = permeability, m^2; σ = interfacial tension between the displacing and displaced slugs, mN/m; θ = contact angle

The numerator and denominator term of (Eq. 8) denotes the gravity and capillary forces respectively. Ultra-low IFT and higher permeable conditions will induce higher gravity forces. As per the conventional belief, lower the IFT, the higher the capillary number and bond number and therefore, higher the oil recovery. However, it is not the case always. Sometimes ultra-low IFT conditions may be detrimental or insignificant or will need to be supplemented by other factors to have an enhanced oil recovery. This will be discussed in this chapter.

In miscible CO_2 flooding, there is a complete removal of interface which means theoretically capillary number should be at infinity and S_{or} should be zero (Eqs. 6 and 7). However, to have the miscibility, certain pressure is required which is known as minimum miscibility pressure (MMP). Unless, the reservoir oil, injection gas composition meets the miscibility requirements, the flooding will be operated in immiscible mode. Recently, surfactants are used to reduce the IFT and MMP of crude oil-CO_2 system. The IFT role on MMP reduction and oil recovery is also discussed.

1.2 Macroscopic Sweep

Macroscopic sweep efficiency gives a measure of how much oil could be contacted volumetrically by the injected fluids to the total oil available before the flooding. Macroscopic sweep efficiency can be further decomposed into vertical and areal sweep efficiency (Eq. 9).

$$E_V = E_A * E_I \tag{9}$$

where

E_A = areal sweep efficiency, the swept area divided by the total reservoir area; E_I = vertical sweep efficiency, the pore space swept by the injection fluid to the total pore spaces in all the layers behind the areal location of the front

Sweep is not only a field-scale phenomenon and it could be important at the core-scale. At the core scale, the mobilized oil needs to be pushed forward for which a favorable mobility ratio needs to be maintained between the mobility buffer, injection chemical slugs and oil/water bank. Mobility buffer efficiency along with the volumetric sweep efficiency and microscopic displacement efficiency determines the overall recovery factor [7]. Sweep is a function of time on both the field-scale [1] and core-scale [4, 5]. Viscous fingering due to the viscosity contrast between the displacing and displaced fluids, excessive channeling of the injection fluids through the high permeable streaks are the main reasons for the poor sweep efficiency during water flooding. Gravity override due to the density differences between the displacing and displaced slugs also leads to poor sweep efficiency in low dense EOR methods such as CO_2 EOR and steam flooding. Mobility ratio (M) is a ratio of the mobility of the displacing slugs to the mobility of the displaced slugs (Eq. 10). Mobility is the ratio of relative permeability of the fluid to its viscosity.

$$M = \left(\frac{K_{rw}}{\mu_w}\right)_{Sor} * \left(\frac{\mu_o}{K_{ro}}\right)_{Siw} \tag{10}$$

where

M = end point mobility ratio, no unit; $\left(\frac{K_{rw}}{\mu_w}\right)_{Sor}$ = mobility of the displacing slug at residual oil saturation; $\left(\frac{K_{ro}}{\mu_o}\right)_{Siw}$ = mobility of the displaced slug at immobile water saturation.

The above definition of mobility ratio holds for water flood in which only the oil flows ahead of the waterfront and only water flows behind the front. However, there will be a saturation variation with respect to space and time in most of the immiscible floods. Mobility ratio at average saturation can be calculated using the average saturation (Eq. 11).

$$M_s = \left(\frac{K_{rD}}{\mu_D}\right)_{S_D} * \left(\frac{\mu_d}{K_{rd}}\right)_{S_d} \tag{11}$$

where

M_s = mobility ratio at average saturation; $\left(\frac{K_{rD}}{\mu_D}\right)_{S_D}$ = mobility of the displacing-phase at the breakthrough saturation; $\left(\frac{K_{rd}}{\mu_d}\right)_{S_d}$ = mobility of the displaced phase at average saturation ahead of the flood front.

For more details about the mobility ratio, the readers can refer to Craig [8] and Green and Willhite [1]. Mobility ratio influences the core-scale linear displacement, areal sweep efficiency and vertical sweep efficiency. To have a better sweep efficiency, mobility ratio needs to be reduced which is accomplished through the injection of the viscous fluids such as polymer and foam solutions. These fluids can impart higher resistance to the flow of the displacing slugs by generating higher apparent viscosity. Surfactants are used in foam flooding and the prime expectation from these surfactants is to have a good foam stability (and not the low IFT), so that the apparent viscosity needed for the enhanced sweep could be achieved. The basic understanding is that foam flooding used for conformance control or used to reduce the adverse mobility of low dense EOR fluids such as steam cannot reduce Sor. However, to have an overall good recovery factor, Sor reduction through enhanced microscopic displacement is needed (Eq. 1). The surfactant blends and surfactant system combined with alkali can lead to higher interfacial activity during foam-based flooding. The role of IFT on Sor reduction during foam-based flooding is discussed in this chapter.

2 Conventional Surfactant Flooding

During surfactant flooding (also called as low-tension or micellar flooding), the surfactants are added to the injection water to reduce the IFT. Reduction in IFT leads to the reduction of capillary pressure and an increase in capillary number (Eqs. 6 and 7). Increase in the capillary number leads to Sor reduction. The example calculations taken from Peter [6] illustrates the importance of having ultra-low IFT for residual oil mobilization during the dynamic surfactant injection at 1ft/day. Please note the words "dynamic surfactant injection", "forced surfactant flooding" and "conventional surfactant flooding" bears the same meaning i.e., the flooding is aided by the imposed flux. These words should not be confused with the surfactant aided spontaneous imbibition where there will be no imposed fluxes. The entire discussion in Sect. 2 pertain only to forced displacement. The discussion about spontaneous imbibition is deferred to Sect. 3.

Problem 1 An oil droplet that got trapped due to snap off at the pore-scale needs to be mobilized. Rock and fluid properties are reported in Table 1.

Determine (a) the pressure gradient that is needed to be release the trapped oil (b) the pressure gradient that could be generated during water flood (c) the new trapping pressure gradient during surfactant flooding who IFT is assumed to be (1) 0.01 mN/m (2) 0.1 mN/m (3) 1 mN/m (Fig. 1).

(a) Calculated trapping pressure gradient using (Eq. 4) is 169.75 psi/ft. During flooding, the gradient higher than 169.75 psi/ft needs to be exerted so that oil will get mobilized.
(b) Calculated pressure gradient that could be generated during forced water flood at 1 ft/day using (Eq. 5) is 0.079 psi/ft. The trapped oil cannot be mobilized because the pressure gradient generated during water flood is less than that required for oil mobilization
(c) Using (Eq. 4), the trapping gradient during surfactants injections with the IFT of 0.01 mN/m, 0.1 mN/m and 1 mN/m are 0.057 psi/ft, 0.57 psi/ft and 5.7 psi/ft respectively.

Table 1 Rock and Fluid properties for sample problem 1

Rock type	Water-wet
$\theta°$	0
IFT between water and oil, mN/m	30
Pore-throat radius, microns	50
Pore-body radius, microns	250
Length of oil blob, microns	250
Darcy velocity, ft/day	1
Water viscosity, cP	1
Permeability, mD	2000

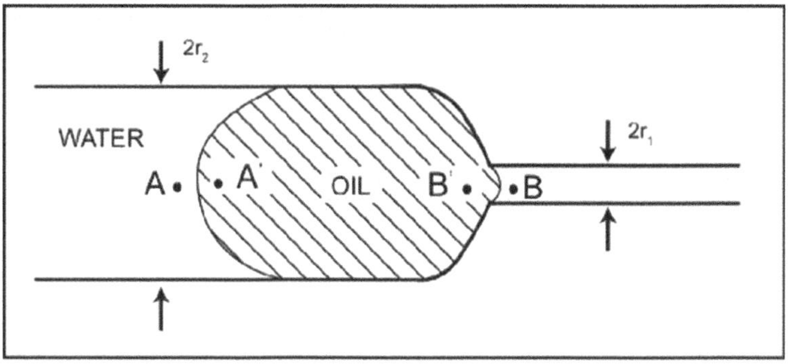

Fig. 1 Trapped oil blob (from [6])

If the surfactant addition to the displacing water reduces the IFT from 30 to 0.01 mN/m, trapping capillary pressure gradient would be reduced from 169.75 to 0.057 psi/ft. The value of 0.057 psi/ft is lower than 0.079 psi/ft, the calculated viscous driving force that could be generated during the water flood. Therefore, trapped oil could be mobilized during the surfactant injection at 1ft/day during the ultra-low IFT conditions of 0.01 mN/m. At the moderate to high IFT conditions of 0.1 to 1 mN/m, the surfactant injection could reduce the capillary pressure gradient from 169.75 psi/ft. to 0.57–5.7 psi/ft. Since these values are more than the viscous pressure gradient generated during the water flood, residual oil cannot be mobilized at 1 ft/day. Operating the flood with the moderate IFT of 0.1 mN/m but at the higher flux of 10 ft/day could have the driving viscous gradient (0.79 psi/ft) exceeding the capillary trapping gradient (0.57 psi/ft). Since the driving viscous force is higher than the trapping capillary force, residual oil could be mobilized. One can expect to have relatively low IFT reduction by surfactant when the flux rate is higher. However, high fluxes are seen only around the wellbore and therefore residual oil in farthest portion of the reservoir would remain immobilized. This signifies the importance of having ultra-low IFT condition during surfactant injection.

Please note the performed calculation is a simplified one because IFT value with vary with respect to time and location due to surfactant's adsorption. Also, formation of micro-emulsion will lead to higher apparent viscosity which may influence the calculation and interpretation. Nevertheless, the performed calculation illustrates the importance of having ultra-low IFT conditions for having a higher microscopic displacement efficiency at 1ft/day. An important point to note here is that there should be an imposed force for the surfactant flooding to be effective in reducing the capillary pressure gradient to be comparable with viscous flood gradient. Reservoirs shouldn't be highly heterogeneous with the extensive presence of high conductivity fractures, and permeability shouldn't ultra-low so that the forced viscous based displacement can be imposed. In the matrix-fracture system characterized by the very low permeable matrix and in very tight rocks, spontaneous imbibition will be a dominant mechanism.

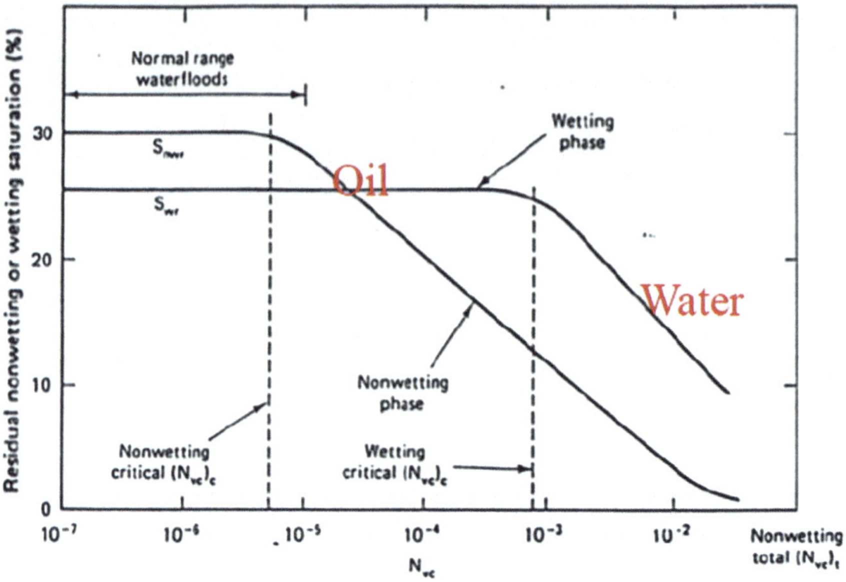

Fig. 2 Residual non-wetting or wetting phase as a function of capillary number (From [9, 10])

The relation between the capillary number and Sor are represented through capillary desaturation curve (CDC). In CDC curve, there will be critical capillary number above which the oil mobilization will be significant (shown as dashed vertical lines in Fig. 2).

At capillary number less than critical capillary number (such as in water flood), capillary force dominates so the injection fluids imbibe into the finer pores and oil gets snapped off in the larger pores. After the critical capillary number, viscous forces relatively dominate the capillary forces and therefore could results in the rapid oil mobilization especially from larger pores [11]. Relative domination of viscous force over capillary force is achieved by the reduction in IFT during the surfactant injection. IFT lowering requirement is dependent on several parameters during forced displacement. In this section, we shall see the effect of rock wettability, oil connectivity, oil viscosity on IFT reduction requirement during oil recovery.

2.1 *Effect of Wettability on IFT Requirement*

Wettability of the rock is an important parameter which influences the requirement of IFT reduction during flooding. In oil-wet rocks, residual oil can occur (1) in the small pore as continuous oil (2) as trapped large globules in many pores (3) as trapped discontinuous oil droplets at the pore throat (4) as pendulum rings along the pore walls. Several studies conducted in the past revealed that when compared to water-wet media, significant reduction in IFT is needed to displace the residual oil occurring in the form of discontinuous droplets or pendulum rings in the oil-wet media [12–16]. Critical capillary number (denoted as Nvc in Fig. 2) that quantifies that onset of rapid oil mobilization is higher when the non-wetting phase displaces the wetting phase. To displace most of the trapping wetting phase by non-wetting phase, a very high capillary number of 2 is needed (Dombrowski and Brownell's curve in Fig. 3) whereas for the efficient displacement of non-wetting phase by the wetting phase, a relatively lower capillary number suffices (Fig. 3). Therefore, it can be said more energy needs to be expended for displacing the wetting phase by the non-wetting phase which can be achieved at the ultra-low IFT conditions.

One important point is that in some oil-wet media, a critical capillary number may appear to occur early, or a visible transition may not be seen.

However, a careful look into Du prey's curve in Fig. 3 and CDC curve for Indiana limestone in Fig. 4 reveals that reduction in Sor with respect to increase in capillary number (or therefore decrease in IFT) is much lesser because of the complex nature of the oil wet rock. This indicates in oil-wet media, recovering the residual oil is relatively a difficult process. For example, in most of the sandstone rocks, increasing the capillary number from 10^{-5} to 10^{-2} could result in the Sor ~5% (Fig. 4). However, for Indiana limestone, Sor could not be reduced to more than 18% with the similar increase in capillary number (Fig. 4). IFT reduction alone may not suffice always in oil-wet rocks because of strong adhesion forces and wider pore size distribution. A strong wettability alteration may need to be coupled with IFT reduction in case of oil-wet rocks.

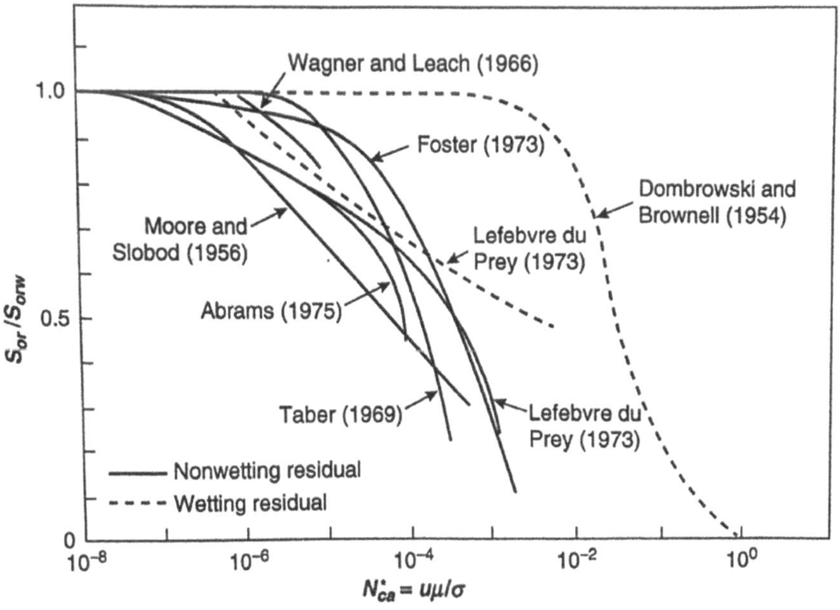

Fig. 3 CDC curve generated by different EOR researchers (From [1])

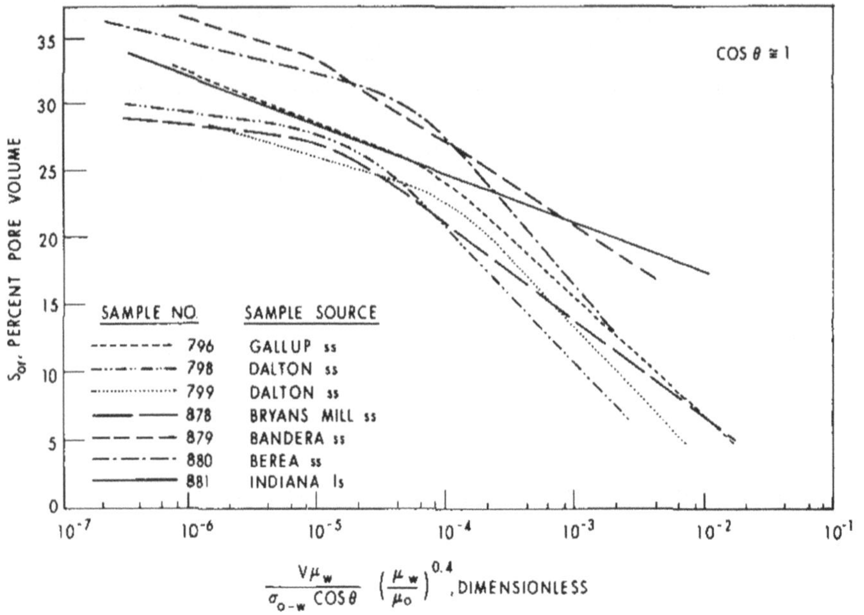

Fig. 4 CDC curve for various sandstone and limestone rocks (From [11])

2.2 *Effect of Oil Continuity on IFT Requirement*

IFT lowering is also dependent on whether the oil exists in a disconnected form or in a continuous form. Aspect ratio between the pore body and pore throat gives a measure of radial capillary variation that the flowing fluid will experience in porous media (Eq. 4). When the oil gets snapped off due to high aspect ratio of above 3, they exist as single smaller droplet in a single pore [17]. Therefore, the trapping capillary pressure is expected to be higher as per (Eq. 4). On a contrary at lower aspect ratio of 3, residual oil exists as large continuous clusters [17]. At the aspect ratio of 2 or less, no significant trapping occurs [17] because capillary pressure tends to be lower as per (Eq. 4). More energy needs to be expended in mobilizing the disconnected oil when compared to recovering the oil that is not trapped. This can be understood by comparing the CDC curve for both the connected and disconnected oil (Fig. 5).

Critical capillary number needed for mobilizing the disconnected oil is relatively higher. Therefore, IFT lowering requirement should be higher for recovering the disconnected oil.

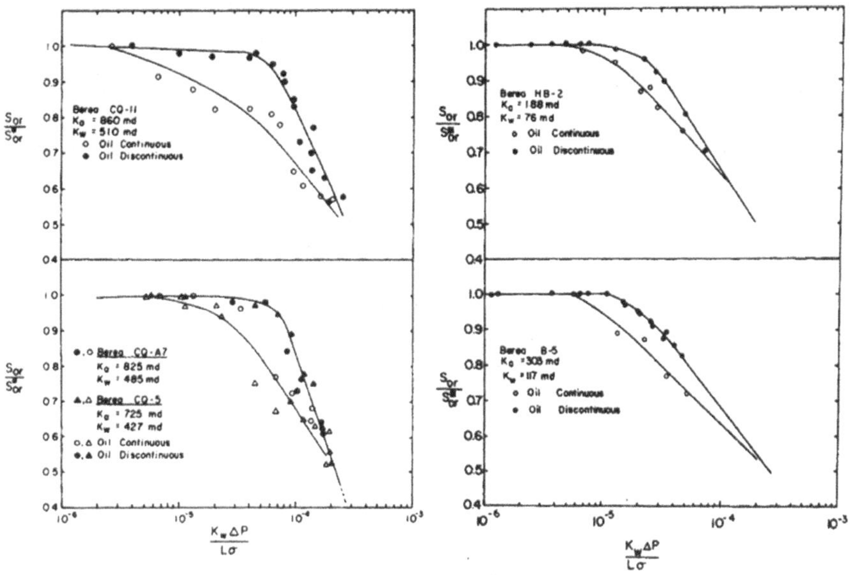

Fig. 5 CDC curve for Berea stones saturated with disconnected and connected oil (From [18])

2.3 *Effect of Oil Viscosity on IFT Requirement*

Displaced oil viscosity is usually not considered in convention capillary number definitions (Eqs. 6 and 7). Therefore, CDC curve generated for the oil of various viscosity looks scattered (Fig. 6).

A closer look into the Fig. 6 reveals at the similar values of capillary number, triangle and diamond symbols representing the higher viscous oil are located at the region of higher Sor. This signifies at the similar IFT level, the higher the oil viscosity, the lesser the recovery. The question arises whether the lowering of IFT will have a beneficial effect on heavy oil recovery during non-thermal EOR methods? To answer this, Zhang et al. [19]'s work is considered.

Zhang et al. [19] performed a systematic study to analyze the relative importance of mobility control and IFT reduction during 1500 cP heavy oil recovery in 3–3.8 Darcy sand pack. Floodings were performed using Alkali surfactant polymer (ASP), alkali surfactant (AS), alkali (A), alkali polymer (AP). Injection rate was 12 mL/hr, an indication that it is a forced displacement. The IFT values and recovery performance of AS and ASP systems are shown in Figs. 7 and 8 respectively. The recovery performance of AP system is shown in Fig. 9.

IFT of AS system is ultra-low on the order of 10^{-3} mN/m (Fig. 7). Adding polymer to alkali-surfactant system increase the IFT by almost an order at all alkali concentration (Fig. 7). Analyzing the poorer recovery performance of ultra-low IFT AS system with the relatively high IFT ASP system at lower NaOH concentration (Fig. 8), detrimental effect of ultra-low IFT is clear. At 0.1% NaOH, AP and ASP

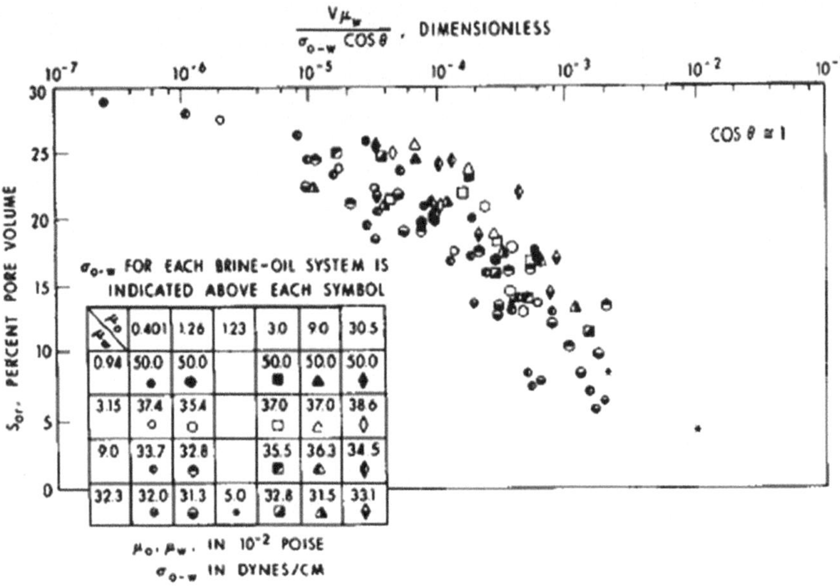

Fig. 6 Effect of oil viscosity on CDC generated using conventional capillary number (From [11])

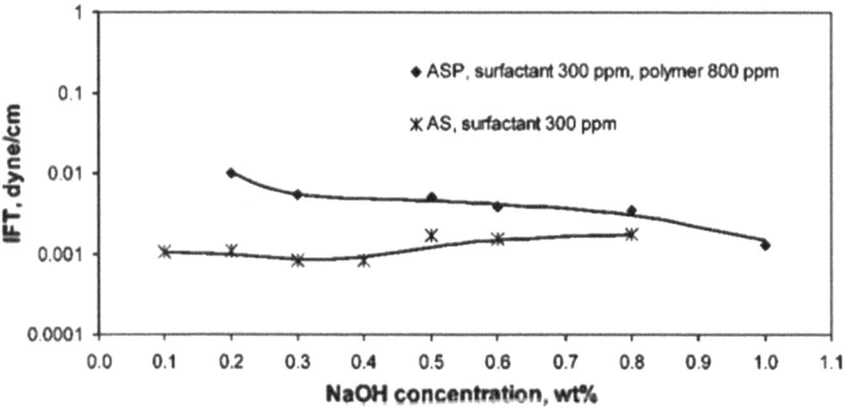

Fig. 7 IFT values as a function of NaOH concentration for ASP and AS systems (From [19])

system corresponded to the recovery factors are 15% and 28% (See Figs. 8 and 9). Maximum pressure drops experienced during these flooding is 9.8 kPa are 6.4 kPa respectively [19]. Higher pressure drop is exhibited by AP flooding. Higher pressure drops means a higher apparent viscosity (Eq. 5) and therefore a lower mobility ratio (Eqs. 10 and 11) and a better recovery. Therefore, it can be said mobility control is more important to arrest the fingering issues in the unstable immiscible floods and the ultra-low IFT conditions should be avoided during heavy oil recovery.

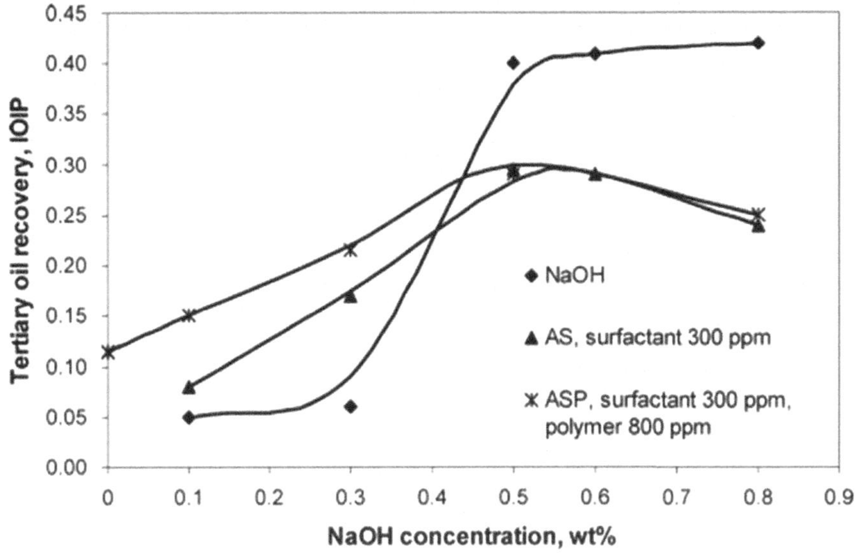

Fig. 8 Recovery performance of ASP and AS systems (From [19])

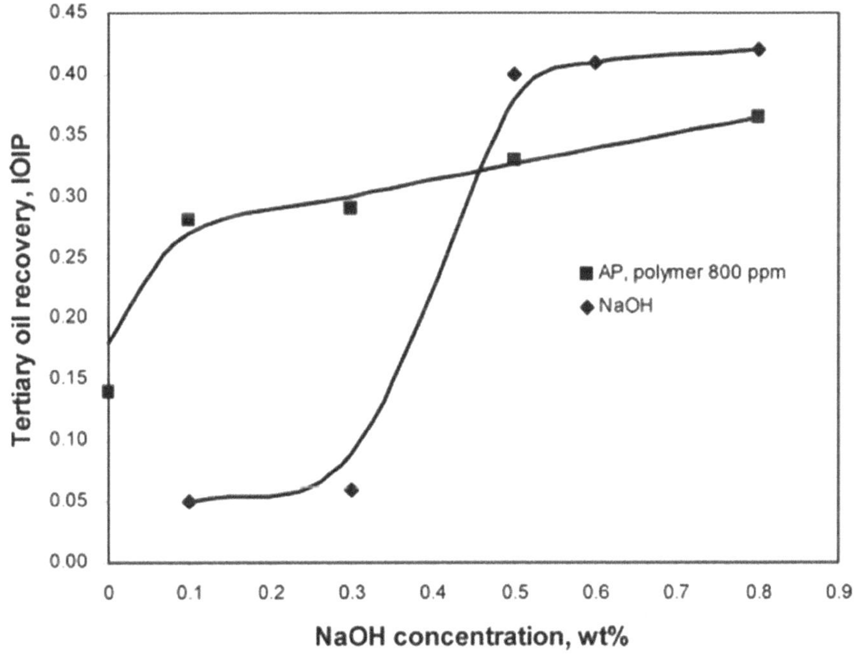

Fig. 9 Recovery performance of AP and A systems (From [19])

3 Spontaneous Imbibition

The spontaneous imbibition is the process in which the wetting phase is imbibed into the pores by the driving capillary force. In the case of tight shale reservoirs, imbibition cannot be a forced one due to permeability constraints, and therefore, spontaneous imbibition must be the main driving mechanisms. Capillary forces and gravity forces are the important forces that governs the imbibition process. The relative importance of these two forces can be quantified through the bond number (Eq. 8). There are two modes of spontaneous imbibition. (1) Counter-current imbibition (2) Co-current imbibition. Capillary force that usually acts as a trapping force in forced displacement is the driving force during spontaneous imbibition. Capillary force drives the aqueous injection fluid to imbibe into the matrix pore (with out an imposed force) and the oil would be expelled out in the opposite direction in a process called counter-current imbibition (Fig. 10a). If the matrix has enough height, gravity forces can cause the fluids to separate so that the low-density oil can be driven out upwardly at the low IFT conditions through the process called co-current imbibition (Fig. 10b).

In this section, I shall discuss the role of IFT on these processes during oil recovery from very tight shale rocks, low permeable limestone and a relatively permeable Berea core.

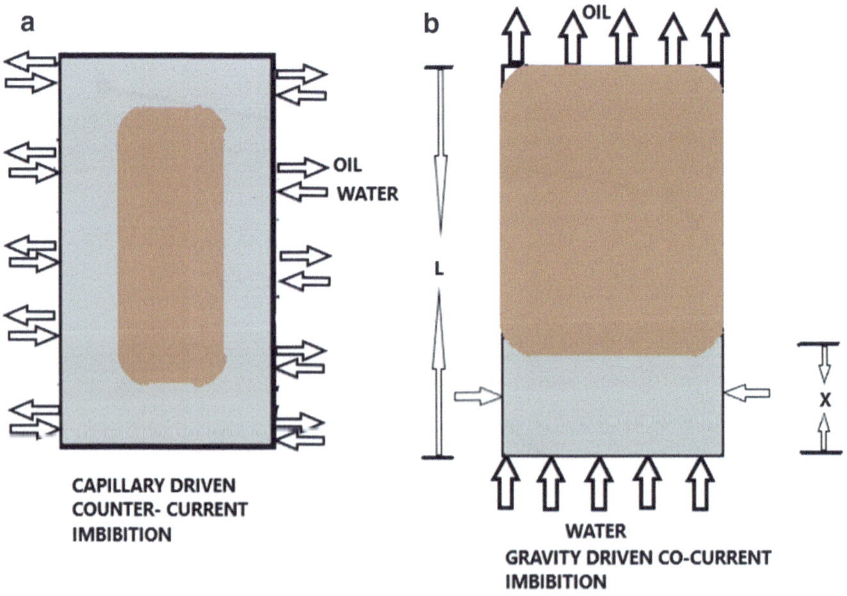

Fig. 10 Schematic showing the **a** Counter-current imbibition **b** Co-current imbibition

3.1 IFT Role in Tight Shale Rocks

Tight shale rocks are usually characterized by a very low permeability (For ex.10^{-6} Darcy) and oil-wet nature. For oil-wet rocks, the contact angle is usually higher than 90° and therefore, the capillary pressure for the oil-wet rocks would be negative as per the (Eq. 3). and it is imperative to have a water-wet surface for the capillary imbibition. Therefore, wettability alteration plays a crucial role. Along with the wettability alteration to water wet, high IFT is important to ensure the good capillary driven counter-current imbibition and therefore a higher recovery rate. Figure 11 compares the simulation effect of ultra-low IFT versus IFT conditions on the recovery factor from the shale core whose wettability got altered from the oil-wet to water-wet.

While 20 mN/m surfactant system could achieve the recovery factor of 40% in 100 days, 1 million days would be needed for 0.008 mN/m system to achieve the similar recovery factor. The higher the IFT, the higher the capillary pressure (Eq. 3) which is beneficial to induce capillary driven counter-current flow during spontaneous imbibition. As can be seen from the schematic depicting the typical capillary dominated flow (Fig. 10a), more oil will be expelled because almost all the faces are open to flow during the counter-current imbibition process. This is the reason for the higher imbibition recovery rate with high IFT system.

While the counter-current capillary imbibition needs high IFT, low IFT should favor gravity driven co-current imbibition (Eq. 8). However, gravity forces, the numerator term in Eq. (8) is also dependent on permeability. Lower the permeability, the lower the gravity forces, and therefore lower the co-current imbibition recovery. Even in the presence ultra-low IFT conditions, gravity forces would be of diminished relevance to separate the two fluids of different density if the media are not permeable. For example, Bond number calculated for ultra-low IFT systems

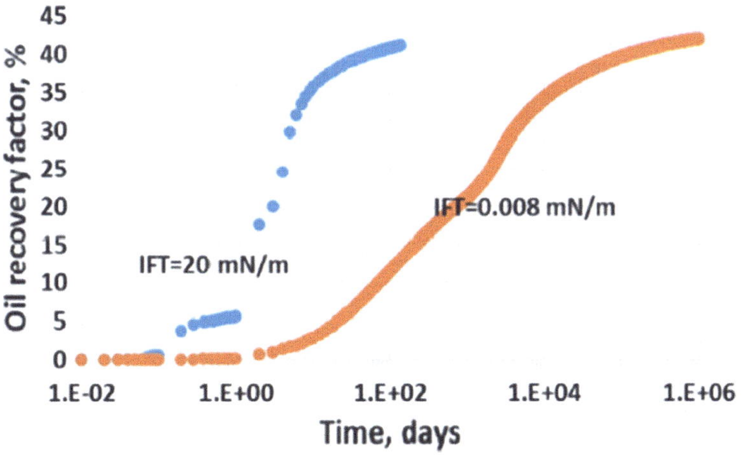

Fig. 11 Effect on IFT on oil recovery rate during spontaneous imbibition from 3.3×10^{-4} mD tight shale (From [20])

(0.008 mN/m) in the shale rock with the permeability of $3.3*10^{-4}$ mD is extremely low ~$7.35*10^{-11}$. Simulation studies performed by Sheng [20] revealed that recovery factor of such ultra-low IFT system will be 0.01 after 138 days. To have a reasonable recovery factor of 22% with ultra-low IFT system through gravity drive, 1.33 million days will be needed [20]. Therefore, the existence of ultra-low IFT conditions on the order of 10^{-3} mN/m will be totally unfavorable for recovering the oil through either co-current or counter-current imbibition from the tight shale rocks. It can be said that maintaining high IFT on the order of more than 10 mN/m and relying mainly on the capillary drive counter-current imbibition is the feasible recovery mechanism in extremely tight shale rocks.

3.2 *IFT Role in Low Permeable Limestone and a Relatively Permeable Berea*

In the previous para, the detrimental effect of having ultra-low IFT condition during oil recovery from extremely low permeable shale oil reservoir was discussed. The IFT role on oil recovery on the relatively permeable medias are discussed by choosing a limestone core with the low permeability of 15 mD and a Berea core with the medium permeability of 100 mD. 15 mD is also relatively low permeable. Therefore, capillary driven counter-current imbibition should dominate. The low IFT systems will ruin the much-needed capillary forces and therefore, the recovery rate is expected to be lower when compared with the high IFT system (Fig. 12).

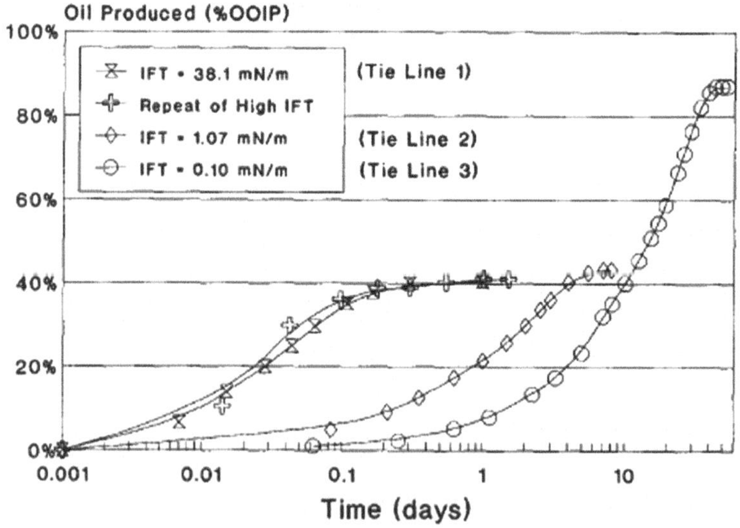

Fig. 12 Effect on IFT on oil recovery rate during spontaneous imbibition from 15 mD limestone (From [21])

As the permeability increases the gravity forces should increase relative to the capillary force as per the bond number (Eq. 8) and therefore co-current imbibition can occur. The lower the IFT, the higher the gravity forces as per the bond number (Eq. 8). Therefore, one would expect the lowest IFT conditions to be of benefit during the co-current process. However, comparing to counter-current imbibition in which all the faces are open to flow (Fig. 10a), a pure vertical co-current imbibition has relatively lesser spaces for fluid intake (Fig. 10b). The higher the fluid intake, the higher the recovery rate. So, the early time recovery should be less with the lowest IFT system which relies mostly on gravity aided co-current imbibition. This is reflected in the relatively lower recovery rate of lowest IFT system (0.1 mN/m) when compared to the one order higher IFT system (1.07 mN/m) during the first two days (Fig. 13).

Therefore, an optimal IFT system which shouldn't be too low nor too high is needed in a relatively high permeable media so that both co-current and counter-current imbibition takes place for a better recovery rate. In other words, both gravity and capillary driven mechanism will be of use, when using an optimal IFT system. In terms of total oil recovery, the system with lowest IFT will be beneficial in both the low permeable limestone and the relatively high permeable Berea (Figs. 12 and 13). It could be attributed to the reduced Sor at the lowest IFT conditions. However, to achieve those recovery, several weeks will be needed especially for the low permeable system. For the medium permeable rocks such as Berea, a low IFT on the order of 10^{-1} mN/m could be fine for having a good recovery rate and ultimate recovery. But an ultra-low IFT on the order of 10^{-3} may hamper the capillary driven counter-current flow which is also needed along with co-current imbibition.

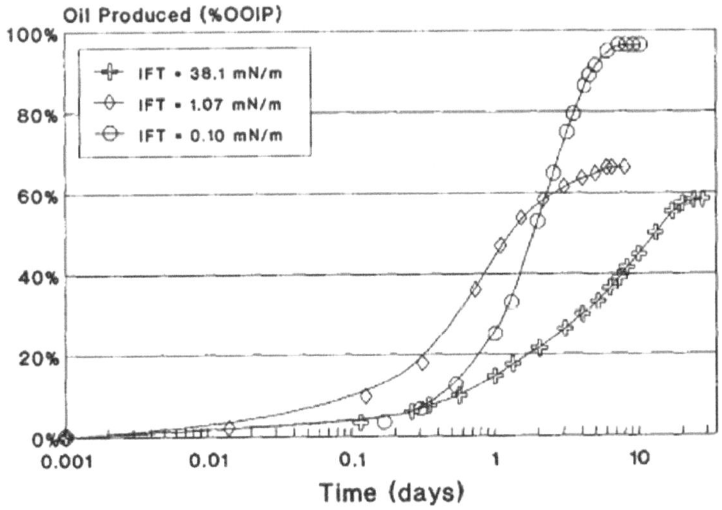

Fig. 13 Effect on IFT on oil recovery rate during spontaneous imbibition from 100 mD Berea (From [21])

4 CO_2 Flooding

Carbon dioxide flooding is one of the most implemented EOR process. Viscosity reduction, oil-swelling and miscibility are some of the recovery mechanisms associated with CO_2 flooding. Miscibility is the most influencing parameter. Basically, miscibility means the two phases are distinguishable and they can flow as a single phase without any interfacial or relative permeability effects. Miscibility between crude oil and CO_2 occur through multiple contacts in the form of vaporizing drive i.e., the intermediate components of crude oil get vaporized into the CO_2 phase. Minimum miscibility pressure (MMP) is the important parameter that governs the efficiency of miscible EOR process. It is defined as the minimum pressure in which the both the injection fluid and crude oil become miscible and any further increase in the pressure will not lead to significant addition in oil recovery.

Miscibility between CO_2 and oil is dependent on reservoir pressure, reservoir temperature, and crude composition etc. For an isothermal reservoir, the only concern is the reservoir pressure [22]. As pressure increases, more CO_2 can be solubilized. The larger the depth, the larger the pressure. Therefore, deeper reservoirs are typical candidate for miscible flooding in general. As per the EOR screening criteria [23], the depth should be greater than 2500 ft for CO_2 to be miscible with crude oil possessing the gravity greater than 40 API °. For 22 API ° oil, the depth should be greater than 4000 ft. CO_2 flooding conducted in miscible mode contributes to higher recovery than the immiscible CO_2 flood. This is because, when the displacing and displaced slugs are miscible, the IFT should be zero theoretically and capillary number should be infinite (Eq. 6). Although, 100% microscopic displacement efficiency though expected theoretically could not be achieved, a very high recovery percent could be expected in the miscible mode. Several low depth reservoirs are the ideal candidates for achieving the MMP associated with CO_2 flooding due to pressure constraints. Efforts were made to reduce the IFT between CO_2 and oil by adding the surfactants to CO_2. In this section, the role of surfactant in reducing the MMP requirement during CO_2 flooding is discussed.

4.1 *Role of IFT in MMP Reduction and Oil Recovery*

One of way enabling the miscibility is by imposing a pressure so the components between the immiscible fluids get exchanged. Another way is to lower the IFT between two immiscible fluids so that molecules between them come closer. A complete removal of interface means two fluids could become miscible. Surfactants are active interfacial agents and their addition to CO_2 can reduce the IFT between CO_2 and oil after a threshold pressure (Fig. 14).

In general, an increase in pressure leads to the decrease in IFT between the systems. However, there should be a threshold pressure beyond which the interfacial activity of surfactant begins to get active because of enhanced solubility. Generally, the higher

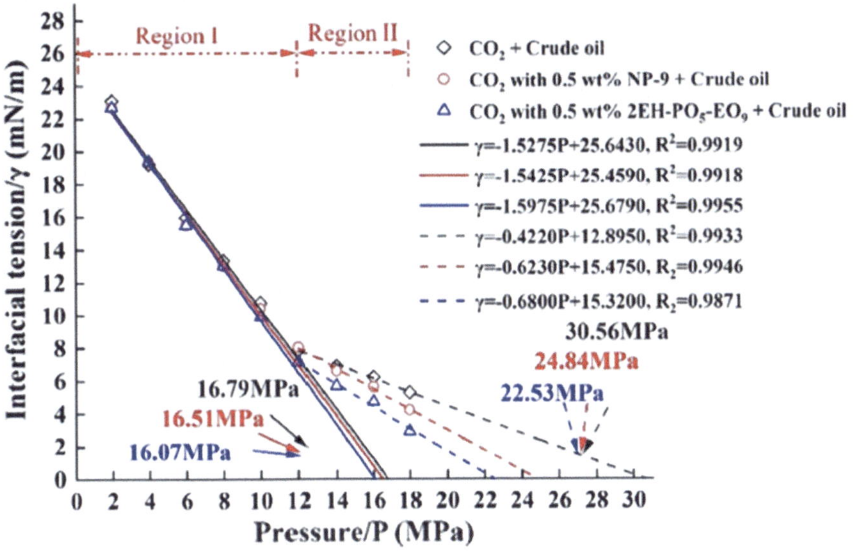

Fig. 14 Effect of pressure on the IFT between crude oil and CO_2 in the absence and presence of CO_2 soluble surfactants (From [24]). NP-9 and 2 EH-PO5-EO9 in Fig. 14 refers to non-ionic and anionic surfactants

the pressure, the higher the solubility. At lower pressures, the surfactants addition doesn't change the MMP of CO_2. MMP of CO_2-crude oil, CO_2 with non-ionic surfactant- crude oil, CO_2 with anionic surfactant-crude oil are 16.79 MPa, 16.51 MPa and 16.07 MPa respectively. Therefore, at lower pressure one cannot expect to have an added benefit of surfactant' interfacial activity during CO_2 flooding because of their limited solubility. At higher pressure greater than 12 MPa, the surfactant begins to show an enhanced interfacial activity. Miscibility between crude oil and CO_2 occurs at the highest pressure of 30 MPA. Miscibility between crude oil and CO_2 can be reduced in the presence of surfactant because of the additional interfacial activity. In the presence of non-ionic surfactant, miscibility between crude oil and CO_2 can occur at 24.84 MPa. In the presence of anionic surfactant, miscibility between crude oil and CO_2 got reduced to 22.53 MPa. Please note that the experiments were carried out by Zhang et al. [24] by dissolving the powdered surfactant in CO_2 without brine. When interacted with formation brine, the foam may be formed which would be of benefit to have a conformance control and thereby an enhanced sweep and overall recovery efficiency. However, the role of IFT on foam-based conformance control during CO_2 flooding is not discussed in this chapter.

The oil soluble surfactant that could reduce the MMP of CO_2 (from 27.3 to 21.2 MPa) could increase the oil recovery factor marginally (Fig. 15). These experiments were performed by Guo et al. [25] using slim tube.

Although, the recovery factor improvement due to surfactant addition is not very significant, the reduction in MMP due to surfactant addition may draw attention

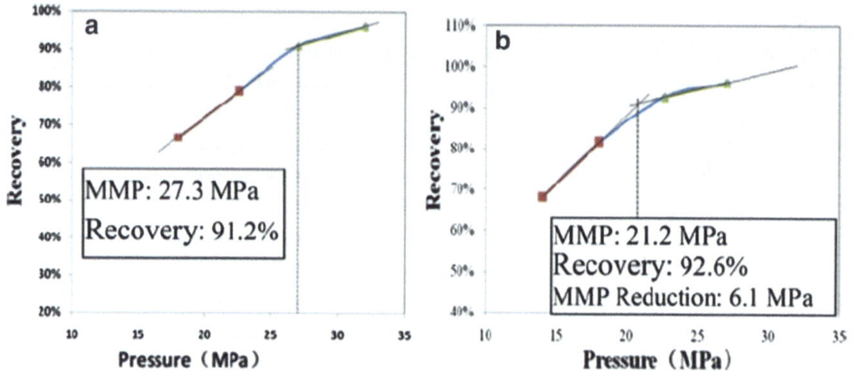

Fig. 15 **a** MMP and the associated recovery of pure CO_2 displacement **b** MMP of oil soluble surfactant pre-slug CO_2 displacement (From [25])

among the EOR community. Slim-tube experiments is a good representative of microscopic efficiencies without macroscopic effects such as gravity override. In miscible conditions, theoretically, IFT is zero, capillary number should be infinite (Eqs 6 and 7) and up to 97% recovery could be expected during MCM process [26].

Oil soluble surfactants may hinder the contact between CO_2 and oil; therefore, miscibility of the process may not be efficient or intact (Fig. 16a).

CO_2 soluble surfactants could be a better option [24] because they can facilitate direct miscible contact between the oil and CO_2 without any hinderance (Fig. 16b). Also, in heterogenous formation, the usage of CO_2 soluble surfactants means the underride problem associated with oil soluble surfactants could be eliminated and IFT

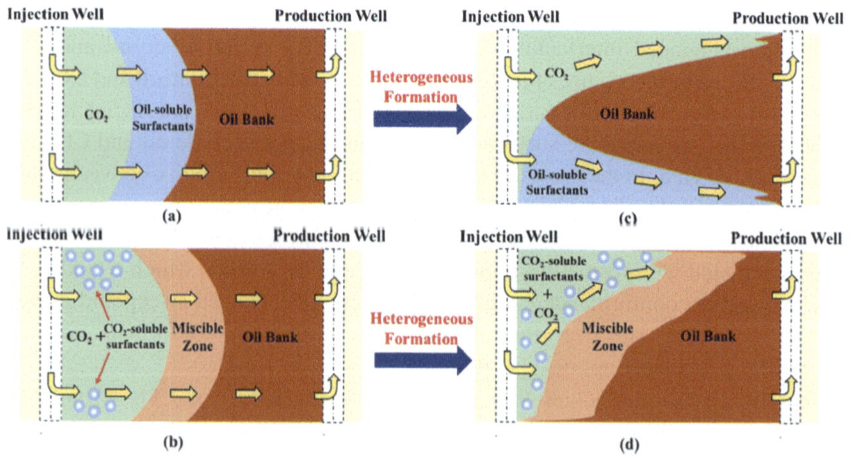

Fig. 16 Schematic depicting the possible advantage of CO_2- soluble surfactant (b and d) over oil-soluble surfactants (a and c) during CO_2 EOR (From [24])

free regimes can be facilitated in CO_2 contacted area (Fig. 16c and d). More works need to be done in this area before the possible consideration of MMP reduction through low IFT surfactants.

5 Foam Flooding Applications

A relatively large volume of gas dispersed in the small volume of liquid is called foam [1]. Generally, such dispersions are quite unstable and tends to break. Stability of foam could be improved by adding surfactants to the liquid [7]. The resistance generated during foam flow is much higher than the resistance generated during the flow of surfactant or gas. The general requirement from mobility control fluids is that they should possess more resistance to flow so that areal, vertical and linear sweep could be improved. While polymer solutions are the widely used mobility control agent, foam are used as mobility control agent in low dense, low viscous gas based EOR methods such as steam flooding [27, 28] and CO_2 flooding [29]. Because the foam has higher apparent viscosity in high permeable media, when compared to low permeable media [30], diversion of the injection fluid from high permeable streaks/fractures to low permeable matrix would be aided by foaming solution. Naturally fractured reservoirs characterized by fracture and matrix are one of the candidate reservoirs for foam-based mobility control applications [31–33]. It is the foam stability that is vital for having a higher apparent viscosity and therefore a favorable mobility ratio (Eqs. and 11). In Sect. 5.1, I shall discuss how the ultra-low IFT and foam stability are related? In Sect. 5.2, the role of IFT on oil recovery during foam-based application in NFR cores such as dolomite and limestone are discussed. The role of IFT on oil recove10ry during alkali steam foam flooding is discussed in Sect. 6.1.

5.1 *Ultra-Low IFT and Foam Stability—A Dilemma*

Ultra-low IFT is important for microscopic displacement efficiency and foam stability is important for mobility and conformance control. For a conventional surfactant system, both cannot be attained simultaneously because of the dependence of surfactant system on salinity. To understand this, Yanatatsaneejit et al. [34]'s work is considered. The authors studied the IFT (Fig. 17) and foam stability (Fig. 18) of an anionic surfactant (Alfoterra 145-4PO) with respect to NaCl concentration.

To have ultra-low IFT for a surfactant solution, optimal salinity of 5% NaCl is needed (Fig. 17). It is because solubilization ratio of oil-surfactant and water-surfactant become equal at the optimal salinity which results in the formation of middle-phase microemulsion with excess water and oil as a separate phase. Since both oil and water can be solubilized equally by the surface-active agents at optimal salinity, an ultra-low IFT can be expected at optimal salinity. For more information

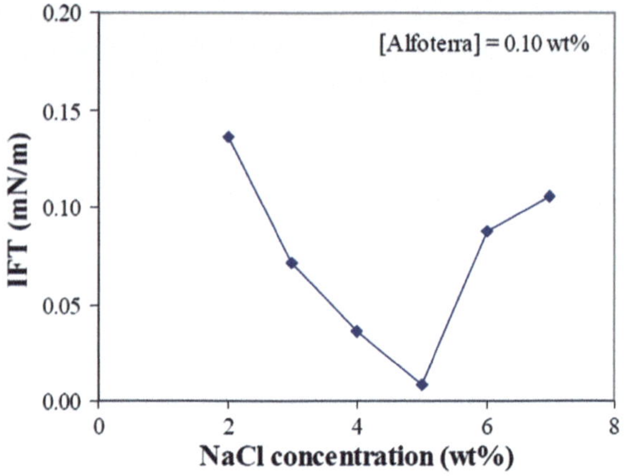

Fig. 17 Effect of NaCl concentration on IFT (From [34])

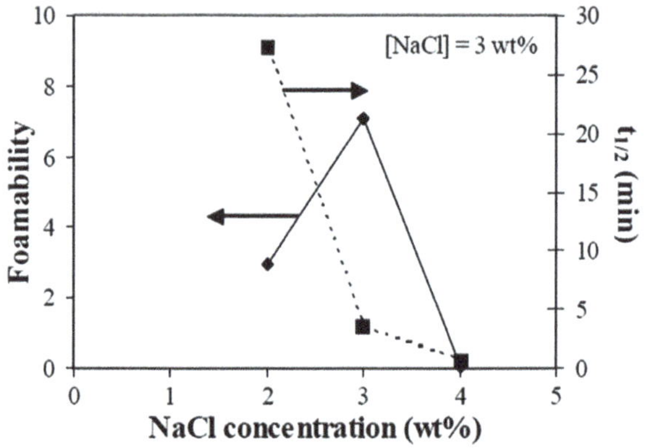

Fig. 18 Effect of NaCl concentration on foamability and foam stability [34]

about the effect of various parameters on IFT and phase behavior, the readers are referred to Green and Willhite [1].

However, for having the better foam stability, the low salinity of 2% NaCl is preferred (Fig. 18). At 2% NaCl, the coalescence time (represented by $t_{1/2}$ in Fig. 18) is higher, which means more time would be needed form an unstable larger foam bubble. Contrarily at high salinity, coalescence time decreases because the negative charge in anionic surfactant gets neutralized by the more amount of positive Na+ ions. Therefore, the repulsive forces between the surfactant head group reduces drastically which cause causes the bubbles to get coalesced to a larger bubble of dry foam in a relatively quicker time. Larger bubble size foam generally tends to be unstable [7]. Therefore, if one wants to have an ultra-low IFT, an optimal salinity that causes the foam instability would be needed.

5.2 IFT Role During Foam Flooding in Naturally Fractured Carbonates

Surfactant formulations for foam-based mobility/conformance control applications were chosen based on foam stability. Attention were not given for optimizing the IFT reduction during foam EOR process and therefore high IFT foam has been used in the past for EOR applications [35, 36]. Because of high IFT, flooding lacks Sor reduction potential which leaves lot of residual oil especially in the matrix. Recent researches have emphasized the formulation of surfactant blends that could achieve both ultra-low IFT and good foam stability [31–33]. The recovery performance of high and low IFT foam in fractured dolomite and limestone are compared from the works of Dong et al. [31] and [33].

5.2.1 High IFT Versus Low IFT Foam Performance in Fractured Dolomite Core

A special low IFT foam is prepared using a zwitterionic surfactant (lauryl betaine-LB), an anionic surfactant called internal olefin sulfonate (IOS), and another anionic surfactant called ethoxylated carboxylate (L38). The salinity of the surfactant solution is 32,690 ppm. LB and IOS produced Winsor-3 regime with n-octane and Winsor-1 regime with simulated live oil made by combining the crude and n-octane. Even though, Winsor-1 regime was achieved with live oil, oil solubilization ratio of 6 leads to the much lower IFT of 0.017 mN/m. The positive charge in the zwitterionic surfactant and negative charge in the anionic surfactant induces a strong synergetic interaction which leads to low IFT even at Winsor-1 regime. However, LB/IOS combination precipitated. L38 increased the aqueous stability without affecting the low IFT potential of LB/IOS system at reservoir temperature. The mixture posses good foamability and foam stability. Alpha olefin sulfonate (AOS) with the chain length of 14 to 16 carbons was used as high IFT foaming surfactant. 1.1 wt% of LB/LOS/L38) and 1% AOS are chosen for comparing the recovery performance of low and high IFT foam. Oil viscosity was 0.9 cP. For more details, please refer to Dong et al. [31].

The incremental recovery performance of 1.1 wt% LB/LOS/L38 (low IFT foam) and 1% AOS (high IFT foam) is compared by injecting the pre-generated foam and nitrogen at 4 ft/day into the water flooded oil-wet fractured dolomite core (Fig. 19). Permeability in the matrix and fracture are 140–149 mD and 120,300 to 158,400 mD respectively. 1.1 wt% LB/LOS/L38 and 1.1% AOS corresponds to the IFT value of 0.017 mN/m and 0.491 mN/m respectively (please note that 0.017 mN/m can also be called as ultra-low IFT. However, for the sake of consistency with Dong et al. [31], it is being called low IFT) Half-time of these bulk foams are 42 min and 16 min indicating low IFT foam possess good foam stability as-well.

Since diverting the fluid from fracture to matrix is important initially, both the foam performed similarly during early stage ~2.5 PV injection at the high flux rate

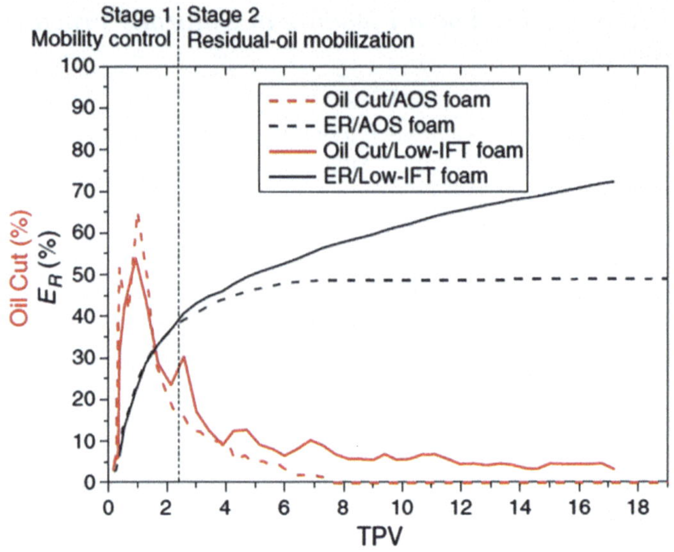

Fig. 19 Recovery performance of low IFT and high IFT foam in oil-wet fractured dolomite [31]

of 4 ft/day. Initially, oil saturation in the matrix should be higher than water flooded Sor and therefore, most of the mobile oil in the matrix gets recovered by the foaming solutions regardless of their IFT reduction potential. In the later stage after 2.5 PV injection, high IFT foam fails to recover the additional residual oil trapped by the capillarity. Despite injecting many PV, a plateau in the recovery profile is seen signifying that the well-swept oil cannot be mobilized at high IFT conditions. Low IFT foam which can contribute to IFT reduction of more than one-order effectively mobilizes well-swept, capillary-trapped residual oil from the matrix. Another advantage with low IFT foam is that lower forced entry pressure is needed for low IFT solutions when compared to gas or high IFT solutions. This could let the low IFT solutions to effectively mobilize more oil from the matrix. More than 20% higher Sor reduction was achieved during low IFT foam injection when compared to high IFT foam. Therefore, it can be said, the microscopic displacement efficiency which is the main constituent of overall recovery factor (Eq. 1) increased in the case of low-IFT foam injecting in fractured reservoirs. Another important point to note is that higher Sor reduction with low IFT foam ensures that oil destabilizing effect on foam is lesser. This was reflected in the reported apparent viscosity of 84 cP and 15 cP during low and high IFT foam injection respectively. Therefore, it can be said that foam stability got improved due to low IFT in dolomite rocks and therefore a potential of good sweep with low IFT foam is also a possibility.

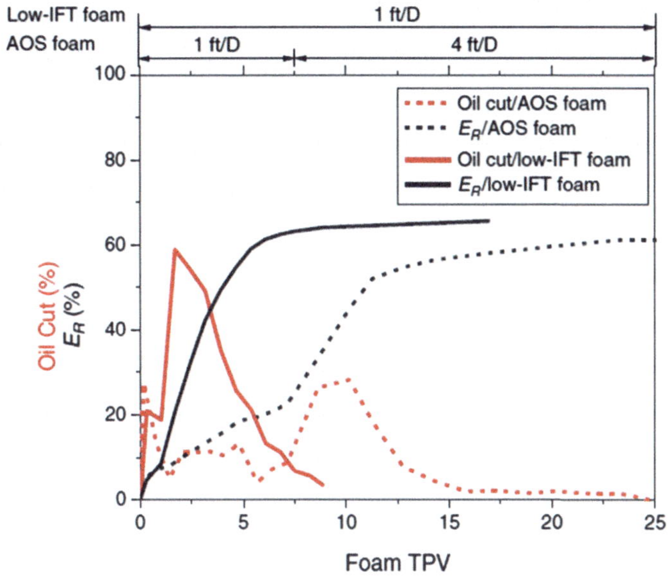

Fig. 20 Recovery performance of low IFT and high IFT foam in oil-wet fractured limestone [33]

5.2.2 Low IFT Foam Performance in Limestone Rocks

In this sub-section, the recovery potential of low IFT foam and AOS foam in the fractured limestone core is compared. Matrix and fracture permeabilities of cores used in the experiments are 5.6–6.4 mD and 85,000–75,322 mD respectively. The concentration, salinity and the IFT values of formulation used in these experiments were same as the one used in the dolomite rocks. The used oil is also the same in both the set of experiments. The recovery performance of low-IFT and high-IFT foam is shown in Fig. 20.

Overall recovery factor is similar during both low (63.8%) and high IFT flooding (61.1%). However, to achieve an incremental oil recovery of 50%, low IFT foam requires 5 PV whereas high IFT foam requires 11 PV of injection. Further, high IFT foam if injected at 1 ft/day can contribute to incremental recovery factor of just 20% after 5 PV. However higher injection rate of 4 ft/day provided the additional viscous force therefore, recovered further oil (Eq. 6). In the early stage mobility control is more important, and therefore it can be said that low IFT foam can effectively contribute to mobility control at low flux of 1ft/day whereas the high IFT foam requires higher flux rate. In the presence of oil, low IFT foam possess more stability than high IFT foam [31]. The main reason is because low IFT ensures additional oil mobilization which means the effect of oil on destabilizing the foam is less.

Comparing the performance of low IFT foam in dolomite and limestone (Figs. 19 and 20), we can see in dolomite, low IFT foam give more than 70% recovery whereas in limestone only around 60% recovery is achieved. Moreover, a plateau in the recovery is seen in limestone formation but not for dolomite formation. This suggests that relative permeability of the oil is getting very low in limestone—a indication of trapping (Fig. 21).

A darker spot in the end of core for low IFT foam indicates that (1) Sor in the matrix is getting mobilized during low IFT foam when compared to high IFT foam (2) Mobilized oil due to low IFT foam injection is getting trapped at the end. Therefore, incremental curve become flattened in the case of low IFT foam. But what causes the trapping in limestone? Why low IFT conditions is detrimental in limestone? Limestone is unstable geochemically and can leads to dissolutions and ion exchange (Fig. 22).

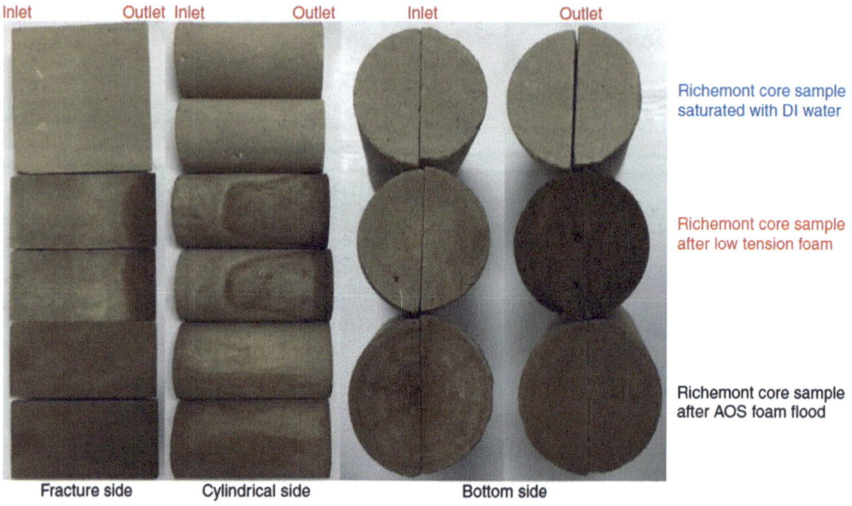

Fig. 21 Cores used during low and high IFT foam flooding [33]

Fig. 22 Changes in concentration of brine and surfactant solutions after contacting with limestone and dolomite rock samples. BR—Brine with Limestone; BS—Brine with Dolomite; SR—Surfactant with Limestone; SS—Surfactant with Dolomite [33]

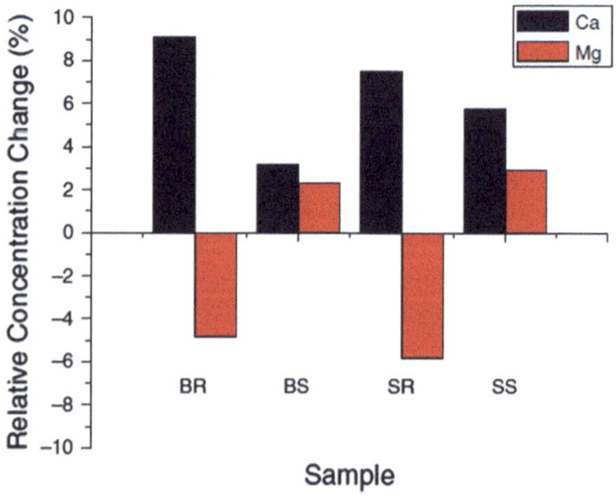

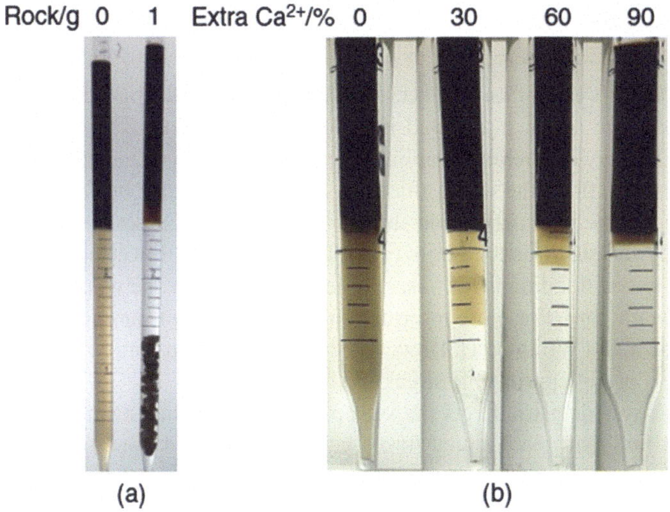

Fig. 23 Phase behavior studies showing the detrimental effect of limestone on low IFT formulations [33]

The concentrations of both the calcium and magnesium ions changes significantly in both the brine and surfactant solutions upon contacting with the limestone rocks when compared to the dolomite rocks (Fig. 22). In limestone rocks, more ions are transferred to surfactant solutions making the system to be over-optimum which in turn leads to a very high viscous micro-emulsion (Winsor 2) that get trapped. A phase behavior study clearly indicates adding limestone (Fig. 23a) and calcium (Fig. 23b) ions will convert lower phase microemulsion into upper phase microemulsion.

Please note in lower-phase micro-emulsion some oil is solubilized and were yellow in color and it is because of this oil solubilization, low IFT was achievable. But the transferring of divalent ions from limestone makes the created low IFT conditions to become highly viscous and therefore, futile. This is the reason why incremental recovery curve during low IFT foam injection is flattening in the case of limestone whereas for dolomite, a steady increase in seen.

This analysis indicates that low IFT foam formulated using surfactants carrying different charge could be expensive but a good option for NFR reservoirs. However, caution need to be exercised when using low IFT foam for EOR applications in different carbonate reservoirs because of their mineralogy etc. Geochemically unstable limestone rocks can make the low IFT injection system to become over optimal which can affect the linear sweep. Both oil mobilization and linear displacement are important, and this issue may be aggravated at the field scale and therefore a case by case investigation about the potential of low IFT foam for various carbonate reservoir is warranted.

6 Steam Flooding

Steam flooding is one of most successful EOR method. During steam flooding, the high temperature steam is injected into heavy oil reservoir. The heat reduces the viscosity of heavy oil thereby enabling its mobility. Steam channelling is one of the main issues that affects the sweep efficiency during the steam flood. Steam channeling occurs around the high permeable streaks. Only oil in high permeable zone would be swept by the steam leaving lots of bypassed oil in the lower permeable region. Generally, steam flooding is not susceptible to fingering as one would think because of its healing nature. For more information about the steam stability, the reader can refer to Green and Willhite [1]. However, the low viscous nature of steam will make the channels or high permeable streaks a short circuit.

Gravity override also affects the sweep efficiency during steam flood. Steam is low dense fluid and gravitational forces causes the steam to override on the top of the reservoir. Oil gets swept by the steam selectively only in the top portion of the reservoir reducing the overall sweep efficiency. The oil remaining in the lower part of the reservoir remain unswept (Fig. 24).

Gravity override will be more severe when the reservoir has non-zero vertical permeability, or not having enough horizontal permeability/ high dip angle [37]. It is important that mobility of steam needs to be reduced to order to increase the sweep efficiency.

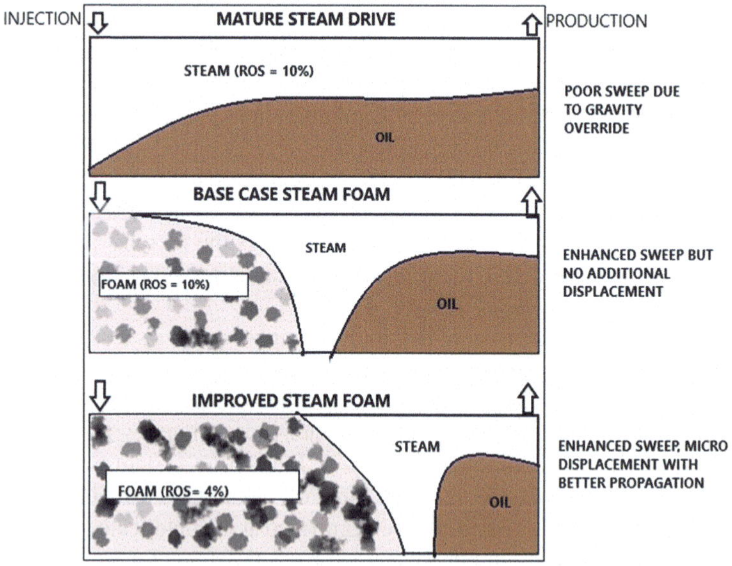

Fig. 24 Schematic showing advantage of using improved steam foam for an enhanced oil recovery (From [28])

6.1 Steam Foam Flooding

Foam generated by the injection of surfactant and steam has gained attention for improving the sweep efficiency. Steam foam used for plugging the high permeable strata and addressing the gravity override was patented by Needham [38] and Dilgren et al. [39]. Since then several researches were done in that area. The prime expectation from foam-steam is to have an enhanced sweep efficiency. Shell company conducted two steam-foam pilots in Kern river field [40]. Foam was generated by continuous injection of 50% quality steam containing 0.5% of Alpha olefin sulfonate (AOS) 1618 and 4 wt% NaCl in the aqueous phase and 0.06 mol% of N_2 in vapor phase. Injected foam generated the apparent viscosity by a factor of 20 to 60 near the injectors and allowed the steam to contact oil in the lower portion of the reservoir thereby improving the vertical sweep efficiency of the project. However, Sor to steam foam is around 10% (Fig. 25) which is similar to Sor values reported during steam flood in Kern river [28, 41] (Fig. 24).

This implies that microscopic displacement efficiency could not be improved with AOS foam during steam flood. To have a high recovery factor, both microscopic displacement and macroscopic sweep are important (Eq. 1). Overall, recovery efficiency during steam flood could be improved if Sor could be reduced in the steam-foam contacted area (Fig. 25).

6.2 IFT Role on Sor Reduction During Alkaline Steam Foam Flooding

Lau and Borchardt [28] undertook an interesting work for Shell company after realizing the performance (microscopic recovery performance in particular) of steam-foam formulation based on alpha olefin sulfonate (AOS) could be improved. Superior formulation consisting of alkaline enhanced steam foam reduce the Sor to very low or even zero whereas for steam foam without alkali, Sor was around 10% [28, 41,

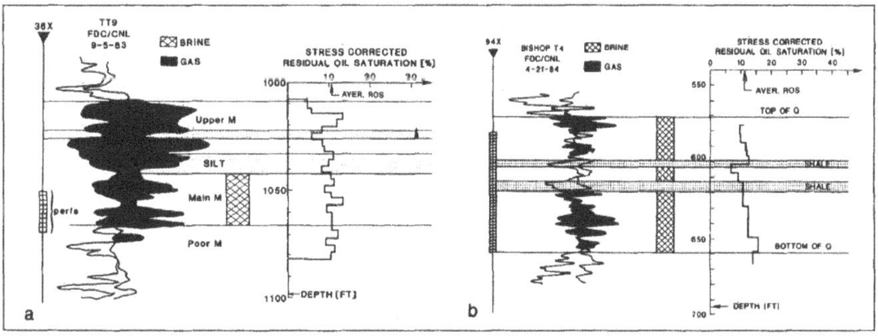

Fig. 25 Average Sor during steam foam flood in Kern river (From [40])

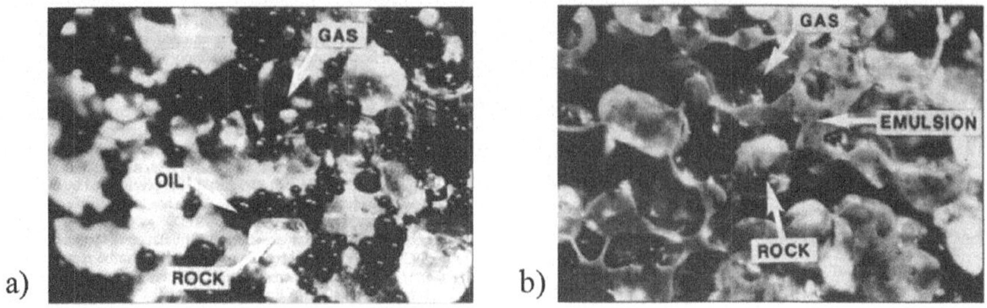

Fig. 26 **a** Larger oil droplets getting trapped in nitrogen-foam. **b** Smaller oil droplets getting mobilized in alkali nitrogen foam [28]. Please note Lau and Borchardt [28] observed the similar behavior when steam was the vapor phase

42]. Alkali generated the in-situ surfactants that lead to low IFT conditions. However, the IFT value is 0.1 mN/m which is not ultra-low. As per (Eq. 6), to have the higher capillary number (and therefore to have low Sor), IFT should be very low. Therefore, to have a higher microscopic displacement, ultra-low IFT conditions are not needed in the case of alkali steam foam flooding. Flowing vapor phase enhances emulsification of the oil beyond that could be achievable by surfactant and alkali alone. It was observed that in the absence of alkali, crude oil remained as the large oil droplets which is too big to propagate through the pore throat (Fig. 26a). Therefore, Sor was higher in the conventional steam-based foam flooding. In the presence of alkali, oil become emulsified into oil-in-water emulsion which shrink its size low enough (Fig. 26b) so that it could propagate through the pore-throat with out getting trapped. As more oil is getting emulsified in to in-situ generated surfactant solutions, there will be a reduction in IFT to some level. Therefore, low IFT conditions might have been needed. However, unless the oil is getting trapped, ultra-low IFT needed to reduce the capillary pressure (Eq. 3) in order to increase the capillary number (Eq. 6) and Sor reduction (Fig. 2) is not needed.

The emphasis that ultra-low IFT is not a requirement for higher microscopic efficiency during alkaline steam foam process can also be understood by looking into the activity maps drawn between Sor and surfactant and alkali concentration (Fig. 27). The region of ultra-low IFT band for kern river is also shown in Fig. 27. Lots of alkaline steam foam experiments corresponds to low Sor of 3 to 4% at the wide range of alkaline and surfactant concentrations which doesn't correspond to ultra-low IFT regimes.

7 Conclusions

Surfactant system has been used for many upstream applications in oil industry such as drilling, stimulation and EOR. For optimal EOR applications, the conventional belief is that a potential surfactant should have capability to reduce IFT between

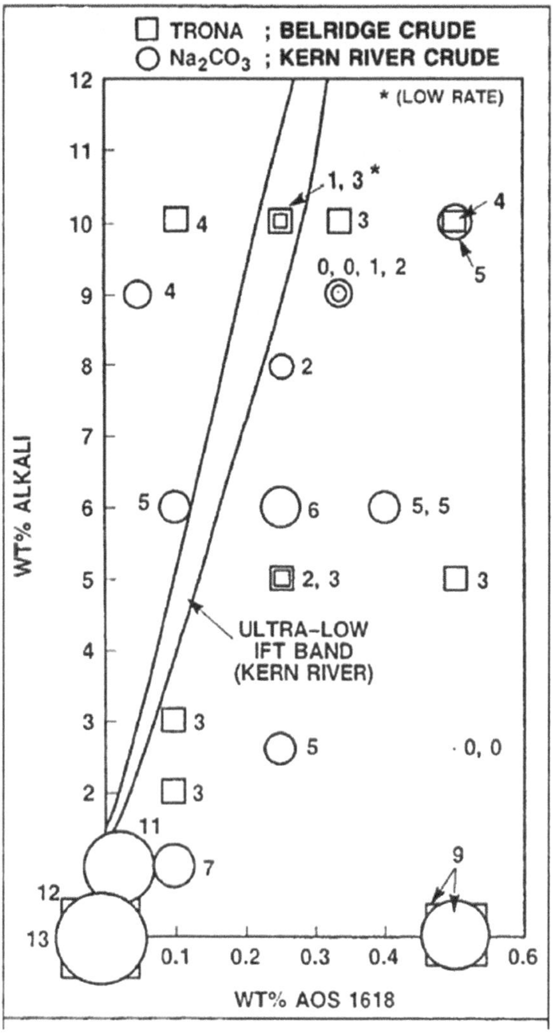

Fig. 27 Sor to alkali steam foam. Area of circle and square proportionate to the Sor [28]

the displacing solutions and displaced oil to an ultra-low level (in the order of 10^{-3} mN/m). Although this is generally true for the forced surfactant injection, the effect of rock-wettability and oil-viscosity needs to be considered to see if there could be a need or precedence for additional recovery mechanism such as mobility control, wettability alteration etc. The effect of rock permeability should also be considered to see if one cannot impose a forced displacement but must rely on capillary driven spontaneous mechanisms such as counter-current, or co-current. An optimal level of IFT which should not be ultra-low should be framed accordingly. CO_2 soluble surfactant appears to reduce the IFT and MMP requirement for a miscible CO_2 flooding. However, more researches are warranted in this area to see if a possible modification to EOR technical screening criteria can be made. Ultra-low IFT foam generated using surfactant blend could be beneficial for both sweep and Sor reductions in naturally fractured carbonates if the ion-exchange, the rock dissolution and other geo-chemical instabilities of the rocks will not induce over-optimal conditions to the displacing systems. In this regard, dolomite rocks are the more preferred candidate than geo-chemically unstable limestone rocks for the ultra-low IFT foam flooding. During steam-based foam flooding at alkaline conditions, having an ultra-low IFT is not a requirement if the in-situ surfactant generation and the subsequent emulsification will ensure that there will be no significant trapping of oil at the pores.

Acknowledgements I am thanking my wife and son (Azad Mathar) for offering their time, patience and support while I was writing this chapter. I am also thanking Dr. Japan Trivedi and University of Alberta for offering me the post-doc position.

References

1. Green D, Willhite P (2018) Enhanced oil recovery, 2nd edn. SPE Textbook Series, Richardson, Texas
2. Taber JJ (1981) Research on enhanced oil recovery: past, present and future. In: Shah DO (eds) Surface phenomenon in Enhanced oil recovery. Springer, Boston, MA. https://doi.org/10.1007/978-1-4757-0337-5_2
3. Al-Quaimi B, Rossen WR (2018) Capillary desaturation curve for residual non-wetting phase in natural fractures. SPE J 23(03):788–802. SPE 189448-PA. https://doi.org/10.2118/189448-PA
4. Azad MS, Trivedi JJ (2020a) Extensional effects during viscoelastic polymer flooding: understanding unresolved challenges. SPE J 25(04):1827–1847. SPE 201112-PA. https://doi.org/10.2118/201112-PA
5. Azad MS, Trivedi JJ (2020b) Does polymer's viscoelasticity influence heavy oil sweep efficiency and injectivity at 1 ft/day? SPE Reservoir Eval Eng. 23(02):446–462. SPE 193771-PA. https://doi.org/10.2118/193771-PA
6. Peter EJ (2012) Advanced petrophysics 2, vol 2. Live OAK book company, Texas
7. Lake LW (1989) Enhanced oil recovery. Prentice-Hall, Englewood Cliffs, New Jersey
8. Craig FW (1980) The reservoir engineering aspects of water flooding, vol 3. Henry L Doherty Memorial Fund of AIME. SPE of AIME. Dallas
9. Delshad M (2020) Fundamentals of Enhanced oil recovery. A short virtual course presented in conjunction with SPE IOR symposium
10. Lake LW (1984) A technical survey of micellar polymer flooding. Presented at SPE EOR, A symposium for the independent producer, Southern Methodist University, Dallas, Texas, Nov 1984
11. Abrams A (1975) The influence of fluid viscosity, interfacial tension, and flow velocity on residual oil saturation left by waterflood. SPE J 15(05):437–447. SPE 5050-PA. https://doi.org/10.2118/5050-PA
12. Dombrowski HS, Brownell LE (1954) Residual equilibrium saturation of porous media. Ind Eng Chem 46(06):1207–1219. https://doi.org/10.1021/ie50534a037
13. Du Prey EJL (1973) Factors affecting liquid-liquid relative permeabilities of a consolidated porous medium. SPE J 13(01):39–47. https://doi.org/10.2118/3039-PA
14. Kremesec VJ, Treiber LE (1978) Effect of system wettability on oil displacement by micellar flooding. JPT 30(01):52–60: https://doi.org/10.2118/6001-PA
15. Slattery JC (1974) Interfacial effects of the entrapment and displacement of residual oil. AIChE J 20:1145–1154

16. Stegemier GL (1976) Mechanisms of entrapment and mobilization of oil in porous media. Paper presented at the AIChE 81st National Meeting, Kansas City, Apr 12–14
17. Chatzis I, Morrow NR, Lim HT (1983) Magnitude and detailed structure of residual oil saturation. SPE J 23(02):SPE 10681-PA. https://doi.org/10.2118/10681-PA
18. Chatzis I, Morrow NR (1984) Correlation of capillary number relationship for sandstone. SPE 24(05):555–562. SPE-10114-PA. https://doi.org/10.2118/10114-PA
19. Zhang H, Dong M, Zhao S (2010) Which one is more important in chemical flooding for enhanced court heavy oil recovery, lowering interfacial tension or reducing water mobility? Energy Fuels 24:1829–1836
20. Sheng JJ (2017) What type of surfactants should be used to enhance spontaneous imbibition in shale and tight reservoirs? J Petrol Sci Eng 159:635–643. https://doi.org/10.1016/j.petrol.2017.09.071
21. Schechter DS, Denqen Z, Orr FM (1991) Capillary imbibition and gravity segregation in low IFT Systems. Paper presented at SPE Annual Technical Conference and Exhibition. Dallas, Texas. 6–9th October. SPE 22594-MS. https://doi.org/10.2118/22594-MS
22. Jarrell PM, Fox CE, Stein MH, Webb SL (2002) Practical aspects of CO_2 flooding, vol 22. Henry L Doherty Memorial Fund of AIME. SPE of AIME. Texas
23. Dickson JL, Dios AL, Wylie PL (2010) Development of improved hydrocarbon recovery screening methodologies. paper presented at SPE improved oil recovery symposium, Tulsa, Oklahoma, USA. 24–28th April
24. Zhang C, Xi L, Wu P et al (2020) A novel system for reducing CO_2-crude oil minimum miscibility pressure with CO_2-soluble surfactants. Fuel 281:118690
25. Guo P, Hu Y, Qin J, Li S, Jiao S, Chen F, He J (2017) Use of Oil-soluble surfactant to reduce the minimum miscibility pressure. Pet Sci Technol 35(4):345–350
26. Metcalfe RS, Yarboroough L (1979) The effect of phase equilibria on the CO_2 displacement mechanism. SPE J 19(4):242–252. SPE 7061-PA
27. Duerksen JH (1986) Laboratory study of foaming surfactants as steam-diverting additives. SPE Reservoir Eng J 1(01):44–52. SPE-12785-PA. https://doi.org/10.2118/12785-PA
28. Lau HC, Borchardt JK (1991) Improved steam-foam formulations: concepts and laboratory results. SPE Reservoir Eng 6(04):470–476. https://doi.org/10.2118/18783-PA
29. Heller JP, Cheng LL, Kuntamukkula MS (1985) Foamlike dispersions for mobility control in CO_2 Floods. SPE J 25(04):603–613. SPE-11233-PA. https://doi.org/10.2118/11233-PA
30. Farajzadeh R, Lotfollahi M, Eftekhari AA et al (2015) Effect of permeability on implicit-texture foam model parameters and the limiting capillary pressure. Energy Fuels 29(5):3011–3018. https://doi.org/10.1021/acs.energyfuels.5b00248
31. Dong P, Puerto M, Jian G et al (2018) Low-IFT foaming system for enhanced oil recovery in highly heterogeneous/fractured oil-wet carbonate reservoirs. SPE J 23(06):2243–2259. https://doi.org/10.2118/184569-PA
32. Dong P, Puerto M, Kun M et al (2019) Ultra-low-interfacial-tension foam-injection strategy in high-temperature ultra-high salinity fractured oil-wet carbonate reservoirs. SPE J 24(06):2822–2840. https://doi.org/10.2118/190259-PA
33. Dong P, Puerto M, Jian G et al (2020) Exploring low-IFT foam EOR in fractured carbonates: success and particular challenges of sub-10-md limestone. SPE J 25(02):867–882. https://doi.org/10.2118/191725-PA
34. Yanatatsaneejit U, Chavadej S, Rangsunvigit P et al. (2004) Effect of interfacial tension and foam characteristics on diesel removal in froth flotation operation. On-Line Number 256. J-STAGE. www.jstage.jst.go.jp
35. Chen Y, Elhag AS, Poon B et al. (2014). Switchable Non-ionic to cationic ethoxylated amine surfactants for CO_2 enhanced oil recovery in high-temperature, high-salinity carbonate reservoirs. SPE J 19(2):249–259. SPE-154222-PA. https://doi.org/10.2118/154222-PA
36. Li RF, Yan W, Liu S et al (2010) Foam mobility control surfactant enhanced oil recovery. SPE J 15(04):928–942. https://doi.org/10.2118/113910-PA
37. Hirasaki GJ (1989) The steam-foam process. J Petrol Technol 41(05):449–456. https://doi.org/10.2118/19505-PA

38. Needham RB (1968) Plugging of high permeable earth strata. US Patent No. 3412793
39. Dilgren RE, Hirasaki GJ, Hill HJ (1978) Steam-channel-expanding steam foam drive. US Patent No. 4086964
40. Patzek TW, Koinis MT (1990) Kern river steam-foam pilots. J Petrol Technol 42(04):496–503. https://doi.org/10.2118/17380-PA
41. Grease GR, Schore RA (1980) Steam flood performance in kern river field. Paper presented at the SPE Enhanced Oil Recovery Symposium, Tulsa, Oklahoma. https://doi.org/10.2118/8834-MS
42. Lau HC (1986) Alkali-enhanced steam foam oil recovery processes. US Patent No. 4609044

Alternative Understanding of Surfactant EOR Based on Micellar Solubilization and In Situ Emulsification

Yujun Feng and Guangzhi Liao

Abstract It is well recognized that ultra-low interfacial tension (IFT) from formation of middle-phase microemulsion is a must in surfactant-based EOR process. However, high concentration of surfactant or surfactant mixtures is generally needed, and it remains unknown how surfactant micelles evolve when they contact oil in porous media. More importantly, few case stories of such microemulsion flooding were reported yet so far. In China, low concentration (<0.3wt%) surfactant slug is always employed with polymer slugs to form binary systems, and more than 15% oil recovery factor has been obtained from field trials. In this chapter, we proposed alternative mechanisms by reviewing our preliminary laboratory results of micellar solubilization of oils and in-situ emulsification with either model surfactant or practically used commodity surfactants. The results show that higher oil recovery factors can be obtained without reaching ultra-low IFT. These findings may provide new guidelines to design surfactant-containing flooding systems for chemically enhanced oil recovery.

Keywords Surfactant EOR · Enhanced oil recovery · Micellar solubilization · In-situ emulsification · Swollen micelles

1 General Overview of Chemical EOR in China

Oil for chemicals, chemicals for oil. Crude oil is not only playing a leading role as the source of primary energy, but also providing the major feedstocks for specialty chemicals such as surfactants and polymers. Some of these chemicals, mainly petroleum

Y. Feng (✉)
Polymer Research Institute, State Key Laboratory of Polymer Materials Engineering, Sichuan University, Chengdu 610065, People's Republic of China
e-mail: yjfeng@scu.edu.cn

G. Liao
PetroChina Exploration and Production Company, Beijing 100007, People's Republic of China

T. Solling et al. (eds.), *Surfactants in Upstream E&P*, Petroleum Engineering,
https://doi.org/10.1007/978-3-030-70026-3_5

sulfonate (PS), heavy alkylbenzene sulfonates (HABS), partially hydrolyzed polyacrylamide (HPAM) and its derivatives, have been employed in turn to extract additional oil during the tertiary oil recovery, or so-called enhanced oil recovery (EOR) process. Based on the chemicals used, chemical EOR (cEOR) can be subdivided into three modes of flooding [1]: polymer, surfactant-polymer (SP) and alkali-surfactant-polymer (ASP). In addition, foam flooding and the emerging nanomaterials flooding can also be classified as cEOR [2], but they will not be touched here as they are beyond the topic of this chapter in which we will focus on surfactant-based EOR.

All these three cEOR technologies have been practically implemented in China oil industry [3], totally around 14.67 million tons of incremental oil were obtained by cEOR in 2019, and PetroChina tops the list: it produced 12.01 million tons of incremental oil by EOR, accounting for 12% of its total output, among which 6.29 million tons came from polymer flooding, 4.4 million tons was contributed by ASP and SP flooding. 14% (OOIP, the same below) and 20% of oil recovery factors have been got in polymer flooding and ASP flooding in Daqing Oilfield, and average ultimate oil recovery factor of 18% is expected in SP projects newly launched in Liaohe, Xinjiang and Dagang oilfields. In addition, PetroChina got more than 10 million tons oil annually by cEOR for over the last 18 years, and accumulatively 270 million tons of oil were produced chemically by the end of 2019.

Although ASP outperformed polymer and SP flooding in oil recovery factor, its notorious problems have been evidenced practically, such as poor demulsification of the produced fluids, scaling of formation, and corrosion of pipelines [4, 5]. To minimize these negative effects in ASP process, but remain the advantages of the surfactants, SP mode has been taken as the alternative for ASP in Shengli Oilfield of Sinopec since early 1990s [6], and 1.4 million tons of oil was produced in 2019 in the high-temperature and high-salinity reservoirs there with oil recovery factor of 10 to 12%. In Liaohe, Xinjiang and Dagang Oilfields of PetroChina, the SP flooding has been launched in 38 injection wells, and 0.35 million tons of oil were produced in 2019 with average oil recovery factor around 15% [7]. In CNOOC, the offshore SP flooding was performed after polymer flooding, and the offshore field trial was just completed in Bohai Bay last year [8, 9].

In both SP and ASP flooding, surfactant is a must in the formulation of the injectant. However, the traditional mechanisms of surfactant EOR seem unable to interpret the findings experienced in SP flooding in China, and some new hints from field cases may need new mechanisms for rational explanation.

2 Traditional Mechanisms of Surfactant-Based EOR

Generally, surfactant cannot be used alone in cEOR process, but is always co-injected with polymer, or the mixture of alkali and polymer to formulate SP and ASP chasing fluids. In both systems, polymer acts as a rheology modifier to thicken and to promote the displacing fluid in a stable, uniform manner for mobility control. In this way, the "viscous fingering" effect can be minimized, oil can be swept out both aerially and vertically [10]. In the ASP ternary mode, the alkaline chemical such as sodium carbonate or sodium hydroxide would lower the adsorption level of the surfactant, and react with acidic components in the crude to generate in situ surface active petroleum soaps, which aid the lead surfactant to decrease oil-water interfacial tension (IFT) synergistically [11], and further produce an oil-rich colloidal dispersion to get ultralow IFT over a much wider range of salinity than in its absence [12].

Concerning the functions of surfactants in cEOR process, mainly two mechanisms, i.e., decreasing IFT and shifting reservoir wettability, have been widely recognized, and were summarized in recent reviews [12–17]. We will not repeat in detail these mechanisms, but just recall the key points to compare what we have found in the later subsections of this chapter.

Up to date, the general rule of thumb to maximize oil recovery factor in surfactant EOR is that IFT should be reduced to ultralow level ($\leq 10^{-3}$ mN/m) to overcome the capillary forces that trap oil in a reservoir. The relationship between the capillary force and the viscous force results in a dimensionless capillary number

$$N_{\mathrm{ca}} = \frac{\eta v}{\sigma \cos\theta} \tag{1}$$

where η denotes the viscosity of the displacing fluid, v represents the injection velocity, σ refers to the interfacial tension between oil and water, and θ stands for the contact angle.

N_{ca} is normally in the range of 10^{-7} to 10^{-6} for a typical reservoir after water flooding. Increasing N_{ca} to between 10^{-4} and 10^{-3} reduces the oil saturation to 90%, and residual oil saturation approaches zero if the capillary number reaches 10^{-2}. From Eq. (1), it is easy to understand that N_{ca} cannot be practically increased more than 1000 times by increasing viscosity or velocity, thus the only thing can do is to decrease the initial IFT value of 20 to 30 mN/m to the range of 10^{-2} to 10^{-3} mN/m by adding surfactant into the driving fluid.

Then, the question is how to get ultralow IFTs essentially with surfactant solutions. Extensive experimental data all confirmed ultralow IFT can be obtained only when Winsor III (middle-phase) microemulsion is formed. Such a microemulsion solubilize both oil and water, and IFT varies as a function of salinity for anionic surfactants. The "optimal salinity" determined from phase behavior is the salinity at which equilibrium IFTs between the microemulsion phase and excess-oil or excess-water phase become equal (and thus the sum becomes a minimum). An empirical correlation between the "solubilization parameters" and the IFTs of the microemulsion with the respective excess phases was established by Huh [18]:

$$\sigma = \frac{\mathrm{C}}{\frac{V_o/V_s}{V_w/V_s}} \tag{2}$$

where V_0, V_W and V_S are the volumes of solubilized oil, solubilized water and surfactant solution, respectively, and C is a constant for a typical crude oil system and its value is approximately 0.3 mN/m.

Because of solubilization, oil is swollen so that oil phase saturation becomes larger, and the resulting oil relative permeability is increased. Thus it is easier for the oil to be produced. So far, the overwhelming majority of surfactant EOR experiments has been dedicated to checking if middle-phase microemulsions can be formed, and salinity scan is an indispensable step used routinely to screen phase behvior of surfactant formulations before conducting more time-consuming coreflood tests.

The second widely-accepted mechanism related to surfactant EOR is wettability alteration [19–21]. In the oil industry, the term "wettability alteration" usually refers to the process of making the reservoir rock more water-wet. As the recovery efficiency of a flooding process is a function of the displacement efficiency and sweep efficiency. These efficiencies are a function of the residual oil saturation (water flood and chemical flood) and mobility ratio, respectively, and the residual oil saturation to waterflood is a function of wettability. The mobility ratio or relative permeability ratio becomes progressively larger as the wettability changes from water-wet to oil-wet. When a formation is strongly oil-wet, it can have both a waterflood residual oil saturation and unfavarable mobility ratio. In addition, an oil-wet formation will have capillary resistance to imbibition of water. Carbonate reservoirs are characterized as intermediate to oil- wet. This leads to a tendency of the oil to adhere strongly on the rock surface inhibiting the oil recovery with the non-wetting phase injection such as water flooding.

In chemical flooding, surfactants will adsorb on the rock surface causing changes in the wetting behavior. While the wettability of reservoir is shifting from strongly oil-wet to neutral wet state, capillary forces that retain oil in porous medium are reduced and then eliminated.

cEOR process may change the crude oil/brine/rock properties by two mechanisms: Coating and cleaning. Coating refers to the process of covering the oil-wet layer by water-wet materials. For example, zirconium nanoparticles are hydrophilic and when they adsorb on the rock surface and form nanotexture coating the oil-wet surfaces, wettability changes to more water-wet. Cleaning mechanism is normally associated with surfactant induced wettability alteration. Cationic surfactants, for instance, desorb the oil-wet layer and thus render the surface to a more water-wet state.

In the EOR industry of China, surfactant has never been solely used except changing wettability to improve water injectivity. In any other cases, surfactant has been always co-injected with polymer in SP flooding, or with alkali and polymer in ASP flooding. As listed in Table 1 [7], much unlike high concentration (generally higher than 1.0 wt%, even as high as 15 wt% surfactant and cosurfactant in early work) of surfactants and cosolvents or cosurfactants used in formulating Winsor III microemulsions, the surfactant concentration of SP flooding in the three oilfields is generally below 0.4% for the cost consideration, and IFT is always measured with

Table 1 Summary of three ongoing surfactant-polymer flooding projects in PetroChina (data cited from Guo et al. 2020, plus updated information)

Oilfield/Block	Injector/Producer	Well spacing (m)	Depth (m)	T (°C)	K^a (mD)	Salinity and hardness of formation brine		Salinity and hardness of injected fluid	
						TDS (mg/L)	Ca^{2+}+Mg^{2+} (mg/L)	TDS (mg/L)	Ca^{2+}+Mg^{2+} (mg/L)
Liaohe, J16	23/34	150	1255–1460	55	2232	2467	10.4	2748.9	0
Xinjiang, 7ZQ	8/13	150	1146	40	94.8	14,250	151.2	3681.1	43.3
Dagang, GX3Q	7/14	120–150	1300	53	1100	14,350	77.0	6726.0	53.0

Oilfield/Block	Chemicals				Inject. rate (PV/y)	Designed $IORF^d$ (%)	Predicted IORF (%)	Current IORF (%)
	Pre-slug	Main slug	Vice slug	Post-slug				
Liaohe, J16	0.1 PV 0.2%P^b_{3000}	0.9 PV (0.25%P_{2500} + 0.2% S^c)	0.2 PV (0.2%P_{2500} + 0.2% S)	0.1 PV 0.14%P_{2500}	0.15	15.5	19.0	16.3
Xinjiang, 7ZQ	0.06 PV 0.18%P_{1000}	0.62 PV (0.1%P_{1000} + 0.2% S)	/	0.1 PV 0.14%P_{1000}	0.10	15.4	18.0	13.7
Dagang, GX3Q	0.1 PV (0.2%P_{2500} + 0.4% S)	0.15 PV (0.2%P_{2500} + 0.35% S)	/	0.15 PV 0.15%P_{2500}	0.12	15.9	16.5	11.4

Note a. "K" refers to permeability of the target reservoirs

b. the subscript "3000" indicate the molecular weight ($\times 10^4$) of the polymer used; the same for P_{2500} and P_{1000}. The polymers used in Liaohe and Xinjiang are HPAM, and that used in Dagang is acrylamide-based associative polymer

c. the surfactant used in Liaohe is amphoteric surfactant, that used in Xinjiang is an alkyl aryl sulfonate (KPS), and that used Dagang is petroleum sulfonate

d. "IORF" stands for "incremental oil recovery factor", expressed in percentage of oil in original place (OOIP) tensiometer rather than estimation from Huh equation mentioned above. Theoretically, with such low concentrations, middle phase microemulsions cannot be formed, and no microemulsions were observed practically. Another curiosity is what happens when surfactant solution just contacts oil, and then how to evolves when more surfactant solutions contact with oil, or more oil was solubilized in the displacing fluid. However, the role of surfactant micelles in the EOR process and the interaction of micelles with oil after the surfactant solution contacted the crude oil, remain largely unknown. In the field trials, we repeatedly observed that if oil has been sufficiently emulsified, the oil recovery efficiency can be largely improved, this is greatly contradictory to the early conclusion that "the formation of stable macroemulsions in the oil fields is considered undesirable and can cause severe problems" [22].

To unravel the mechanisms behind these new findings, we recently carried out micellar solubilization and in situ observation of emulsion in porous media, and the preliminary results are summarized in the following sections.

3 Solubilization of Oil by Surfactant Micelles

As mentioned above, the overwhelming mechanism of surfactant EOR lies on the formation of middle-phase microemulsions to attain ultra-low oil-water interfacial tension. However, residual oil can also be recovered using low concentration surfactant solutions without microemulsion formation, and the interaction between surfactant solution and crude oil at very early contact has not been revealed yet. While many studies have been focusing on microemulsions already formed, the role of surfactant micelles in surfactant flooding before microemulsion formation has not been clarified yet. We hypothesize micelle solubilization of oil as an alternative EOR mechanism. In fact, low surfactant concentrations to recover oil without forming middle-phase emulsions was also reported. Rosen et al. [23] found that residual oil can be mobilized at surfactant concentration as low as 0.05 wt%, or even below that (0.01 wt%). Wang and coworkers [24] reported that 0.4 wt% petroleum sulfonate solution combined with polymer solutions can increase oil recovery up to 18.1%. In fact, such binary formulations with less than 0.4 wt% surfactant were extensively used in China's petroleum industry to annually produce more than 1 million tons of incremental oil [25]. Therefore, it is necessary to re-gain insight into surfactant flooding mechanisms, specifically those using low surfactant concentrations.

One of the most important functions of micelles is their solubilization capacity, and the structure of micelles leads to anisotropic water distributions, and nonpolar molecules are solubilized in the micellar core, forming "swollen micelles". It is well reported that dyes, drugs, oils, and benzene can be solubilized inside the hydrophobic core of micelles. Although extensive research has been focusing on solubilizing hydrophobic compounds in micellar systems, few studies have studied the interaction between crude oil and surfactant micelles and the role of micelles in enhancing oil recovery. This knowledge gap limits the understanding of the surfactant flooding recovery mechanism at the very early stage of surfactant–oil contact, as well as the

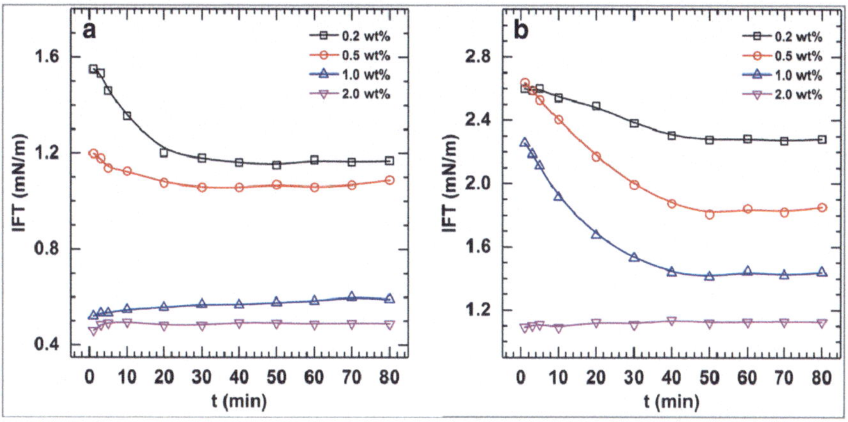

Fig. 1 Variation of IFT with concentration of aqueous SDBS solutions and **a** 1-hexene or **b** anisole at 40 °C as a function of time [26]

reason for the increase in the oil recovery factor by low concentration surfactant solutions.

To address this concern, we investigated how representative polar and nonpolar compounds in crude oil, anisole and 1-hexene, are solubilized in the micelles of the model surfactant sodium dodecylbenzene–sulfonate (SDBS). SDBS was selected because its structure was similar to that of other alkyl benzene sulfonates widely applied in chemical EOR processes, but they are a complex mixture containing both solvents and a series of derivatives with different hydrophobic tail length. The interaction between surfactant micelles and these additives was investigated with dynamic light scattering, UV-Vis spectroscopy, 1H NMR spectroscopy, cryo-TEM, confocal microscope and small angle neutron scattering. The temperature of all experiments was maintained at 40 °C to simulate the reservoir condition of the Xinjiang Oilfield of PetroChina. The detailed experimental procedures can be found in our newly published work [26].

It is one of the most important criteria in surfactant-based EOR process that if IFT can be decreased to ultralow level. So the IFT between SDBS solution and oil phase (anisole and 1-hexene) was firstly examined. After determining IFT, we measured the solubilization capacity and size of SDBS micelles in the presence of anisole or 1-hexene, then studied the location of anisole (or 1-hexene) solubilization in SDBS micelles, and finally the micellar morphology change with increasing anisole (or 1-hexene) concentration was visualized.

3.1 IFT Between Anisole (1-Hexene) and SDBS Solution

Depicted in Fig. 1 is the dynamic IFT between anisole (or 1-hexene) and SDBS solutions at 40 °C. One can immediately find that all IFTs only reach the order of 10^0 magnitude mN/m, whatever the surfactant concentration is.

At 0.2 and 0.5 wt% SDBS concentration, the IFT value starts at 1.6 or 1.2 mN/m and slightly decreases to 1.2 and 1.1 mN/m, respectively, over 80 min of equilibrium (Fig. 1a). When higher than 1.0 wt%, the IFT only reaches 5×10^{-1} mN/m, still much higher than ultra-low level. Similarly, the IFT between anisole and SDBS solution is also falling in the magnitude 10^0 mN/m within the SDBS concentration from 0.2 to 2.0 wt%.

Based on the mechanism outlined above, the oil recovery efficiency cannot be improved without ultralow IFTs. Such a premise stimulates us to see if the micellar solubilization can increase oil recovery factor.

3.2 Solubilized Amount of Anisole and 1-Hexene in SDBS Micelles

The capacity of SDBS micelles to solubilize anisole or 1-hexene was determined by UV-Vis absorbance, and the results are displayed in Fig. 2. In all cases, the absorbance of SDBS aqueous solutions keeps almost unchanged before a critical value, after which the absorbance increases steeply. The insets in both Fig. 2a and b demonstrated the appearance of solutions changes from clear to cloudy when increasing the concentration of 1-hexene or anisole.

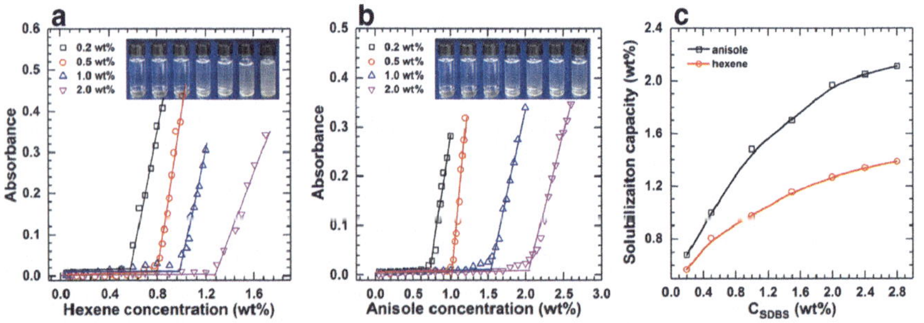

Fig. 2 Solubilization of **a** 1-hexene and **b** anisole at 40 °C in SDBS solutions with different concentrations. **c** Solubilization capacity as a function of SDBS concentration at 40 °C. The insets show pictures of 0.2 wt% SDBS solutions solubilizing different concentrations of anisole (or 1-hexene), ranging from 0.2 to 1.5 wt%. Reproduced from Ref. [26]

The maximum solubilization capacity of SDBS micelles is determined from the threshold values, i.e., the intersection of the two straight lines at low and high additive content. When below this turning point, micelles can further solubilize additive molecules to form "swollen micelles"; however, when above this critical value, the micelles are already saturated with additive molecules, and the solutions begin to become turbid, indicative of formation of emulsions or rupture of the swollen micelles.

Furthermore, the solubilization capacity of SDBS solutions is correlated with its concentration (Fig. 2c). One can find that the solubilization capacity of anisole and 1-hexene in SDBS micelles increases with SDBS concentration. The increase in SDBS concentration offers the extent of micellization and the size of micelles available to hold anisole or 1-hexene molecules.

Based on the data listed in Fig. 2c, 100 g of the 0.2 wt% SDBS solution can solubilize 0.68 g anisole or 0.57 g 1-hexene. This means, micelle solubilization cannot be ignored during cEOR process, even at low surfactant concentrations. Such findings underpin our hypothesis that surfactant micelles can solubilize oils at low surfactant concentrations without forming middle-phase microemulsions. Taking 1-hexene as an example, one can estimate that 1 ton of 0.2 wt% SDBS solution (2 kg surfactant powder) is capable of solubilizing 5.7 kg alkane. Similarly, in the field application, the surfactant solutions can also solubilize both polar and nonpolar components in crude oil, neither high concentrations of surfactant solutions are needed, nor the ultralow IFTs are obtained.

3.3 *Size Change of SDBS Micelles in the Presence of Additives*

Exhibited in Fig. 3a is the impact of 1-hexene concentration on micelle size of 0.2 wt% SDBS solutions. In the absence of 1-hexene, a single family of small particles was observed with a diameter around 3.5 nm. This is in good line with literature data of typical spherical SDBS micelles. Upon addition of 0.5 and 1.0 wt% 1-hexene, the micelles size sharply increased to 9.7 and 113.9 nm, respectively. The particle diameter does not increase significantly when the content of 1-hexene is lower than 0.57 wt%; on the contrary, when above 0.57 wt%, the SDBS micelles were saturated with 1-hexene, the particle size significantly increases up to 600 nm. Similarly, as shown Fig. 3b, the size of SDBS micelles increases from 3.5 to 13.1 nm when the anisole concentration increases from zero to 0.5 wt%, but sharply to 105.3 nm when 1.0 wt% anisole is added into the solution. Further increase to 1.5 wt% anisole, the size grows as large as 700 nm in diameter.

When comparing Figs. 3c, 3D with Fig. 2a and b, one can find the additive concentrations corresponding to size change are exactly the threshold concentrations at which maximum solubilization capacity occurs. This means, particle diameters only increase slightly below the maximum solubilization capacity; once the maximum

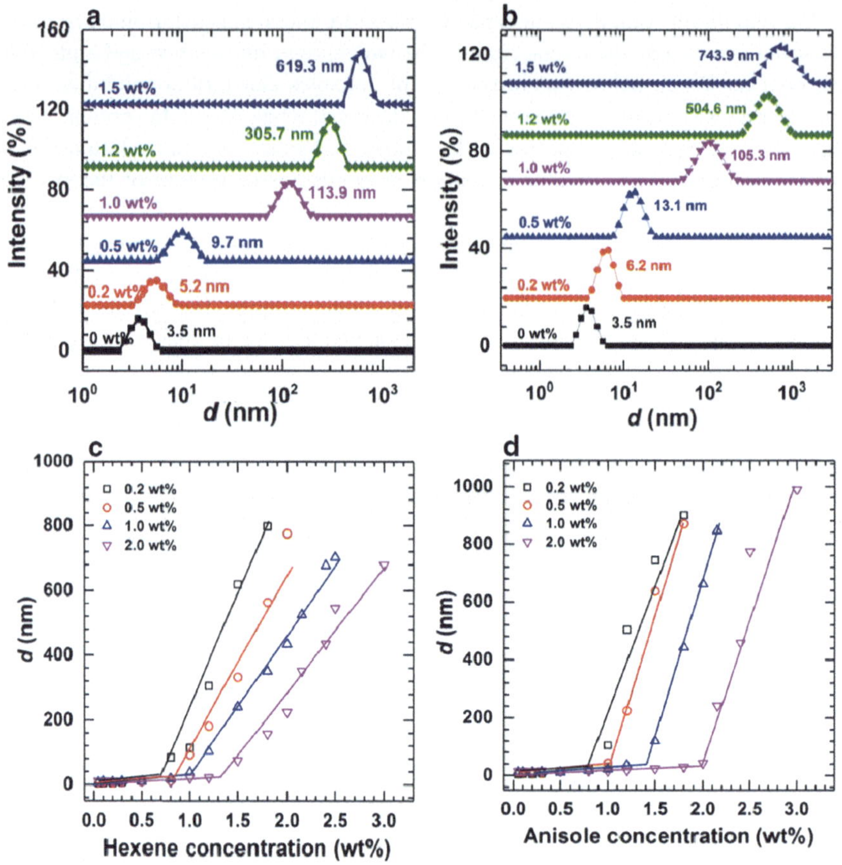

Fig. 3 Influence of **a** 1-hexene and **b** anisole concentration on micellar size in 0.2 wt% SDBS solutions at 40 °C; Variation of micellar diameter with addition of **c** 1-hexene and **d** anisole to aqueous SDBS solutions with different concentrations at 40 °C. Reproduced from Ref. [26]

solubilization capacity is exceeded, the size increases to several hundred nanometers, suggesting the formation of emulsion droplets, which was further verified by the confocal microscope observation in the later subsection.

Such a change in curvature can be explained as follows. The mean micelle diameter remained relatively low (<50 nm) at low 1-hexene concentrations, only slight changes in diameter were observed. When the 1-hexene addition exceeds the maximum solubilization capacity of SDBS, a sharp increase in diameter is observed. At this point, the system did no longer contain swollen micelles, but emulsion droplets formed, as imaged by CLSM in Sect. 3.5. Similar results were observed with SDBS solutions in the presence of anisole, presented in Fig. 3d. To recapitulate, the 1-hexene and anisole solubilization increases with SDBS concentration, meaning that the higher the SDBS concentration, the higher the additive concentration required for a sharp increase in particle diameter. The increase in micelle diameter shows that the surfactant solutions have the ability to solubilize oils. This implies that for EOR processes, micellar solubilization can improve oil recovery.

3.4 *Solubilization Site of Anisole and 1-Hexene in SDBS Micelles*

Investigation of the solubilization site can provide more information about micellar solubilization. In this work, we applied ^1H NMR chemical shifts to distinguish where exactly anisole and 1-hexene are solubilized in SDBS micelles.

From Figs. 4a and b where the ^1H NMR spectra of 0.2 wt% SDBS with and without anisole (or 1-hexene) addition are present, we can find the α-CH_2 and bulk-CH_2 protons maintains almost unchanged in their chemical shifts for this system, showing that 1-hexene has little or no effect on the hydrophilic part and hydrocarbon chain of SDBS. From Fig. 4c, we derive the fact that the shift ($\Delta\delta$) of SDBS protons does not change as the 1-hexene concentration increases. It shows that 1-hexene is mainly located in the hydrophobic core of the SDBS micelles. Similar experiments were performed on the SDBS/anisole system, increasing anisole concentrations shifted the α-CH_2 and bulk-CH_2 signals to the higher magnetic field (Figs. 4d and e). Figure 4f shows the relationship between shift extent ($\Delta\delta$) of the SDBS protons and anisole

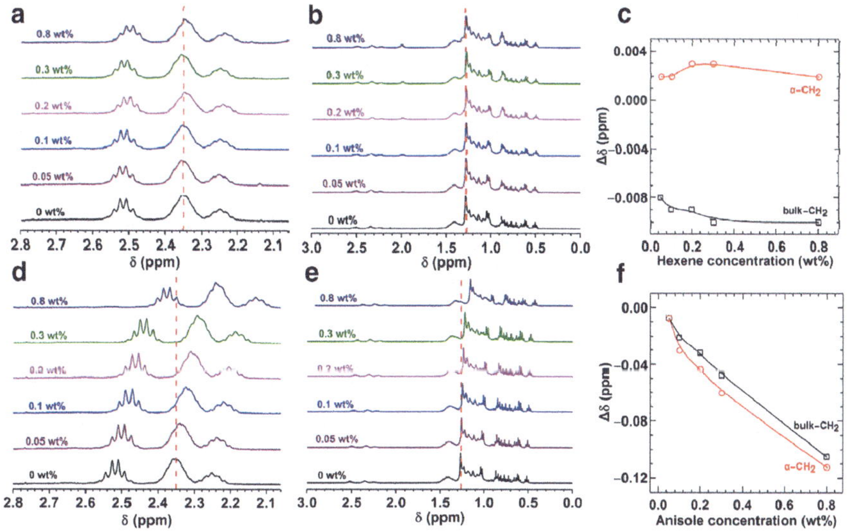

Fig. 4 α-methylene resonances of 0.2 wt% SDBS in the presence of **a** 1-hexene **d** anisole; the bulk methylene resonances of 0.2 wt% SDBS in the presence of **b** 1-hexene **e** anisole; plot of observed chemical shifts of 0.2 wt% SDBS in D_2O as a function of **c** 1-hexene and **f** anisole concentration. All experiments were conducted at 40 °C. Reproduced from Ref. [26]

concentration. The negative value of Δδ represents the upfield shift, while the positive one denotes the downfield shift. As the anisole concentration increased, the Δδ value of SDBS protons also increased. Anisole possesses a hydrophilic polar –OMe group and a hydrophobic benzene ring. Therefore, anisole molecules penetrate into the palisade layer in such a way that the –OMe group is fixed to the surface of the surfactant micelles and the benzene ring resides inside the palisade layer.

The extent of SDBS resonance shifts in the presence of anisole and 1-hexene is closely related to the relative hydrophilic/hydrophobic behavior of the solubilizate. From the results in Fig. 4, anisole, being more hydrophilic than 1-hexene, resides at the palisade layer, and 1-hexene, being more hydrophobic, is solubilized in the micelle core.

3.5 *Morphology of Micelle Assemblies*

The above results revealed that micelles grow in the presence of anisole or 1-hexene, it is interesting to see how the morphology of these assembles change, and what morphologies can be obtained when anisole or 1-hexene is solubilized in SDBS micelles. Thus, we used SANS and cryo-TEM to monitor the morphology evolution of micelles in the presence of anisole or 1-hexene. The SANS profile can give an indication on particle size and shape, at low q values (typically <0.1 $Å^{-1}$), the scattering may scale as $I(q) \sim q^{-x}$.

Plotted in Fig. 5 are the variations of scattering intensity, I, against wave-vector (q) for 0.2 wt% SDBS solutions with or without 1-hexene (or anisole). From Fig. 5a, a near-zero gradient indicates essentially spherical particles for SDBS solution with and without 1-hexene, all indicative of spherical micelles. For 0.2 wt% SDBS, the diameter was 3.2 nm, which is consistent with the aforementioned DLS result (3.5 nm). Furthermore, the micellar aggregation number was calculated to be $N = 41$ for 0.2 wt% aqueous SDBS solution. When 0.2 and 0.5 wt% 1-hexene was separately added, the diameter increased up to 4.8 nm and 9.4 nm, respectively, clearly indicating the formation of swollen micelles. It can be seen that the increase of scattering intensity is due to the increase of 1-hexene, which leads to the increase of micelle size and does not change the micelle morphology.

The curves in Fig. 5b show different changes with increasing anisole concentration. At 0.2 wt% anisole, the curve is similar to that where no anisole was added, only the scattering intensity increases, indicating that anisole entered the micelle and swelling occurred. The diameter of 5.6 nm, slightly smaller than that (6.2 nm) measured by DLS. However, when 0.5 wt% anisole was added, the scattering peak was not very obvious and became very narrow in scattering curves, with a slight shift of the peak to lower q. These features revealed the particle separation is increased and the effect of interaction reduces, indicative of the formation of long scattering objects. The slope of the scattering curve in the low q region is only used to confirm the micelle shape. This suggests that the transition of spherical to rod-like micelles with increasing anisole concentration. The curves at high-q regions overlap perfectly

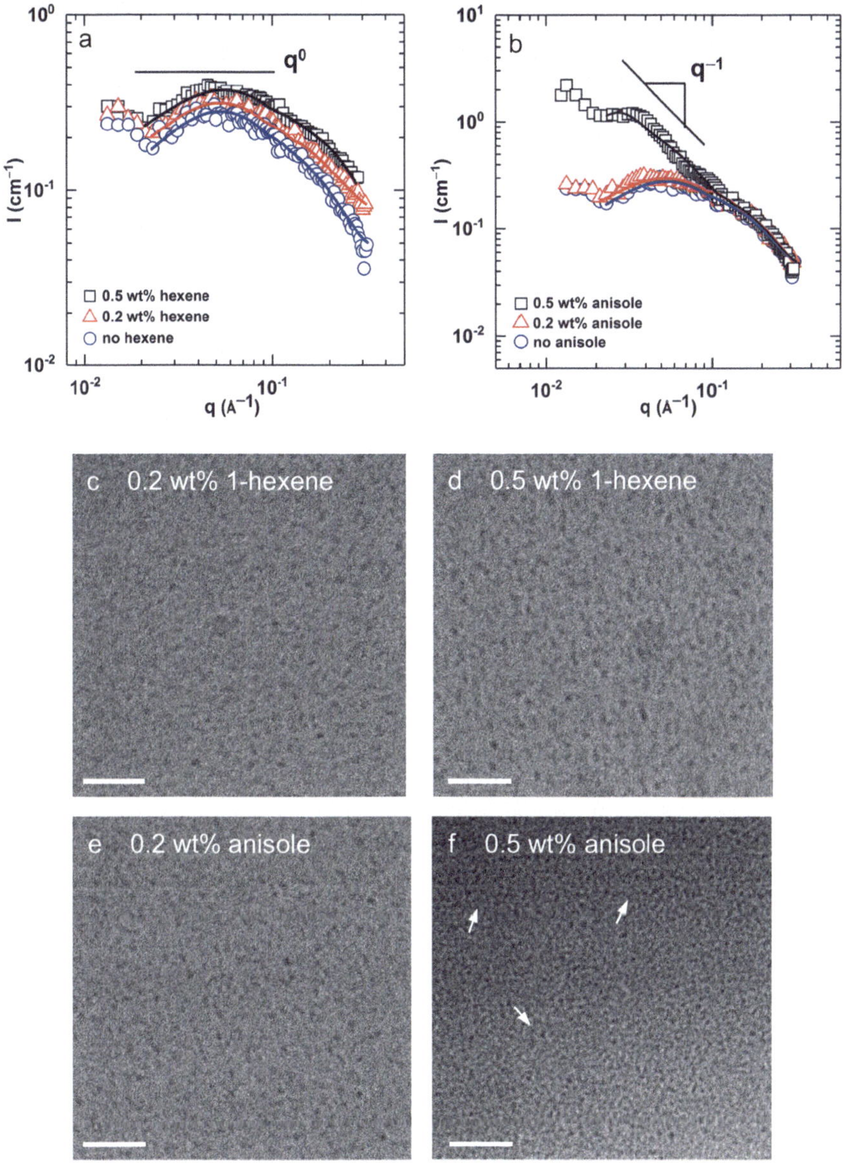

Fig. 5 Morephology change of SDBS micelles. Scattering intensity obtained from SANS experiments plotted as a function wave-vector for 0.2 wt% aqueous SDBS solution in the presence of **a** 1-hexene and **b** anisole in D_2O at 40 °C. Solid lines represent best fitting curves. Cryo-TEM images of 0.2 wt% SDBS micelles co-solubilized with **c** 0.2 wt% 1-hexene; **d** 0.5 wt% 1-hexene; **e** 0.2 wt% anisole; **f** 0.5 wt% anisole. The scale bar is 50 nm. Confocal fluorescence images of micelles (0.2 wt% SDBS) solubilized with **g** 1.5 wt% 1-hexene; (H) 1.5 wt% anisole. The scale bar is 50 μm. Reproduced from Ref. [26]

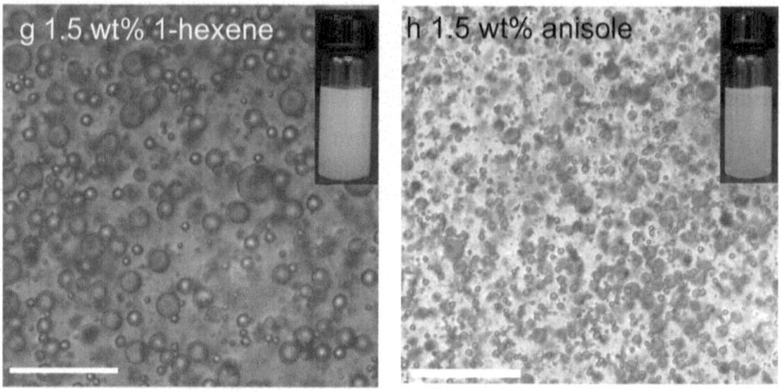

Fig. 5 (continued)

for all anisole concentrations, showing that the smallest dimension of the micelles remains approximately the same. The cylinder model was appropriate to describe the system, giving a "rod" length of around 11.4 nm, and a diameter of roughly 5.8 nm. These findings imply that the diameter will not grow any further when more than 0.2 wt% anisole is added, but the swollen micelles will grow along the uni-dimension to become a cylinder.

To support the SANS results, cryo-TEM was used to investigate the microstructure of SDBS micelles. The cryo-TEM image of 0.2 wt% SDBS with 0.2 wt% 1-hexene is shown in Fig. 5c. The shape of the micelles was spherical with a diameter of roughly 7 nm, which is slightly larger than those measured by DLS (5.2 nm) and SANS (4.8 nm). Increasing the concentration of 1-hexene further to 0.5 wt%, as shown in Fig. 5d, the micelles still remained spherical, with a diameter of ~10 nm. Similar experiments were performed with the micellar SDBS system using anisole (Fig. 5e), when 0.2 wt% anisole was added, the micelles were spherical and the diameter was ~8 nm. Changes in the shape of micelles with 0.5 wt% anisole are presented in Fig. 5f. The spherical micelles coexist with short rod-like assemblies (as indicated by the arrows), with diameters of roughly 7 nm and lengths of 13 nm.

Addition of the nonpolar 1-hexene did not induce any effect on the micelle shape, even at a high concentration of 0.5 wt%. The effect of anisole, on the other hand, is noticeable. Solubilization of anisole with rod-like structures led to a lengthening of the micelles, from 3 to 13 nm. Embedding of anisole decreased the repulsion forces between surfactant headgroups, resulting the sphere–rod transitions. Increasing the anisole concentration increases the number of anisole molecules per micelle. Anisole penetrates into the surfactants between the similarly charged headgroups and minimizes the headgroup repulsion, promoting the formation of rod-like micelles. In contrast, 1-hexene shows no transition effect, as 1-hexene resides in the micelle core without influencing the system.

The solution became more turbid with increasing additive amounts; therefore, a confocal fluorescence microscope was used to observe the morphology. Figure 5g shows emulsion droplets of ~600 nm diameter for the 0.2 wt% SDBS solution containing 1.5 wt% 1-hexene, exceeding the maximum SDBS solubilization capacity. Similarly, with increasing anisole concentration, emulsion droplets with a diameter as high as 700 nm were formed (Fig. 5h). At additive concentrations higher than the maximum solubilization capacity, such as 1.5 wt%, the solution became turbid. Based on the DLS results (Fig. 3a, b) and CLSM images where diameters of several hundred nanometers were observed, it can be inferred that emulsion droplets did form, indicating that high additive concentrations induce the micelle-to-emulsion transition. These observations may contribute to understand the EOR mechanism of surfactant flooding better; in other words, before oil contacts with surfactant solution slug to form emulsion, the presence of swollen micelles can enhance the recovery efficiency by solubilizing the oils.

Based on above findings, the relationship between micellar solubilization and additive concentration were realized. At low additive concentrations, the micelle diameter is barely affected. However, spherical and rod-like micelles are formed, because solubilization sites of different polarity are present. As the additive concentration increases and exceeds the maximum solubilization capacity (as shown by the dotted red line), micelles are unable to solubilize the additives anymore, causing swollen micelles to transfer into emulsion droplets, resulting in a significant increase in diameter.

3.6 Solubilization Behavior of Some Other Surfactants

Apart from the pair of SDBS and 1-hexene or anisole, we also investigated the solubilization capacity of HABS, and alkyl aryl sulfonate [27–29]. The additives solubilized include paraffin oil and main component of crude oil. Similar to SDBS system, the nonpolar additive is always solubilized in the hydrophobic core, while the polar one is located in the palisade, and the size of the micelles gradually increases until emulsions are formed.

We picked out some recent results [28] to highlight oil recovery efficiency can be improved without attaining ultralow IFT. The surfactant used, raw naphthenic aryl sulfonates (NAS), is a typical alkyl aryl sulfonated derived from the local oil cracking components in Xinjiang Oilfield, PetroChina. It is highly diversified in components dominated with benzodicyclohexane sulfonate, supplemented by phenanthrene sulfonate, acenaphthene sulfonate, alkylbenzene sulfonate, indan sulfonate, and alkyl naphthalene sulfonate [30].

Displayed in Fig. 6a are interfacial tensions between paraffin oil or crude oil and with different NAS concentrations solutions. One can find that the IFT between saline water and paraffin oil or crude centers around 4 mN/m, and decreases sharply to the minimum, 9×10^{-2} mN/m, when increasing NAS solution to 0.1%, after which, the IFT goes back slowly but still remains in the magnitude of 10^{-1} mN/m even up to 1.0 wt% NAS solution. As for the IFT between crude oil and NAS solutions, it goes down to the lowest value, 2×10^{-2} mN/m, when increasing NAS concentration to 0.2 wt%, then bounds back to 10^{-1} mN/m level up to 1.0 wt% NAS concentration.

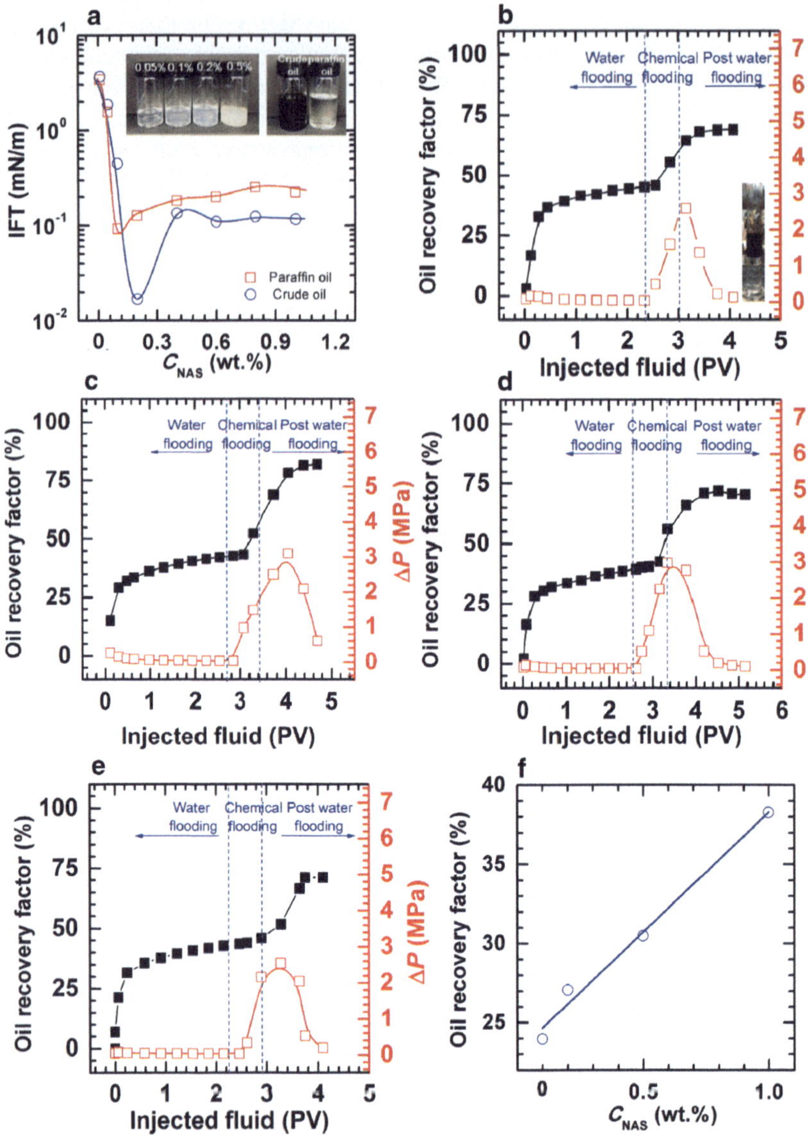

Fig. 6 **a** The comparison of dynamic interfacial tension of NAS solution/paraffin oil and NAS/crude oil at different NAS concentration. The left inset in it indicates the mixture of NAS solution and paraffin oil with different NAS concentration, and the right demonstrates the difference between crude oil and paraffin oil. Cumulative oil recovery and flooding pressure plotted as a function of cumulative injected volume for core flooding with mixture of 0.24% polymer solution and concentration of NAS at (B) 0.1%, **c** 0.2%, **d** 0.5% and **e** 1.0%. **e** The increased oil recovery factor changes as a function of NAS concentration. The saline water is 14,000 mg/L NaCl solution, and 0.2 PV 0.22% polymer solution was used as post-flush fluid. The temperature for all the experiments is 40 °C. Reproduced from Ref. [28]

Table 2 Core flooding results of NAS-KYPAM flooding (The core is 30 cm in length, and 4 cm in diameter; Salinity is 14,000 mg/L NaCl, and T = 40°C)

Core No.	ϕ^{a} (%)	K^{b}(mD)	SP slug (0.5 PV)	Post slug (0.2 PV)	E_{w}^{c}	E_{t}^{c}	E_{sp}^{c}
1	9.8	190.8	0.1% NAS + 0.24% KYPAM[d]	0.22% KYPAM	44.9	68.9	24.0
2	9.9	116.9	0.2% NAS + 0.24% KYPAM	0.22% KYPAM	44.0	71.1	27.1
3	9.7	148.7	0.5% NAS + 0.24% KYPAM	0.22% KYPAM	40.0	71.9	30.5
4	10.0	154.0	1.0% NAS + 0.24% KYPAM	0.22% KYPAM	42.2	81.6	38.3

Note a. "ϕ" means porosity; b. "K" demotes permeability; c. "E_w", "E_t" and "E_{sp}" refer to oil recovery factors of water flooding, total oil recovery factor and oil recovery factor of NAS-KYPAM SP flooding; d. "KYPAM" is an HPAM polymer with improved salt tolerance

All in all, at all the surfactant concentrations, the IFT has not reached the ultralow level.

Although the IFT is not ultralow, as can be seen from Figs. 6b–e and Table 2, the oil recovery factor of the NAS-KYPAM flooding with NAS concentrations of 0.1, 0.2, 0.5 and 1.0% is all higher than 24%. The viscosity of SP slugs and polymer slugs is around 60 mPa s at 40 °C, and the viscosity of crude oil is 16 mPa·s at same temperature. All pressure peaks reach almost 3 MPa. As indicated in the inset in Fig. 6a, there was no emulsion found during flooding process. Quite surprisingly, as exhibited in Fig. 6f, the oil recovery factor increases linearly with increasing surfactant concentration from 0.1 to 1.0% [28]! These preliminary results imply that ultralow IFT may be not a must, and in some extent, the oil recovery factor can be increased by elevating surfactant concentration to some extent.

4 In Situ Emulsification in Porous Media

Apart from the widely accepted mechanisms mentioned above, theoretically in situ emulsification must naturally occur when the chasing surfactant solution contacts crude oil in porous media under pumping pressure, and oil droplets must be carried within emulsions to be transported out to the surface. Several laboratory studies [31–35] also reported that emulsions were generated during the surfactant EOR process. For instance, Chen et al. [36] found during a core flooding, part of driving fluid propagates as an emulsion after contacting with oil, implying that in situ emulsion was formed during the displacing process, unlike the separate two-phase flow of water and oil during water flooding. Moreover, Shah and coworkers [22] confirmed that both micro- and macroemulsion could be generated in porous media when surfactant solution is mixed with oil, and the former is favourable because it is helpful to decrease IFT, while the latter is undesirable, but it is also useful to block big channels so as to divert the follow-up chasing fluid to go to more permeable zones to sweep residual oil thereof. Practically, the SP field trail experience in some oilfields of PetroChina also supported that the higher the oil recovery efficiency can be significantly improved if the emulsion occurs during EOR process. However, how emulsions are formed and how important the emulsion to increase oil recovery factor remains unanswered so far, mainly due to the unavailability of the direct visualization of flow process in porous media. However, no direct proofs are reported yet to unravel how emulsion is formed in porous media and how important to increase oil recovery factor due to unavailability to visualize the emulsification process, thus it is desirable to verify visibly the formation of emulsion in porous media and the contribution of emulsification to EOR process.

Thanks to the transparency and designability of channel structure, microfluidic chips become ideal tools to mimic the underground porous media. In this subsection, direct visualization of emulsion formation in a microfluidic apparatus will be presented. In situ O/W and W/O emulsification processes were monitored with microfluidic setup equipped with two types of microfluidic chips with heterogeneous and homogeneous pore geometries to simulate the underground oil reservoir environment. SDBS was selected as a model surfactant, and its aqueous solution was injected into the paraffin oil-saturated microchip to mimic the displacing process. A series of tests were conducted by varying SDBS concentration, electrolyte content, injection rate, and pore-scale. Snapshots were captured for qualitative and quantitative analysis of in situ emulsification during the surfactant flooding. To differentiate oil and water phase, 1×10^{-4} mol L^{-1} Fluorescein aqueous solution was introduced to generate light green fluorescence, and 3×10^{-5} mol L^{-1} Nile Red was added to label oil phase to emit dark green color. The detailed experimental procedures can be found in Ref. [37].

4.1 IFT Between Paraffin Oil and SDBS Solution

As reported previously [38, 39], IFT plays a vital role in generating and manipulating microscale droplets, and the droplet size and its generation mechanism strongly depends on the equillibrium of viscosity, pressure and IFT. Therefore, the measurement of IFT can provide a guideline for screening and opitimizing injection fluids in the subsequent microfluidic experiments. Dynamic IFTs between a series of SDBS solutions and paraffin oil with and without NaCl were monitored and plotted in Fig. 7. One can find that all the IFT values only reach the 10^{-1}–10^{0} mN m^{-1} level regardless of the surfactant or electrolyte concentration, two order of magnititude higher than the ultra-low level (10^{-3} mN m^{-1}). While optimal salinity is believed helpful to form middle-phase microemulsions, here the IFT values are also as high as 10^{-1}–10^{0} mN m^{-1} within NaCl concentration ranging from 0.2 to 1.4 wt%.

Without ultralow IFT, oil recovery factor generally cannot be increased based on the traditional mechanisms aforementioned. So, it is necessary to see if in situ emulsification can get more oil to be produced under such a circumstance.

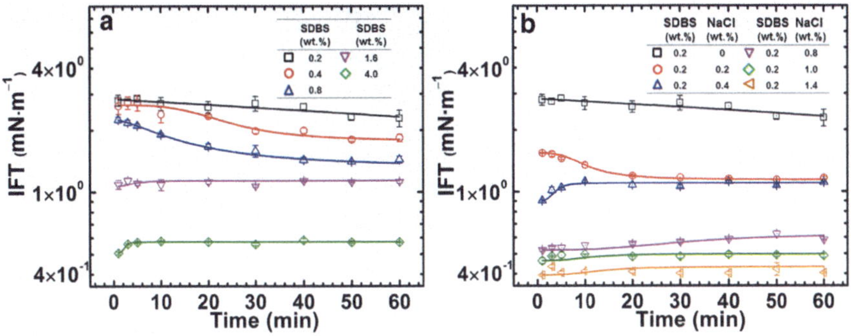

Fig. 7 Dynamic interfacial tension between paraffin oil and **a** different concentration of SDBS solutions without salt, and **b** 0.2% SDBS solutions with different content of NaCl at 40 °C. Reproduced from Ref. [37]

4.1.1 Emulsification in Heterogeneous Chip

We first visualize the in situ emulsification process during flow of SDBS solutions in the micromodel with random geometry, and to reveal the factors affecting the emulsification process through qualitative and quantitative analysis of the pore-scale snapshots.

(1) **Macroscopic behaviour of emulsions**

To preliminarily check if emulsification occurs in the micromodel, eluates in each test were collected to see the appearance, and analysed by the Digimizer software so as to calculate the volume of each phase. The concentration of SDBS, NaCl and the parameters associated with displacement are summarized in Table 3, and the images of eluate collected from Runs 1–5 in Table 3 are displayed in Fig. 8a–e, respectively. It is worth noting that the injection volume is 900 μl in all groups, and all the eluate in each experiment was collected in a glass tube with the total volume approximately equal to the injection volume. As depicted in Fig. 8, opalescent phase was found in the intermediate layer of the eluate except the blank control with only pure water (Fig. 8a), above and below which are paraffin oil and aqueous phase, respectively.

Table 3 Properties of injected fluid and the corresponding displacement results [37]

Run	C_{SDBS} (wt%)	C_{NaCl} (wt%)	Injection rate (μl min^{-1})	Emulsion volume (μl)	Oil volume (μl)	Total volume (μl)
1	0	0	10	0	279.7	279.7
2	0.2	0	10	50.6	293.8	344.4
3	1.6	0	10	63.2	296.4	359.6
4	0.2	1	10	67.9	221.0	288.9
5	1.6	0	1	19.3	362.3	381.6

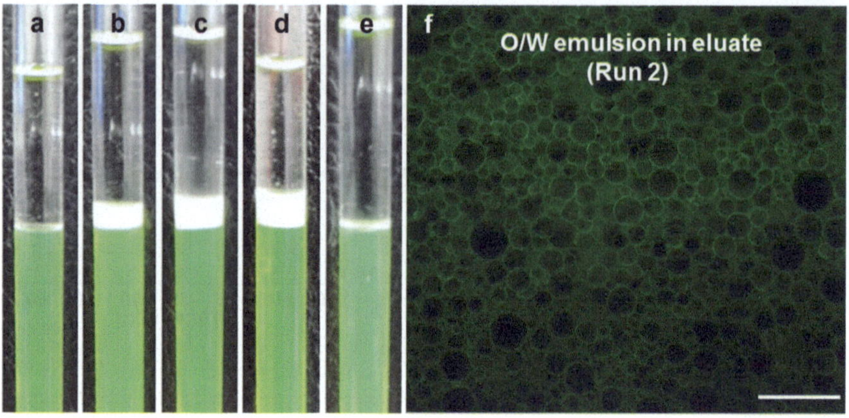

Fig. 8 Snapshots of eluate collected from the outlet of the heterogeneous model using **a** water (Run 1, Table 1), **b** 0.2% SDBS in water (Run 2, Table 1), **c** 1.6% SDBS in water Run 3, Table 1), **d** 0.2% SDBS aqueous solution containing 1% NaCl (Run 4, Table 1) as displacing phase at flux of 10 μl min^{-1}, and (E) 1.6% SDBS aqueous solution as displacing fluid at flux of 1 μl min^{-1} (Run 5, Table 1). All the experiments were conducted at 40 °C. The inner diameter of the glass tube is 5 mm. Fluorescent microscopic image **f** of the emulsion in eluate from the glass tube (Run 2, Table 1). λ_{ex} = 420–485 nm. The scale bar is 500 μm. Reproduced from Ref. [37]

It should be noted that the oil and aqueous phases were labeled by fluorescent dyes, thus appear pale yellow and green color under normal light.

To unravel the composition, the opalescent layer from Run 2 (Fig. 8b) was extracted with a syringe and observed using a fluorescent microscope. Since both oil and water phases were labeled by fluorescent dyes, Fig. 8f exhibited that the darker droplets were surrounded by bright green water phase, which confirmed that the intermediate phase is indeed an O/W emulsion. One can find that with the gentle shaking of the tube, visible emulsion droplets rise and fall in aqueous phase and finally stabilize in the top layer of the water phase but lie below the oil one. The reason for this phase distribution can be explained as follows: relatively low density (0.81 g cm^{-3}) of paraffin oil makes the emulsion droplets difficult to disperse in continuous phase evenly, but move to the top layer of the water phase under the action of buoyancy; at oil-water interface, in combination with the hydration caused by the exposed hydrophilic group of SDBS and the infiltration of water into oil droplets, the density of emulsion droplets is compensated to be greater than that of pure oil, which prevents the oil droplets from further rising to the oil phase. Besides, the electrostatic repulsion among sulfonate groups in SDBS molecules stops the droplets from coalescence with each other, so that the O/W droplets eventually were stabilized on the top layer of the water phase but lie below the oil phase. This kind of distribution brought us a great convenience to calculate the volume of each phase, so as to evaluate the emulsification degree and oil displacement performance.

One can find from Fig. 8b and c that the increase of SDBS concentration is beneficial to the emulsion output, and a 1.4% increase in SDBS concentration produces an additional 25% of the emulsion (Run 2 and Run 3 in Table 3), suggesting higher surfactant concentration can get more oil to be emulsified. The presence of electrolyte also promotes the production of emulsion, but has an adverse effect on the extraction of pure oil in this system (as exhibited in Fig. 8d and Run 4 in Table 3). Noticeably, as can be seen in Fig. 8e, when injection rate is reduced from 10 to 1 μl min^{-1}, there was a significant decrease in emulsion yield but remarkable increase in pure oil output, indicating that the reduction of injection rate is not conductive to in situ emulsification but beneficial to the production of pure oil. In short, the yield of pure oil and emulsion depends on fluid composition and injection rate, and the reasons will be given in the next section.

(2) **Mechanism of O/W emulsification**

Exhibited in Fig. 9 are the screenshots of the emulsification process in heterogeneous chip using 1.6% surfactant solution as the injection fluid (Run 3 in Table 3). It should be noted that the surfactant concentration used here is quite high. However, with higher surfactant concentration the emulsification is highly visible. The dotted rectangle indicates the exact location where the emulsification occurs, and the insets refer to the evolution of the emulsion droplets over time. One can find that when the residual oil arrived at the pore throat (0 ms), as the displacement pressure is smaller than the capillary resistance, the front edge of the residual oil is gradually extruded within 50 to 170 ms, and eventually snapped off at 200 ms; however, a large amount of trailing edge residual oil was still trapped in the throat. As depicted in the image captured at 240 ms, the detached residual oil turns into spherical shape quickly,

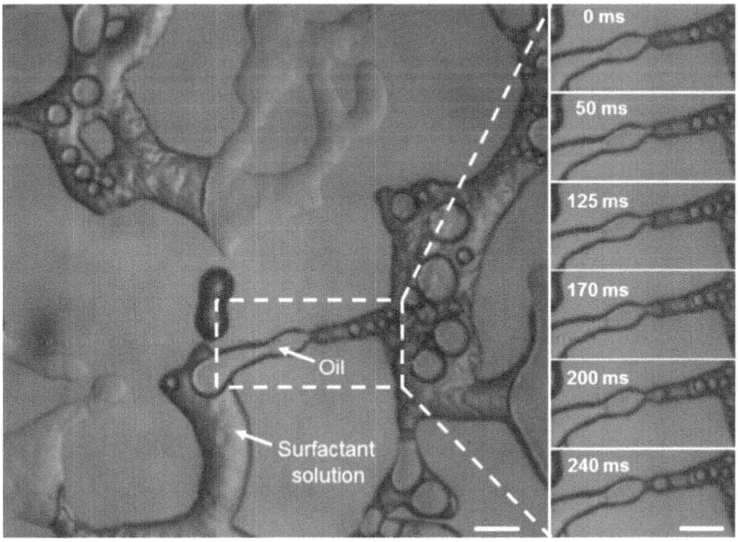

Fig. 9 Screenshots of emulsification process in heterogeneous chip. 1.6% SDBS solution was injected continuously at a flux of 10 μl min^{-1} at 40 °C. The scale bar is 200 μm. Reproduced from Ref. [37]

forming pore-throat scale emulsion and transporting downstream. It is worth noting that droplet coalescence was rarely seen during the observation, indicating emulsion formed in situ was typically stabilized by SDBS.

(3) **Surfactant concentration dependence of emulsification**

The above section has confirmed that emulsification dose occur in porous media. In this section, we continue to examine the factors affecting in situ emulsification. To investigate the surfactant concentration dependence of emulsification, three injection tests (Runs 1–3, Table 3) are compared here. Initially pure water was injected into the oil-saturated microchip as a blank control. Pore-scale snapshots were captured after 1 h of injection when the flow reached steady state, and fluorescent images were taken to distinguish oil from aqueous phase. There was still a large amount of residual oil in pore space after waterflooding. No obvious emulsification occurred due to absence of surfactant, where oil was expelled from the outlet in a continuous oil riband. As displayed in Fig. 8a, only oil and water phases were present in the container, which further proved that no emulsification occurred without surfactant.

Further microfluidic experiments were compared with 0.2 and 1.6% SDBS solution as the injection fluids and the experimental protocol was basically the same as that of the control case. Snapshots were captured after 1 h of injection when the flow reached steady state, and fluorescent images are presented to distinguish between paraffin oil, aqueous phase and pore body.

In the lower concentration group (Run 2, Table 1), a coarse O/W emulsion was created, whose particle size is larger than channel diameter, thus the droplets squeeze each other with poor mobility. When SDBS concentration is increased to 1.6% (Run 3, Table 3), smaller droplets were formed, with a size comparable to that of the pore throat. To determine more precisely the size of emulsion droplets, the snapshots were further digitized using Digimizer program. Taking the sample from Run 3, Table 3 as an example, the image was digitized to calculate the diameter of all individual droplets of the dispersed phase. It is notable that the pore-level snapshots captured in this section do not cover the entire area of the micromodel; hence, the calculated size and size distribution of emulsion droplets are neither exact nor real average of the size and size distributions observed in each run. However, a relative comparison of these data can provide some information insight into the effect of process variables on emulsification considering similar location and timestamp associated with these snapshots.

In the area near the inlet, emulsion droplets with an average diameter of 113.2 μm were formed in 1.6 wt% SDBS solution, which were just around half those formed from its 0.2 wt% counterpart (205.7 μm in average diameter). Moreover, the standard deviation of diameter in 1.6% group is also significantly smaller than that in 0.2% group (37.3 μm vs. 142.7 μm), implying the uniformity of size distribution in the former case.

As the surfactant concentration increases from 0.2 to 1.6%, the capillary number increases accordingly from 1.7×10^{-5} to 3.6×10^{-5} due to the reduction of oil-water IFT, resulting in a smaller capillary resistance, and the large size residual oil bank can be pushed more easily by displacing fluid to undergo snapping action, thus more smaller drops can be produced finally. In addition, a higher concentration of surfactant solution is more conducive to modify the wettability of the reservoir which will be verified later, thus reducing the adhesion of oil to the channel surface and making it easier flow to be emulsified.

When pure water was used as displacing fluid, the pressure drop increased in the initial stage with the advancement of water injection, peaking at ~10.1 kPa, and then eventually declined until stabilized around ~3.6 kPa. The peak pressure reflected the highest resistance reached in the microchannel, which corresponded to the minimum total mobility. With more and more oil displaced, the flow resistance decreased, and then the pressure drop declined. In the case of 0.2% SDBS solution used as displacing fluid, the pressure drop increased once the syringe pump was started, and reached the maximum (~18.0 kPa) after several fluctuations. Then it declined and stabilized at ~12.0 kPa, significantly higher than that of blank control, indicating the presence of big emulsion droplets, which can compensate the pressure decline by blocking the flow paths. It should be noted that a series of distinct pressure fluctuations occurred at the stable stage in this group (during the period of injection volume from 480 to 680 μl), which could be attributed to the passing of big droplets at the pore throat. However, when 1.6% SDBS solution was used, the extra pressure was weakened, which could be explained by the poor blockage effect caused by smaller emulsion droplets. As we know, smaller droplets are not easily intercepted by pore throat, and they cause less resistance to the fluid flow. Similarly, at the stable stage in this group, a slight pressure fluctuation occurred at the injection volume of ~600 μl, which could be attributed to the transportation of aggregated small droplets in the flow path. Obviously, smaller droplets are more likely to pass through the pore throat under displacement pressure, so the pressure fluctuation caused by them is slight.

In addition, we found the emulsion particle size is also associated with the propagation distance in the 0.2 wt% SDBS group. As shown in Fig. 10, the snapshots A, B and C were captured at the up-, middle- and down-streams, respectively. The O/W droplets have a large size at the beginning of flow stage (Fig. 10a), but over the propagation, as depicted in Fig. 10b and c, smaller and more homogeneous O/W droplets were found.

Histograms reflecting particle size distribution are presented in Fig. 10d, e and F, and the red box in the inset indicates the observation area. One can find that with the extension of propagation distance, the size of droplets gradually evolved from a random distribution (Fig. 10d) to a quasi-normal one (Fig. 10f), and the standard deviation decreased from 142.7 to 40 μm, which is related to the number of snapping action experienced during the propagation. As aforementioned, snapping represents the main mechanism for the formation of small size emulsion, it is obvious that with the increase of migration distance, more time of snapping action is consumed and consequently smaller size droplets are produced.

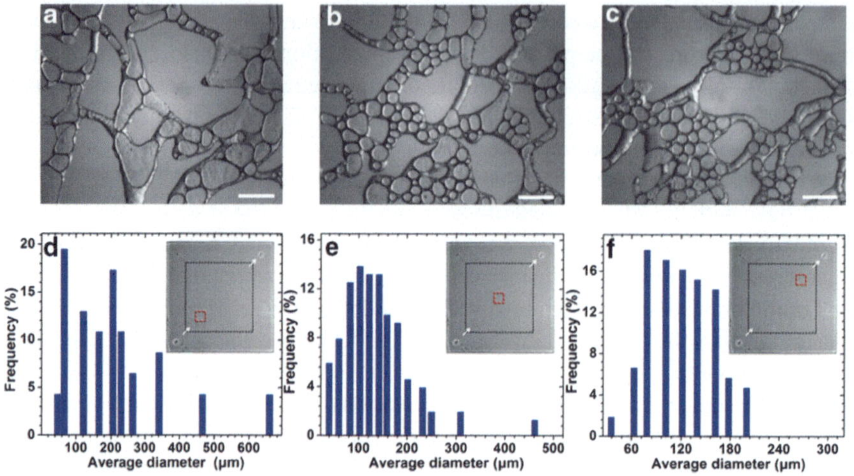

Fig. 10 Snapshots (**a**, **b** and **c**) and corresponding droplet size distribution (**d**, **e** and **f**) for O/W emulsion captured near the inlet (**a** and **d**), the centre (**b** and **e**) and the outlet (**c** and **f**) of the micromodel after 1 h of injection of 0.2% SDBS solution. SDBS solution was injected continuously at a flux of 10 μl min^{-1} at 40 °C. The red rectangle in insets indicates the viewing position. The scale bar is 500 μm. Reproduced from Ref. [37]

4.1.2 Influence of Electrolyte on Emulsification

As depicted in Fig. 7, the IFT between paraffin oil and SDBS solution is influenced significantly by NaCl added in aqueous solution, so we further studied the effect of electrolyte on emulsification.

Figure 11 shows the snapshots captured at different regions and their statistical histograms of particle size. As given in Fig. 11a, after introduction of 1.0% NaCl into 0.2% SDBS aqueous solution, smaller particles (118.6 μm) have been created at the upstream of the flow path in comparison with that without electrolyte (205.7 μm). As exhibited in Fig. 11b–f, at the midstream and downstream of the flow channel, the droplets are also smaller than that in control group, and gradually evolves to a quasi-normal size distribution, indicating good emulsification. Noticeably, the oil saturation decreased significantly in the dominant channel, and the largest amount of emulsion presented in the effluent. As shown in Table 3, an additional emulsion yield of 24% was obtained over water flooding. However, the oil productivity in this group was the lowest in all tests, which could be attributed to the further dredging of the dominant seepage channel caused by the transition from residual oil to smaller oil droplets; that is, as the IFT decreased to 0.5 mN m^{-1} in this group, the capillary number increased to 8.0×10^{-5}, a smaller capillary resistance could be achieved, the residual oil in main flow path was more likely to be pushed to pass through the constriction, and snapped into small droplets with high mobility and be extracted quickly, resulting in a more unimpeded flow of displacing fluid in dominant channel, which is detrimental to reach a satisfactory sweep efficiency. The pressure data can

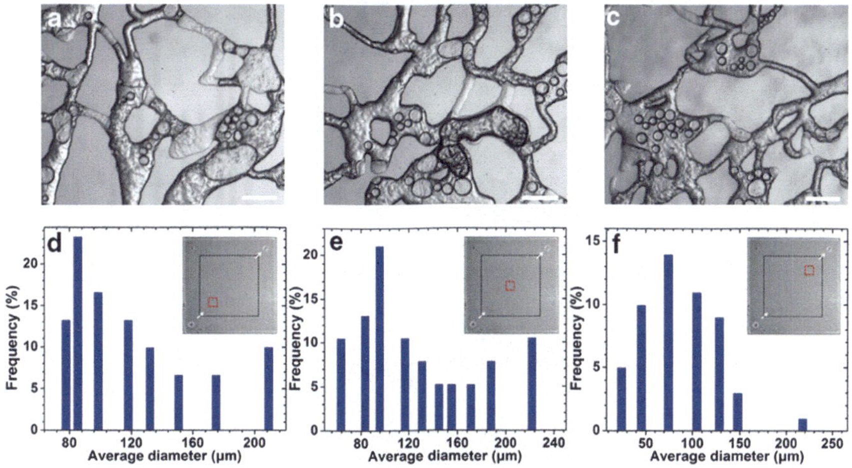

Fig. 11 Snapshots (**a**, **b** and **c**) and droplet size distribution (**d**, **e** and **f**) for O/W emulsion taken near the inlet (**a** and **d**), the centre (**b** and **e**) and the outlet (**c** and **f**) of the microchip after 1 h of injection of 0.2% SDBS aqueous solution containing 1% NaCl. SDBS solution was injected continuously at a flux of 10 μl min^{-1} at 40 °C. The red rectangle in inset indicates the viewing position. The scale bar is 500 μm. Reproduced from Ref. [37]

be used to prove the above inference. The pressure drop in this case is stable at ~2.2 kPa, even lower than that of the blank control (~3.6 kPa), indicating that the displacing fluid flows more smoothly in the dominant paths.

4.1.3 Influence of Injection Rate on Emulsification

The macroemulsion is kinetically stable and not formed spontaneously, which means it requires external energy to be produced. In microchip, this external energy is provided by displacement pressure, which is proportional to the injection rate. Here, 1.6% SDBS solution was injected continuously at 1 and 10 μl min^{-1}, respectively, to explore the influence of injection rate on emulsification. As shown in Fig. 12a, with the same injection time, the droplets formed in low flux group were larger than those in control group (224.2 μm at 1 μl min^{-1} vs. 118.1 μm at 10 μl min^{-1}). Oil droplets formed in Run 5 (Table 3) are squeezed too densely to distinguish between oil and pore body under normal light, so fluorescent images are given here in Fig. 12b. Moreover, the droplets produced in Run 5 have a discrete size distribution with a size range from ~120 to ~400 μm (Fig. 12c). In sharp contrast, as depicted in Fig. 12d, e and f, emulsions formed in Run 3 (Table 3) have smaller size with a more uniform size distribution (size range from ~60 to ~180 μm).

In addition, even if we extend the injection time to keep the injection volume consistent with high injection rate, the produced droplets in Run 5 are still larger than those in Run 3. This behaviour is closely related to the displacement pressure.

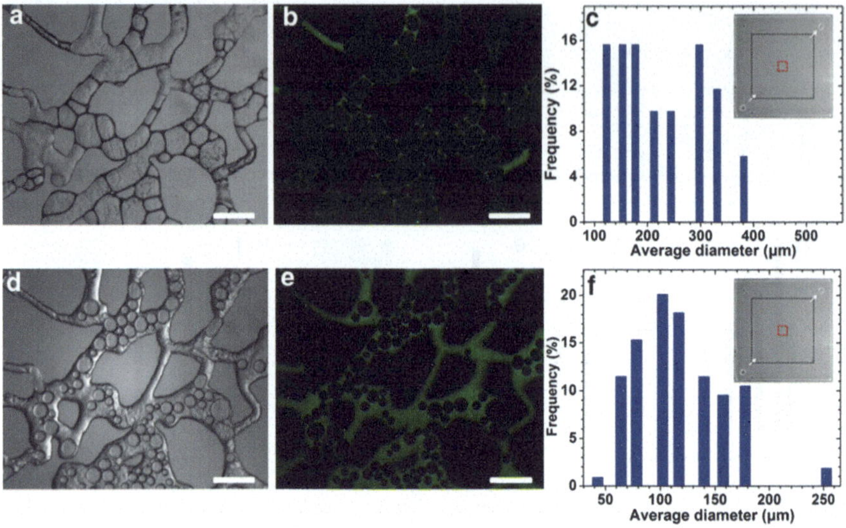

Fig. 12 Snapshots (**a**, **b**, **d** and **e**) and droplet size distribution (**c** and **f**) for O/W emulsion captured in the flow process of 1.6% surfactant solution was injected continuously at 1 μl min^{-1} (**a**, **b** and **c**) and 10 μl min^{-1} (**d**, **e** and **f**). Snapshots were captured near the centre of the micromodel under normal light (**a** and **d**) and excitation light (**b** and **e**). All the experiments were carried out at 40 °C. The red rectangle in inset indicates the viewing position. The scale bar is 500 μm. λ_{ex} = 420–485 nm. Reproduced from Ref. [37]

In low flux group, the displacement pressure is too low to drive residual oil to go through the narrow throat, most residual oil maintains its natural volume, or has experienced a time-consuming snapping action to form relatively large oil droplets at the wider throat. When the residual oil arrived at the pore throat (0 ms), the front edge of the residual oil is gradually extruded, and eventually snapped off at 450 ms, the detached residual oil turns into spherical shape quickly, forming emulsion with relatively big particle size.

Oil droplets of this size range are difficult to pass through the narrower throat, resulting in a significant decrease in emulsion yield, to only one third of that of high flux group. Surprisingly, the amount of oil displaced in low flux group reached the maximum in all cases, with the yield increasing by 30% compared to waterflooding (Run 1, Table 3) and 22% to high flux group (Run 3, Table 3), indicating an improved oil displacement performance, which could be explained by the blocking mechanism.

In Run 5, the coarser emulsion formed in situ acts as dynamic mobility control agents, by blocking the dominant seepage channel and diverting the displacing fluid toward unswept regions of porous media; that is, when the flow was hindered in the direction pointed by red arrow, it turned to flow to the orientation that green arrow pointed. In this way, the macroscopic sweep efficiency improves, thus more oil can be extracted. This finding is in good line with previous reports which demonstrated the major advantage of macroemulsion-based EOR was thought to be that the emulsion blocks the high-permeability paths and force more displacing fluid into unswept regions. However, it is worth noting that these researches were based on injection of pre-prepared O/W emulsion, which should be distinguished from the emulsion formed in situ we reported here.

Due to the reduction of injection rate in Run 5, the comparison of the pressure data with other groups is meaningless, but some insight into the flow of the emulsion can be obtained from the curve. The pressure drop showed a distinct oscillation at the injection volume of ~440 μl, which corresponded to the passing of oil droplets at pore throat. As we know, there is a minimum pressure needed for an oil droplet to pass through a pore throat, which is determined by size of droplet and pore throat and the frictional force between them. When droplets block the channel, the fluid flow is impeded and the pressure increases until it can drive the droplets to pass the pore throat by deformation.

4.1.4 Emulsification in a Homogeneous Chip

It is well recognized that W/O emulsion is also beneficial for improving oil recovery, though to a lesser, through decreasing the effective viscosity of oil [40, 41]. In above experiments, in addition to the widespread presence of O/W emulsions, W/O emulsions were also observed in the regions far from the dominant seepage channel. Taking Run 3 (Table 3) as an example, under fluorescent excitation, bright green water droplets appeared in the dark green oil, indicative of the existence of W/O emulsion. However, in the heterogeneous chip used above, due to the irregularity of channel, the area where W/O emulsification occurs was random, which prevented us from observing the W/O emulsification process continuously. Therefore, we used a microchip with well-defined structure to investigate the W/O emulsification.

Similarly, the injection started with oil-saturated microchannel. The injection rate was controlled at a low level (0.1 μl min^{-1}) due to the small size of the channel (100 μm in length and 40 μm in depth). First, pure water injection test was performed as a blank control. After 3 h of injection, the trapped oil in branch channel maintained its initial morphology, no bright green water droplet was found in oil phase under fluorescent excitation, indicating no emulsification occurred due to the absence of surfactant.

Then, separate injection experiments were carried out with 1.6 wt% SDBS solution as displacing fluid and the experimental protocols were the same as that of blank control. W/O emulsification started rapidly after SDBS solution was injected into the model. W/O drops appeared at the front of branch channel after 3 min of injection, and reached the first turning of branch channel in 20 min. Whereafter, the emulsified range gradually expanded, accounting for 36% of whole branch channel at 40 min, 53% at 90 min, and reached the dead end at 180 min. The bright green fluorescent patches in oil phase further confirmed the occurrence of W/O emulsification. By diffusing along the wall, the aqueous phase penetrated into oil phase and accumulated at the dead end of the channel, squeezed the space occupied by oil phase, so as to achieve the purpose of oil displacement.

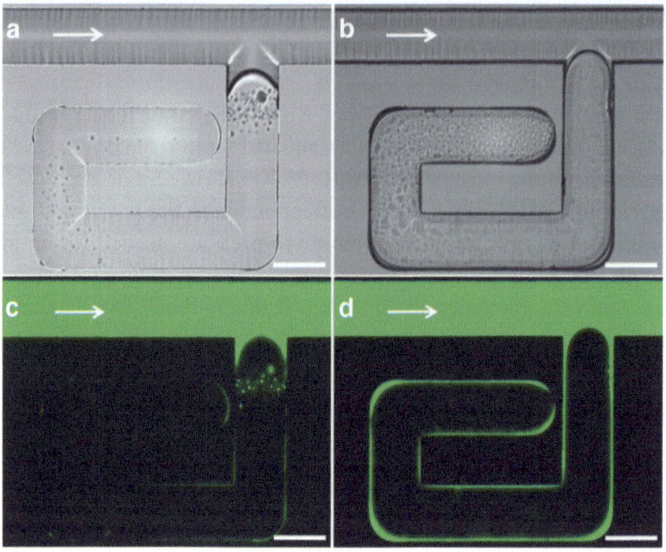

Fig. 13 Comparison of the W/O emulsion produced by (**a** and **c**) 0.2% and (**b** and **d**) 1.6% surfactant solution flood. Surfactant solution was injected continuously at 0.1 μl min^{-1} at 40 °C. Snapshots were taken under normal light (**a** and **b**) and excitation light (**c** and **d**). The scale bar is 100 μm. $\lambda_{ex} = 420$–485 nm. Reproduced from Ref. [37]

The influence of surfactant concentration on emulsification was also investigated. As shown in Fig. 13, after 3 h of injection, there was a significant contrast between the two groups. The emulsification in 0.2% group was limited to the front of the branch channel, accounting for 6% of the whole region (Fig. 13a), while the 1.6% group had higher emulsification degree with higher density of W/O droplets, and the penetration region diffused to the dead end after 3 h of injection (Fig. 13b).

In addition, the wettability of the channel surface was significantly modified by SDBS solution. As shown in Fig. 13c and d, SDBS solution diffused along the surface of the channel, and formed an isolated layer between trapped oil and channel surface, which was beneficial to reduce the adhesion between oil and channel surface, thereby increasing the mobility of the trapped oil. By comparing Fig. 13c and d, one can find that this phenomenon is more obvious in 1.6% group, indicating that the wettability modification is also concentration dependent. The evaluation of wettability of the channel surface can be explained by the adsorption of surfactant molecules to the solid-liquid interface. As the surfactant solution flows through the channel, the surfactant molecules begin to be adsorbed with their hydrophilic heads in the aqueous phase and hydrophobic tails on the channel surface, thus changing the wettability. When the surfactant concentration is high enough, the channel surface can be modified to completely hydrophilic [42].

Combined with the reduction of oil viscosity and space replacing effect caused by W/O emulsification, as well as the decrease of adhesion between oil and channel surface caused by wettability alteration, the trapped oil is more likely to be pushed to the main flow route and carried away by displacing fluid. Furthermore, by measuring the volume occupied by aqueous phase (considering the practical difficulties, we can only calculate the volume occupied by the continuous water phase and ignore the water droplets in oil), we can roughly calculate the oil recovery factor. After 12 h injection of 1.6% SDBS solution, almost 14% trapped oil was displaced. It should be mentioned that after being exposed to excitation light for a long time, the fluorescence of W/O droplets gradually weakens and even disappears. In addition, since the above process was performed so slowly in the low-concentration group that exceeded the observation time scale, we only estimated the oil recovery performance in high-concentration counterpart.

5 Conclusions and Perspectives

The use of surfactant in oil production can be traced back to one hundred years ago, and the surfactant EOR has been extensively studied during the last half century. However, field trials with microemulsion flooding have less documented because the high cost of the high concentration surfactant solutions impeded its wider applications. In addition, during the displacement of crude oil, the surfactant solution slug gradually interacts with crude oil, and it is impossible to form middle-phase microemulsions upon initial contact, or at the early stage, because the surfactant concentration and time are not sufficient to form microemulsions. To the best of laboratory knowledge, weeks of time are needed to form middle-phase microemulsions under mild mixing of all the components in tubes under the simulated oil reservoir temperature. On the other hand, micelle solubilization can immediately occur once the surfactant solution contacts oil. Moreover, not all displacement fluids can form middle-phase microemulsions with any salinity range. According to all literature results reported so far, optimal salinity is always necessary to form middle-phase microemulsions, and a negative gradient must be designed for core flooding with middle-phase microemulsions. Lastly, from produced fluids in field trails with surfactant-polymer flooding, we have not observed the presence of middle-phase microemulsion or even microemulsion (but macroemulsions were found indeed), which in turn indicates that middle-phase microemulsions in porous media underground were not formed during SP flooding with low surfactant concentration.

On the contrary, surfactants were utilized with much lower concentration in both SP and ASP flooding with polymer or the mixture of polymer and alkali, and almost 8 million tons of incremental oil were produced annually by these two technologies. In both processes, phase diagram was not used for screening surfactants.

Undoubtedly, ultralow IFT is still the dominating mechanism for cEOR, and phase diagram is still a useful tool for screening surfactants for cEOR. Nevertheless, the new findings in the practical field trials drive us to understand the mechanisms alternatively.

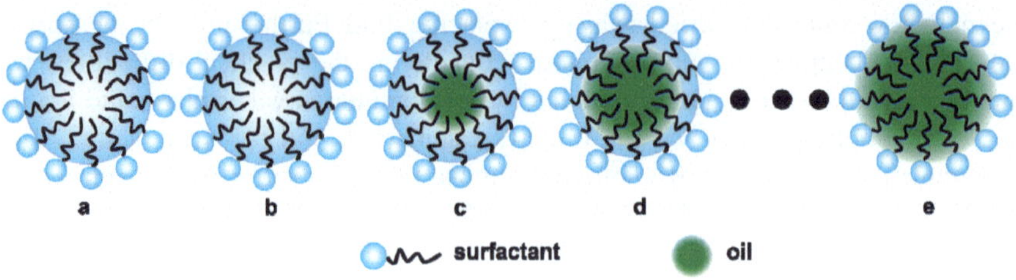

Fig. 14 The schematic illustration of the transition from micelles (**a**) to emulsion (**b**, **c**, **d**) as oil loading increases. Reporduced from [43]

Unlike the previous microemulsion flooding mechanism (the optimal salinity required to form middle-phase microemulsion), we found micelle solubilization is also helpful for surfactant improved oil recovery before forming into microemulsion, and noted that lower concentrations of surfactants solutions are also helpful in displacing oil.

Both O/W and W/O emulsions are formed in porous media during the surfactant flooding. Increasing surfactant concentration, migration distance, injection rate, or addition of electrolyte tends to form smaller O/W particles through snapping action at pore throat, and vice versa. Smaller size endows oil with a better mobility to go through the pore throat, and up to 24% extra emulsion can be achieved through emulsification entrainment; bigger droplets can block the dominant paths, thus improving sweep efficiency and increasing oil recovery factor up to 30% compared to waterflooding. Furthermore, W/O emulsification was found to be a time-dependent process influenced by surfactant concentration, and oil was recovered by diffusing surfactant solution into oil phase and replacing the oil-occupied space in porous media.

Combined both micellar solubilization and in situ emulsification, we may draw the following scheme to show how surfactant micelles evolve into swollen micelles and finally emulsion droplets when gradually contacting oil in the porous media (Fig. 14).

It is woth emphasizing that the results summarized in this chapter are just very preliminary, and much work needs to be done in the future, mainly, solubilization capacity of different surfactants, quantitative of solubilization amount, solubilization of oil with commodity surfactants, in situ emulsion in cores should be further examined to verify the newly-proposed mechanisms in this chapter.

Acknowledgements The financial support from the joint funding of Natural Science Foundation of China (U1762218, 21173161) is gratefully acknowledged. The authors also thank Miss Lingyan Gong, Mr. Xuezhi Zhao, Miss Fuxing Zhan for their extensive laboratory experiments.

References

1. Sheng JJ (2010) Modern chemical enhanced oil recovery: theory and practice. Gulf Professional Publishing, Boston
2. Gbadamosi AO, Junin R, Manan MA, Agi A, Yusuff AS (2019) An overview of chemical enhanced oil recovery: recent advances and prospects. Int Nano Lett 9:171–202
3. Guo H, Dong J, Wang Z, Liu H, Ma R, Kong D, Wang F, Xin X, Li Y, She H (2018) EOR survey in China-Part 1. Paper SPE 190286 presented at SPE Improved Oil Recovery Conference, 14-18 April, Tulsa, USA
4. Guo H, Li Y, Wang F, Yu Z, Chen Z, Wang Y, Gao X (2017) ASP flooding: theory and practice progress in China. J Chem. Article ID 8509563
5. Denney D (2013) Progress and effects of ASP flooding. J Petrol Technol 65:77–81
6. Gao C, Shi J, Zhao F (2014) Successful polymer flooding and surfactant-polymer flooding projects at Shengli Oilfield from 1992 to 2012. J Petrol Explor Prod Technol 4:1–8
7. Sun C, Guo H, Li Y, Song K (2020) Recent advances of surfactant-polymer (SP) flooding enhanced oil recovery field tests in China. J Chem. Article ID 8286706
8. Zhang J, Zhang F, Kang X, Li B (2019) Development and application of chemical EOR technologies in China offshore oil fields. In: Samsuri A (ed) Enhanced oil recovery processes: new technologies. IntechOpen
9. Lu Q, Ning Y, Wang J, Yang X (2015) Full field offshore surfactant-polymer flooding in Bohai Bay China. Paper SPE174591 presented at SPE Asia Pacific Enhanced Oil Recovery Conference, 11–13 Aug 2015, Kuala Lumpur, Malaysia
10. Sorbie KS (1991) Polymer-improved oil recovery. CRC Press, Boca Raton
11. Sheng JJ (2014) A comprehensive review of alkaline–surfactant–polymer (ASP) flooding. Asia-Pac J Chem Eng 9:471–489
12. Hirasaki GJ, Miller CA, Puerto M (2011) Recent advances in surfactant EOR. SPE J 16:889–907
13. Sheng JJ (2015) Status of surfactant EOR technology. Petroleum 1:97–105
14. Kamal MS, Hussein IA, Sultan AS (2017) Review on surfactant flooding: phase behavior, retention, IFT, and field applications. Energy Fuels 31:7701–7720
15. Pal S, Mushtaq M, Banat F, Al Sumaiti AM (2018) Review of surfactant-assisted chemical enhanced oil recovery for carbonate reservoirs: challenges and future perspectives. Pet Sci 15:77–102
16. Sheng JJ (2013) Review of surfactant enhanced oil recovery in carbonate reservoirs. Adv Pet Explor Dev 6:1–10
17. Jamaloei BY (2009) Insight into the chemistry of surfactant-based enhanced oil recovery processes. Recent Pat Chem Eng 2:1–10
18. Huh C (1979) Interfacial tensions and solubilizing ability of a microemulsion phase that coexists with oil and brine. J Colloid Interface Sci 71:408–426
19. Mohammed M, Babadagli T (2015) Wettability alteration: a comprehensive review of materials/methods and testing the selected ones on heavy-oil containing oil-wet systems. Adv Coll Interface Sci 220:54–77
20. Wang Y, Xu H, Yu W, Bai B, Song X, Zhang J (2011) Surfactant induced reservoir wettability alteration: recent theoretical and experimental advances in enhanced oil recovery. Pet Sci 8:463–476
21. Spinler EA, Baldwin BA (2000) Surfactant induced wettability alteration in porous media. In: Schramm LL (ed) Surfactants: fundamentals and applications in the petroleum industry, Cambridge University Press, pp 159–202
22. Sharma MK, Shah DO (1985) Macro- and microemulsions in enhanced oil recovery. In: Shah DO (ed) Macro- and microemulsions. ACS Symposium Series No. 272, American Chemical Society, Washington, DC, pp 149–172
23. Rosen MJ, Wang H, Shen P, Zhu Y (2005) Ultralow interfacial tension for enhanced oil recovery at very low surfactant concentrations. Langmuir 21:3749–3756
24. Wang H, Cao X, Zhang J, Zhang A (2009) Development and application of dilute surfactant–polymer flooding system for Shengli oilfield. J Petrol Sci Eng 65:45–50

25. Zhu Y, Jian G, Liu W, Cheng L, Hou Q, Li J (2013) Recent progress and effects analysis of surfactant-polymer flooding field tests in China. Paper SPE 165213 presented at SPE Enhanced Oil Recovery Conference held in Kuala Lumpur, Malaysia, 2–4 July, 2013
26. Gong L, Liao G, Luan H, Chen Q, Nie X, Liu D, Feng Y (2020) Oil solubilization in sodium dodecylbenzenesulfonate micelles: New insights into surfactant enhanced oil recovery. J Colloid Interface Sci 569:219–228
27. Zhao X, Gong L, Liao G, Luan H, Chen Q, Feng Y (2021) Micellar solubilization of crude oil components by heavy alkylbenzene sulfonate surfactant for enhanced oil recovery. J Mo Liq 329:115519
28. Zhan F, Gong L, Luan H, Chen Q, Liao G, Feng Y (2020) Enhancing oil recovery based on micelle solubilization without ultra-low interfacial tension. J Surfactants Deterg https://doi.org/10.1002/jsde.12488
29. Zhao X, Gong L, Liao G, Luan H, Chen Q, Feng Y (2019) Micelle solubilization of crude oil by alkyl aryl sulfonate surfactants for enhanced oil recovery. Under preparation
30. Luan H, Gong L, Yue X, Nie X, Chen Q, Guan D, Que T, Liao G, Su X, Feng Y (2019) Micellar aggregation behavior of alkylaryl sulfonate surfactants for enhanced oil recovery. Molecules 24, article ID 4325
31. Bryan J, Wang J, Kantzas A (2009) Measurement of emulsion flow in porous media: improvement in heavy oil recovery. J Phys Conf Ser 147:012058
32. Mandal A, Bera A (2015) Modeling of flow of oil-in-water emulsions through porous media. Pet Sci 12:273–281
33. Bennlon DB, Chan M, Sarloglu G, Courtnage D, Wansleeben J, Hirata T (1993) The in situ formation of bitumen-water stable emulsions in porous media during thermal stimulation, Paper CIM 93-46 presented at the CIM Technical Conference, Calgary, May 9-12, 1993
34. Cuthiell D, Green K, Chow R, Kissel G, McCarthy C (1995) The in situ formation of heavy oil emulsions. Paper SPE 30319 presented at the International Heavy Oil Symposium, Calgary, June 19–21, 1995
35. Liu ZY, Li YQ, Luan HX, Gao WB, Guo Y, Chen YH (2019) Pore scale and macroscopic visual displacement of oil-in-water emulsions for enhanced oil recovery. Chem Eng Sci 197:404–414
36. Chen T, Chakrabarty T, Cullen MP, Thomas RR, Sieben MC (1991) Laboratory and field evidence of emulsion flow in porous media. Paper CIM/AOSTRA 91-78 presented at the CIM/AOSTRA Technical Conference, Banff, April 21-24, 1991
37. Zhao X, Feng Y, Liao G, Liu W (2020) Visualizing in-situ emulsification in porous media during surfactant flooding: a microfluidic study. J Colloid Interface Sci 578:629–640
38. Peng L, Yang M, Guo S, Liu W, Zhao X (2011) The effect of interfacial tension on droplet formation in the flow-fousing microfluidic device. Biomed Mcirodevices 13:559–564
39. Mastiani M, Seo S, Jimenez SM, Petrozzi N, Kim MM (2017) Flow regime mapping of aqueous two-phase system droplets in flow-focusing geometries. Colloids Surf A 531:111–120
40. Zhou YZ, Yin DY, Wang DQ, Gao X (2018) Emulsion particle size in porous media and its effect on the displacement efficiency. J Dispersion Sci Technol 39:1532–1536
41. Pei HH, Zhang GC, Ge JJ, Jin LC, Liu XL (2011) Analysis of microscopic displacement mechanisms of alkaline flooding for enhanced heavy-oil recovery. Energy Fuels 25:4423–4429
42. Xu JH, Li SW, Tan J, Wang YJ, Luo GS (2006) Preparation of highly monodisperse droplet in a T-junction microfluidic device. AIChE J 52:3005–3010
43. Gong L, Liao G, Chen Q, Luan H, Feng Y (2019) Swollen surfactant micelles: properties and applications. Acta Phys Chim Sinica 35:816–828

Biosurfactants and Its Application in Oil Recovery

Abhijit Samanta

Abstract This chapter discusses about the application and potential role of biosurfactants in oil recovery technology. The mechanism of mobilizing of trapped crude oil within the pores of reservoir rocks has been discussed on the basis of reduction of interfacial tension (IFT), wettability alteration, emulsion formation, and biodegradation of oil. In situ production and ex situ injection of biosurfactants for different suitable microbial strains have been reviewed and laboratory based evaluation of microbial enhanced oil recovery (MEOR) have been discussed. Internationally major field trials of MEOR by worlds leading oil producing nations have been reported. Thus, this chapter tries to concentrate on nearly all the concern issues about the past, latest trends and potential aspects of biosurfactants applications in oil recovery.

Keywords Biodegradation · Interfacial tension · Wettability alteration · Emulsification index · Microbial enhanced oil recovery

1 Introduction

The modern world and global intensive economic growth is very much dependent on energy resources and the petroleum remains the main power source, which brings to drive the development enormously for the past century [77, 82, 106]. It has been predicted that the global energy consumption will increase by 1.7% per annum in the quantity of oil barrels produced between the years of 2000 and 2030, while projected oil utilization is to be reached about 15.3 billion tons yearly. Hence, if this intensity of demand will carry, the availability of crude oil in the reservoirs can fulfill these requirement by around 40 years [33, 100, 106]. The implicational of conventional oil recovery methods or primary recovery process is able to produce only about 5–15% of initial reservoir oil [51, 94]. The secondary recovery is the step to use an attempt to enhance the extraction of remaining trapped oil inside reservoir by injecting water or gas [28]. Additionally, the improvement of oil production efficiency can be possible

A. Samanta (✉)
School of Engineering and Applied Sciences, The Neotia University, Sarisha, West Bengal 743368, India

T. Solling et al. (eds.), *Surfactants in Upstream E&P*, Petroleum Engineering,
https://doi.org/10.1007/978-3-030-70026-3_6

by application of enhanced oil recovery (EOR) techniques [21, 35, 53, 99, 101]. Among different EOR technologies, surfactant flooding process i.e. the reduction of interfacial tension (IFT) or wettability alteration by use of surfactants is implemented globally [68, 76, 98].

Surfactants have both the hydrophilic and hydrophobic groups and have capacity to decrease the surface tension of air/water and oil/water interfaces, consequently it reduces the capillary forces between water and oil within the reservoir pores, which leads to recover the trapped oil easily [54, 55, 61, 86, 127]. It has been reported that use of commercial synthetic surfactants with high concentration as well as have high cost effect were required to achieve satisfactory results [26, 35]. Moreover these have significant acute toxic effects on the environment, which has motivated researchers to focus on biosurfactants or natural surfactants [30]. Therefore, the development of efficient applications of biosurfactants from natural recourses in EOR is growing rapidly and becoming an important technology [71, 85]. The microbial enhanced oil recovery (MEOR) is an alternatively more cost-effective and ecofriendly technology among different EOR methods used in past few decades using in situ and ex situ application of biosurfactants [37, 60, 81, 90, 97].

2 Biosurfactants

Recent advancements in sustainable development have focused towards environmental and biodegradable compounds to enhance the oil recovery process from the reservoirs contaminated by remaining oil. Therefore this has attracted researchers to give more attention to the discovery of biosurfactants from natural recourses. Biosurfactants are heterogeneous group of surface active reagents derived naturally from biological entities, especially microorganism. Biosurfactants are produced from different microorganisms like fungi, bacteria, and yeast, which are known for biosurfactant producing strains [30]. The bacteria like *Pseudomonas*, *Bacillus*, *Candida sphaerica* UCP 0995 and *Acinetobacter* of genera are the excellent producer of biosurfactant [3, 108, 109]. It has been reported that the bacterial species like *Pseudomonas aeruginosa*, *Bacillus subtilis*, *Bacillus licheniformis*, *Candida albicans* and *Acinetobacter calcoaceticus* are extensively studied for biosurfactant production [32, 49, 63, 116, 45]. Biosurfactants are also produced from fungi, *Candida bombicola* and yeast, *Pseudozyma rugulosa* [17, 41].

Primarily biosurfactants are neutral or anionic in natural, while cationic biosurfactants consist of amine groups. Generally hydrophobic moiety contains a long-chain hydrophobic moiety (fatty acid) and the hydrophilic moiety is mainly amino acid, carbohydrate, cyclic peptide, phosphate carboxyl acid or alcohol. Usually the molecular mass of biosurfactants is ranges from 500 to 1500 Da [18]. These are classified into two types based on their biochemical characteristics viz. low molecular weight (LMW) and high molecular weight (HMW) biosurfactants. The LMW biosurfactants efficiently reduce the surface as well as interfacial tensions where the HMW biosurfactants act as emulsion stabilizing compounds. On the other hand according to their applications in environmental biotechnology, the biosurfactants are classified into glycolipids, lipopeptides, fatty acids/phospholipids/neutral lipids, polymeric surfactants and particulate surfactants (Table 1) [31, 34, 74, 93, 95, 117].

Among the above biosurfactants, the LMW biosurfactants include glycolipids, lipopeptides, phospholipids and Surface active antibiotics whereas polymeric and

Table 1 Classification of biosurfactants, respective producer microorganisms and applications

Type of biosurfactants	Producer microorganisms	Applications in bioremediation and hydrocarbon contaminated sites
Glycolipids		
Rhamnolipids	*Pseudomonas aeruginosa*, *Pseudomonas chlororaphis*, *Acinetobacter calcoaceticus*, *Serratia rubidaea* SNAU02	Enhancement of degradation, dispersion and emulsification of hydrocarbons; metals removal from contaminated soil
Trehalose lipids	*Rhodococcus erythropolis*, *Arthrobacter sp.*, *Mycobacterium tuberculosis*, *Gordonia amarae*, *Nocardia sp.*, *Corynebacterium sp.*	Improvement of the bioavailability of hydrocarbons
Sophorolipids	*Torulopsis bombicola*, *Torulopsis petrophilum*, *Aspergillus flavus*, *Trichodermaviridis*, *Fusarium sp. S33*, *Trichosporon asahii*, *Mucor mucedo*, *Rhizopus oryzae*	Enhancement of oil recovery and removal of heavy metals from sediments
Lipopeptides		
Surfactin/iturin/fengycin	*Bacillus subtilis*, *Bacillus mojavensis*	Development of the biodegradation of hydrocarbons and chlorinated pesticides; increasing the effectiveness of phytoextraction
Viscosin	*Pseudomonas fluorescens*	Improvement of oil recovery
Lichenysin	*Bacillus licheniformis*	Improvement of oil recovery
Subtilisin, Plipastatin	*Bacillus subtilis*	Improvement of oil recovery
Putisolvin	*Pseudomonas putida*	Improvement of oil recovery
Fatty acids/phospholipids/neutral lipids		

(continued)

Table 1 (continued)

Type of biosurfactants	Producer microorganisms	Applications in bioremediation and hydrocarbon contaminated sites
Phospholipids	*Acinetobacter* sp., *Thiobacillus thiooxidans*, *Rhodococcus erythropolis*	Enhancement of the bacterial tolerance towards heavy metals
Spiculisporic acid	*Penicillium spiculisporum*	Enhancement of removal of the heavy metal ions from ionic liquids; preparation of organogels, superfine microcapsules
Corynomycolic acid	*Corynebacterium lepus*	Improvement of hydrocarbon removal
Polymeric surfactants		
Emulsan	*Acinetobacter calcoaceticus*	Increasing the stability of oil/water emulsions
	Pseudomonas fluorescens	
Alasan	*Acinetobacter radioresistens*	
Biodispersan	*Acinetobacter calcoaceticus* A2	Enhancement of dispersion of limestone in water
Liposan	*Candida lipolytica*, *Candida tropicalis*	Increasing the stability of hydrocarbon-in-water emulsions
Mannoprotein	*Saccharomyces cerevisiae*	
Particulate biosurfactants		
Vesicles	*A. calcoaceticus*	Under investigation
Emulcyan	*Phormidium J-1*	

particulate biosurfactants belong to the HMW biosurfactants [105]. Chemical structures of most used biosurfactants are depicted in Fig. 1.

3 Role of Biosurfactants in MEOR Mechanism

The residual oil saturation (S_{or}) in oil recovery process is very much dependent on the capillary number (N_{cap}), so the oil recovery is also dependent on the number of capillaries from various aspects. The capillary number (N_{cap}) is defined as equation given below (Eq. 1).

$$N_{cap} = \frac{\text{viscous force}}{\text{capillary force}} = \frac{\upsilon\mu}{\sigma cos\theta} \quad (1)$$

where σ is the IFT, υ is the superficial velocity, μ is the velocity of displacing (wetting) fluid and θ is the contact angle.

The IFT plays a significant role on oil recovery process along with other parameters. The change of IFT can be done by injection or production of biosurfactants during MEOR process by indigenous or exogenous metabolisms of different microorganisms [7, 12, 46, 60, 88, 101, 16].

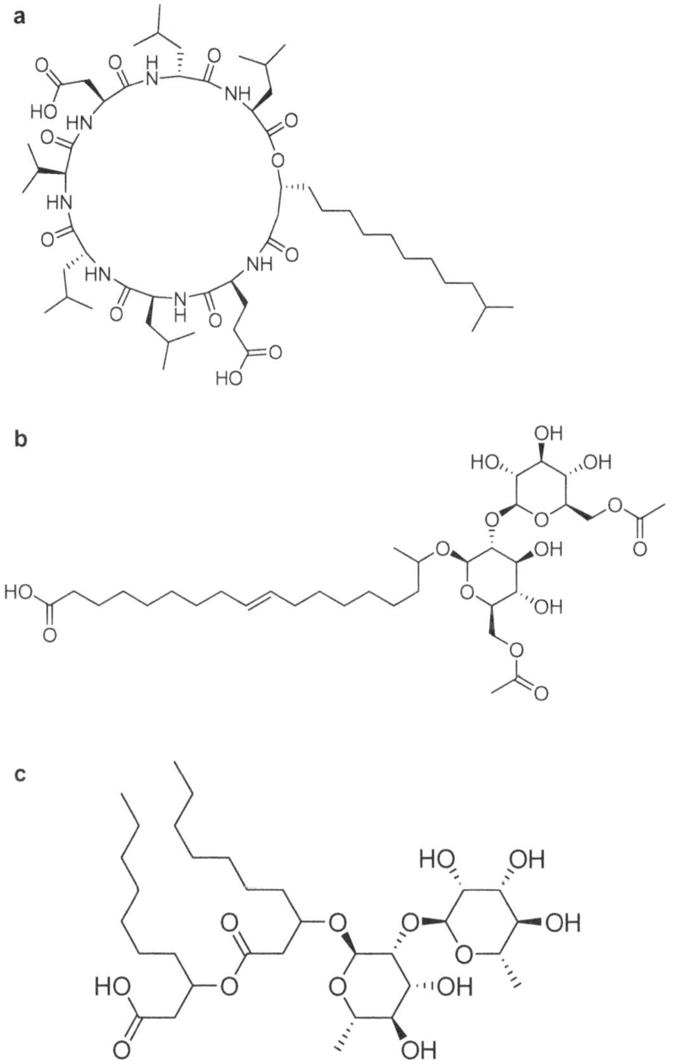

Fig. 1 Chemical structure of most studied biosurfactant—**a** surfactin, **b** sophorolipids, **c** rhamnolipids, and **d** emulsan

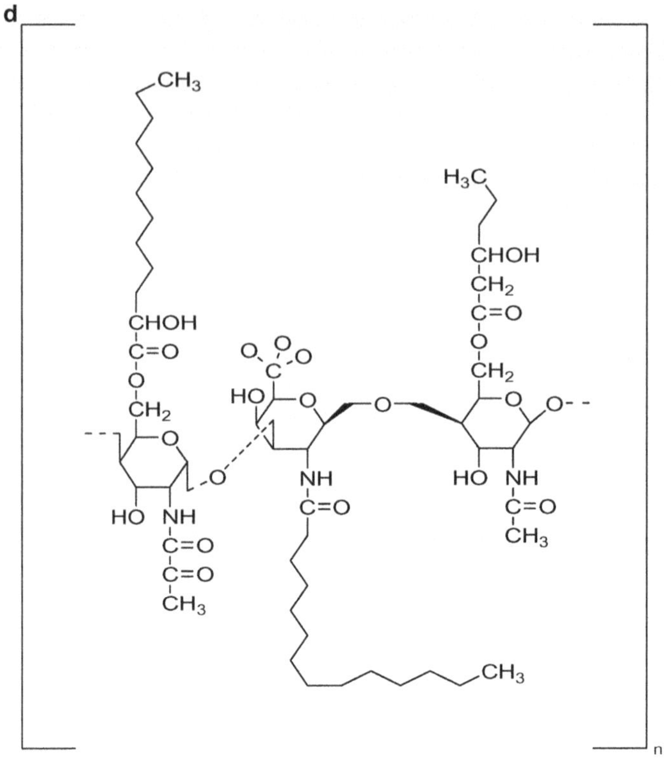

Fig. 1 (continued)

Biosurfactants confer extensive range of properties due to presence of both hydrophilic and hydrophobic moiety, which include lowering of surface and interfacial tension of liquids, creating aggregation of micelles and forming microemulsions between two different phases. The role of biosurfactants in MEOR is based on two fundamental mechanisms. Firstly, these are altering the wettability of reservoir rocks by decreasing the IFT of the oil/water interface and consequently increasing the permeability, driving force, fluidity and sweep efficiency. Hence the trapped oil in reservoir pore moves easily by increasing the displacement efficiency, reservoir pressure, viscosity reduction and mobility control. Therefore employment of biosurfactants in oil recovery process can significantly reduce the capillary forces between oil/water and oil/rock that leads to move hindered oil droplet through rock pores (Fig. 2). The second mechanism is based on the formation of emulsion at oil/water interface, which increases the surrounding area of contact of oil droplets [48, 100] and stabilises the desorbed oil in water and hence allows removal of oil along with the injection water [67, 85].

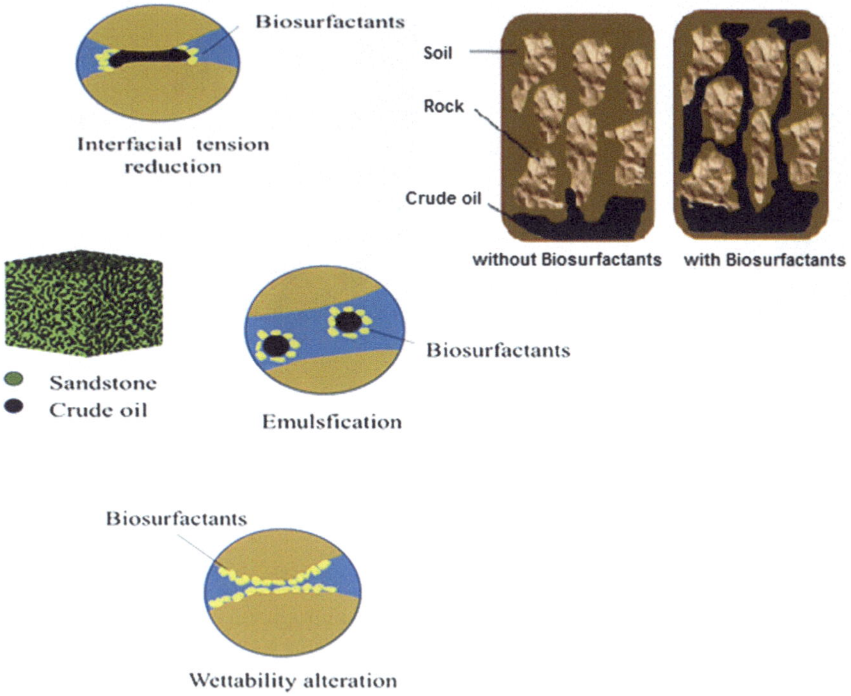

Fig. 2 Schematic illustration of mechanism role of biosurfactants in MEOR (adopted with author permission from Santos et al. [100] and from Niu et al. [81] Copyright (2020), with permission from Elsevier)

Another mechanism of enhanced oil recovery can be explained by two ways based on biodegradation of hydrocarbons by biosurfactants. Firstly the bioavailability of hydrophobic biomass to microbes was increased and the surface tension of the medium around bacterium was reduced, subsequently it caused the reduction of the IFT between oil molecules and bacterium cell wall. On the other hand, the interaction between cell surface and biosurfactant leads to change in the membrane, increase in hydrophobicity by facilitating hydrocarbon adherence and lowering the lipopolysaccharide index of the cell wall without destruction of the membrane. Therefore it blocks the hydrogen bridge formation and allows hydrophobic and hydrophilic interactions, which cause molecular rearrangements and decrease IFT of the liquid by expanding its surface area as well as supporting bioavailability and consequent biodegradability. The action of biosurfactants on contaminated oil droplets by bacterium cell is illustrated in Fig. 3.

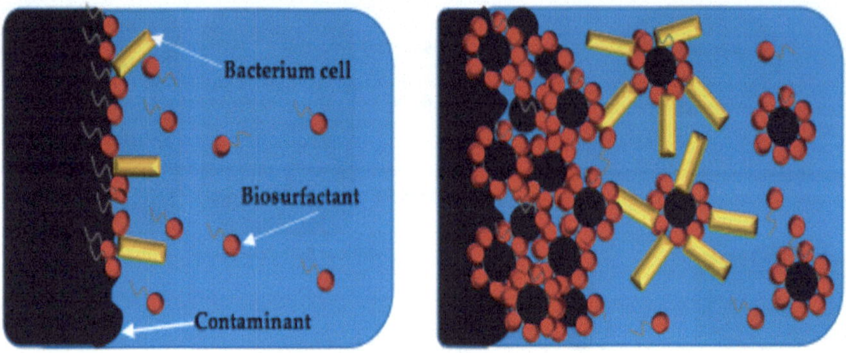

Fig. 3 Illustration of biosurfactant action on petroleum (adopted with author permission from Santos et al. [100])

3.1 Oil/Water Interfacial Tension Reduction by Biosurfactants

The biosurfactant molecules are aligned themselves upright on the air/water interface in aqueous medium with their hydrophilic moiety leaning towards the water. This has the effect to decrease the surface tension and to form emulsion. The concentration above which surfactants start to form micelles in aqueous medium is known as critical micelle concentration (CMC). Micelles of biosurfactant molecules contain a fat loving liophilic center that is why biosurfactants can easily dissolve oils (Fig. 4). The point corresponds to CMC has been indicated by the lowest value of surface tension of a tensioactive reagents in Fig. 5 [25].

The aggregation size, configuration and shape of micelles are dependent on the nature of biosurfactant and number of monomers, e.g. saponins from *S. officinalis* and

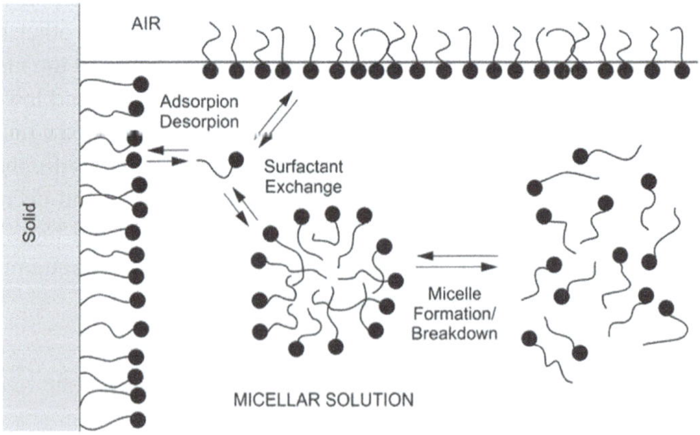

Fig. 4 View of different regions of aqueous surfactant solution and micelle formation

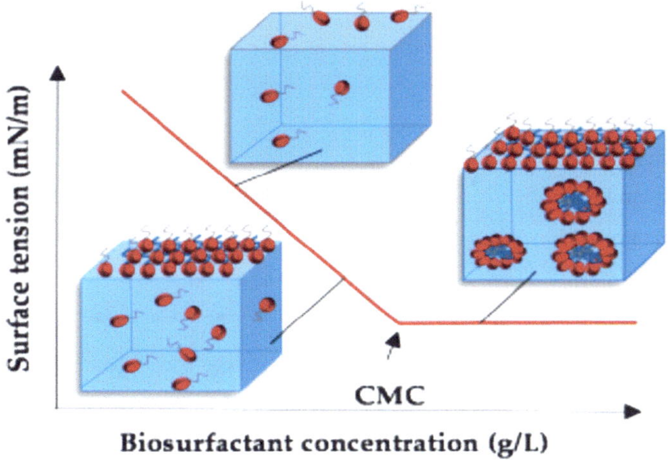

Fig. 5 Change in surface tension with increase in biosurfactant concentrations (adopted with author permission from Santos et al. [100])

soya bean form aggregates of only two molecules, whereas fifty molecules require for the micelles formation by saponins from *Q. saponaria.* It has been found that elongated or smooth filamentous micelles are formed by *Saponaria* and *Q. saponins,* while circular micelles are created by saponins of *G. Max*. These differences are due the structural variation of aglycone, the compound after the glycosyl group of saponins is replaced by a hydrogen atom. Micelles formation is also dependent on other physical and chemical parameters like salt concentration, temperature and pH etc.

The quality of a biosurfactant can be determined by its CMC, the point of maximum drop of surface tension. The CMC values were obtained from the meeting point of surface tension curves by the linear adjustment of before and after the change in surface tension [64]. It has been documented that the rhamnolipid biosurfactant produced from *P. aeruginosa* has ability to decrease the surface tension from 72 to 35.25 mN/m and CMC was found at 127 mg/L as shown in Fig. 6 [27]. The CMC values of biosurfactant containing four homologs produced from the bacteria *P. aeruginosa* are varying from 5 to 200 mg/L [42, 75, 115]. The CMC value of saponins extracted from *Q. saponaria* is equal to 0.5 to 0.8 g/l at 25 °C temperature and decreases with addition of salt concentration [83].

The measurement of contact angle changes and IFT reduction is challenging during microbial growing and surfactant production. In Fig. 7, an experimental arrangement was planned to examine the contact angle and IFT changes through biosurfactant production under emulating reservoir pressure and temperature conditions [87]. The wettability alternation has monitored and viewed through two quartz crystal windows. The IFT was measured by pendant drop method (Fig. 7a), whereas the sessile drop method was used to determine the contact angle (Fig. 7b). A high-pressure syringe pump was fitted to create pressure between the view cell and

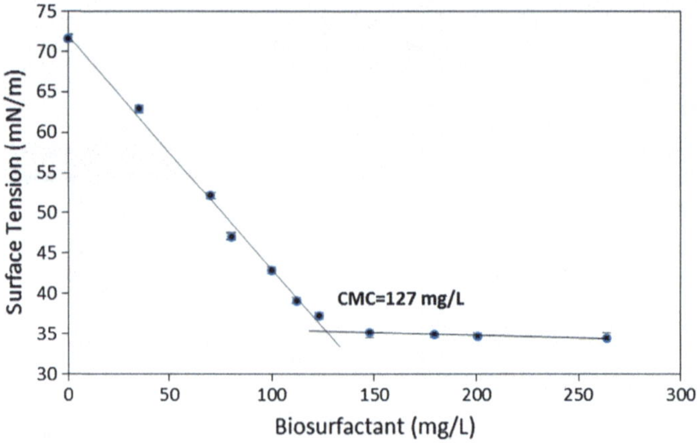

Fig. 6 Determination of CMC of rhamnolipid (adopted from Câmara et al. [27])

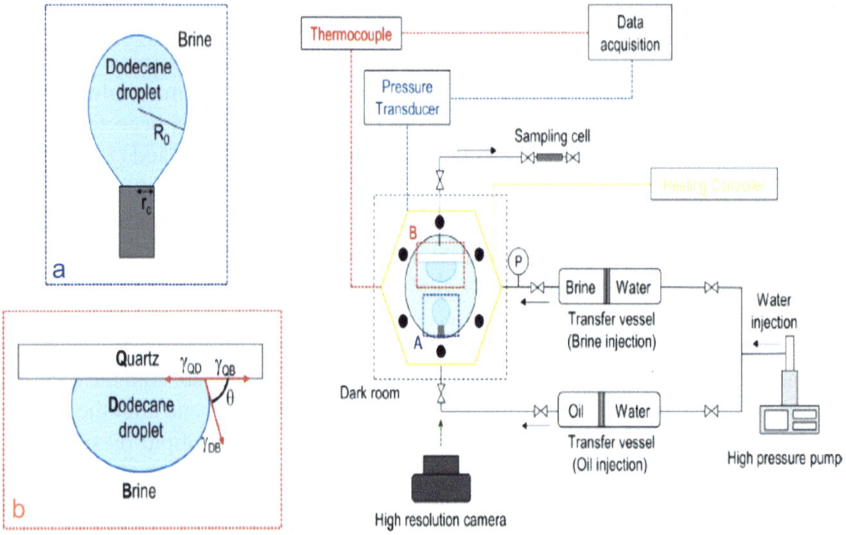

Fig. 7 Illustration of the experimental arrangement for measurement of **a** IFT and **b** contact angle during microbial growth (adopted with author permission from Park et al. [87])

transfer vessels for fluid injection. An electric heater was inserted inside the cell for controlling the temperature. To capture the droplet images, a high resolution digital-single-lens-reflex (DSLR) camera was used.

The effectiveness on enhanced oil recovery of biosurfactants has been examined by calculation of the capillary pressure against the residual oil saturation. This was investigated by measurement of contact angle and IFT changes on dodecane-brine-quartz system during the production of surfactin from *B. subtilis*. Figure 8 highlights

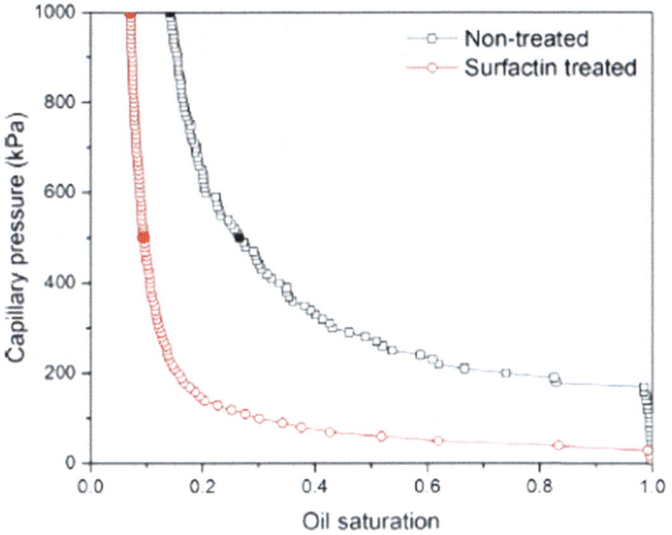

Fig. 8 Capillary pressure versus residual oil saturation before and after surfactin addition (adopted with author permission from Park et al. [87])

that the residual oil saturation was reduced under a certain capillary pressure for surfactin, which causes the alteration of interfacial properties. Due to lower capillary factor (defined as $\gamma \cos \theta$), lowering of residual oil saturation and alteration of interfacial properties, more trapped reservoir oil can be easily displaced or swept by water injection. It has been found that surfactin has the ability to decrease the capillary factor from 33 to 7.5 mN/m at 500 kPa water injection pressure and the residual oil saturation was reduced from 0.27 to 0.09 after modification of water/oil/rock interface.

The wettability alteration is another significant mechanism for oil recovery from reservoirs [5]. The wettability alteration is measured by contact angle measurements of a droplet of water on the reservoir cores by biosurfactants treatment. Figure 9 show that initially a water droplet forms 71.6° of contact angle on a cleaned core surface (Fig. 9a) and it changes to 45° on surface of the biosurfactants treated core (Fig. 9b). This indicates the biosurfactants have the capacity to alter the wettability of reservoir rocks.

4 Emulsification Index by Biosurfactants

Besides lowering of the surface tension of aqueous medium, surfactant is also able to form the emulsions between oil and water, hence increase the solubility of hydrocarbons. Generally, the reduction of surface tension has been benchmarked for a effective that below is 35 mN/m, but several investigation have been reported that

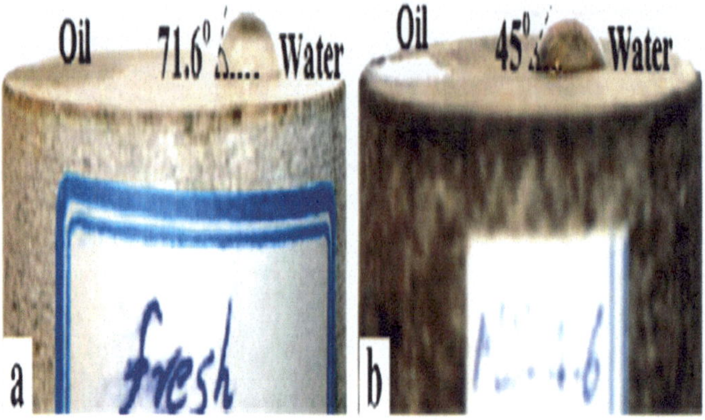

Fig. 9 Change of contact angle of a water droplet on the **a** cleaned and **b** biosurfactant treated dolomite core surface (adopted from Ghojavand et al. [39] Copyright (2020), with permission from Elsevier)

some biosurfactants have high emulsification capacity with hydrophobic compounds whose medium surface tension were above 35 mN/m. The stability of emulsion is very important for environmental application of biosurfactants.

The emulsion efficiency of biosurfactants can be evaluated through its emulsification index (E_{24}). The emulsification index (E_{24}) of biosurfactant is generally determined by adding a small amount of oil (crude petroleum) and added to cell-free culture broth in a graduated tube. The mixture is then shaken with a vortex for 2 min and kept for 24 h to determine the emulsion stability. Therefore E_{24} can be measured by following equation [123]:

$$E_{24} = \frac{H_a}{H_b} \times 100\% \tag{2}$$

[H_a = height of the emulsion layer and H_b = total height of the mixture].

The emulsion stability is dependent on the biosurfactant concentrations. For example, E_{24} of the rhamnolipid was reported for concentration ranges from 50 to 260 mg/L, at 30 °C in Fig. 10, where the maximum value of E_{24} was found as ~69% for crude oil [27]. This is consequence with others reported as ~85% [10] and ~68% [9]. E_{24} of biosurfactants produced by six microorganisms (S1, S2, S3, S4, S5 and S6) separated from oilfield wastewater in Daqing oil field was also measured after 72 h fermentation with various hydrocarbons such as xylene, n-pentane, kerosene and crude oil [62]. The results are shown in Fig. 11.

The oil–water emulsion formation reduces the interfacial energy and adsorbs in reservoir rocks which assists to increase the oil recovery [29]. It has been found from Fig. 6 that the emulsifying capacity of ~125 mg/L of rhamnolipid has a tendency to stabilize and attain at maximum value 69% for 260 mg/L concentration. Hence,

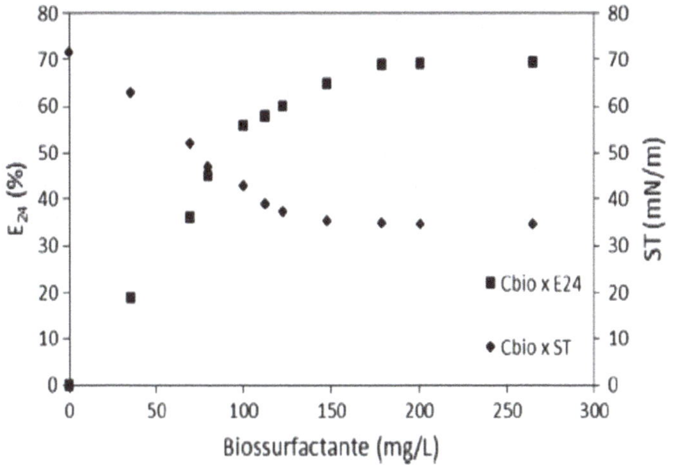

Fig. 10 Effect of biosurfactant concentrations on E_{24} and surface tension (ST) (adopted from Câmara et al. [27])

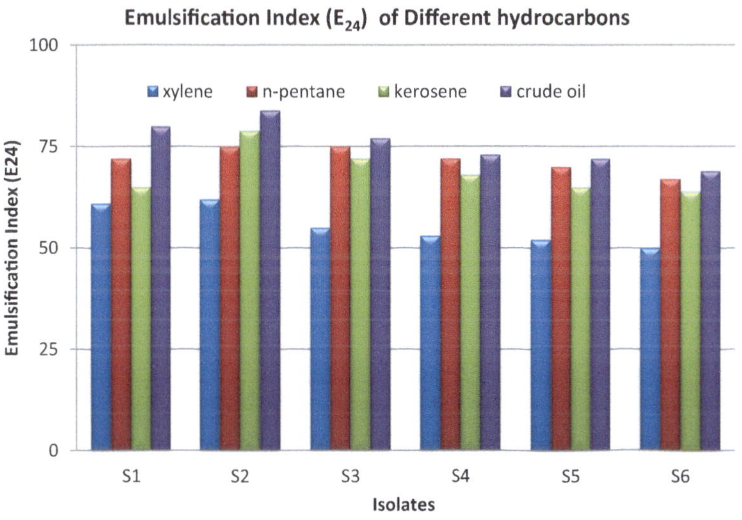

Fig. 11 Emulsification index (E_{24}) of different hydrocarbons by biosurfactant producing microorganisms

the rhamnolipid has a good emulsifying capacity as well as capable of decreasing the surface tension of oil–water interface. These properties of biosurfactants make possible to be suitable reagent for MEOR.

5 Use of Biosurfactants in Enhanced Oil Recovery (EOR)

The primary and key processes that govern the performance of MEOR are transportation of injected microorganisms and nutrients within reservoir, production of biosurfactant and consequently reduction oil/water IFT [2, 101]. It has been proven that the biosurfactants play a major role in MEOR mechanism, which includes both in situ production [125, 126] and ex situ injections [8]. Introduction of effective microorganisms inside the reservoirs is essential for in situ application. However most of the biosurfactant producing microorganisms are aerobic. In case of the in situ biosurfactant production inside the reservoir rock the oxygen-depleted conditions are essential, which may weaken the efficiency of oil recovery. Moreover the oxygen injection in the reservoirs is cost effective, time dependent and uncertain. Therefore anaerobic biosurfactant producing microorganisms are crucial for in situ application. On the other hand in the ex situ biosurfactant flooding, at first the biosurfactants are formed by culture medium in bioreactor and then introduced into the reservoir wells. This process involves effective bioprocessing technology and high cost for transportation and purification [4, 126]. Therefore the in situ biosurfactant flooding is comparatively beneficial technique for MEOR applications. Three main strategies of biosurfactant application in MEOR are shown in Fig. 12 [6].

Surfactin and rhamnolipid are the two extensively studied biosurfactants amongst a large variety of biosurfactants. The most surface active biosurfactant is surfactin which be able to decrease the air/water IFT from 72.8 to 27.9 mN/m and successfully applied in MEOR [89]. The glycolipid biosurfactants have been progressively developed and broadly utilized in MEOR due to its higher yield. For example, glycolipid produced from the bacteria *Pseudomonas sp.* has been widely used in oil recovery

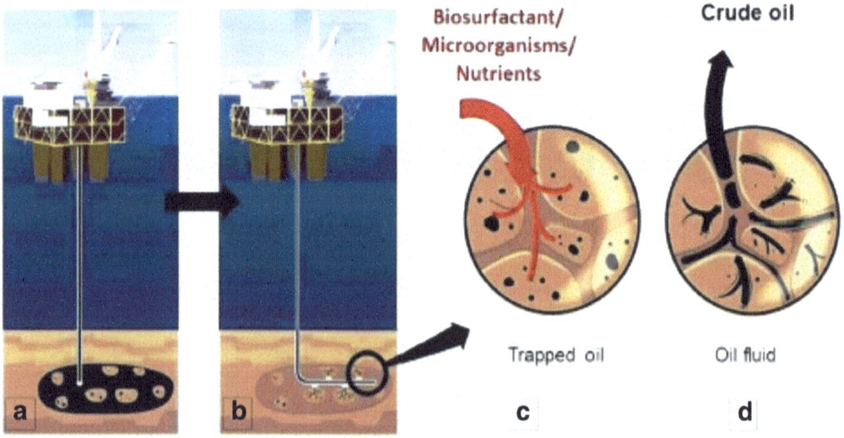

Fig. 12 Process of MEOR using biosurfactants—**a** primary oil recovery, **b** reduction of pressure in oil well, **c** main strategies of biosurfactants application in oil recovery, and **d** oil well pressure restored to facilitate oil extraction (adopted with author permission from [6])

Table 2 Various effective biosurfactants for MEOR application

Microbial source	Biosurfactant	Literatures
Bacillus sp.	Surfactin	[90]
	Rhamnolipid	[40]
	Lichenysin	[15, 19, 70, 124]
Pseudomonas sp.	Rhamnolipid	[11, 20]
Bombicola	Sophorolipid	[96]
Acinetobacter sp.	Emulsan	[111]
	Alasan	[78]
Rhodococcus sp.	Viscosin	[80, 106]
	Trehaloselipids	[112]

experiments. Other biosurfactants such as lichenysin and lipid emulsan have also been found extremely effective in MEOR. Another biosurfactant *rhamnolipids* produced by *B. subtilis* reduces the IFT and also useful to enhance the oil recovery process [40]. A list of various potential biosurfactants and their sources for MEOR application is given in Table 2.

A lipopeptide biosurfactant surfactin has been prepared in a fermenter under the controlled environment and successfully used in MEOR. Further surfactin production is increasing by developing bioprocessing technology in three stages viz. optimization of nutrition parameters, inoculums age and environmental parameters like temperature, pH, agitation and aeration rate [102]. The produced surfactin was recovered and purified by ultrafiltration method [103]. In situ biosurfactant production and application by microorganisms inside a reservoir constitutes an effective mechanism in MEOR. The *Clostridium sp.* at reservoir temperature conditions and *Bacillus licheniformis* under both pH and temperature are potentially effective in MEOR [13, 107]. The microbial strain *B. subtilis* has the capability to produce in situ biosurfactants and to degrade oil inside the reservoirs. Also, the biosurfactant from *Pseudomonas putida* has the capacity to decrease hydrocarbon/water IFT at different pH environment [50]. Experimentally it has been investigated that in situ treatment of *B. subtilis* with different hydrocarbon mixtures in sand-packs are capable to increase the oil by 13–18% for Arabian Light oil, 16–24% for viscous paraffin, 6–25% for heating oil, and 15–17% for heavy crude oil. The crude biosurfactants can be synthesized by high performance thin layer chromatography (HPTLC) and gel filtration chromatography by *Bacillus circulans,* which increase surface activity and a decrease in critical micelle concentration (CMC) values [72, 73]. Several laboratory experiments have been conducted on sand pack column to investigate dynamics of oil recovery and the effect of biomass and nutrient concentrations. Numerous experimental studies have been performed for the potential of in situ biosurfactant production using selected microorganisms and significant amounts of residual oil have been recovered from mature reservoirs. The percentages of oil recovery along with the emulsification index (E_{24}) are represented in Fig. 13 to illustrate the efficiency of biosurfactants producing microorganisms [101]. The total recovery factor was found as 50.45 ±

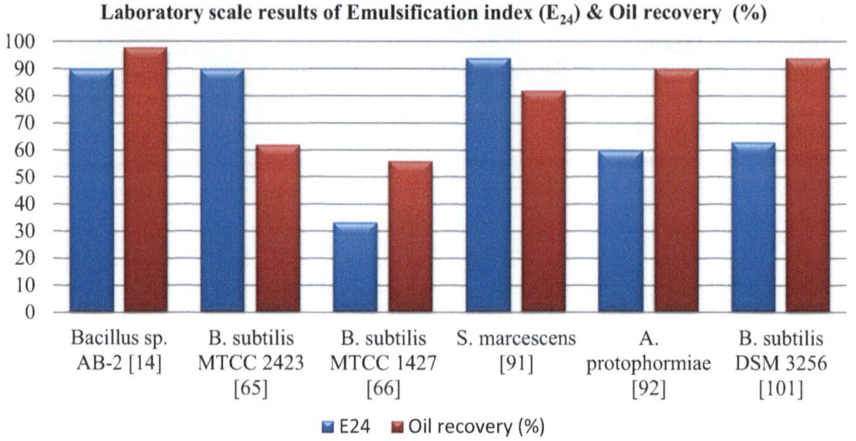

Fig. 13 Results of laboratory scale experimental results of emulsification index (E_{24}) and oil recovery (%)

0.79% by rhamnolipid from *P. aeruginosa*, of which recovery correspond to MEOR was evaluated as 11.91 ± 0.39% (Fig. 14).

One of the most important limitations of biosurfactants application in MEOR is its high production costs, where nearly half of the total expenses account for the matrix costs and separation costs. However utilization of biosurfactants in oil recovery is successful because of low purity is needed, not required successive separation and purification procedure. Recently various sugarcane producing countries like Brazil,

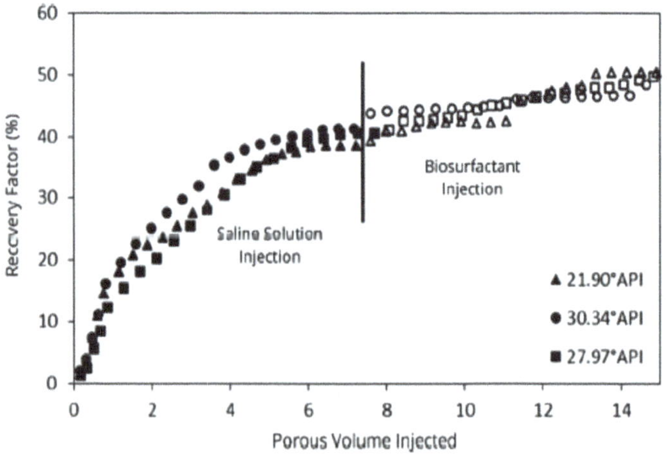

Fig. 14 Recovery factor versus porous volume of biosurfactant solution injected (adopted from Câmara et al. [21])

China, and India have employed cheaper renewable agricultural residues as carbon resources in biosurfactant matrix which can decrease the cost matrix [1, 38].

6 Field Applications

The idea that microbes could able to remove oil from porous media is not proposed recently, though biosurfactants effects in MEOR have been extensively studied and developed in recent years. Beckman was proposed the concept of MEOR in the year of 1926. In year 1946, for the first time Zobell has proposed a series of mechanism of MEOR, which included reduction of oil–water interfacial tension, increase in rock permeability and reduction of viscosity of crude oil by microbial strains [128, 129]. First field test of MEOR was conducted by Yarbrough in the year 1954 [22]. In the 1970s, MEOR technology became one of promising and economically-feasible scientifically confirmed tertiary oil recovery process. Various oil producing nations like China, U.S., Canada, Russia, Australia, Romania, Poland, Hungary, Czech Republic, Great Britain, Germany, Norway, and Bulgaria have supported and promoted numerous research projects and field trials. The laboratory studies have been carried out since the beginning of 1980s to give various references for consequent field applications to connect the comparability between laboratory data and field trials. The suitable strain, strain concentration, nutrient compositions, treatment periods etc. are the important screening parameters for the field trials. A numbers of filed applications have been trialed around the worldwide in past few decades and achieved varying degree of success [52, 101, 104]. It has been found that greater than 90% of field tests have specified by encouraging results to enhance the oil recovery in MEOR technology.

The United States has conducted field trials in the Lisbon oil fields, Arkansas involving in situ biosurfactants production with injection of *Clostridium acetobutylicum* [97]. The in situ production of biosurfactants was also investigated in the Bebee field in Oklahoma using *Bacillus* strain, which confirmed that an effectual enhancement oil recovery required nine times lower biosurfactant concentration than minimum concentration [125]. In several oil fields in Romania, the average oil production was improved about 100% and 200% after MEOR application [97]. The injection of anaerobic fermentation and hydrocarbon degrading microbes in the Piedras Coloradas oil field of Argentina for twelve months has broadly increased oil production by 66% among the six production wells. Similarly, the collective oil production was increased ~27,984 m^3 after nineteen months of field test by the injection of facultative anaerobes and nutrients in the Vizacheres oil field in Argentina. The crude oil production was increased from ~3.6 to ~5.9 tons/day by endogenous MEOR technology in the oil reservoirs at Saskatchewan, Canada [113].

China has widely implemented MEOR field trials in a number of oil reservoirs such as Changqing, Daqing, Jilin, Liaohe, Qinghai, Shengli and Xinjiang. Almost 2500 wells in 23 oil fields together with Jilin, Shengli, Zhongyuan and Daqing have been extensively tested [119]. Among which ~2150 wells were treated with the microbial huff and puff, where crude oil production was improved by ~219,000 tons in Shengli Oilfield and ~64,000 tons in Daqing Oilfield [36]. China is the leader in successful MEOR field application in past decades. The Oil and Natural Gas Corporation (ONGC) Limited, India in association with The Energy and Resources Institute (TERI, New Delhi) and the Institute of Reservoir Studies (IRS), Ahmedabad, has conducted some field trials on the basis of rigorous anaerobic microbes segregated from the oil reservoir. A total of 12 wells in 4 field trials have confirmed three times enhancement of oil recovery [101]. Therefore use of biosurfactants in field tests is one of the sacrificial agents in the MEOR method. The outcome from different field trials have shown that the total cost reduced by 30%, which is indicating that applications of biosurfactants are as good as compatible with other oil recovery methods. Some field experimentations in different oil producing nations are tabulated in Table 3.

7 Conclusion

Use of biosurfactants in MEOR compare to other EOR technologies is the most advantageous and eco-friendly. In situ production and ex situ incorporation of biosurfactants by microorganisms could improve the ultimate oil recovery. Various field applications are reviewed in this chapter which confirmed that biosurfactants successfully enhance the tertiary oil recovery. Moreover, in situ application of biosurfactants is low cost than conventional oil recovery technologies. This operation could make active to the marginal reservoirs, fault blocks in reservoirs, and economically chemical flooding restricted reservoir. It has been concluded that field trials of biosurfactants flooding have positively leads inactive oil fields to recover additional oil the in last two decades. It has also been bring into notice that use of biosurfactants in MEOR could affect the performance of neighboring wells, therefore to get better effectiveness, understanding of reservoir properties, microbial activity in porous medium are essential before field applications. Further research should expansively necessitate on the selection of exact microbial strains, production of novel and high efficient biosurfactants, development of possible microbial products in reuse and cost-effective approaches, and suitable mathematical modeling for optimization and maximization of oil production. Hence the advancement in oil biotechnology research is increasingly gaining recognition and appreciation among new researchers in recent years.

Table 3 Various MEOR field trials globally

Country	Microbial systems	Literatures
Argentina	Hydrocarbon degrading anaerobic-facultative microorganisms	[69, 110]
China	Culture of *Arcobacter*, *Bacillus*, *Bacteroides*, *Eurobacterium*, *Pseudomonas*, *Thauera sp.*	[36, 59, 120, 121]
Canada	Pure culture of emulsifying bacteria *Leuconostoc mesenteroides*	[44, 113]
Germany	*Clostridium sp.*	[118]
Hungary	*Clostridium sp.*	[47]
India	Thermophillic and halophillic anaerobes—*Clostridium* sp., *Thermoanaerobacterium* sp. and *Thermococcus sp.*	[89]
Netherlands	*Bacillus polymyxa*, *Clostridium gelatinosum* and *Betacoccus dextranicus*	[57]
Peru	Petroleum Hydrocarbon-degrading microorganisms—*Mycobacterium*, *Streptococcus*, and *Rhodococcus*	[24, 114, 122]
Romania	Modified mixed developed cultures—Clostridium, Bacillus, Pseudomonas, and other gram-negative rods	[58, 56]
Russia	Sulfate-reducing and fermentative microorganisms—*Clostridium tyrobutiricum*	[43, 79]
USA	Culture of *Bacillus*, *Clostridium acetobutvlicum*, *Leuconostoc mesenteroides*, *Pseudomonas*, Petroleum Hydrocarbon-degrading bacteria, gram-negative rods	Oppenheimer and Heibert [84], [23, 57]

References

1. Al-Bahry SN et al (2013) Biosurfactant production by Bacillus subtilis B20 using date molasses and its possible application in enhanced oil recovery. Int Biodeterior Biodegrad 81:141–146
2. Al-Sulaimani H, Al-Wahaibi Y, Al-Bahry S, Elshafie A, Al-Bemani A, Joshi S (2012) Residual-oil recovery through injection of biosurfactant, chemical surfactant, and mixtures of both under reservoir temperatures: induced-wettability and interfacial-tension effects. SPE Reserv Eval Eng 15(2):210–216
3. Al-Wahaibi Y, Joshi S, Al-Bahry S et al (2014) Biosurfactant production by *Bacillus subtilis* B30 and its application in enhancing oil recovery. Colloids Surf B Biointerfaces 114:324–333
4. Albino JD, Nambi IM (2010) Partial characterization of biosurfactants produced under anaerobic conditions by *Pseudomonas* sp. ANBIOSURF-1. Adv Mater Res 93:623–626
5. Alkan H et al (2019) Investigation of spontaneous imbibition induced by wettability alteration as a recovery mechanism in microbial enhanced oil recovery. J Petrol Sci Eng 182
6. De Almeida DG, Soares Da Silva RCF, Luna JM, Rufino RD, Santos VA, Banat IM, Sarubbo LA (2016) Biosurfactants: promising molecules for petroleum biotechnology advances. Front Microbiol 7:1718
7. Alvarez V, Jurelevicius D, Marques J et al (2015) Bacillus amyloliquefaciens TSBSO 3.8, a biosurfactant-producing strain with biotechnological potential for microbial enhanced oil recovery. Colloids Surf B Biointerfaces 136:14–21
8. Amani H, Müller MM, Syldatk C, Hausmann R (2013) Production of microbial rhamnolipid by *Pseudomonas aeruginosa* MM1011 for ex situ enhanced oil recovery. Appl Biochem Biotechnol 170:1080–1093
9. Amani H, Sarrafzadeh MH, Haghighi M, Mehrnia MR (2010) Comparative study of biosurfactant producing bacteria in MEOR applications. J Petrol Sci Eng 75:209–214
10. Amani H (2015) Study of enhanced oil recovery by rhamnolipids in a homogeneous 2D micromodel. J Pet Sci Eng 128:212–219
11. Arino S, Marchal R, Vandecasteele JP (1996) Identification and production of a rhamnolipidic biosurfactant by a Pseudomonas species. Appl Microbiol Biotechnol 45:162–168
12. Armstrong RT, Wildenschild D, Bay BK (2015) The effect of pore morphology on microbial enhanced oil recovery. J Petrol Sci Eng 130:16–25
13. Arora P, Ranade DR, Dhakephalkar PK (2014) Development of a microbial process for the recovery of petroleum oil from depleted reservoirs at 91–96°C. Bioresour Technol 165:274–278
14. Banat IM (1993) Isolation of thermophilic biosurfactant producing Bacillus sp. Biotechnol Lett 15:591–594
15. Banat IM (1995) Biosurfactant production and possible uses in microbial enhanced oil recovery and oil pollution remediation: a review. Bioresour Technol 51:1–12
16. Banat IM, Franzetti A, Gandolfi I et al. (2010) Microbial biosurfactants production, applications and future potential. Appl Microbiol Biotechnol 87:427–444
17. Bhardwaj G, Cameotra SS, Chopra HK (2013) Biosurfactants from fungi: a review. J Petrol Environ Biotechnol 4:1–6
18. Bognolo G (1999) Biosurfactants as emulsifying agents for hydrocarbons. Colloids Surf, A Physicochem Eng Asp 152(1–20):41–52
19. Bonmatin JM., Laprevote O, Peypoux F (2003) Diversity among microbial cyclic lipopeptides: iturins and surfactins. Activity-structure relationships to design new bioactive agents. Comb Chem High Throughput Screen 6(6):541–556
20. Bordoloi NK, Konwar BK (2008) Microbial surfactant-enhanced mineral oil recovery under laboratory conditions. Colloids Surf B Biointerfaces 63(1):73–82
21. Brown LR (2010) Microbial enhanced oil recovery (MEOR). Curr Opin Microbiol 13:316–320
22. Bryant RS, Douglas J (1988) Evaluation of microbial systems in porous media for EOR. SPE Reserv Eng 3(2):489–495

23. Bryant RS, Stepp AK, Bertus KM, Burchfield TE, Dennis M (1994) Microbial enhanced waterflooding field tests. Society of Petroleum Engineers
24. Bybee K (2006) MEOR in northwest Peru. J Petrol Technol 58(1):48–49
25. Campos JM, Stamford TLM, Sarubbo LA, Luna JM, Rufino RD, Banat IM (2013) Microbial biosurfactants as additives for food industries. Biotechnol Prog 29:1097–1108
26. Cerón-Camacho R, Martínez-Palou R, Chávez Gómez B, CuéllarCésar FHB, JorgeAburto JCC (2013) Synergistic effect of alkyl-O-glucoside and -cellobioside biosurfactants as effective emulsifiers of crude oil in water. A proposal for the transport of heavy crude oil by pipeline. Fuel 110:310–317
27. Câmara JMDA, Sousa MASB, Barros Neto EL, Oliveira MCA (2019) Application of rhamnolipid biosurfactant produced by *Pseudomonas aeruginosa* in microbial-enhanced oil recovery (MEOR). J Petrol Explor Prod Technol 9:2333–2341
28. Daghlian Sofla SJ, Sharifi M, Hemmati Sarapardeh A (2016) Toward mechanistic understanding of natural surfactant flooding in enhanced oil recovery processes: the role of salinity, surfactant concentration and rock type. J Mol Liq 222:632–639
29. Darjani S, Koplik J, Pauchard V (2017) Extracting the equation of state of lattice gases from random sequential adsorption simulations by means of the Gibbs adsorption isotherm. Phys Rev E 96(5):052803-1–052803-7
30. Deleu M, Paquot M (2004) From renewable vegetables resources to microorganisms: new trends in surfactants. C R Chim 7(6–7):641–646
31. Desai D, Banat IM (1997) Microbial production of surfactants and their commercial potential. Microbiol Mol Biol Rev 61(1):47–64
32. El-Sheshtawy HS, Aiad I, Osman ME, Abo-ELnasr AA, Kobisy AS (2016) Production of biosurfactants by *Bacillus licheniformis* and *Candida albicans* for application in microbial enhanced oil recovery. Egypt J Petrol 25:293–298
33. Elraies KA, Tan IM (2012) The application of a new polymeric surfactant for chemical EOR. In: Romero-Zerón L (ed) Introduction to enhanced oil recovery (EOR) processes and bioremediation of oil-contaminated sites. InTech, Rijeka, pp 45–70
34. Fenibo EO, Douglas SI, Stanley HO (2019) A review on microbial surfactants: production, classifications, properties and characterization. J Adv Microbiol (3):1–22
35. Fernandes PL, Rodrigues EM, Paiva FR et al (2016) Biosurfactant, solvents and polymer production by *Bacillus subtilis* RI4914 and their application for enhanced oil recovery. Fuel 180:551–557
36. Gao C (2018) Experiences of microbial enhanced oil recovery in Chinese oil fields. J Petrol Sci Eng 166:55–62
37. Gao CH, Zekri A (2011) Applications of microbial-enhanced oil recovery technology in the past decade. Energy Sources Part A Recover Util Environ Eff 33:972–989
38. Geetha SJ, Banat IM, Joshi SJ (2018) Biosurfactants: production and potential applications in microbial enhanced oil recovery (MEOR). Biocatal Agric Biotechnol 14:23–32
39. Ghojavand H, Vahabzadeh F, Shahraki AK (2012) Enhanced oil recovery from low permeability dolomite cores using biosurfactants produced by a Bacillus mojavensis (PTCC 1696) isolated from Masjed-I Soleyman field. J Petrol Sci Eng 81:24–30
40. Gudiña EJ, Rodrigues AI, Alves E et al. (2015) Bioconversion of agro-industrial by-products in rhamnolipids toward applications in enhanced oil recovery and bioremediation. Bioresour Technol 177:87–93
41. Hirata Y, Ryu M, Oda Y, Igarashi K, Nagatsuka A, Furuta T, Sugiura M (2009) Novel characteristics of sophorolipids, yeast glycolipid biosurfactants, as biodegradable low foaming surfactants. J Biosci Bioeng 108:142–146
42. Hörmann B, Müller MM, Syldatk C, Hausmann R (2010) Rhamnolipid production by Burkholderia plantarii DSM 9509T. Eur J Lipid Sci Technol 112:674–680
43. Ibragimov KM, Abdullayeva F, Guseynova NI (2015) Experience of microbial enhanced oil recovery methods at Azerbaijan fields. In: SPE annual Caspian technical conference & exhibition. Society of Petroleum Engineers, Baku, p 9

44. Jackson SC et al (2012) Field implementation of DuPont. In: SPE annual technical conference and exhibition. Society of Petroleum Engineers, San Antonio, p 10
45. Jagtap S, Yavankar S, Pardesi K, Chopade B (2010) Production of bioemulsifier by Acnetobacter species isolated from healthy human skin. Indian J Exp Bio 48(1):70–76
46. Jahanbani Veshareh M, Ganji Azad E, Deihimi T et al (2018) Isolation and screening of Bacillus subtilis MJ01 for MEOR application: biosurfactant characterization, production optimization and wetting effect on carbonate surfaces. J Petrol Explor Prod Technol 9:233–245
47. Jaranyi I (1968) Beszamolo a nagylengyel terzegeben elvegzett koolaj mikrobiologiai Kiserletkrol. M. All. Foldtani Intezet Evi Jelentese A 423–426
48. Joseph PJ, Joseph A (2009) Microbial enhanced separation of oil from petroleum refinery sludge. J Hazard Mater 161:522–525
49. Joshi SJ, Al-Wahaibi YM, Al-Bahry SN, Elshafie AE, Al-Bemani AS, Al-Bahri A, Al-Mandhari MS (2016) Production, characterization, and application of *Bacillus licheniformis* W16 biosurfactant in enhancing oil recovery. Front Microbial 7:1853
50. Kanna R, Gummadi SN, Kumar GS (2014) Production and characterization of biosurfactants by Pseudomonas putida MTCC 2467. J Biol Sci 14:436–445
51. Kaster KM, Hiorth A, Kjeilen-Eilertsen G, Boccadoro K, Lohne A, Berland H, Stavland A, Brakstad OG (2012) Mechanisms involved in microbially enhanced oil recovery. Transp Porous Media 91(1):59–79
52. Ke CY, Lu GM, Li Y, Bin et al (2018) A pilot study on large-scale microbial enhanced oil recovery (MEOR) in Baolige Oilfield. Int Biodeterior Biodegrad 127:247–253
53. Khajepour H, Mahmoodi M, Biria D, Ayatollahi S (2014) Investigation of wettability alteration through relative permeability measurement during MEOR process: a micromodel study. J Petrol Sci Eng 120:10–17
54. Khan MY, Samanta A, Ojha K, Mandal A (2009) Design of Alkaline/surfactant/polymer (ASP) slug and its use in enhanced oil recovery. Petrol Sci Technol 27:1926–1942
55. Kumar A, Mandal A (2017) Synthesis and physiochemical characterization of zwitterionic surfactant for application in enhanced oil recovery. J Mol Liq 243:61–71
56. Lazar I, Petrisor IG, Yen TE (2007) Microbial enhanced oil recovery (MEOR). Petrol Sci Technol 25(11–12):1353–1366
57. Lazar I (1991) Ch. A-1 MEOR field trials carried out over the world during the last 35 years. In: Developments in petroleum science. Elsevier, pp 485–530
58. Lazar I et al (1991) Preliminary results of some recent MEOR field trials in Romania. In: Developments in petroleum science. Elsevier, pp 365–385
59. Le J et al (2014) A field test of activation indigenous microorganism for microbial enhanced oil recovery in reservoir after polymer flooding. Acta Petrol Sin 35(1):99–106
60. Le JJ, Wu XL, Wang R et al (2015) Progress in pilot testing of microbial-enhanced oil recovery in the Daqing oilfield of North China. Int Biodeterior Biodegrad 97:188–194
61. Li N, Zhang G, Ge J et al (2012) Ultra-low interfacial tension between heavy oil and betaine-type amphoteric surfactants. J Dispers Sci Technol 33:258–264
62. Liu J, Chen Y, Xu R, Jia Y (2013) Screening and evaluation of biosurfactant producing strains isolated from oilfield wastewater. Indian J Microbiol 53:168–174
63. Liu B, Liu J, Ju M, Li X, Yu Q (2016) Purification and characterization of biosurfactant produced by *Bacillus licheniformis* Y-1 and its application in remediation of petroleum contaminated soil. Mar Pollut Bull 107:46–51
64. Ma KY, Sun MY, Dong W et al (2016) Effects of nutrition optimization strategy on rhamnolipid production in a *Pseudomonas aeruginosa* strain DN1 for bioremediation of crude oil. Biocatal Agric Biotechnol 6:144–151
65. Makkar RS, Cameotra SS (1997) Biosurfactant production by a thermophilic Bacillus subtilis strain. J Ind Microbiol Biotechnol 18:37–42
66. Makkar RS, Cameotra SS (1998) Production of biosurfactant at mesophilic and thermophilic conditions by a strain of Bacillus subtilis. J Ind Microbiol Biotechnol 20:48–52
67. Mao X, Jiang R, Xiao W, Yu J (2015) Use of surfactants for the remediation of contaminated soils: a review. J Hazard Mater 285:419–435

68. Marhaendrajana T, Kurnia R, Irfana D et al (2018) Study to improve an amphoteric sulfonate alkyl ester surfactant by mixing with nonionic surfactant to reduce brine–waxy oil interfacial tension and to increase oil recovery in sandstone reservoir: T-KS field, Indonesia. J Petrol Explor Prod Technol 9:675–683
69. Maure MA, Dietrich FL, Diaz VA, Arganaraz H (1999) Microbial enhanced oil recovery pilot test in Piedras Coloradas field. Society of Petroleum Engineers, Argentina
70. McInerney MJ, Javaheri M, Nagle DP Jr (1990) Properties of the biosurfactants produced by Bacillus licheniformis strain JF-2. J Ind Microbiol 5:95–102
71. Mohan PK, Nakhla G, Yanful EK (2006) Biokinetics of biodegradation of surfactants under aerobic, anoxic and anaerobic conditions. Water Res 40:533–540
72. Mukherjee S, Das P, Sen RK (2009a) Rapid quantification of a microbial surfactant by a simple turbidometric method. J Microbiol Methods 76:38–42
73. Mukherjee S, Das P, Sivapathasekaran C, Sen R (2009b) Antimicrobial biosurfactants from marine Bacillus circulans: extracellular synthesis and purification. Lett Appl Microbiol 48:281–288
74. Muthusamy K, Gopalakrishnan S, Ravi TK, Sivachidambaram P (2008) Biosurfactants: properties, commercial production and application. Curr Sci 94(6):736–747
75. Müller MM, Hörmann B, Syldatk C, Hausmann R (2010) Pseudomonas aeruginosa PAO1 as a model for rhamnolipid production in bioreactor systems. Appl Microbiol Biotechnol 87:167–174
76. Nabipour M, Ayatollahi S, Keshavarz P (2017) Application of different novel and newly designed commercial ionic liquids and surfactants for more oil recovery from an Iranian oil field. J Mol Liq 230:579–588
77. Najafi-Marghmaleki A, Kord S, Hashemi A, Motamedi H (2018) Experimental investigation of efficiency of MEOR process in a carbonate oil reservoir using *Alcaligenes faecalis*: impact of interfacial tension reduction and wettability alteration mechanisms. Fuel 232:27–35
78. Navon-venezia S, Zosim Z, Gottlieb A, Legmann R, Carmeli S, Ron EZ et al (1995) Alasan, a new bioemulsifier from Acinetobacter radioresistens. Appl Environ Microbiol 61:3240–3244
79. Nazina TN et al (2013) Microorganisms of the carbonate petroleum reservoir 302 of the Romashkinskoe oilfield and their biotechnological potential. Microbiology 82(2):190–200
80. Neu T, Härtner T, Poralla K (1990) Surface active properties of viscosin: a peptidolipid antibiotic. Appl Microbiol Biotechnol 32:518–520
81. Niu J, Liu Q, Lv J, Peng B (2020) Review on microbial enhanced oil recovery: mechanisms, modeling and field trials. J Petrol Sci Eng 192:107350
82. Okoliegbe IN, Agarry OO (2012) Application of microbial surfactant (a review). Scholarly J Biotechnol 1:15–23
83. Oleszek W, Hamed A (2010) Saponin-based surfactants. In: Surfactants from renewable resources, Chap 12, pp 239–249
84. Oppenheimer CH, Hiebert FK (1987) Microbial enhanced oil production field tests in Texas. Paper presented at the DOE/NIPER symposium on applications of microorganisms to petroleum technology, Bartlesville, OK, Aug 1987, pp 12–13
85. Pacwa-Płociniczak M, Płaza GA, Piotrowska-Seget Z, Cameotra SS (2011) Environmental applications of biosurfactants: recent advances. Int J Mol Sci 12(1):633–654
86. Pal N, Verma A, Ojha K, Mandal A (2020) Nanoparticle-modified gemini surfactant foams as efficient displacing fluids for enhanced oil recovery. J Mol Liq 310:113193
87. Park T, Jeon MK, Yoon S, Lee KS, Kwon TH (2019) Modification of interfacial tension and wettability in oil-brine-quartz system by in situ bacterial biosurfactant production at reservoir conditions: implications for microbial enhanced oil recovery. Energy Fuels 33:4909–4920
88. Passman FJ, Skovhus TL, Whitby C et al (2013) Prospects for using native and recombinant rhamnolipid producers for microbially enhanced oil recovery. Int Biodeterior Biodegrad 81:133–140
89. Patel J, Borgohain S, Kumar M, Rangarajan V, Somasundaran P, Sen R (2015) Recent developments in microbial enhanced oil recovery. Renew Sustain Energy Rev 52:1539–1558

90. Pereira JFB, Gudiña EJ, Costa R et al (2013) Optimization and characterization of biosurfactant production by Bacillus subtilis isolates towards microbial enhanced oil recovery applications. Fuel 111:259–268
91. Pruthi V, Cameotra SS (1997a) Production, properties of a biosurfactant synthesized by A. protophormiae. An Antarctic strain. World J Microbiol Biotechnol 13:137–139
92. Pruthi V, Cameotra SS (1997b) Production of a biosurfactant exhibiting excellent emulsification and surface activities by S. marcescens. World J Microbiol Biotechnol 13:133–135
93. Purwasena IA, Astuti DI, Syukron M, Amaniyah M, Sugai Y (2019) Stability test of biosurfactant produced by *Bacillus licheniformis* DS1 using experimental design and its application for MEOR. J Petrol Sci Eng 183:106383
94. Rabiei A, Sharifinik M, Niazi A, Hashemi A, Ayatollahi S (2013) Core flooding tests to investigate the effects of IFT reduction and wettability alteration on oil recovery during MEOR process in an Iranian oil reservoir. Appl Microbiol Biotechnol 97(13):5979–5991
95. Rosenberg E, Ron EZ (1999) High- and low-molecular-mass microbial surfactants. Appl Microbiol Biotechnol 52:154–162
96. Saborimanesh N, Mulligan CN (2015) Effect of sophorolipid biosurfactant on oil biodegradation by the natural oil-degrading bacteria on the weathered biodiesel, diesel and light crude oil. J Biorem Biodegrad 6(6):1–8
97. Safdel M, Anbaz MA, Daryasafar A, Jamialahmadi M (2017) Microbial enhanced oil recovery, a critical review on worldwide implemented field trials in different countries. Renew Sustain Energy Rev 74:159–172
98. Samanta A, Bera A, Mandal A, Ojha K (2012) Mobility control and enhanced oil recovery using partially hydrolyzed polyacrylamide (PHPA). J Petrol Explor Prod Technol 2:67–74
99. Samanta A, Mandal A, Sarkar A (2011) Design of alkali-surfactant-polymer flooding for enhanced oil recovery. LAP LAMBERT Academic Publishing GmbH & Co. KG, Germany. ISBN 978-3-8465-4147-0
100. Santos DKF, Rufino RD, Luna JM, Santos VA, Sarubbo LA (2016) Biosurfactants multifunctional biomolecules of the 21st century. Int J Mol Sci 17(3):401
101. Sen R (2008) Biotechnology in petroleum recovery: the microbial EOR. Prog Energy Combust Sci 34(6):714–724
102. Sen R, Swaminathan T (2004) Response surface modeling and optimization to elucidate the effects of inoculum age and size on surfactin production. Biochem Eng J 21:141–148
103. Sen R, Swaminathan T (2005) Characterization of concentration and purification parameters and operating conditions for the small-scale recovery of surfactin. Proc Biochem 40:2953–2958
104. She H, Kong D, Li Y, Hu Z, Guo H (2019) Recent advance of microbial enhanced oil recovery (MEOR) in China. Geofluids 1–16
105. Shekhar S, Sundaramanickam A, Balasubramanian T (2015) Biosurfactant producing microbes and their potential applications: a review. Crit Rev Environ Sci Technol 45:1522–1554
106. Silva RCFS, Almeida DG, Rufino RD, Luna JM, Santos VA, Sarubbo LA (2014) Applications of biosurfactants in the petroleum industry and there mediation of oil spills. Int J Mol Sci 15:12523–12542
107. Sivasankar P, Suresh Kumar G (2014) Numerical modelling of enhanced oil recovery by microbial flooding under non-isothermal conditions. J Petrol Sci Eng 124:161–172
108. Sobrinho HBS, Luna JM, Rufino RD, Porto ALF, Sarubbo LA (2013) Application of biosurfactant from *Candida sphaerica* UCP 0995 in removal of petroleum derivative from soil and sea water. J Life Sci 7:559–569
109. Sobrinho HBS, Luna JM, Rufino RD, Porto ALF, Sarubbo LA (2013) Assessment of toxicity of a biosurfactant from *Candida sphaerica* UCP 0995 cultivated with industrial residues in a bioreactor. Electron J Biotechnol 16(4):1–2
110. Strappa LA, De Lucia JP, Maure MA, Llopiz MLL (2004) A novel and successful MEOR pilot project in a strong water-drive reservoir Vizcacheras field, Argentina. In: SPE/DOE symposium on improved oil recovery. Society of Petroleum Engineers, Tulsa, p 24

111. Suthar H, Hingurao K, Desai A, Nerurkar A (2008) Evaluation of bioemulsifier mediated microbial enhanced oil recovery using sand pack column. J Microbiol Methods 75(2):225–230
112. Tango MSA, Islam MR (2002) Potential of extremophiles for biotechnological and petroleum applications. Energy Sources 24:543–559
113. Town K, Sheehy AJ, Govreau BR (2010) MEOR success in Southern Saskatchewan. SPE-158022-PA 13(5):773–781
114. Tremblay J, Yergeau E, Fortin N, Cobanli S, Elias M, King TL et al (2017) Chemical dispersants enhance the activity of oil-and gas condensate-degrading marine bacteria. ISME J 11:2793–2808
115. Varjani SJ, Upasani VN (2017) Critical review on biosurfactant analysis, purification and characterization using rhamnolipid as a model biosurfactant. Bioresour Technol 232:389–397
116. Veshareh MJ, Azad EG, Deihimi T, Niazi A, Ayatollahi S (2019) Isolation and screening of *Bacillus subtilis* MJ01 for MEOR application: biosurfactant characterization, production optimization and wetting effect on carbonate surfaces. J Petrol Explor Prod Technol 9:233–245
117. Vijayakumar S, Saravanan V (2015) Biosurfactants-types, sources and applications. Res J Microbiol 10:181–192
118. Wagner M, Lungerhansen D, Nowak U, Ziran B (1993) Microbially improved oil recovery from carbonate. Biohydrometal Technol 2:695–710
119. Wang W (2012) Laboratory research and field trials of microbial oil recovery technique. Oil Drilling Product Technol 34(1):107–113
120. Weidong W et al (2014) MEOR field test at block Luo801 of Shengli oil field in China. Petrol Sci Technol 32(6):673–679
121. Xiang T et al (2004) Mechanism of indigenous microbial enhancement of oil recovery and pilot test. Acta Petrol Sin 25(6):63–67
122. Xu X, Liu W, Tian S, Wang W, Qi Q, Jiang P et al (2018) Petroleum hydrocarbon-degrading bacteria for the remediation of oil pollution under aerobic conditions: a perspective analysis. Front Microbiol 9:2885
123. Xu W (2005) Experimental investigation of dynamic interfacial interactions at reservoir conditions. MS thesis, Department of Petroleum Engineering, Louisiana State University
124. Yakimov MM, Amro MM, Bock M, Boseker K, Fredrickson HL, Kessel DG et al (1997) The potential of Bacillus licheniformis strains for in situ enhanced oil recovery. J Petrol Sci Eng 18:147–160
125. Youssef N, Simpson DR, Duncan KE, McInerney MJ, Folmsbee M, Fincher T, Knapp RM (2007) *In situ* biosurfactant production by *Bacillus* strains injected into a limestone petroleum reservoir. Appl Environ Microbiol 73:1239–1247
126. Youssef N, Simpson DR, McInerney MJ, Duncan KE (2013) *In-situ* lipopeptide biosurfactant production by *Bacillus strains* correlates with improved oil recovery in two oil wells approaching their economic limit of production. Int Biodeterior Biodegrad 81:127–132
127. Zhang QQ, Cai BX, Xu WJ et al (2015) Novel zwitterionic surfactant derived from castor oil and its performance evaluation for oil recovery. Colloids Surf A Physicochem Eng Asp 483:87–95
128. Zobell CE (1946) Action of microorganisms on hydrocarbons. Bacteriol Rev 10(1–2):1–49
129. Zobell CE (1947) Marine bacteriology. Annu Rev Biochem 16:565–586

Novel Surfactants for Enhanced Oil Recovery

Satyajit Chowdhury, Saket Srivastava, Srawanti Medhi, and Jitendra S. Sangwai

Abstract After the primary and secondary methods of hydrocarbon production, the reservoir depletes and often relies on enhanced oil recovery techniques (EOR, a tertiary method) to reduce the residual oil saturation (S_{or}) to a minimum value. Amongst many methods, the chemical EOR (cEOR) technique of oil recovery is widely implemented. The cEOR technique aims to optimize mobility ratio and reduce interfacial tension (IFT) and the viscosity of in situ oil. A subclass of cEOR is surfactant flooding which uses the principle of IFT reduction to facilitate additional oil gain. This process incorporates the use of surfactants which promotes favorable wettability and forms in situ oil–water (o/w) emulsions driving the oil towards the producing well. This chapter provides insight into different classes and types of surfactants used in cEOR methods. The discussions on different surfactants are carried out broadly in terms of its effect on IFT, temperature stability, adsorption on rock matrix and effectiveness in saline environments.

Keywords Chemical enhanced oil recovery · Recovery mechanism · Surfactant types

1 Introduction

The primary mechanism to recover oil from a reservoir is dependent on the intrinsic pressure of a reservoir. After a certain production limit, the reservoir flow declines and further aid is necessary to maintain a prescribed level of flow rate. Hereafter,

S. Chowdhury · J. S. Sangwai (✉)
Gas Hydrate and Flow Assurance Laboratory, Petroleum Engineering Program, Department of Ocean Engineering, Indian Institute of Technology Madras, Chennai 600036, India
e-mail: jitendrasangwai@iitm.ac.in

S. Chowdhury · S. Medhi
Assam Energy Institute, A center of Rajiv Gandhi Institute of Petroleum Technology, Sivasagar 785697, India

S. Srivastava
Department of Petroleum Engineering and Earth Sciences, University of Petroleum and Energy Studies, Dehradun 248007, India

T. Solling et al. (eds.), *Surfactants in Upstream E&P*, Petroleum Engineering,
https://doi.org/10.1007/978-3-030-70026-3_7

a secondary mechanism is adopted where the immiscible displacement of oil is achieved by waterflooding or the reservoir pressure is maintained by injecting gas [9]. Re-energizing the reservoir by external pressure maintenance mechanism helps to regain better production rates. However, after these efforts turn inadequate a tertiary method is implemented to recover the residual oil, mostly after a waterflood. Enhanced oil recovery (EOR) or tertiary techniques aim to reduce the residual oil saturation (trapped oil) and offer an additional incremental oil gain. A broader classification of different EOR techniques is listed in Fig. 1.

Chemical EOR (cEOR) is one of the most efficient tertiary recovery techniques which increases the recovery factor by improving both macroscopic (volumetric) as well as microscopic displacements. Yielding a favorable mobility ratio contributes to improving volumetric displacement. The objective of polymer EOR is to decrease the viscous fingering (Fig. 2) effect by reducing the mobility ratio between displacing and displaced fluid [38].

Pore-scale displacement is associated with micro displacement. Dominant capillary forces do not allow entire oil in contact with the injected fluid to be produced. Hence, a fraction of oil is inherently trapped by capillary forces. A dimensionless number known as capillary number (ratio of inertial by capillary force) is often used to characterize the efficiency of oil recovery from a pore volume. Capillary number (Ca) is expressed as:

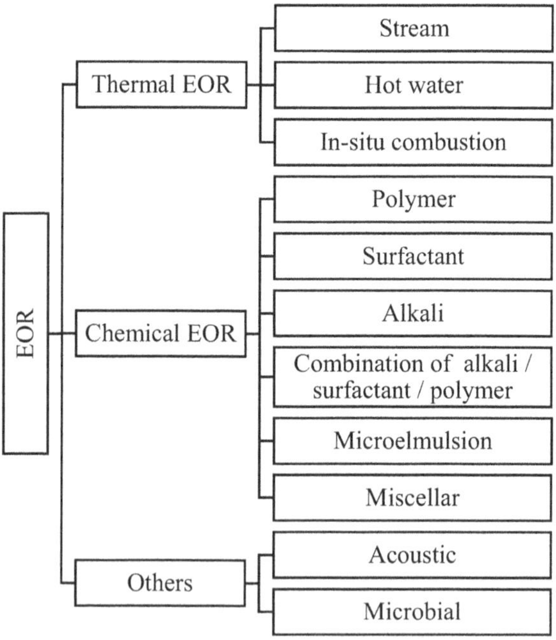

Fig. 1 Broad classification of EOR methods

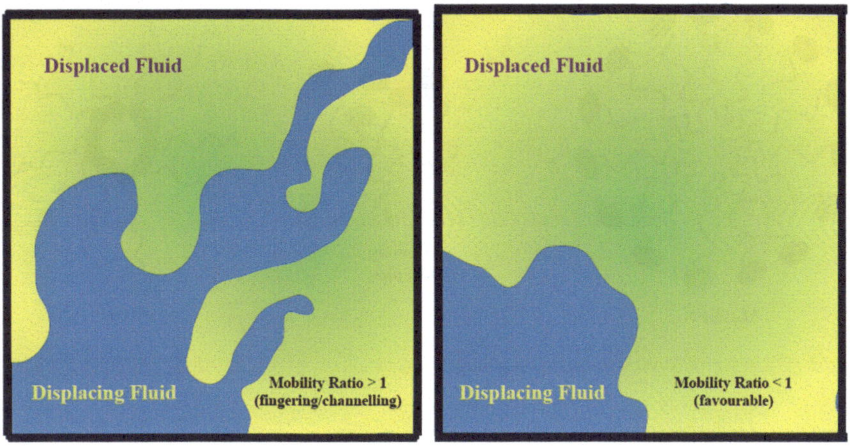

Fig. 2 Viscous fingering effect due to varied mobility ratio

$$Ca = \frac{Viscous\,forces}{Surface\ or\ interfacial\,forces} = \frac{\mu V}{\gamma \cos \theta} \tag{1}$$

where, μ is the viscosity of the injected aqueous phase, V is the velocity and γ is the oil/water interfacial tension (IFT). *Ca* can be increased by reducing the magnitude of interfacial tension between oil and water. Higher *Ca* minimizes the residual oil saturation trapped inside the pore volume. During water flooding operation *Ca* is typically in a range of 10^{-7} to 10^{-6} which results to a significant amount of residual oil saturation. An increase in *Ca* to 10^{-4} to 10^{-3} reduces the trapped oil by 90% [13]. Residual oil saturation ideally reaches a value of zero if *Ca* is increased to 10^{-2} which can be achieved by lowering of IFT between oil and aqueous phases, alteration of wettability and interfacial rheological properties [21, 42]. Both of these phenomena can be accomplished by the use of surfactants along with the displacing fluid (brine). Surface-active-agents or surfactants are fundamentally organic compounds that have a solvent loving and solvent fearing group known as hydrophilic and hydrophobic groups respectively in case the solvent is water [33]. The addition of surfactants brings down the IFT to lower values and weakens the capillary forces within the pore network [10]. Surfactant on mixing with aqueous solution tends to adsorb between the interface of oil and water and reduces the free energy of the given solution (Fig. 3). With the eventual increase in the concentration of surfactant, micelles are formed which are highly dynamic. Critical micelle concentration (CMC is the minimum concentration of surfactant required to initiate the formation of micelles; which is one of the most vital properties of a surfactant.

The solubility of a surfactant is a function of temperature. The minimum temperature beyond which micelles are formed is called a Krafft point which is usually dictated by the melting point of the hydrocarbon tail [29]. It is important to note that due to the immobile hydrocarbon tail below the Krafft point, the solubility of the surfactant is significantly less and bears no CMC. In that case, the solution

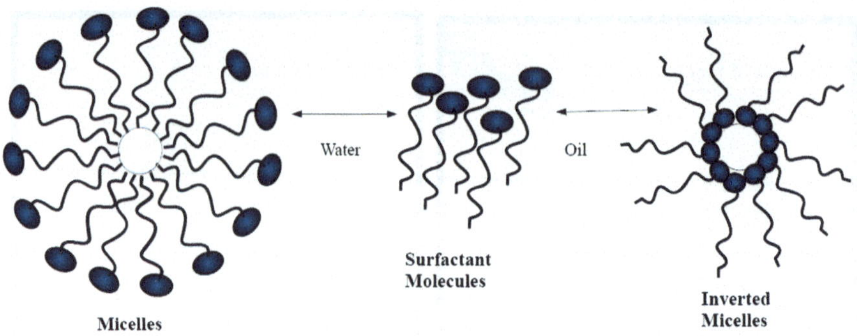

Fig. 3 Oil in water (o/w) emulsion (micelles) and water in oil (w/o) emulsion (inverted micelles)

appears cloudy and phase separation becomes unavoidable [22]. The mark of CMC also affects the physical properties of an oil–water solution with a surfactant, e.g.: trend of conductance of a surfactant solution turns non-linear beyond CMC [44]. Furthermore, surface tension decreases with an increase in surfactant concentration till CMC, past which its value remains constant [5]. Hydrophile-lipophile balance (HLB) is another parameter to profile surfactants. It specifies the inclination of a surfactant to form oil in water (o/w) or water in oil (w/o) emulsion. The value of HLB varies from 0 to 20, where 0 corresponds to absolute hydrophobic behavior, and 20 to complete hydrophilic molecule of surfactant. Furthermore, lipophilic surfactants are characterized by a low HLB value typically less than 9 whereas, a value greater than 11 signifies that the surfactant is hydrophilic [35].

1.1 General Overview on the Effect of Surfactant on IFT and Wettability Alteration

The natural existing oil reservoirs can be classified as oil-wet, water wet and mixed wet based on the contact angle measurements (typically measure for water droplet). On oil-wet conditions, this contact angle (of water) is greater than 90°, while a contact angle less than 90° exhibits water wet conditions. In oil-wet reservoirs, the matrix retains the oil in micropores due to the presence of capillarity resulting in the trapping of oil. The water-wet reservoirs, on the other hand, get easily filled with water in microchannels which leaves the oil phase to occupy more permeable channels with larger pore throat size. Hence, the water wet reservoirs support spontaneous imbibition allowing easy flow of oil towards the producing well. Production is difficult to achieve from a reservoir bearing oil-wet conditions owing to which there lies a constant attempt to change its wettability to water-wet conditions. This process is known as wettability alteration.

The interfacial tension of the formed droplet during the liquid–liquid interaction at the time of surfactant injection serves as an important factor for the stability of the emulsion. The value of the interfacial tension declines faster in the case of droplets in the size of microscale as compared to milliscale droplets. Smaller droplet sizes lead to a smaller diffusion boundary layer thickness, which reduces the time required for the surfactant molecule to diffuse into the depletion region away from the aggregated bulk. The surfactant in the outer phase catalyzes the reduction of the interfacial tension, but the decaying rate of the interfacial tension is independent of the phase to which surfactant is added for milliscale droplets. The stability of the emulsion is also deeply affected by the surface gradient of the fluid–fluid interface which infers the response of the interface deformation. It can also be linked to the probability of the droplet coalescence. Surfactants, when used in EOR techniques, lower the IFT between oil–water systems and mobilizes the oil to form oil pools or banks. It primarily helps to reduce the capillary forces and allows the oil to gain mobility and avoid retrapping. Mechanism of emulsion formation and its associated behavior with pore throat size are described in Figs. 4, 5 and 6.

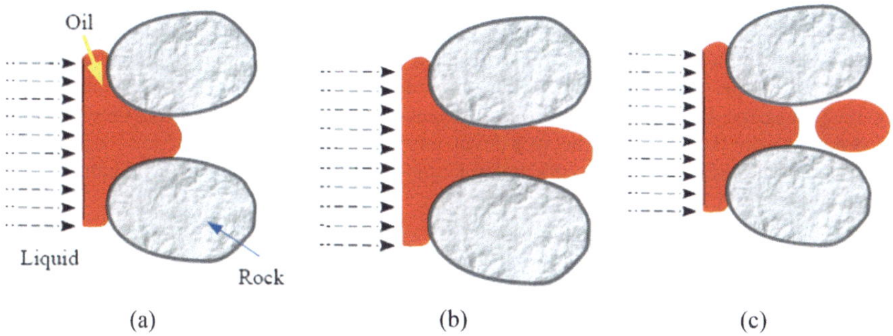

Fig. 4 Snapping off mechanism of emulsion formation: **a** residual oil trapped inside rock grains, **b** trapped oil elongates as it comes in contact with the surfactants present in the displacing fluid or injected fluid, **c** oil gets snapped off and forms emulsion. Adapted with permission from [46] as per open access policy. Copyright 2019 Society of Chemical Industry and John Wiley & Sons Ltd.

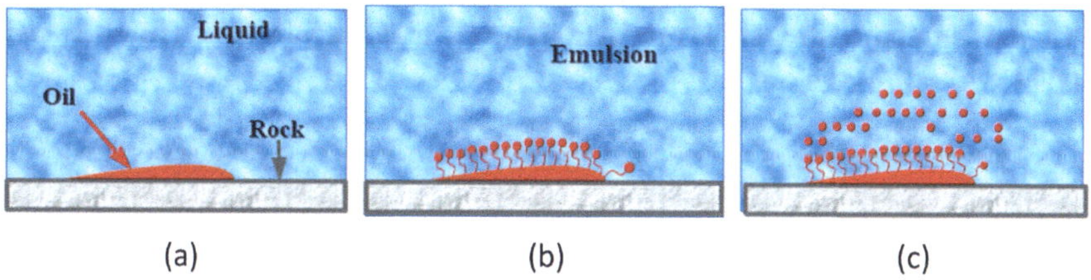

Fig. 5 Shear based emulsion formation: **a** oil film trapped as residual, **b** emulsion droplet formation by shearing action of liquid (displacing fluid) which has emulsifying properties due to the presence of surfactants, **c** oil–water emulsion formed due to shearing mechanism, gets dispersed in the liquid phase and additional emulsions being formed reducing the oil film. Adapted with permission from [46] as per open access policy. Copyright 2019 Society of Chemical Industry and John Wiley & Sons Ltd.

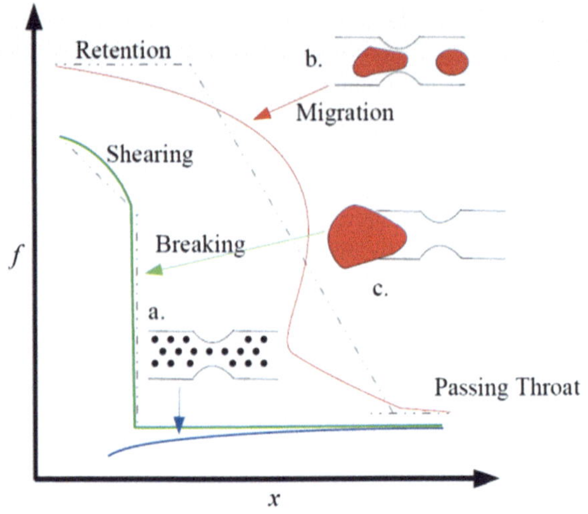

Fig. 6 The resistance factor 'f' for different droplet size emulsions: **a** the emulsion droplet size is far smaller than the throat diameter, **b** the emulsion droplet size is slightly larger than the throat diameter but smaller than the pore diameter, **c** the emulsion droplet size is significantly larger than the pore throat diameter. 'f' is the resistance factor and is defined as the ratio of the pressure difference between the inlet and outlet of a throat, and x is the length of pore throat. Adapted with permission from [46] as per open access policy. Copyright 2019 Society of Chemical Industry and John Wiley & Sons Ltd.

Crude oil having long hydrocarbon chains typically possess a negative charge which gets adsorbed on the surface of minerals bearing a positive charge. The interaction between the cationic monomer and the anionic groups leads to the formation of ion pairs which are non-soluble in aqueous and soluble in the oleic phase. This ion pair causes desorption at the oil, water and rock interface which facilitate easy entry of water eventually changing the surface to water-wet. Anionic surfactant cannot desorb the negatively charged groups from the reservoir rock surface. But they form water-wet bilayer between the surface of the rock and oil, which generate weak capillary forces through hydrophobic interaction and facilitate easy incursion of water and expulsion of oil.

2 Types of Surfactants and Their Role in EOR

The nature of the surfactant is characterized by the hydrophilic group which can be cationic, anionic, non-ionic, or amphoteric. These hydrophilic groups are positively and negatively charged in the case of cationic and anionic surfactants, respectively. Non-ionic surfactants do not ionize and are rather soluble in water through hydrogen bonding [45]. Amphoteric surfactants have a hydrophilic group that bears both negative and positive charges.

2.1 Cationic Surfactants

Cationic surfactants are those which comprise a positive hydrophilic head in combination with a halide group. Due to its positive change, it is habitually attracted to negative charges. Hence, its application is limited. Clays being one of the most abundant and dominant compositions of reservoir rocks are negatively charged. Upon interaction with a cationic surfactant, it adsorbs these surfactants into its matrix exposing the hydrophobic part to the hydrocarbon fluid. Additionally, by the means of these cationic surfactants, a water-wet reservoir can turn into an oil-wet. Hence, cationic surfactants are very effective to alter reservoir rock wettability. However, the cationic surfactant has found wide applications in carbonate formation as they have relatively small adsorption in calcite minerals. Studies have found that oil recovery from limestone (a carbonate rock) is higher subjected to spontaneous imbibition (Fig. 7). It is generally observed that oil recovery is significantly less for anionic surfactants such as sodium dodecyl sulfate (SDS) from calcite rich cores. As compared to the family of C_nTABs (Fig. 4a), heavy-headed phosphonium surfactants ($C_{10}TPPB$, $C_{12}TPPB$) contribute to higher oil recovery. For C_nTAB, as the number of carbon atom increases, CMC and IFT decreases [19].

The wettability alteration due to surfactant is due to the interaction between rock type, a cationic surfactant, salinity, and crude oil composition. The presence of salt with a cationic surfactant like CnTAB also largely affects the competence of wettability alteration after sometime when it reaches its limiting value of concentration, consequently increasing repulsive forces which accounts for higher contact angle of the droplet. On the contrary, the diffusion of surfactants into the aqueous phase is inhibited at higher salt concentrations owing to which it displays adverse effects on wettability [14, 17]. Furthermore, from a rheological standpoint, low-frequency oscillations show viscoelastic liquids behavior and viscoelastic solid behavior is displayed at higher frequencies [20]. Higher elastic properties imply extensional flow which is useful to sweep oil from trapped pores [2]. Some cationic surfactants like hexadecyl pyridinium chloride monohydrate [$C_{21}H_{38}ClN.H2O$] (CPC) shows higher adsorption in carbonate formation (Fig. 8) especially those rich in silica and aluminum [24]. Nonetheless, due to the high capital expenditure during the synthesis process of cationic surfactants owing to the high-pressure hydrogenation process, their use is limited.

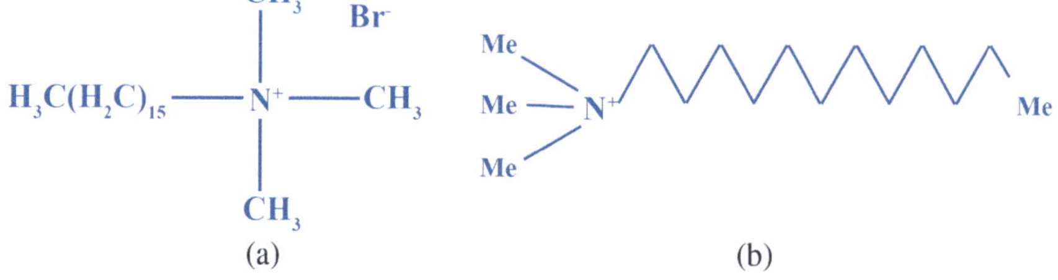

Fig. 7 **a** Structure of cetyl trimethyl ammonium bromide (CTAB), **b** structure of dodecyl trimethyl ammonium (DTAB)

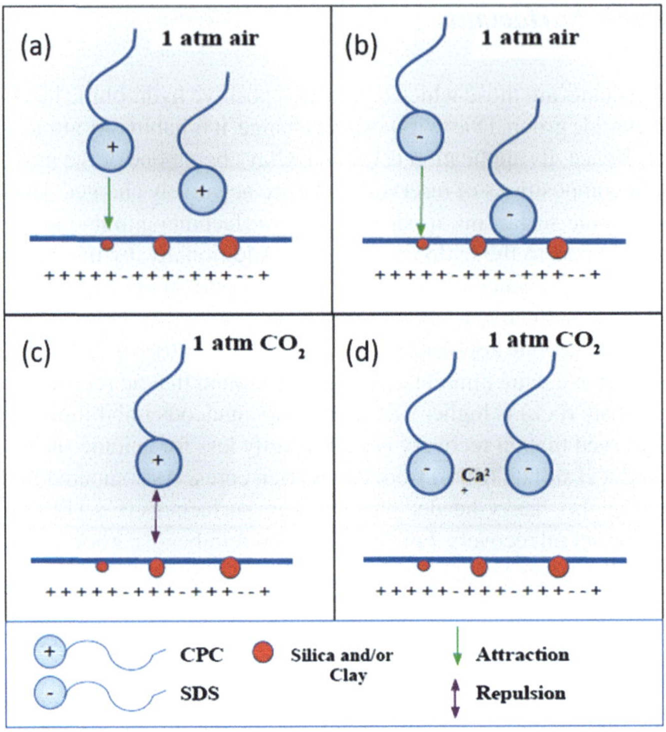

Fig. 8 Mechanisms for CPC/SDS adsorption on natural carbonates in the presence/absence of CO_2: **a** CPC adsorption on natural carbonates in DI water equilibrated with 1 atm air, **b** SDS adsorption on natural carbonates in DI water equilibrated with 1 atm air, **c** CPC adsorption on natural carbonates in DI water equilibrated with 1 atm CO_2, **d** SDS adsorption on natural carbonates in DI water equilibrated with 1 atm CO_2. (CPC: Hexadecylpyridinium chloride monohydrate [$C_{21}H_{38}ClNH_{20}$]). Adapted with permission from [24]. Copyright 2013 Elsevier

2.2 *Anionic Surfactant*

Anionic surfactants are those which comprise of a negative charge on their hydrophilic end. The functional group end contains phosphate, carboxylate, sulfonate, etc. Anionic surfactants are widely used as they are easily available and undergo very low adsorption especially when implemented in sandstone reservoirs (negatively charged sandstone). They are efficiently used to reduce IFT and bears stability under high-temperature conditions [18]. Anionic surfactant dominates the market owing to its low cost of manufacture and highly potent detergency.

Alkyl aryl sulfonates (Fig. 9) were first introduced around the 1930s and soon captured the market. It is produced by the reaction between propylene tetramer and benzene. The higher member of this family of surfactants are hydrophobic and interact well with crude oil. Higher recovery can be attained when used with optimum salinity. On the contrary, these are non-biodegradable. One of the prime benefits

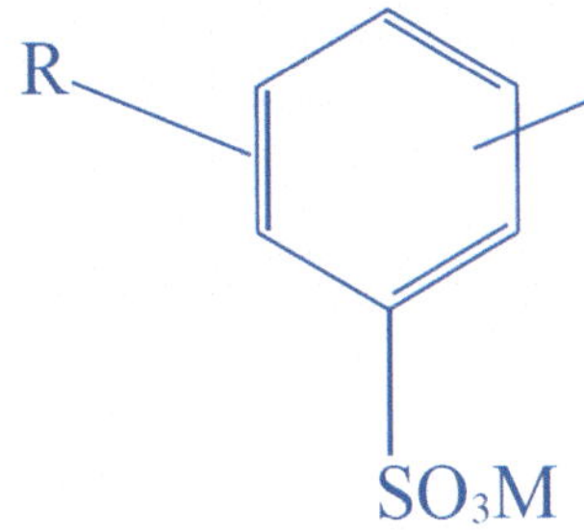

Fig. 9 Structure of alkyl aryl sulfonates

of alkyl aryl sulfonate surfactant is its ability to adapt as per reservoir conditions. Additionally, they are stable for a range of crude oils at different temperatures. However, at higher salt concentrations especially in the presence of divalent salts, this class of anionic surfactants is not stable [12].

Linear alkyl benzene sulfonate (Fig. 10) is a subclass of anionic alkyl aryl sulfonates. Alkyl chains with higher carbon number are not synthesized or used, as increasing the number of carbon atoms slows the biodegradability of the surfactant. The sulfonate group is essential for temperature resistance. Linear alkyl benzene sulfonate is toxic and becomes more poisonous with increasing salinity and hardness of water [30]. On the other hand, increasing the value of the carbon in the alkyl chain reduces the value of CMC [32].

This linear variant of surfactant significantly reduces the value of IFT especially with crudes with a carbon number from 7 to 10. Alcohol propoxy sulfate (APS) is the structure of anionic surfactant consisting of a hydrophobe tail that is a higher carbon branched alcohol along with 7 propylene oxide groups.

The alcohol propoxy sulfate (APS) (Fig. 11) is used alongside internal olefin sulfonate (IOS) to increase the hydrocarbon sweep up to 30% [43]. It is widely used in reservoirs with light oil and low temperature due to the high stability of foam

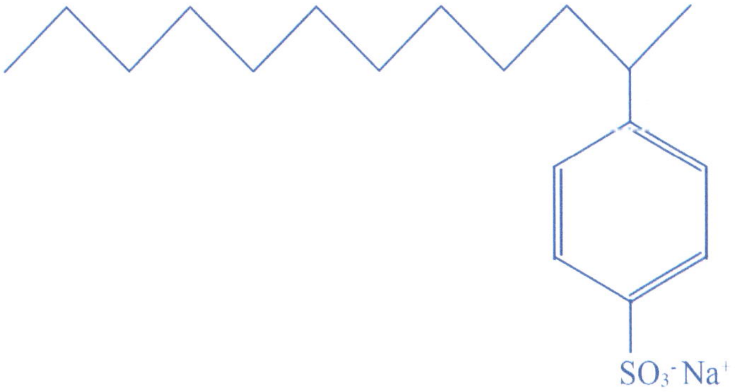

Fig. 10 Molecular structure of alkylbenzene sulfonate

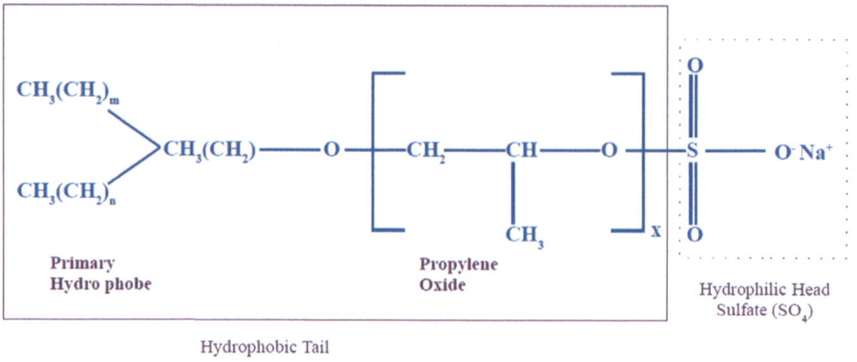

Fig. 11 Structure of alcohol propoxy sulfate

in an aqueous medium [23]. The surfactant is stable at high saline formations and its optimum salinity is at 4500 ppm, but the presence of a sulfate group makes it vulnerable to drastic temperature changes and high-temperature conditions. Alkyl ether sulfates are anionic surfactants that show high aqueous stability. However, the sulfate group makes it vulnerable to drastic temperature changes or high formation temperature and gets easily precipitated in high saline conditions. The addition of the ethoxy group to the initial structure increases the stability of the surfactant in high temperature and saline environment, still, its effectiveness declines at a temperature higher than 120 °C and salinity greater than 180,000 ppm [31]. N-ethoxy sulfonate gives an upper hand in terms of stability at higher reservoir temperatures due to its improvised design and does not precipitate in the presence of divalent ions like Ca^{2+} and Mg^{2+} [36]. Sodium Dodecyl Sulfate (SDS) (Fig. 12) is the most popular surfactant from the sulfate family group. This surfactant improves the EOR process alongside nanoparticles (NPs). The cloud point of SDS is over 100 °C [8]. Mixing NPs with SDS prevents the problem of NP adsorption on the reservoir rock.

Alpha-olefin sulfonate family of anionic surfactants are very efficient in reservoirs with high divalent ion concentration and has a high biodegradation rate. The most common of these surfactants are sodium C14-16 olefin sulfonate; C14-16 alkane hydroxy and C14-16 alkene sodium salts; sodium alpha-olefin sulfonate. Other advantages of alpha olefin sulfonate include stability over a wide range of

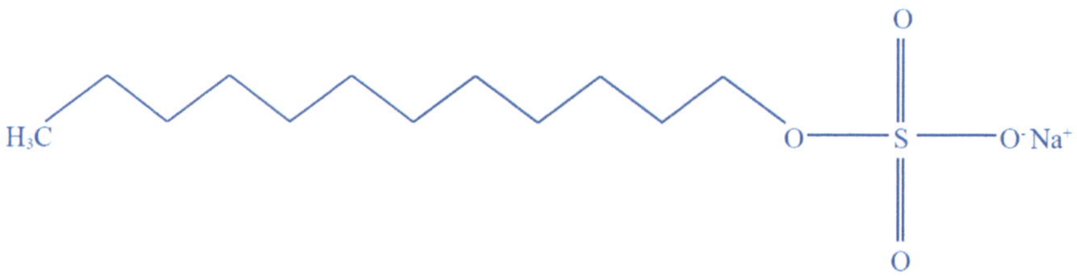

Fig. 12 Structure of sodium dodecyl sulfate (SDS)

pH, good foaming ability even in the case of hard water, and acid resistance, hence it is used often with polymeric fluids. Branched alkyl benzene sulfonate is another anionic surfactant in which the existence of the benzene group makes it long term stable against high reservoir temperature conditions but turns it non-biodegradable in nature. Alpha olefin sulfate is very similar to alpha-olefin sulfonate however it has thermal sensitivity due to the C–O–S bond [12].

Docusate sodium is a double tail anionic surfactant. It is used worldwide in ASP flooding of conventional reservoirs. Increasing the concentration of docusate sodium drastically increases the viscosity of the displacing fluid which makes it favorable to deploy for recovery in case of thin small laminated reservoirs [6]. Ethoxy or propoxy glycidyl sulfonates is an upgraded member from the family of sulfonates surfactant that is resistant to a high range of salinity variations. It is also used to form highly stable macroemulsion with high-temperature tolerance [12]. Alkyl ether sulfates, on the other hand, show excellent aqueous stability but the sulfate group present in it makes it sensitive to drastic temperatures. Moreover, it is not stable in high salinities. Internal olefin sulfonates (IOS) has two hydrophobic tails. The presence of sulfonate unit in this surfactant makes it stable for a long term in high-temperature reservoir conditions (up to 200 °C) [31]. The characteristic of aqueous stability and stable foaming at low concentration makes it widely used by researchers for core flooding operations. IOS has optimum salinity at 4500 ppm and enhances the recovery up to 30% on being introduced to sandstone for tertiary recovery operations. The only drawback is its instability in regions of high divalent ion concentration. Sulfonated, ethoxylated alcohol also known as alkyl phenol is designed for reservoirs with high salinity or divalent ion concentration [40]. Sodium petroleum sulfonate is a family of anionic surfactant and precipitates even in low salt concentration in water [7]. TDA-9PO-Sulfate (Tridecyl-9 propylene oxide sulfate) has a commercial name 'Petrostep S13-C', which is an anionic surfactant with a high tolerance level of salt [37]. Alkyl alcohol propoxylated sulfate is a very strong anionic surfactant that can reduce the interfacial tension even when employed at low concentrations. Carboxybetaine based anionic surfactant and amphoteric surfactant show the potential to replace all the existing surfactants as they are highly compatible with drastic reservoir conditions [16].

2.3 *Nonionic Surfactants*

Nonionic surfactants do not undergo ionization, rather hydrogen bonding and van der Waals forces primarily influence its solubility. The inferior dissolution of nonionic surfactant transpires due to elevated thermal energy and poor hydrogen bonding at higher temperatures, which consequently forms a turbid solution. The temperature where this type of surfactant develops turbidity is marked as a cloud point [45]. Fundamentally, the cloud point depends on factors like the number of ethylene oxide units, the concentration of the surfactant used and branched chains. Although nonionic surfactants bear high salinity tolerance, yet the magnitude of IFT reduction achieved by its use is lower as compared to cationic and anionic surfactants [1]. Alkyl ethoxy carboxylates are the most common nonionic surfactants at normal pH, although they behave as anionic surfactant at alkaline pH conditions. The character of resistivity to high temperature and high salinity condition is due to the ethoxy group along with the carboxylate compound. The ethoxy unit present in the surfactant increases interaction energy (as a result of the hydrogen bond formation with water molecule) while the carboxylate group increases long term stability in the high-temperature conditions. Alkyl polyglycoside is a non-ionic surfactant used under high-temperature reservoirs even at a high concentration of divalent ions. Due to the structure of the glucose, it can work in high temperatures and be stable at a maximum salinity of 180,000 ppm [34]. Nonylphenol ethoxylate belongs to the family of alkyl phenol ethoxylate. The cloud point for this class of nonionic surfactants is over 100 °C. The surfactant is generally used as a co-surfactant along with the cationic surfactant in carbonate reservoirs that bears high temperature (up to 100 °C) and salinity conditions. The nonylphenol ethoxylate mainly acts on the wettability of reservoirs and can recover 80% of oil in place under the spontaneous imbibition process. Secondary alcohol ethoxylate/propoxylate, another non-ionic surfactant is characterized by high solubility. It is used as a co-surfactant alongside cationic surfactant in carbonate reservoirs having extreme conditions. Moreover, in an inert atmosphere, the surfactant is stable up to temperatures as high as 300 °C [27]. Tridecyl alcohol ethoxylate known by the trade name TDA-30-EO is used as a co-surfactant in high temperature (approximately 100 °C) and salinity of 200,000 ppm [36]. The hydrophilicity of the reservoir can be increased by just adding more numbers of ethoxylate units which increases its solubility in water. Poly ethoxylated alkylphenols are used for EOR in high salinity reservoirs but its low cloud point makes it insoluble even at a moderate temperature range of 35–50 °C [40]. Poly glycol ether, a nonionic surfactant is classified as 'r series' and 'p series' based on the constituent chemical at the time of preparation, namely, ethylene oxide and propylene oxide respectively. They work in extremely high temperatures and have elevated cloud points. This surfactant family consists of completely biodegradable members.

Triphenylmethane also known as TMP Fig. 13, is a nonionic surfactant. Being insoluble in water, TPM shows excellent stability in high salinity (up to 18.6%) and high temperature (>70 °C) [41]. A secondary linear ethoxylated alcohol, known as Tergitol surfactant has foamability characteristics. Tergitol 15-S-12 shows the maximum stability in distilled water as well as in saline water. The surface excess value for the nonionic surfactant increases with the decrease of molecular cross-section of the polar head group. At a particular CMC, the micelle sizes of this surfactant increase with an increase in ethylene oxide number (EON). A higher EON value yields greater HLB value. As compared to water flooding, additional recovery of 26% is observed for Tergitol 15-S-12 [25].

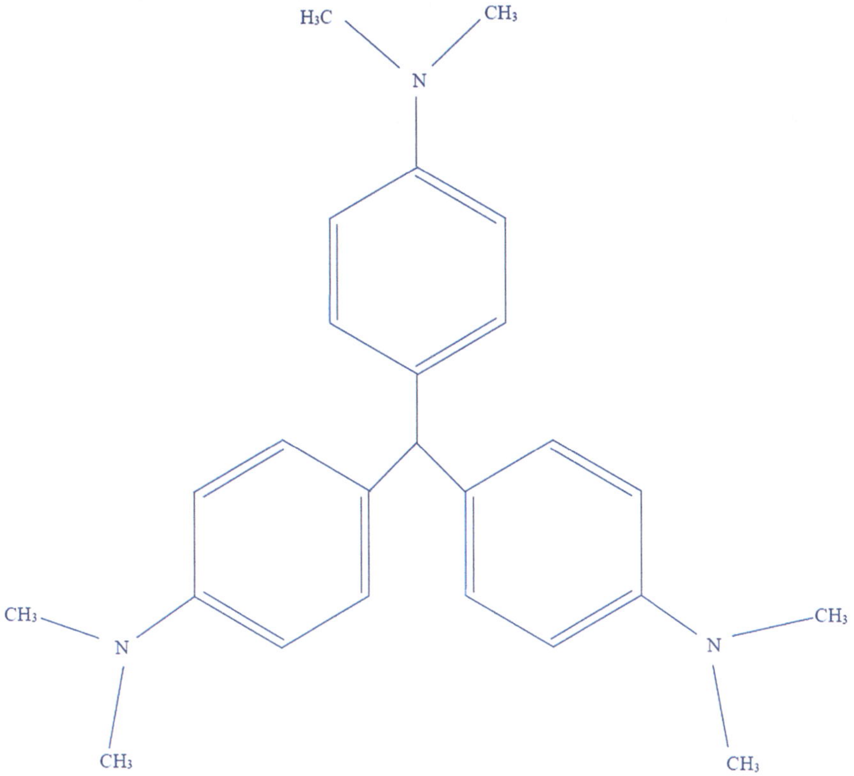

Fig. 13 Structure of triphenylmethane (TPM), a nonionic surfactant

2.4 *Zwitterionic Surfactant*

Zwitterionic or amphoteric surfactant (Fig. 14) is amphiphilic organic compounds that have both negative and positive charges in its hydrophilic head. The presence of dual charge engages its application in both sandstone and carbonate reservoirs. Moreover, due to its net neutral nature, it has a superior tolerance level in high saline environments. The most common surfactant in this family is made from castor oil that can achieve ultra-low interfacial tension value of 5.4×10^{-3} mN/m at very low concentrations in the absence of alkaline mediator. Zwitterionic surfactants have strong thermal stability, electrolytic tolerance, foam capability, and wetting performance which makes it an important chemical for EOR development in the future. It undergoes interaction with anionic surfactants like SDS as well as cationic ones like DTAB [26]. Like any other surfactant, increasing the concentration of the zwitterionic surfactant decreases the value of the surface tension until the point of CMC. For a solution having 5000 ppm concentration, the value of surface tension observed is 33.25 mN/m. Reduced adsorption after the point of CMC will not allow further lowering in the value of the surface tension. The value of the IFT decreases with

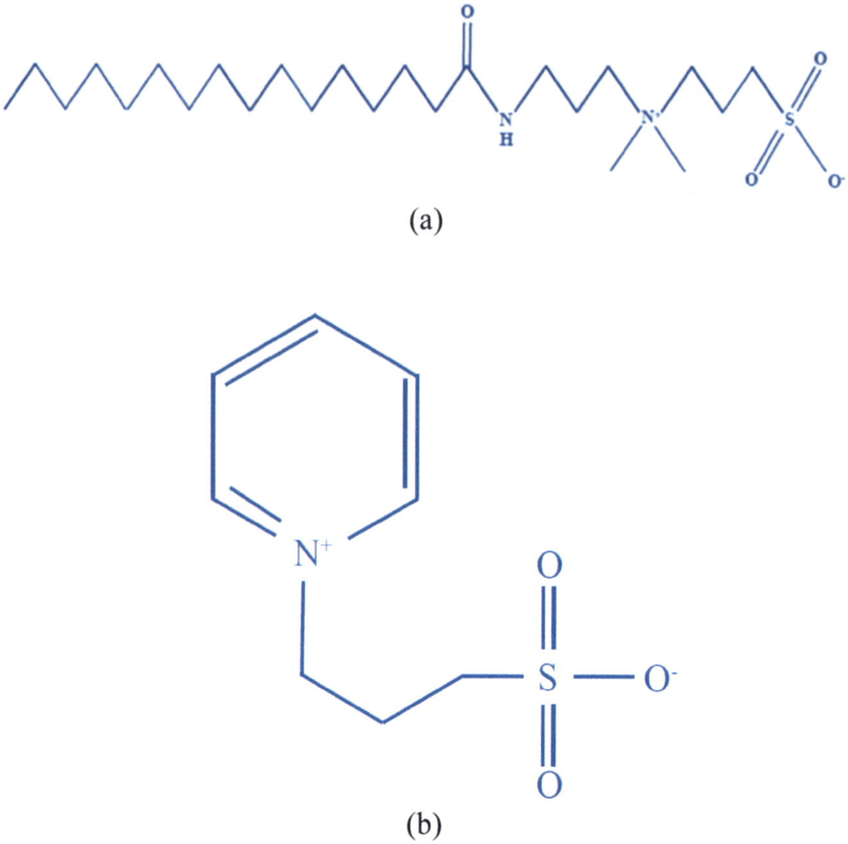

Fig. 14 **a** Molecular structure of amidosulfobetaine-16, **b** general chemical structure of the geminal zwitterionic surfactant 3-(1-pyridinio)-1-propanesulfonate

an increase in the concentration of the surfactant. Sample containing 500 ppm of the zwitterionic surfactant easily pulls down the IFT value to 0.0377 mN/m from 17 mN/m [28].

Zwitterionic surfactant shows no precipitation even at a temperature of 90 °C and a salt concentration of 20%. The solubility of surfactant and salt increase with an increase in temperature [4]. The thermo-gravimetric analysis shows excellent temperature stability up to a temperature of 160 °C. The flooding experiments using the zwitterionic surfactant show additional recovery of up to 30% [3]. A class of zwitterionic surfactants has positively charged ammonium and negatively charged carboxylate ions. The hydrocarbon lengths generally have a carbon chain of 12–14 carbons whose CMC is found to be 60 ppm. At this value of CMC this surfactant effectively sweeps residual oil as it achieves ultra-low interfacial tension profile between the oil–water interface. Moreover, it can alter wetting conditions from oil to water as observed from core flood experiments. The presence of both hydrophobic and hydrophilic groups reduces the amount of surfactant loss due to adsorption during recovery operation and also induces characteristics of good thermal stability with excellent salt tolerance [15]. Carbonate reservoirs are excellent candidates for carboxybetaine based zwitterionic surfactants. Along with excellent salt tolerance capability and high thermal stability, it also shows a low degree of adsorption in a positively charged carbonate reservoir. However, few surfactants of this class show contrary results. Zwitterionic surfactants derived from sulfobetaine shows excellent pH tolerance. They are further divided into two classes based on the saturated or unsaturated nature of their tail. Amidosulfobetaine zwitterionic surfactant with a saturated tail shows poor solubility in saline and freshwater reservoirs, whereas, with an unsaturated tail it shows excellent solubility in both saline and freshwater reservoirs. The latter surfactant with unsaturated tail shows no sign of decomposition at high temperatures (up to 225 °C), thereby showing excellent thermal stability. The CMC of this amidosulfobetaine unsaturated surfactant is 1.02×10^{-4} mol L^{-1} in saltwater and 2.05×10^{-4} mol L^{-1} in freshwater. The value of the IFT for this surfactant lies close to 10^{-1} mN m^{-1} and remains tolerant to changes in temperature and salinity. This surfactant shows the best result when deployed as a co-surfactant along with polymeric surfactant [15]. The novel alkyl-hydroxyl-sulfobetaine (C_{16}HSB and C_{18}HSB) shows high IFT reduction when used with HPAM. Additionally, it has significant stability and shows the potential for a new opportunity in high temperature and high salinity reservoirs [11]. Gubert amine-oxide is another zwitterionic surfactant that shows low IFT at low CMC value [39].

3 Conclusion

This chapter laid an overview of common, as well as novel surfactants that has emerged in surfactant cEOR. Most of the recent research and findings have been generally on anionic surfactants as they are found to be more suitable for sandstone reservoirs; which are relatively more abundant and accessible. Anionic surfactants mostly work by reducing IFT at o/w interfaces. However, for carbonate reservoir, the cationic surfactants are preferred over anionic surfactants due to their ability to alter wettability. Nonionic surfactant emerges as an apt candidate for reservoirs that bear high salinity with extreme formation water hardness. Zwitterionic surfactants are fairly explored later and they yield low IFT values and also own the potential to alter the wettability of reservoir rock. They are also known to have high thermal stability and tolerate strong electrolytes due to their net charge neutrality.

Acknowledgements This work is supported by Science and Engineering Research Board (SERB), Department of Science and Technology (DST), Government of India through project TAR/2019/000348. The authors would like to thank the funding agency for supporting this research.

References

1. Abrahamsen A (2012) Apply chemical EOR and on the Norne field C-segment. Norwegian University of Science and Technology
2. Afsharpoor A, Balhoff MT, Bonnecaze R, Huh C (2012) CFD modeling of the effect of polymer elasticity on residual oil saturation at the pore-scale. J Petrol Sci Eng 94–95:79–88
3. Amirpour M, Shadizadeh SR, Esfandyari H, Ahmadi S (2015) Experimental investigation of wettability alteration on residual oil saturation using nonionic surfactants: capillary pressure measurement. Petroleum 1:289–299
4. Bharmoria P, Gupta H, Mohandas VP, Ghosh PK, Kumar A (2012) Temperature invariance of NaCl solubility in water: inferences from salt-water cluster behavior of NaCl, KCl, and NH_4Cl. J Phys Chem B 116:11712–11719
5. Chakraborty T, Chakraborty I, Ghosh S (2011) The methods of determination of critical micellar concentrations of the amphiphilic systems in aqueous medium. Arab J Chem 4:265–270
6. Dang CTQ, Nguyen NTB, Chen Z, Nguyen HX, Bae W, Phung TH (2012) A comprehensive evaluation of the performances of alkaline/surfactant/polymer in conventional and unconventional reservoirs. Presented in the SPE Asia Pacific oil and gas conference and exhibition, Perth, 22–24 Oct 2012
7. Dwarakanath V, Shong RG, Winslow G, Malik T, Solano M (2017) Enhanced oil recovery compositions and methods thereof. International Patent No. WO 2017/040903 Al
8. Esmaeilzadeh P, Hosseinpour N, Bahramian A, Fakhroueian Z, Arya S (2014) Effect of ZrO_2 nanoparticles on the interfacial behavior of surfactant solutions at air-water and n-heptane-water interfaces. Fluid Phase Equilib 361:289–295
9. Fanchi JR (2016) Improved recovery, pp 272–281
10. Gbadamosi AO, Junin R, Manan MA, Agi A, Yusuff AS (2019) An overview of chemical enhanced oil recovery: recent advances and prospects. Springer, Berlin Heidelberg
11. Guo S, Wang H, Shi J, Pan B, Cheng Y (2015) Synthesis and properties of a novel alkyl-hydroxyl-sulfobetaine zwitterionic surfactant for enhanced oil recovery. J Petrol Explor Prod Technol 5:321–326
12. Hirasaki GJ, Miller CA, Puerto M (2008) Recent advances in surfactant EOR. Presented in the SPE international petroleum technology conference, Denver, 21–24 Sept 2008
13. Howe AM, Clarke A, Mitchell J, Staniland J, Hawkes L, Whalan C (2015) Visualising surfactant enhanced oil recovery. Colloids Surf A Physicochem Eng Asp 480:449–461
14. Jiang N, Li P, Wang Y, Wang J, Yan H, Thomas RK (2005) Aggregation behavior of hexadecyltrimethylammonium surfactants with various counterions in aqueous solution. J Colloid Interface Sci 286:755–760
15. Kamal MS, Shakil Hussain SM, Fogang LT (2018) A zwitterionic surfactant bearing unsaturated tail for enhanced oil recovery in high-temperature high-salinity reservoirs. J Surfactants Deterg 21:165–174
16. Kamal MS, Sultan AS, Hussein IA (2015) Screening of amphoteric and anionic surfactants for cEOR applications using a novel approach. Colloids Surf A Physicochem Eng Asp 476:17–23
17. Khatory A, Kern F, Lequeux F, Appell J, Porte G, Morie N, Ott A, Urbach W (1993) Entangled versus multiconnected network of wormlike micelles. Langmuir 9:933–939
18. Kronberg B, Holmberg K, Lindman B (2014) Surface chemistry of surfactants and polymers. Wiley, Chichester
19. Kumar S, Panigrahi P, Saw RK, Mandal A (2016) Interfacial interaction of cationic surfactants and its effect on wettability alteration of oil-wet carbonate rock. Energy Fuels 30:2846–2857
20. Kuperkar K, Abezgauz L, Danino D, Verma G, Hassan PA, Aswal VK, Varade D, Bahadur P (2008) Viscoelastic micellar water/CTAB/NaNO3 solutions: rheology, SANS and cryo-TEM analysis. J Colloid Interface Sci 323:403–409
21. Latil M (2015) Enhanced oil recovery. In: Petroleum engineer's guide to oil field chemicals and fluids. Elsevier, pp 477–565
22. Lee KS, Lee JH (2019) Hybrid chemical EOR using low-salinity and smart waterflood. In: Hybrid enhanced oil recovery using smart waterflooding. Elsevier, pp 65–110

23. Levitt DB, Jackson AC, Heinson C, Britton LN, Malik T, Dwarakanath V, Pope GA (2006) Identification and evaluation of high-performance EOR surfactants. Presented in the SPE/DOE symposium on improved oil recovery, Tulsa, Apr 2006
24. Ma K, Cui L, Dong Y, Wang T, Da C, Hirasaki GJ, Biswal SL (2013) Adsorption of cationic and anionic surfactants on natural and synthetic carbonate materials. J Colloid Interface Sci 408:164–172
25. Mandal A, Bera A, Ojha K, Kumar T (2012) Characterization of surfactant stabilized nanoemulsion and its use in enhanced oil recovery. Presented in the SPE international oilfield nanotechnology conference and exhibition, Noordwijk, 12–14 June 2012
26. McLachlan AA, Marangoni DG (2006) Interactions between zwitterionic and conventional anionic and cationic surfactants. J Colloid Interface Sci 295:243–248
27. Mohan K (2009) Alkaline surfactant flooding for tight carbonate reservoirs. Presented in the SPE annual technical conference and exhibition, New Orleans, 4–7 Oct 2009
28. Mukherjee I, Moulik SP, Rakshit AK (2013) Tensiometric determination of Gibbs surface excess and micelle point: a critical revisit. J Colloid Interface Sci 394:329–336
29. Nakama Y (2017) Surfactants. In: Cosmetic science and technology. Elsevier, pp 231–244
30. Oya M, Takemoto Y, Ishikawa Y (2008) Large decrease in acute aquatic toxicity of linear alkylbenzene sulfonate in hard water and seawater by adding adsorbent. J Oleo Sci 57:15–21
31. Patil PD, Rohilla N, Katiyar A, Yu W, Nelson C, Falcone S, Rozowski P (2018) Surfactant based EOR for tight oil unconventional reservoirs through wettability alteration: novel surfactant formulations and their efficacy to induce spontaneous imbibition. Presented in the SPE/AAPG/SEG unconventional resources technology conference, Texas, 23–25 July 2018
32. Pena ALML (2015) Experimental evaluation of linear alkylbenzene sulphonates for enhanced oil recovery. Instituto Superior Técnico, pp 1–9
33. Sagir M, Mushtaq M, Tahir MS, Tahir MB, Shaik AR (2020) Surfactants for enhanced oil recovery applications. Springer International Publishing, Cham
34. Santa M, Alvarez-Jürgenson G, Busch S, Birnbrich P, Spindler C, Brodt G (2011) Sustainable surfactants in enhanced oil recovery. Presented in the SPE enhanced oil recovery conference, Kuala Lumpur, 19–21 July 2011
35. Schramm LL, Kutay SM (2000) Emulsions and foams in the petroleum industry. In: Surfactants. Cambridge University Press, pp 79–118
36. Sharma G, Mohanty KK (2013) Wettability alteration in high-temperature and high-salinity carbonate reservoirs. SPE J 18:646–655
37. Sharma MM, Gao B (2014) Salt-tolerant anionic surfactant compositions for enhanced oil recovery (EOR) applications. International Patent No. WO 2014/055158 Al
38. Sheng JJ (2015) Status of surfactant EOR technology. Petroleum 1:97–105
39. Shi C, Li Z, Zhang H, Chen Y, Zhang M, He M (2020) Synthesis of Guerbet amine-oxide for enhanced oil recovery. J Chem 2020:1–6
40. Shupe RD, Maddox JJ (1976) Surfactant oil recovery process usable in high temperature, high salinity formations. International Patent No. 4018278
41. Siggel L, Santa M, Gmbh WH, Hansch M, Nowak M, Ranft M, Weiss H, Schreiner E, Oetter G, Tinsley J, Se B (2012) A new class of viscoelastic surfactants for enhanced oil recovery. Presented in the SPE improved oil recovery symposium, Tulsa, 14–18 April 2012
42. Spildo K, Johannessen AM, Skauge A (2012) Low salinity waterflood at reduced capillarity. Presented in the SPE improved oil recovery symposium, Tulsa, 14–18 April 2012
43. Szlendak SM, Nguyen N, Nguyen QP (2013) Laboratory investigation of low-tension-gas flooding for improved oil recovery in tight formations. SPE J 18:851–866
44. Taylor DW, Nieman TA (1984) Determination of critical micelle concentrations by bipolar pulse conductance with an exponential dilution flow system. Anal Chem 56:593–595
45. Zhao G, Khin CC, Chen SB, Chen BH (2005) Nonionic surfactant and temperature effects on the viscosity of hydrophobically modified hydroxyethyl cellulose solutions. J Phys Chem B 109:14198–14204
46. Zhou Y, Yin D, Chen W, Liu B, Zhang X (2019) A comprehensive review of emulsion and its field application for enhanced oil recovery. Energy Sci Eng 7:1046–1058

Low Salinity Surfactant Flooding: Role of Surfactant and Salt

Abhijit Kakati and Jitendra S. Sangwai

Abstract The need for effective enhanced oil recovery (EOR) methods in terms of economics and technical feasibility are growing rapidly along with the steeply growing demand for crude oil in the energy sector. Such demands driving researchers to innovate novel EOR solutions and also to explore ways to enhance the effectiveness of conventional EOR methods. This chapter summarises advancement in one such hybrid EOR method developed by combining novel low salinity water flooding with conventional surfactant flooding. The synergistic benefits of low salinity water and different low salinity surfactant formulations in terms of improving reservoir properties and oil recovery efficiency are summarized. This chapter also aims to provide a very detailed discussion on the complex pore level mechanism of oil recovery through the hybrid low salinity surfactant flooding process.

1 Introduction

Surfactant flooding is one of the oldest and the most widely used chemical enhanced oil recovery (EOR) method practiced by the oil industry. It involves the injection of an aqueous solution of surfactant into oil reservoirs generally tailored with water flooding. A surfactant solution in the reservoirs can produce ultra-low oil–water interfacial tension, increase the capillary number to a great extent, causes water wetness of the reservoir rock and in-situ emulsification of oil and water which enhances pore level displacement efficiency of the crude oil [1, 2]. Surfactant flooding, although it has distinctive working mechanisms, it is generally applied in combination with other chemical EOR methods like polymer flooding and alkaline flooding [3].

A. Kakati
School of Petroleum Technology, Pandit Deendayal Energy University, Gandhinagar, Gujarat 382007, India

J. S. Sangwai (✉)
Petroleum Engineering Programme, Department of Ocean Engineering, Indian Institute of Technology Madras, Chennai, Tamil Nadu 600036, India
e-mail: jitendrasangwai@iitm.ac.in

T. Solling et al. (eds.), *Surfactants in Upstream E&P*, Petroleum Engineering,
https://doi.org/10.1007/978-3-030-70026-3_8

Synergism between different EOR methods is always served as an optimization route for maximizing oil recovery and reducing the cost of EOR operations. One such novel EOR method whose synergism has been studied with other chemical EOR methods in recent times is the 'low salinity water flooding'. This method is based on designing of the injection water chemistry, ideally one with injection water salinity much lower than that of the reservoir connate water [4, 5]. The method is found to be effective in increasing oil recovery from the laboratory as well pilot sale applications [6–12] however, it is also proven to be an excellent partner in producing synergistic benefits with other chemical EOR methods. Many laboratory investigations have shown that low salinity water flooding, when applied in combination, can be effective in enhancing oil recovery as well as in mitigation of many problems that arise during chemical EOR such as polymer flooding and surfactant flooding [13–16]. The synergistic benefits of low salinity water flooding with chemical EOR methods lead to hybrid EOR concepts such as low salinity surfactant flooding and low salinity polymer flooding.

This chapter summarizes the development in the field of low salinity surfactant flooding including the effect of low salinity surfactant formulation on reservoir properties, their oil recovery efficiency obtained from laboratory studies and the pore level physicochemical mechanism involved.

2 Low Salinity Water Flooding

The petrophysics and surface chemistry research group at the University of Wyoming while studying wettability effects on waterflooding observed that change in injection brine salinity affects oil recovery [17]. Advancement of their research on the impact of brine salinity on oil recovery established that reducing injection brine salinity can increase oil recovery by waterflooding [11, 18, 19]. Their research was actively followed by researchers of British Petroleum through numerous laboratory and single-well tests which finally led to the registration of LoSal™ [7, 12, 20, 21]. Meanwhile, researchers from industry and academia actively participated in investigating the low salinity waterflooding process. Most of the laboratory core flooding investigation showed positive results whereas a few reported no incremental recoveries. Apart from laboratory evaluations, low salinity water flooding has been tested in pilot-scale in many parts of the world e.g. Alaska, North Sea, Wyoming, Syria, and Kuwait, etc. [7, 9, 10, 12, 21, 22, 23]. The majority of the projects showed significant potential for low salinity water flooding. The incremental oil recovery of low salinity water flooding from laboratory core flooding test was as high as 36% of OOIP whereas pilot-scale tests showed the maximum incremental recovery is 15% [7]. The difference in magnitude of incremental recovery arises due to the fact that, many pore volumes of low salinity water were injected in laboratory experiments which is not realistic in field applications.

Apart from investigating the efficiency of the low salinity waterflooding process another effort made by the researcher community is to understand the underlying working mechanism. Various mechanisms of low salinity waterflooding have been identified which are: (1) formation fine migration; (2) in-situ soap generation; (3) multi-component ion-exchange; (4) mineral dissolution; (5) salting-in effect; (5) electrical double layer expansion and (6) mineral dissolution. Although different mechanisms have been identified, extensive active research is going on for the search for mechanism of the low salinity water flooding [4, 24, 25, 26]. Table 1 summarizes the proposed low salinity mechanisms. All the proposed mechanisms may not work together for every reservoir, as it may primarily depends on specific reservoir conditions.

Laboratory investigations have demonstrated that the low salinity effect is significant when certain conditions pre-exits in the reservoir. This prerequisite conditions include: (1) presence of non-swelling clay in the reservoir, (2) presence of polar component in crude oil, (3) presence of high salinity connate water in the reservoir, (4) high divalent ion concentration in the formation water, (5) significant salinity contrast between injection water and formation water [4, 5, 11]. However, some studies reported the low salinity effect even outside the regime of above conditions.

Table 1 Description of low salinity water flooding mechanisms

Low salinity mechanism	Description	Source
Fine migration	• Mixed wet formation fine detached from pore surface due to double layer expansion • Oil droplet adsorbed on clay fine also stripped away along with formation fine	[11]
In-situ soap generation	• Low salinity water increases reservoir pH • Elevated pH causes in-situ soap generation by alkaline reactive components of crude oil	[7]
Multi-component ion exchange	• Exchange of divalent cations between low salinity water and organo-metallic complexes • Resulted in decomplexation of organo-metallic complexes adsorbed on the rock surface • Wettability alteration due to desorption of organic matter	[27]
Salting-in effect	• The solubility of organic material increases drastically as the salinity decreases • Adsorbed organic molecules of crude oil on the clay surface detached and dissolves into water • Desorption of organic material increases water-wetness of the reservoirs rock	[28]
Double layer expansion	• Increasing electrostatic repulsion of between crude oil and sandstone surface • Expansion of intervening water film	[29]
Mineral dissolution	• Dissolution of carbonate minerals like anhydrite, dolomite or calcite and release of particle • Increase in pressure and wettability alteration promote oil recovery	[30]

3 Low Salinity Surfactant Flooding

The role of salinity and salinity gradient on the surfactant flood process has been a topic of interest for many petroleum researchers since many decades. A salinity gradient implies formation or waterflood brine more saline than surfactant slug and surfactant slug more saline than the chase brine. Some studies reported a negative impact of salinity gradient on surfactant flooding in terms of minimizing ion-exchange and maintaining optimal salinity in the mixing zone [31, 32]. On the other hand, there are studies which claimed that a salinity gradient can result in higher oil recovery efficiency as compared to constant salinity [33, 34]. A salinity gradient is also observed to be responsible for lowering surfactant adsorption to nearly half as compared to constant salinity [34, 35]. The salinity level is known to controls the phase behavior of a surfactant/oil/brine system as: (1) low salinity results in type II (−) or under-optimum emulsion where surfactant predominantly portioned into aqueous phase; (2) high salinity results in type II (+) or over-optimum emulsion where surfactant predominately partitioned into oleic phase; (3) optimal salinity resulted in type (III) or middle phase emulsion which causes lowest interfacial tension value as shown in Fig. 1. An increase in salinity shows a steady progression from type II (−) through type (III) to type II (+) phase behavior [36].

In the present time the growing interest of oil companies over low salinity water flooding has driven some researches to explore the synergistic advantages of combining low salinity water flooding with surfactant flooding in the direction of conceptualizing a hybrid EOR method. Alagic and Skauge [37] were the first to

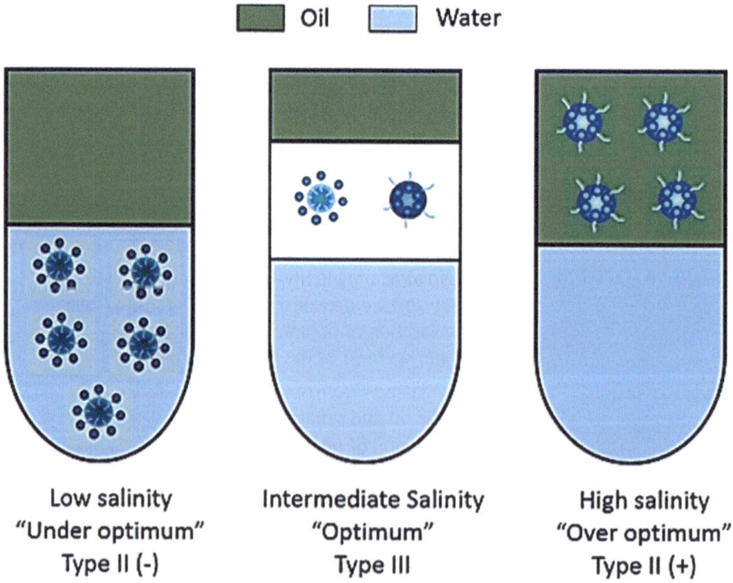

Fig. 1 Variation in types of emulsion phase behavior with salinity

investigate this low salinity surfactant synergism through a series of laboratory core flooding experiments. They observed a higher tertiary oil recovery when applied a pre-flush with low salinity brine and lower oil recovery without low salinity preflush. Meanwhile, others contributed to low salinity surfactant flooding research by investigating the rock-fluid interaction in the presence of low salinity surfactant solution in the reservoir. The low salinity surfactant flooding studies have been performed for various formulation using different surfactants like sodium dodecylbenzene sulfonate (SDBS), sodium dioctylsulfosuc-cinate (Aerosol OT or AOT), etc. Some of these low salinity surfactant formulations were found to have a significant impact in altering reservoir properties which could greatly favor in mobilizing residual oil. The LSS could be applied in two ways: (1) using low salinity water flooding as preflush or (2) optimal salinity surfactant injection [38].

4 Synergistic Effects of Low Salinity and Surfactant on Reservoir Properties

The most important property which has been targeted to change in surfactant EOR is the crude oil–water interfacial tension. Surfactants are capable of reducing the oil–water interfacial tension [39, 40]. A low interfacial tension between crude oil and injected water resulted in higher capillary number and formation of micro-emulsion which facilitate mobilization of entrapped residual oil. Micro-emulsions are thermodynamically stable solutions of swollen micelles and thus its formation helps in the transportation of mobilized oil in the reservoir. The relation between interfacial tension and micro-emulsion formation can be related through Eq. (1).

$$\gamma = k\left(\frac{V_{surf}}{V_{diss}}\right)^2 \tag{1}$$

where, 'γ' is the oil–water interfacial tension, 'V_{surf}' and 'V_{diss}' are the volumes of surfactant and dispersed phase (either oil or water depending on emulsion phase behavior) and 'k' is a constant whose value depends on the surfactant [41]. Thus, the interfacial tension depends on the solubilization capacity of the surfactant between oil and water. The interfacial tension reaches its minimum value when the solubilization ratio approaches unity. Generally, co-surfactants (e.g., long-chain alcohol) are added to increase the solubilization capacity of a surfactant in a micro-emulsion system and results in an ultra-low interfacial tension (~10^{-3} mN/m). The addition of a co-surfactant disturbs the organization of surfactant molecules at the oil–water interface which otherwise increases the stiffness of the interface [42, 43].

Recent studies of low salinity surfactant flooding show that an ultra-low interfacial tension could be achieved between crude oil and aqueous low salinity surfactant solutions without the addition of a co-surfactant [44, 45]. The interfacial tension of a pure hydrocarbon-aqueous surfactant system varies with the ionic strength of the surfactant solution over a pH range of 2 to 10. At low pH, a high ionic strength results in higher interfacial tension; but a high pH and a low ionic strength together can increase interfacial tension of hydrocarbon-aqueous surfactant system. The presence of inorganic divalent cation in the system governs the interfacial tension behavior over varying pH range as well as the absolute value of interfacial tension (Fig. 2). With ionic surfactant in the system even at the same ionic strength, the presence of divalent cations (e.g. Ca^{2+}) can result in a significantly lowers interfacial tension. Particularly, when anionic surfactants are used, the presence of divalent ions in the solution could drop the interfacial tension to a very low value (<0.1 mN/m) [45].

When Tichelkamp et al. [45] extended their measurements of interfacial tension for crude oil-low salinity surfactant systems it was observed that the interfacial tension could reach an ultra-low value when an anionic surfactant is applied with or without divalent cations. However, depending on the nature of the crude oil the presence and absence of cations could play a role. When AOT surfactant is applied a low salinity surfactant solution can reduce the oil–water interfacial tension which could be further reduced by introducing divalent cations into the system irrespective of the crude oil type. However, the low salinity SDBS solution doesn't produce an ultra-low interfacial tension (Fig. 3).

Jha et al. [44] studied the salinity effect on interfacial tension for heptane-aqueous surfactant systems containing SDS and AOT surfactants over a wide range of salinity with varying surfactant concentrations. In-line with the earlier discussed study this

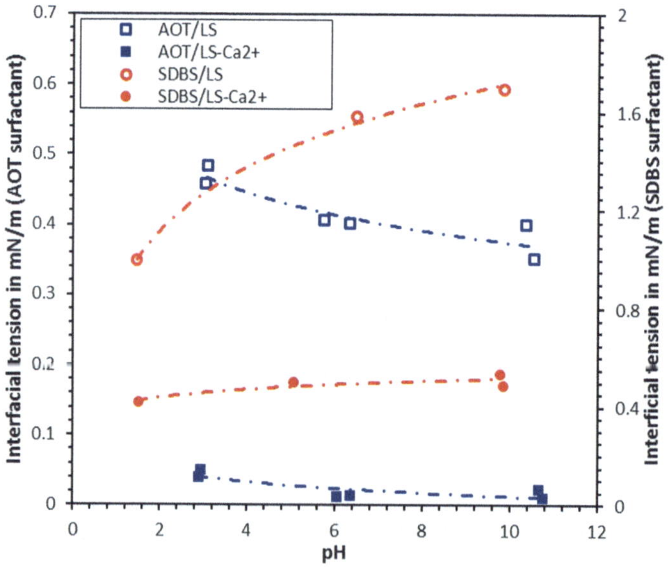

Fig. 2 The interfacial tension between a model oil (heptane + toluene in v/v: 1/1) and two anionic surfactants (SDBS and AOT) solutions as a function of pH (LS stands for low salinity water and LS-Ca^{2+} stands for low salinity water with calcium, in each case ionic strength = 20). ("Adapted with permission from [45]. Copyright 2014 American chemical society")

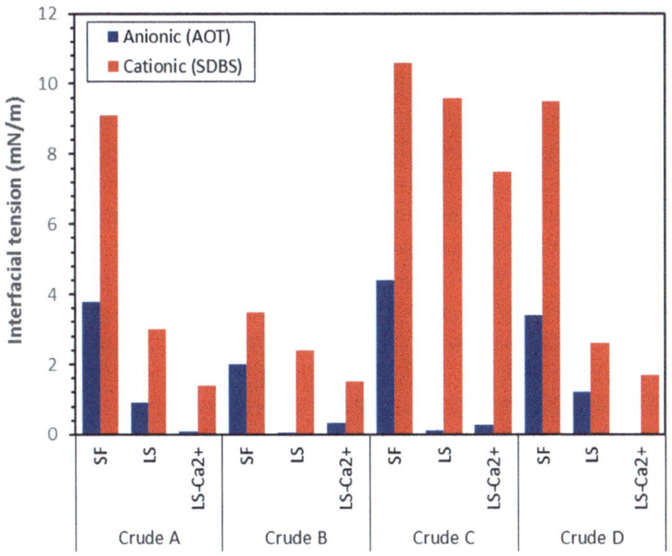

Fig. 3 Interfacial tension between crude oil and aqueous surfactant solutions of two anionic surfactants (SDBS and AOT) with no electrolyte (FS), low salinity water (LS) and low salinity water with calcium ions (LS-Ca^{2+}). For crude A [°API, TAN, TBN] = [33.4, 1.08, 1.16], crude B [°API, TAN, TBN] = [34.5, 0.1, 0.56], crude C [°API, TAN, TBN] = [27.3, 2.46, –] and crude D [°API, TAN, TBN] = [44.3, 0.1, 0.18]. ("Adapted with permission from [45]. Copyright 2014 American chemical society")

study also reported that the addition of salt can increase the ability of a surfactant to reduce the oil–water interfacial tension. However, there is an optimum salinity level for each low salinity-surfactant formulation for achieving a minimum interfacial tension value for an oil–water system. The lowest interfacial tension can be achieved when a combination of a divalent salt and anionic surfactant AOT was used. The relative concentration of divalent and monovalent ions on the interfacial tension has studied by Khanamiri et al. [46]. The optimum ionic strength corresponding to minimum IFT for crude oil-low salinity surfactant system decreases as the divalent to monovalent ratio increases. However, an increase in the relative concentration divalent doesn't consistently result in reduced IFT, instead, there is an optimum value that exists for divalent to monovalent ration to obtain the minimum achievable IFT.

A regression model to predict the change in interfacial tension with varying salinity for a hydrocarbon-aqueous surfactant system has been proposed by Jha et al. [44] given in Eq. (2).

$$\Delta\gamma = -RTaN_S^{\max}\left(\frac{v_- + v_+}{v_-}\right)ln\left(1 + \frac{1}{55.5M \times e^{(\Delta G_{ads}/RT)}}Cv_-\right) \qquad (2)$$

where, '$\Delta\,\gamma$' is the difference in interfacial tension value between the hydrocarbon-low salinity aqueous surfactant solution to that of hydrocarbon-water systems, 'R' is the universal gas constant, 'T' is the temperature, '$v-$' and 'v_+' are the stoichiometric number of anions in the surfactant, 'C' is the bulk surfactant concentration. The model has three adjustable parameters: 'a', N_s^{max} and 'ΔG_{ads}' represents a component specific parameter, maximum surface excess and Gibbs free energy of adsorption obtained from non-linear regression. The values of these parameters for different low salinity surfactant formulations were reported by [44].

The interfacial tension of an oil–water system in presence of surfactant depends on the aggregation behavior of the surfactant molecules at the interface of the two phases. The interfacial tension varies with the concentration of surfactant until a particular concentration level, known as critical micellar concentration (CMC). Above this concentration, the surfactant molecules start the formation of micelles and interfacial tension remains constant even with the addition of more surfactant [47]. The ionic strength of the surfactant solution governs the value of CMC. Generally, lower ionic strength results in a higher CMC value. Therefore, a low salinity surfactant solution results in lower oil–water interfacial tension as it increases the CMC value and delays the micellization process. The surfactant molecules can form different types of aggregates depending on their structure and the type of aggregation governs the value of interfacial tension. The critical packing parameter (CPP) is a number that relates the structure of the surfactant to the shape of the aggregate it forms as given in Eq. (3). [48, 49].

$$\mathrm{CPP} = \frac{v_o}{a_e l_0} \tag{3}$$

where, 'v_0' is the volume of the surfactant tail, 'a_e' is the equilibrium area per molecule at the aggregate interface and 'l_0' is the length of the surfactant tail. The values of CPP for the formation of a spherical, cylindrical, bilayer, and inverted structure are CPP $\leq 1/3$, $1/3 \leq$ CPP $\leq 1/2$, $1/2 \leq$ CPP ≤ 1, CPP ≥ 1, respectively [49, 50]. When the CPP approaches unity, the surfactant molecules aggregates to form a lamellar structure where the surfactant molecules form tight packing leads to a minimum interfacial tension value. For common surfactants, the v_0/l_0 value is constant and thus only reflects the specificity of surfactant in the CPP [48]. In case of an ionic surfactant addition of salt into the system causes charge shielding and resulted in a decrease of the surfactant head group area. Small head group area helps in the formation of bilayer aggregates [44, 45]. This effect is maximum observable at an optimal salt concentration whereby the interfacial tension is minimum. Again, another important factor that governs the shape of surfactant aggregates is the 'surfactant aggregation number' which can be defined as the number of surfactant molecules in a micelle above CMC. A decrease in ionic strength results in a decrease in the aggregation number. Again, a decrease in the aggregation number changes the shape of the micelles from spherical to bilayer [48]. Studies also showed that the divalent cations particularly calcium forms stronger binding with anionic surfactants and leads to extra tight packing at the interface [51]. The lower interfacial tension values with AOT surfactant is the result of higher CPP due to its two-tailed structure in comparison to one-tailed SDBS.

Wetting is another important property that plays a vital role in the oil displacement process during chemical EOR applications. A detailed study on the effect of salinity on wetting of reservoirs rock during surfactant flooding is not reported in the literature. However, there are few studies available which partly studied this effect through oil desorption and contact angle measurements for a very limited number of low salinity surfactant formulations [46, 52].

Figure 4 portrays oil desorption efficiencies of low salinity surfactant formulations from silica and an Al–silicate coated quartz surface. The silica surface represents sandstone and the Al–silicate simulates clay mineral often present in sandstone reservoirs. The oil desorption efficiency increases with an increase in surfactant concentration for both surfaces. The low salinity surfactant solutions of both AOT and SDBS surfactant increases the oil desorption from the mineral surfaces. The AOT showed higher potential for oil desorption compared to the SDBS surfactant. Apart from salinity and surfactant type, the oil desorption depends also on the mineral that constitutes the rock surface. The desorption was much higher for the silica surface in comparison to the Al–silicate surface [52].

Khanamiri et al. [46] studied the efficiency of low salinity surfactant formulations with a varying range of monovalent to the divalent ratio in altering wettability of a mineral surface through contact angle measurements. An alkylbenzene sulfonate

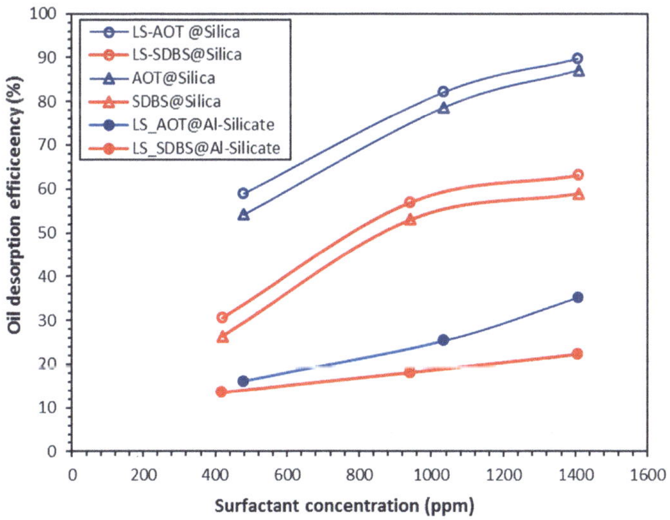

Fig. 4 Oil desorption efficiency form silicate and Al–silicate surfaces as a function of surfactant concentration for different low salinity surfactant and surfactant formulation with AOT and SDBS surfactants obtained from quartz crystal microbalance (QCM) measurements. ("Adapted with permission from Nourani [52]. Copyright 2014 American chemical society")

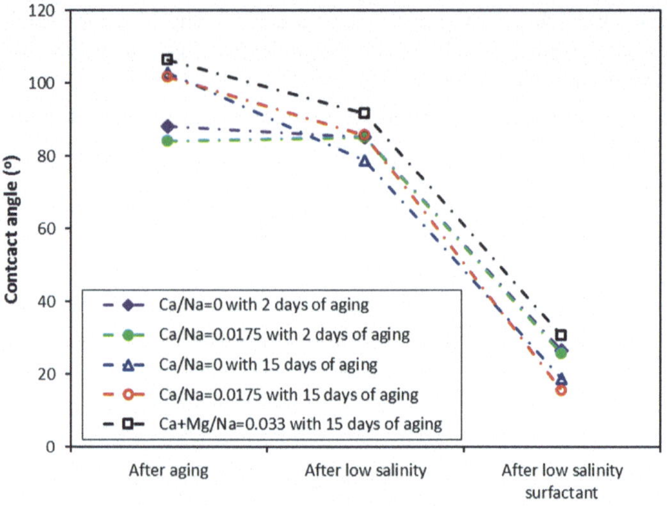

Fig. 5 Contact angles on a silicon dioxide coated crystal of an alkylbenzenesulfonate surfactant (chain length distribution of C_{15}–C_{18}) solution droplet before and treatment in low salinity surfactant. ("Adapted with permission from Khanamiri et al. [46]. Copyright 2016 American chemical society")

surfactant with chain length distribution of C_{15}–C_{18} was used at a concentration of 21.5 weight %. The measured contact angles of the deionized water droplets on the oil aged silica surface (at 70 °C) are shown in Fig. 5. A longer aging time result in more oil wetness of the surface in comparison to a shorter aging time. It can be observed from Fig. 5 that exposure to low salinity water can shift the wettability of a silicate mineral surface towards water wet regime and this effect is highly prominent when further exposed to a low salinity surfactant solution. However, a lower divalent to monovalent ion ratio in the low salinity surfactant formulation can result in more water-wetness of the mineral surface for a sufficiently long-aged surface.

5 Oil Recovery Potential of Low Salinity Surfactant Formulations

The oil recovery efficiency of any novel EOR technique requires its laboratory-scale evaluation before it is tested in the field. The oil recovery efficiency is generally tested in laboratory though core flooding experiments using rock and fluid samples at pressure and temperature conditions representative of a reservoir. Unlike low salinity water flooding, the low salinity surfactant flooding has not been extensively evaluated through core flooding experiments. It has only been studied by a few researchers in the last decade. Also, pilot/field-scale testing has not been performed yet by any oil company. The detail on the limited number of core flooding studies of low salinity water flooding performed by researchers is presented in Table 2. The reported results of core flooding suggested that application low salinity surfactant flooding can increase the oil recovery which is found to vary between 3 to 32% of OOIP over secondary water flooding. The low salinity surfactant flooding has been tested in two different modes: low salinity surfactant is tailored behind a low salinity water flooding in secondary mode and low salinity surfactant flooding after high salinity secondary water flooding. Although incremental oil volume was produced in both cases the former resulted in higher incremental oil recovery as seen from Table 2. Figure 6a presents the recovery profile showing both test mode.

Again the ratio of divalent to monovalent metal cations in both low salinity brine and low salinity surfactant solution observed to influence the incremental recovery by low salinity surfactant injection. An increase in the relative concentration of divalent

Table 2 Summary on low salinity surfactant core flooding studies conducted by different research groups using various formulations along with their obtained recovery efficiencies

Sl. No.	Low salinity surfactant formulation for tertiary flood	Secondary waterflood	Chase brine flood	Crude oil and core used	Incremental oil recovery (% OOIP)	Source
1	5000 ppm NaCl + ENORDET 0242L (an Olefin sulfonate from Shell Chemicals) + 10,000 wt % isoamyl alcohol (IAA)	5000 ppm NaCl	No chase brine flood	Crude with TAN = 2.84 and TBN = 0.95 Berea sandstone	30–32	[2]
2		Synthetic water, TDS = 32321 ppm			20	
3	31,051 ppm NaCl + 500 ppm SDBS	3250 ppm NaCl	No chase brine flood	Crude with API gravity = 33^0 TAN = 1.08 and TBN = 1.16 Berea sandstone	3	[25]
4	3087 ppm NaCl + 136 ppm $CaCl_2$ + 500 ppm SDBS	3087 ppm NaCl + 136 ppm $CaCl_2$			5.1	
5	2825 ppm NaCl + 221 ppm $CaCl_2$ + 30 ppm $MgCl_2$ + 500 ppm SDBS	2825 ppm NaCl + 221 ppm $CaCl_2$ + 30 ppm $MgCl_2$			5.4	
6	31,051 ppm NaCl + 5000 ppm S3 (an alkylbenzenesulfonate surfactant with chain length C_{15} to C_{18})	3087 ppm NaCl + 136 ppm $CaCl_2$	32,500 ppm NaCl		27.1	[26]
7	3087 ppm NaCl + 136 ppm $CaCl_2$ + 5000 ppm S3	3087 ppm NaCl + 136 ppm $CaCl_2$	30,875 ppm NaCl + 1362 ppm $CaCl_2$		27	

'OOIP' stands for original oil in place

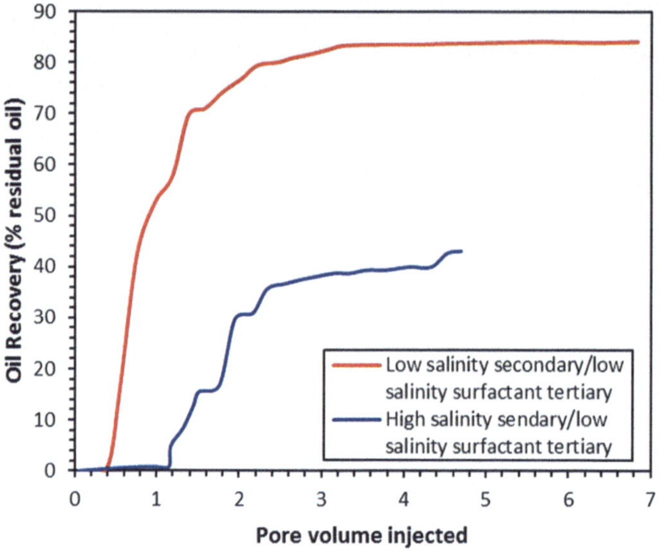

(a)

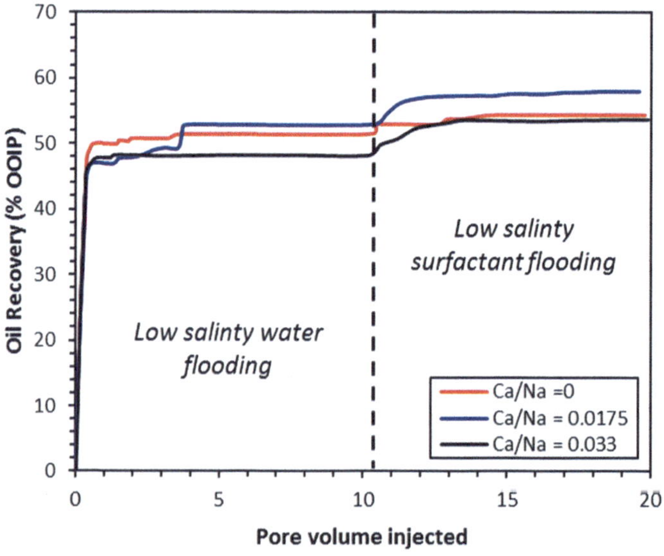

(b)

Fig. 6 Oil recovery profile as function of pore volume injected **a** comparison of low salinity surfactant flooding with low and high salinity secondary flood (Sl. No. 1 and 2 of Table 2) ("Adapted with permission from Alagic and Skauge [37]. Copyright 2010 American chemical society."), **b** comparison of low salinity surfactant flooding for varying divalent to monovalent ratio of metal cation (Sl. No. 3–5 of Table 1). ("Adapted with permission from Khanamiri et al. [14]. Copyright 2016 American chemical society") metal cations in low salinity preflush, as well as low salinity surfactant formulation, resulted in slightly higher incremental recovery (Table 2). The oil recovery profile for different divalent to monovalent cation ratio is presented in Fig. 6b.

When high salinity water is injected as chase brine after low salinity surfactant injection it results in small incremental oil recovery. However, the influence of the composition of the chase brine in terms of divalent to monovalent cation on the oil recovery is negligible.

6 Mechanism of Low Salinity Surfactant EOR

As a hybrid EOR method, low salinity surfactant flooding works through a set of complex mechanisms which are a combination of both low salinity mechanism and surfactant flooding mechanism. The low salinity surfactant solution resulted in a lower oil–water interfacial tension in the oil reservoir in comparison to high salinity surfactant injection due to: (a) higher CMC resulting in delayed micellization; (b) bilayer formation at optimum salt concentration due to charge shielding of surfactant head groups resulting in a tight packing at the interface; (c) lower surfactant aggregation number resulting in lamellar aggregation of surfactant molecules at the interface. A lower interfacial tension between oil and low salinity surfactant solution increases the capillary number (an example shown in Table 3). However, the residual oil saturation after low salinity surfactant flooding (e.g., first case cited in Table 3 resulted in residual oil saturation 6 to 4% of OOIP) is much lower than residual oil saturation which could be achieved according to capillary number theory (40% for the above-mentioned case [53]).

An increase in reservoir pH is a predominant effect associated with low salinity water flooding as evident from various laboratory core flooding reports [6, 11, 15, 27]. Injection of low salinity water into a reservoir that originally contains high salinity formation water causes disturbance to a pre-existing chemical equilibrium established in the crude oil-rock-brine system. High salinity formation water results in the formation of organo-metallic complexes where divalent cation acts as a bridge to

Table 3 Capillary number (N_c) before and after low salinity surfactant injection

Low Salinity surfactant formulation	Secondary water flood composition	N_c during secondary low/ high water flood	N_c during low salinity surfactant flooding
5000 ppm NaCl + ENORDET 0242L (an Olefin sulfonate from Shell Chemicals) + 10,000 wt % isoamyl alcohol (IAA)	5000 ppm NaCl	9.5×10^{-8}	2.2×10^{-4}
	Synthetic water, TDS = 32321 ppm	7.5×10^{-8}	2.2×10^{-4}

("Adapted with permission from Alagic and Skauge [37]. Copyright 2010 American chemical society")

bind negatively charged crude oil molecules to the negatively charged silicate surface of sandstone rock. Low salinity water breaks this kind of complexes and results in the release of divalent cations from the rock or clay surface to re-established chemical equilibrium. The surrounding water molecules dissociate in this process into H^+ and OH^- ions. H^+ ions due to their higher affinity for clay and also to substitute divalent cations adsorbed onto the silicate or clay surface. Thus, OH^- ions remain in solution with increasing pH. With the injection of low salinity surfactant solution, the reservoir pH can further increases as evident from the effluent brine analysis performed by Alagic et al. [54] as shown in Fig. 7. An increase in the pH causes the following effects: (a) saponification of natural alkaline reactive compounds in crude oil, (b) Increasing surface activity of asphaltene. Crude oil consists of saponifiable alkaline reactive components known as petroleum acid. These petroleum acids are also known as nathenic acid which represents an unspecified mixture, of which carboxylic acids are the dominant constituents. Other, components include carboxyphenols, porphyrins, and asphaltene. A reaction between petroleum acids and their subsequent hydrolysis resulted in the generation of insoluble surfactant [55]. The generated surfactant can act in three different ways: reduce interfacial tension, alter reservoir rock wettability and cause emulsification. Again, at high or low pH the functional groups of asphaltene molecules become charged. It results in increased hydrophilic behavior and enhances their surface activity thereby reduces the interfacial tension. The effect of pH on the

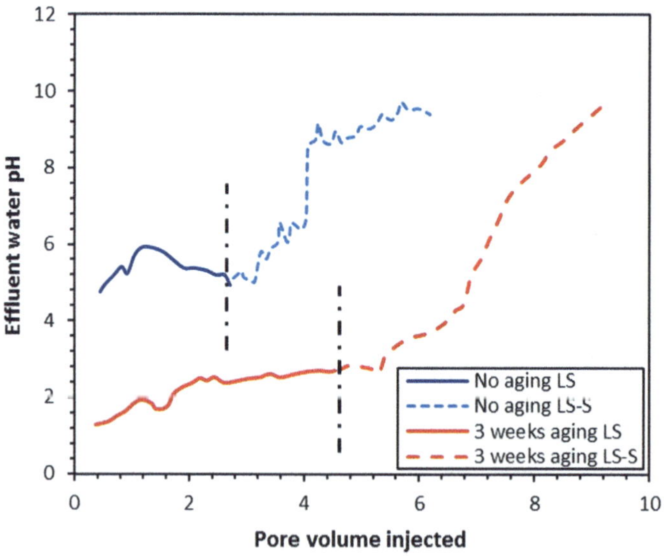

Fig. 7 Variation in effluent pH as function of pore volume injected during low salinity (LS) and subsequent low salinity surfactant (LS-S) flooding for cores (Berea sandstone) with and without aging case. Composition of low salinity water: 5000 pmm; low salinity surfactant solution: 5000 ppm NaCl + 10000 ppm Enordet O242L (an internal olefin sulfonate) +5000 ppm iso-amyl alcohol. ("Adapted with permission from Alagic et al. [54]. Copyright 2011 Elsevier")

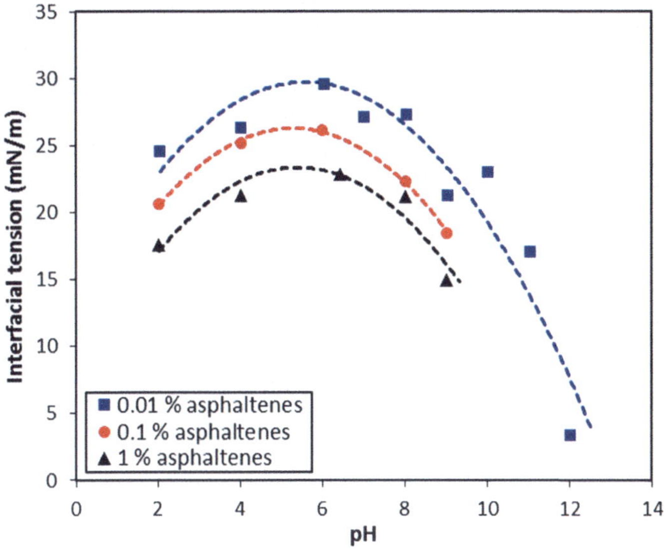

Fig. 8 Oil water between interfacial tension model oil (toluene + asphaltene in 0.01, 0.1, 1.0 wt %) and water as function of pH ("Adapted with permission from Poteau et al. [56]. Copyright 2005 American chemical society")

interfacial tension between a model oil (containing asphaltene) and water is shown in Fig. 8. Though this effect also observed at low pH, it is more prominent at high pH because asphaltene molecules consist of more acidic functional groups as compared to basic functional groups [56]. A high and low pH cause reduce coalescence of water and oil droplets and stabilize oil–water emulsions formed by injected surfactant solution.

A high salinity condition promotes the formation of the water-in-oil emulsion and the surfactant moves over to the oleic phase. An increasing salinity gradient resulted in trapping of surfactant molecules in the oil phase and causes a delayed surfactant breakthrough. On the other hand, a low salinity environment in the reservoir causes surfactants to stay in the aqueous phase and form micro-emulsion by solubilizing oil in water, thus it increases the oil recovery efficiency of the surfactant solution.

Adsorption of surfactant on the rock-forming mineral surface causes an increase in hydrophobicity and thus causes higher oil wetness of a surface [57]. High salinity conditions are known to result in high surfactant adsorption irrespective of the concentration of surfactant. For example, anionic surfactants are widely used for sandstone reservoirs to reduce adsorption due to the similarities in their similar charges; but at high salinity condition the cations in water help in binding of negatively charged anionic surfactant molecules with negatively charged sandstone surface thus promotes its adsorption [58]. Also, high salinity causes precipitation of anionic surfactants and subsequent adsorption of it on the reservoir rock surface [59]. Low salinity surfactant combination thus lowers surfactant adsorption and causes increased water wetness of the reservoir rock. Moreover, the surfactant adsorption also depends on the reservoir pH. An increasing pH causes a sandstone surface to become more negatively charged and thus reduces the adsorption of surfactant [60]. Since low salinity surfactant injection results in increased pH, it further lowers the adsorption of surfactant and promote water wetness. The increased water wetness directly due to low salinity water and as well as due to pH increase, enhances the oil recovery of the low salinity surfactant flooding process. In addition, the wettability is also influenced by the relative concentration of divalent ions in the low salinity surfactant formulation. The presence of divalent ions forms cross-linked aggregations which result in increased elastic modulus of the oil–water interface. Subsequently, a higher elastic modulus of the interface between crude oil and low salinity surfactant solution causes bypassing of the oil by surfactant solution as a result of the smaller strain of the interface [61].

7 Conclusion

Combining novel low salinity water flooding with conventional surfactant flooding can be an effective EOR solution. The synergism of low salinity water flooding and surfactant flooding can greatly impact the reservoir properties and thus improves the oil displacement process. Although a few studies have been published on low salinity surfactant flooding very interesting observations were made in these studies. Low salinity surfactant solutions were able to produce ultra-low oil–water interfacial tension without the application of a co-surfactant. The presence of divalent ions in the low salinity surfactant solution further suppress the interfacial tension values. The pH of the low salinity surfactant solution impacts the interfacial tension values depending on the ionic strength of the solution. Compassion between SDBS and AOT showed low salinity surfactant formulations containing AOT result in much lower interfacial tension as compared to SDBS. Low salinity surfactant solutions can also change the wettability of reservoir rock from oil-wet to water-wet regime. A lower relative concentration of divalent ions can result in a greater shift towards water-wet regime. Laboratory core flooding studies show that a significantly high incremental oil recovery (as high as 32% of OOIP) can be achieved with low salinity surfactant flooding. A complex interplay of physicochemical mechanisms involved in the oil recovery process through low salinity surfactant flooding including interfacial bilyer formation, lamiller surfactant aggregation, insitu soap generation, wetting alteration, oil in water solubilisation etc. However, further research is needed to understand the process and to optimize low salinity surfactant formulations to maximize oil recovery.

References

1. Kamal MS, Hussein IA, Sultan AS (2017) Review on surfactant flooding: phase behavior, retention, IFT, and field applications. Energy Fuels 31:7701–7720
2. Kumari R, Kakati A, Nagarajan R, Sangwai JS (2019) Synergistic effect of mixed anionic and cationic surfactant systems on the interfacial tension of crude oil-water and enhanced oil recovery. J Dispers Sci Technol 40:969–981
3. Hamza MF, Sinnathambi CM, Merican ZMA (2017) Recent advancement of hybrid materials used in chemical enhanced oil recovery (CEOR): a review. IOP Conf Ser: Mater Sci Eng 206:012007
4. Katende A, Sagala F (2019) A critical review of low salinity water flooding: mechanism, laboratory and field application. J Mol Liq 278:627–649
5. Sheng JJ (2014) Critical review of low-salinity waterflooding. J Pet Sci Eng 120:216–224
6. Kakati A, Kumar G, Sangwai JS (2020) Oil recovery efficiency and mechanism of low salinity-enhanced oil recovery for light crude oil with a low acid number. ACS Omega 5:1506–1518
7. McGuire PL, Chatham JR, Paskvan FK, Sommer DM, Carini FH (2005) Low salinity oil recovery: an exciting new EOR opportunity for Alaska's north slope. Presented in SPE western regional meeting, Irvine, CA, USA, 30 March–1 April
8. Needham RB, Doe PH (1987) Polymer flooding review. J Pet Technol 39:1503–1507
9. Robertson EP (2007) Low-salinity waterflooding to improve oil recovery—historical field evidence. Presented in SPE annual technical conference exhibition, Anaheim, California, USA, 11–14 November
10. Seccombe JC, Lager A, Webb K, Jerauld G, Fueg E (2008) Improving wateflood recovery: LoSal™ EOR field evaluation. Presented in SPE symposium on improved oil recovery Tulsa, Oklahoma, USA, 20–23 April
11. Tang GQ, Morrow NR (1999) Influence of brine composition and fines migration on crude oil/brine/rock interactions and oil recovery. J Pet Sci Eng 24:99–111
12. Webb KJ, Black CJJ, Tjetland G (2005) A laboratory study investigating methods for improving oil recovery in carbonates. In the proceddings of internations petroleum technology conference, Doha Qatar, 21–23 November. pp 785–791
13. Kakati A, Kumar G, Sangwai JS (2020) Low salinity polymer flooding: effect on polymer rheology, injectivity, retention and oil recovery efficiency. Energy Fuels 34:5715–5732
14. Khanamiri HH, Enge IB, Nourani M, Stensen JÅ, Torsæter O, Hadia N (2016) EOR by low salinity water and surfactant at low concentration: impact of injection and in situ brine composition. Energy Fuels 30:2705–2713
15. Lee Y, Lee W, Jang Y, Sung W (2019) Oil recovery by low-salinity polymer flooding in carbonate oil reservoirs. J Pet Sci Eng 181:106211
16. Mohsenatabar FA, Zargar G, Kazemzadeh E (2019) An investigation into polymer flooding in high temperature and high salinity oil reservoir using acrylamide based cationic co-polymer: experimental and numerical simulation. J Pet Explor Prod Technol 9:1485–1494
17. Jadhunandan PP, Morrow NR (1995) Effect of wettability on waterflood recovery for crude-oil/brine/rock systems. SPE Reserv Eng 10:40–46
18. Morrow NR, Tang GQ, Valat M, Xie X (1998) Prospects of improved oil recovery related to wettability and brine composition. J Pet Sci Eng 20:267–276
19. Tang GQ, Morrow NR (1997) Salinity, temperature, oil composition, and oil recovery by waterflooding. SPE Reserv Eng 12:269–276
20. Jerauld GR, Mohammadi H (2013) Enhanced oil recovery process using low salinity water. Internatioanl patent no. WO 2013/117741 Al
21. Lager A, Webb KJ, Collins IR, Richmond DM (2008b) LoSal™ enhanced oil recovery: evidence of enhanced oil recovery at the reservoir scale. Presented in SPE symposium improved oil recover, Oklahoma, USA, 19–23 Aprils
22. Ahmetgareev V, Zeinijahromi A, Badalyan A, Khisamov R, Bedrikovetsky P (2015) Analysis of low salinity waterflooding in bastrykskoye field. Pet Sci Technol 33:561–570

23. Skrettingland K, Holt T, Tweheyo MT, Skjevrak I (2011) Snorre low-salinity-water injection-coreflooding experiments and single-well field pilot. SPE Reserv Eval Eng 14:182–192
24. Kakati A, Jha NK, Kumar G, Sangwai JS (7–8 Nov 2017) Application of low salinity water flooding for light paraffinic crude oil reservoir. Presented in SPE symposium: production enhancement and cost optimisation symposium, Kuala Lumpur, Malaysia
25. Kakati A, Sangwai JS (2017) Effect of monovalent and divalent salts on the interfacial tension of pure hydrocarbon-brine systems relevant for low salinity water flooding. J Pet Sci Eng 157:1106–1114
26. Kakati A, Sangwai JS (2018) Wettability alteration of mineral surface during low-salinity water flooding: role of salt type, pure alkanes, and model oils containing polar components. Energy Fuels 32:3127–3137
27. Lager A, Webb KJ, Black CJJ, Singleton M, Sorbie KS (2008) Low salinity oil recovery—an experimental investigation. Petrophysics 49:28–35
28. Rezaeidoust A, Puntervold T, Strand S, Austad T (2009) Smart water as wettability modifier in carbonate and sandstone: a discussion of similarities/differences in the chemical mechanisms. Energy Fuels 23:4479–4485
29. Ligthelm D, Gronsveld J, Hofman J, Brussee N, Marcelis F, Van Der Linde H (2009) Novel waterflooding strategy by manipulation of injection brine composition. Presented in the SPE EUROPEC/EAGE Annual conference and exhibition, Amsterdam, The Netherlands, 8–11 June
30. Pu H, Xie X, Yin P, Morrow NR (2010) Low salinity waterflooding and mineral dissolution. Presented in SPE annual technical conference and exhibition, 19–22 September, Florence, Italy
31. Hill HJ, Lake LW (1978) Cation exchange in chemical flooding—3 Experimental. Soc Pet Eng J 18:445–456
32. Pope GA, Lake LW, Helfferich FG (1978) Cation exchange in chemical flooding—1. Basic theory without dispersion. Soc Pet Eng J 18:418–434
33. Gupta SP, Trushenski SP (1979) Micellar flooding—compositional effects on oil displacement. Soc Pet Eng J 19:116–128
34. Nelson RC (1982) Salinity-requirement diagram—a useful tool in chemical flooding research and development. Soc Pet Eng J 22:259–270
35. Glover CJ, Puerto MC, Maerker JM, Sandvik EL (1979) Surfactant phase behavior and retention in porous media. Soc Pet Eng J 19:183–193
36. Hirasaki GJ, van Domselaar HR, Nelson RC (1983) Evaluation of the salinity gradient concept in surfactant flooding. Soc Pet Eng J 23:486–500
37. Alagic E, Skauge A (2010) Combined low salinity brine injection and surfactant flooding in mixed-wet sandstone cores. Energy Fuels 24:3551–3559
38. Khanamiri HH, Torsæter O, Stensen JÅ (1–4 June 2015) Experimental study of low salinity and optimal salinity surfactant. Prsented in SPE EUROPEC, Madrid, Spain
39. Hirasaki GJ, Miller CA, Puerto M (2011) Recent advances in surfactant EOR. Soc Pet Eng J 16:889–907
40. Sheng JJ (2015a) Status of surfactant EOR technology. Petroleum 1:97–105
41. Pope GA, Baviere MP (1991) Reduction of capillary forces by surfactants. In: Baviere M (ed) Basic concepts in enhanced oil recovery processes. Elsevier Applied Science, London
42. Gräbner D, Hoffmann H (2017) Rheology of cosmetic formulations. In: Sakamoto K, Lochhead R, Maibach H, Yamashita Y (eds) Cosmetic science and technology: theoritical principles and applications. Elsvier Inc. pp 471–488
43. Vaidya S, Ganguli AK (2019) Microemulsion methods for synthesis of nanostructured materials. In: Lipson RH, Nann T (eds) Andrews DL. Elsvier Inc., Comprehensive Nanoscience and Nanotechnology, pp 1–10
44. Jha NK, Iglauer S, Sangwai JS (2018) Effect of monovalent and divalent salts on the interfacial tension of n-heptane against aqueous anionic surfactant solutions. J Chem Eng Data 63:2341–2350
45. Tichelkamp T, Vu Y, Nourani M, Øye G (2014) Interfacial tension between low salinity solutions of sulfonate surfactants and crude and model oils. Energy Fuels 28:2408–2414

46. Khanamiri HH, Nourani M, Tichelkamp T, Stensen JÅ, Øye G, Torsæter O (2016) Low-salinity-surfactant enhanced oil recovery (EOR) with a new surfactant blend: effect of calcium cations. Energy Fuels 30:984–991
47. Kellermann M, Bauer W, Hirsch A, Schade B, Ludwig K, Böttcher C (2004) The first account of a structurally persistent micelle. Angew Chem Int Ed 43:2959–2962
48. Nagarajan R (2002) Molecular packing parameter and surfactant self-assembly: the neglected role of the surfactant tail. Langmuir 18:31–38
49. Wang Z (2010) Interactions between an anionic fluorosurfactant and a PEO-PPO-PEO triblock copolymer in aqueous solutions. J Surfactants Deterg 13:97–102
50. Malcolmson C, Lawrence MJ (1995) Three-component non-ionic oil-in-water microemulsions using polyoxyethylene ether surfactants. Colloids Surf B 4:97–109
51. Sein A, Engberts JBFN, Van Der Linden E, Van De Pas JC (1996) Lyotropic phases of dodecylbenzenesulfonates with different counterions in water. Langmuir 12:2913–2923
52. Nourani M, Tichelkamp T, Gaweł B, Øye G (2014) Method for determining the amount of crude oil desorbed from silica and aluminosilica surfaces upon exposure to combined low-salinity water and surfactant solutions. Energy Fuels 28:1884–1889
53. Garnes JM, Mathisen AM, Scheie A, Skauge A, Hydro N (April 22–25 1990) Capillary number relations for some north sea reservoir sandstones. Presented in SPE/DOE seventh symposium on enhanced oil recovery. Tulsa, Oklahoma
54. Alagic E, Spildo K, Skauge A, Solbakken J (2011) Effect of crude oil ageing on low salinity and low salinity surfactant flooding. J Pet Sci Eng 78:220–227
55. Sheng JJ (2015b) Investigation of alkaline–crude oil reaction. Petroleum 1:31–39
56. Poteau S, Argillier JF, Langevin D, Pincet F, Perez E (2005) Influence of pH on stability and dynamic properties of asphaltenes and other amphiphilic molecules at the oil-water interface. Energy Fuels 19:1337–1341
57. Somasundaran P, Zhang L (2006) Adsorption of surfactants on minerals for wettability control in improved oil recovery processes. J Pet Sci Eng 52:198–212
58. Yekeen N, Manan MA, Idris AK, Samin AM (2017) Influence of surfactant and electrolyte concentrations on surfactant adsorption and foaming characteristics. J Pet Sci Eng 149:612–622
59. Belhaj AF, Elraies KA, Mahmood SM, Zulkifli NN, Akbari S, Hussien OSE (2019) The effect of surfactant concentration, salinity, temperature, and pH on surfactant adsorption for chemical enhanced oil recovery: a review. J Pet Explor Prod Technol 10:125–137
60. Gbadamosi AO, Junin R, Manan MA, Agi A, Yusuff AS (2019) An overview of chemical enhanced oil recovery: recent advances and prospects. Int Nano Lett 9:171–202
61. Khanamiri HH, Torsæter O, Stensen JÅ (2016) Effect of calcium in pore scale oil trapping by low-salinity water and surfactant enhanced oil recovery at strongly water-wet conditions: in situ imaging by X-ray microtomography. Energy Fuels 30:8114–8124

Conformance Control

Combining Particles with Surfactants to Improve Microscopic Displacement and Sweep Efficiency

Mustafa Almahfood and Baojun Bai

Abstract Surfactants and gel treatments are two major types of chemicals that have gained a great deal of attention in the oil industry due to their oil recovery enhancement and water shutoff improvement. The surfactants are mainly used to improve the microscopic displacement efficiency or change the rock wettability from oil wet to oil wet for oil recovery improvement, while gel treatments are often used to significantly reduce the fluid flow through channels or fracture and thus improve water flooding efficiency. The combination of these two methods have a great potential to significantly improve oil recovery in both micro- and macro-scope. This chapter first reviews the fundamentals of surfactants, conventional nanoparticles, polymeric nanoparticles and preformed gel particles that are often used in EOR applications. Then, it describes how the combination of surfactants with these particles can be used to significantly enhance oil recovery in terms of their mechanisms and laboratory experimental results.

Keywords Preformed particle gels · Nanoparticles · Polymeric nanogels

1 Introduction

The majority of oil reservoirs around the globe have already reached or will soon reach at the stage where the oil production rate is approaching the decline phase [83]. As a result, one of the significant tasks that faces the oil business today is the means by which to defer the abandonment of currently flowing fields by producing more oil economically and limiting the amount of produced water. Excess water produc-

M. Almahfood
EXPEC Advanced Research Center, Saudi Aramco, Dhahran, Saudi Arabia

B. Bai (✉)
Missouri University of Science and Technology, Rolla, MO 65401, USA
e-mail: baib@mst.edu

T. Solling et al. (eds.), *Surfactants in Upstream E&P*, Petroleum Engineering,
https://doi.org/10.1007/978-3-030-70026-3_9

tion is viewed as a major issue that prompts early well abandonment and quickens oil reduction for developed wells. EOR applications are commonly executed in oilfields to improve oil recovery and decrease water production. Chemical based EOR strategies can improve oil recovery through four major mechanisms: (1) interfacial tension reduction, (2) wettability alteration towards a water-wet state, (3) conformance control improvement for better sweep efficiency, and (4) emulsifying crude oil [6, 24, 145].

Particles and Surfactants are two major types of chemicals that have been widely investigated to enhance oil recovery. Particles can be classified into conventional particles and deformable particle gels. Conventional particles, especially nanoparticles, have become a recent popular research topic in the industry because of their roles in wettability modification, emulsion stabilization, interfacial tension reduction, and conformance improvement [6, 33]. Particle gels with various sizes from nanometer to millimeter have been also demonstrated their roles in sweep efficiency improvement and microscopic displacement efficiency improvement [8, 15, 16, 30, 34, 50, 53, 117]. Surfactants are the most popular materials that are widely used to reduce interfacial tension and modify rock wettability towards more water-wet for better oil recovery. Overall, both particles and surfactants have been evaluated and applied individually to enhance oil recovery. However, many research studies have indicated that the combination of the two materials can create a synergy effect to greatly improve oil recovery. This chapter will first provide a brief summary about particles and surfactants that are often used in EOR, and then describe the combination of surfactant with nanoparticles, nanogel, and preformed particle gel, individually.

2 Surfactants in EOR

Surface active agents, commonly known as surfactants, are chemical substances that can adsorb on a surface or fluid-fluid interface when present at low concentrations in a system. Surfactants have been widely employed in EOR processes over the years through decreasing oil-water interfacial tension (IFT), adjusting the rock wettability towards more water-wet condition, and emulsifying crude oil [56, 69, 153]. During field operations, surfactant can be injected into a reservoir either through an injection well for flooding purpose or through a production well as a huff-puff process for wettability alteration in carbonate reservoirs.

2.1 Recovery Mechanisms of Surfactants

The main EOR mechanisms of surfactants are interfacial tension reduction and wettability modification towards water-wet state.

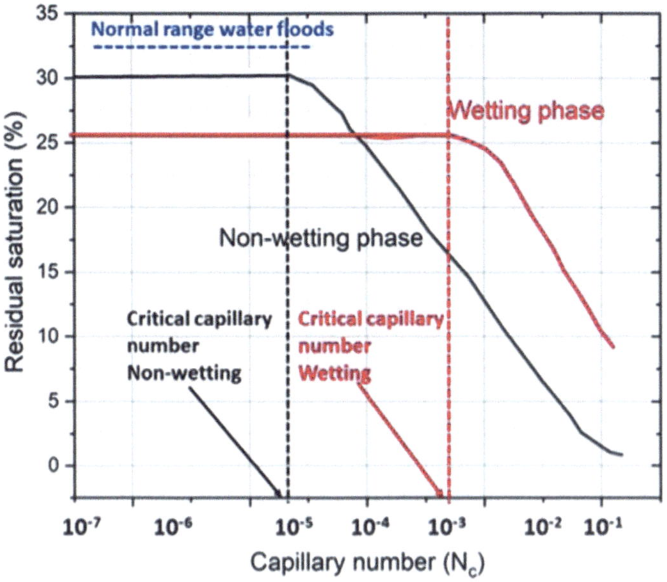

Fig. 1 A typical capillary desaturation curve [120]

Interfacial Tension Reduction: Interfacial tension effect on oil recovery improvement can be explained through the relationship between capillary number and residual oil saturation, which is referred to capillary desaturation curve in the literature. The capillary number is defined as:

$$N_c = \frac{\nu\mu}{\sigma \cos\theta} \tag{1}$$

where ν is the displacing darcy velocity, μ is the displacing fluid viscosity, θ is the contact angle, and σ is the interfacial tension between the displacing and displaced fluids. Many experimental studies have shown that the residual oil saturation can be decreased as the capillary number increases [49, 62, 76, 97, 133]. The relationship between the capillary number and the residual oil saturation is referred to in the literature as capillary desaturation curve. Figure 1 shows a typical capillary desaturation curve.

As shown in Fig. 1, in order to reduce the waterflooding residual oil saturation by half or more, the capillary number has to increase 1000 times from a typical capillary number of 10^{-7} for waterflooding process. Equation 1 shows that the capillary number can be increased by three ways: (1) increasing the injection velocity ν, (2) increasing displacing fluid viscosity μ, and (3) reducing the interfacial tension σ. Increasing the capillary number by 1000 times using the first two ways is not practically feasable. On the other hand, the interfacial tension between displacing and displaced fluids can be reduced from 30 to a value in the order of 10^{-3} mN/m by introducing surfactants to the solution. In other words, the capillary number can be increased by more than 1000 times when surfactants are added to the displacing fluid. As a result of the low IFT, oil droplets can flow easily through the pore throats due to the reduction in the capillary trapping.

Wettability Modification: Compared to sandstone rocks, carbonate rocks tend to be more oil wet, which is one major reason why carbonate reservoirs usually have low oil recovery [74, 96]. As a result, wettability modification has become one of the major targets for carbonate reservoir EOR. After the rock becomes more water wet, water can imbibe into matrix and reduce oil saturation. Many experimental studies have demonstrated the significant contribution of surfactants in wettability alternation projects [1, 39, 121, 139, 140, 160].

3 Particles Used in EOR

3.1 Conventional Nanoparticles

Nanoparticles are defined as solid colloidal particles with sizes ranging from 1 to 100 nm. They are characterized with unique material features due to their extremely small particle size. As materials shrink in size to the nanoscale, their properties are unique and different from those found in bulk materials due to the immense surface area related with nanoparticles. Lately, they have been considered for EOR applications due to their nano-size and environmental friendliness [6].

Types of Conventional Nanoparticles: In the context of EOR, nanoparticles could be subdivided into four main categories: (1) metal oxide, (2) magnetic, (3) silica, and (4) organic particles. Table 1 summarizes the four types of nanoparticles and their associated possible EOR mechanisms.

Application of Conventional Nanoparticle and Associated Mechanisms: A comprehensive understanding of the EOR mechanisms of various nanoparticles is significant when making a selection. Currently, the nanoparticles applied in EOR processes can be divided into three main types: (1) nanofluids, (2) nano-emulsions, and (3) nano-catalysts. Below is a brief description of each application and their associated EOR mechanisms.

Nanofluids: Nanofluids in oil and gas applications refer to the base liquids that contain one added substance with a particle size under 100 nm [44]. Generally, the base liquid can be oil, water, or gas. The major EOR mechanisms related with nanofluids include dis-joining pressure effect, interfacial tension (IFT) reduction, wettability alteration, pore plugging to divert the flow to unswept zones/areas, injection fluid viscosity increase, oil viscosity reduction, and asphaltene precipitation prevention.

Disjoining Pressure: In general, nanoparticles in an aqueous dispersion tend to form a self-assembled structural array at the discontinuous phase such as oil, gas, or polymer. The particles in the three contact region prefer to force themselves into a wedge-shaped film structure and start to force themselves between the solid surface and the discontinuous phase, as delineated in Fig. 2. Particles that are available in the aqueous solution persistently push the particles in the confined region forward

Table 1 Classes of conventional nanoparticles [6]

Category		Type	Possible EOR mechanism	References
Metal oxide		Aluminum oxide (Al_2O_3)	IFT reduction	[109]
			Oil viscosity reduction	[59]
		Copper oxide (CuO)	Heavy oil viscosity reduction	[127]
		Iron oxide (Fe_2O_3)	Viscosity reduction	[151]
				[75]
		Nickel oxide (NiO_3)	Displacing fluid viscosity enhancement	[109]
			Oil viscosity reduction	[108]
		Magnesium oxide (MgO)	Weak recovery agent	[109]
			Causes permeability impairment	[64]
		Tin oxide (SnO_2)	Wettability modification	[104]
				[68]
		Titanium dioxide (TiO_2)	Wettability modification	[43]
			IFT reduction	
		Zinc oxide (ZnO)	Causes permeability impairment	[109]
				[48]
		Zirconium oxide (ZrO_2)	Not common in EOR applications	[109]
Magnetic		Ferro nanofluids	IFT reduction	[75]
				[65]
		Cobalt Ferrite	Oil viscosity reduction	[156]
Silica		SiO_2	Wettability modification	[109]
			IFT reduction	[136]
				[144]
		Alumina Coated	Wettability modification	[135]
		Hydrophobic oxide	Wettability modification	[123]
		Spherical fumed silica	Wettability modification	[161]
		Nano composite	IFT reduction	[107]
		Polysilicon NP	IFT reduction	[119]
		(HLP-LHP-NWP)	Wettability modification	[128]
				[60]
Organic		Carbon nanoparticles	Wettability modification	[158]
				[73]
		Carbonate nanotubes(CNT)	Oil viscosity reduction	[51]
	Polymer	CDGs	Sweep improvement	[42]
				[29]
		Polymer coated	Viscosity reduction	[124]
			Sweep improvement	

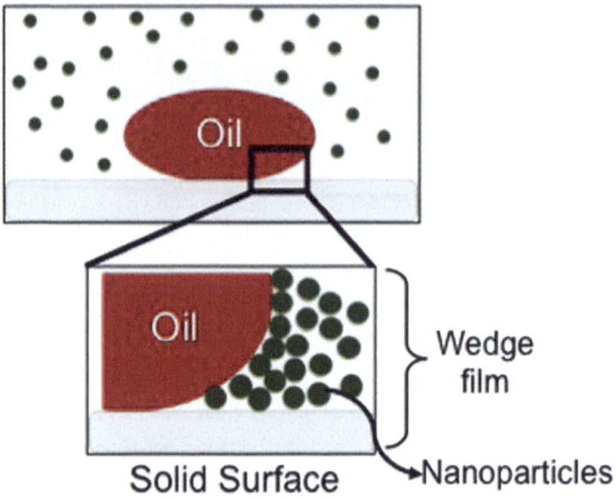

Fig. 2 An illustration of nanofluid wedge film structure (sketched after [78])

and impart a large force known as the disjoining pressure force [32]. The energies that drive this mechanism are Brownian motion and electrostatic repulsion between the particles. The size of the disjoining pressure force is significantly influenced by the size of nanoparticles, temperature, and the saltiness of the base liquid. The force released by a single particle is incredibly feeble, yet when a lot of little particles are available, the force can be as much as 50,000 Pa at the vertex [89]. Hence, the disjoining pressure force causes the system to lose its balance. All together for the system to recover its balance, a portion of its properties, for example, IFT and wettability, would be changed and oil recovery would take place.

IFT Reduction: Interfacial tension (IFT) is viewed as one of the primary parameters estimated to determine liquid distribution and transportation in permeable media. IFT reduction has been regarded as EOR mechanism for nanofluid flooding; however, argument has also existed because the IFT reduction level is usually limited. For example, Hendraningrat et al. [83] estimated the IFT between crude oil and silicon dioxide (SiO_2) nanoparticles utilizing the pendant drop technique. Initially, IFT between the oil and brine solutions was estimated at 19 $\frac{mN}{m}$. Nonetheless, upon adding nanoparticles into the solution, IFT diminished to 8 $\frac{mN}{m}$.

Wettability Modification: Wettability is characterized as the propensity of one liquid to spread on or adhere to a solid surface in the presence of another immiscible fluid. Wettability is viewed as one of the key parameters in multi-phase flow where it influences other reservoir parameters, for example, capillary pressure, relative permeability, and oil recovery efficiency [35]. In addition, it governs the fluid flow in porous media, residual oil saturation and distribution in rocks [13]. Multiple qualitative and quantitative techniques are accessible in the literature for wettability estimations [13]. Qualitative techniques incorporate imbibition tests, microscopic visualization, and wettability assessment utilizing relative permeability curves [35], while quantitative techniques incorporate contact angle estimations and the Amott method [12]. Reservoir rocks can be either water-wet, oil-wet, or intermediate-wet. Figure 3 delineates the contact angle range for each wettability type.

The most elevated portion of oil recovery is corresponding to the propensity towards the most water-wet state in an oil reservoir. Wettability alteration utilizing chemical treatments is an ongoing and developing field of research that is motivated by academic and industrial interests [102]. It is a significant method to enhance the oil recovery from oil-wet or intermediate-wet reservoirs.

Quite a few experimental studies have been conducted to examine the impact of various nanoparticles on wettability alteration. Al-Anssari et al. [4] demonstrated that SiO_2 nanoparticles actuated wettability adjustment on oil-wet and intermediate-wet calcites. These outcomes are consistent with the findings of Roustaei et al. [118]. Wettability modification caused by nanoparticles is influenced by various factors, such as nanoparticle size, concentration, and the saltiness of the base fluid. Hendraningrat et al. [61] showed that the contact angle of the aqueous phase reduced as nanoparticle size decreased. Meanwhile, the incremental oil recovery due to nanoparticles increased as the size of nanoparticles decreased.

The EOR mechanisms of nanofluids are essentially connected together during an oil recovery process. Nonetheless, at least one recovery mechanism might highly affect the recovery. In general, the size, type and concentration of nanoparticles are the fundamental criteria that lead to an alternate nanofluid recovery mechanism. It is well reported in the literature that expanding the nanoparticle size might result in formation damage (pore plugging), which eventually decreases the oil recovery. On the other hand, the smaller size of nanoparticles can increase electrostatic repulsion between nanoparticles, which causes higher disjoining pressure. Additionally, the higher concentration of nanoparticles lead to lower IFT values and higher disjoining pressure.

Nano-emulsion: Smart fluids are prepared by employing nano-sized particles, which have gotten progressively utilized in the oil and gas industry [11]. Nano-emulsions, as illustrated in Fig. 4 are viewed as one of the major types of smart

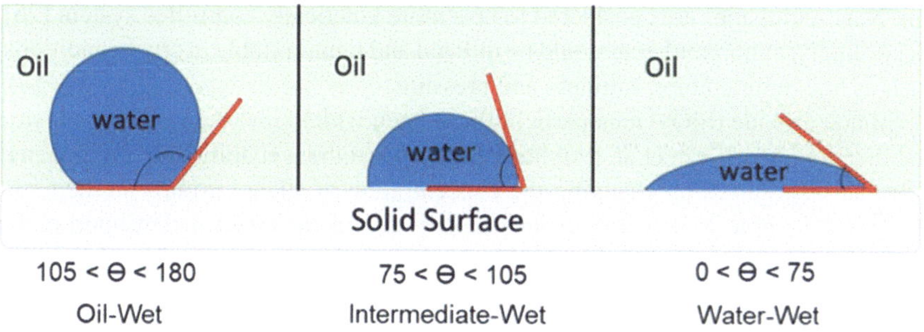

Fig. 3 A schematic of the contact angle of two immiscible fluids with a solid surface. For $105 < \theta < 180$, the liquid drop barely wets the solid wall, leading to poor wettability. For $75 < \theta < 105$, the drop partially wets the solid surface, leading to intermediate wettability. For $0 < \theta < 75$, the liquid drop perfectly wets the solid surface

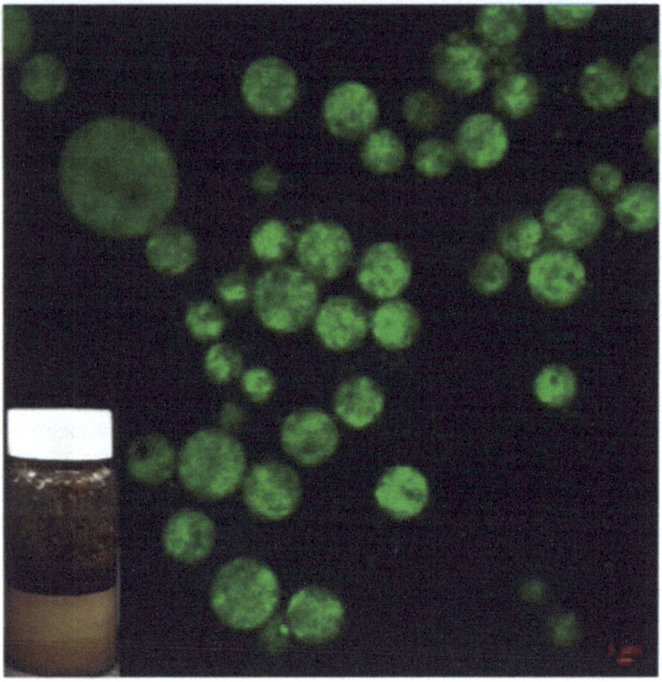

Fig. 4 Confocal microscopy image illustrating nano emulsion from nanofluids (Scale bar is 5 μm) [5]

fluids, which are intended to enhance the oil recovery from hydrocarbon reservoirs. These emulsions stabilized by nanoparticles have been demonstrated to have a greater ability to overcome the limitations and drawbacks of conventional emulsions [88]. Nano- and submicro sized emulsions have attracted a great deal of attention over the years due to their small droplet size, which could be in the range of 50–500 nm. Nano-emulsions are considered to be a more kinetically controlled system [26]. Accordingly, nano-emulsions could be utilized and remain stable in cruel conditions, such as high temperature, saltiness and pressure.

Silicon dioxide (SiO_2) nanoparticles have been widely investigated for emulsion stabilization. Mcelfresh et al. [90] have detailed that the wettability of this nanoparticle can be adjusted by changing the concentration of silanol groups on their surfaces. For example, more stable oil-in-water emulsions can be formed by hydrophilic nanoparticles with a high level of silanol group (>90%). Alternately, when silica nanoparticles are coated with a small amount of the silanol group (<10%) on their surface, water-in-oil emulsions can be formed.

Nano-Catalysts: Nano-catalysts are known as nano-sized particles utilized as catalysts during steam injection into heavy oil reservoirs [58]. In EOR applications, nano-catalysts are basically utilized to reduce the viscosity of heavy oil [134]. It should be noted that this chapter is not intended to profoundly explore nano-catalysts.

3.2 *Particle Gels*

Recently, the study of preformed gels for conformance control improvement has gained a great interest among gel-based conformance control processes. Seright et al. [125, 126] have studied the behavior of preformed bulk gel through fractures and reported that preformed bulk gels showed better placement than in-situ gel systems, and effectively reduced the damage in unswept areas. However, direct pumping preformed bulk gel in fields is usually not feasible because its ultra-high viscosity affects the pumping ability and injectivity. As an alternative, particle gels have attracted a lot of interest in gel treatments. Preformed particle gel systems have been developed and applied for conformance control improvement where they overcome some of the drawbacks inherent in an in-situ gelation system, such as lack of control over the gelation time, gelling uncertainty due to shear degradation, chromatographic fractionation, and dilution by formation water [15–17, 31]. Unlike in-situ gel systems, preformed gels are formed at surface facilities before being injected into the reservoir; thus no gelation occurs in the reservoir.

Currently available particle gel systems can be classified as three types based on the range of their particle sizes: nano-sized and submicro-sized cross-linked polymeric particles known as nano- and submicro-gels [7, 8, 53], microgels [30, 117], and millimeter-sized ($>10\mu m$) preformed particle gels (PPG) [16, 17, 34].

Nanogels: Nanogels as shown in Fig. 5 as an example are recently evolved particles in EOR applications. Nanogels have been perceived as attractive agents for in-depth treatments in heterogeneous reservoirs with low permeability [63, 82]. Like conventional nanoparticles, nanogels are known for their simple injection process due to their small particle size, which is much smaller than the diameter of the pore throats in oil reservoirs [8, 114]. In addition, they are characterized by low viscosity, particularly at low concentrations [7, 98]. Despite their similarities, nanogels contrast from conventional nanoparticles by their ability to deform and swell a few times of their original size when dispersed in salt water. Almahfood et al. [7] and Geng et al. [53] have revealed that the swelling abilitiy of nanogels relies upon brine composition synthesized the particles, saltiness, pH, and temperature. Nanogels are also able to decrease the relative permeability of water more than that of oil due to their hydrophilic polymeric networks that contribute to the permeability reduction [9]. Also, nanogels are known for their ability to adsorb at the oil-water interface to decrease the interfacial tension and stabilize oil-in-water emulsions, which lead to the oil recovery improvement.

Preformed Particle Gels (PPG): PPG technology was initially developed by the Research Institute of Exploration and Development in PetroChina in 1996. It was first applied in Zhongyuan oilfield in 1999. PPGs shown in Fig. 6 as one example are kind of materials that can absorb from a few to several hundred times of their original size. PPGs can overcome some of the drawbacks inherent in an in-situ gelation system, such as lack of gelation time control, gelling uncertainty due to shear degradation, chromatographic fractionation, and dilution by formation water. PPG is formed at a surface facility before injection, so no gelation occurs in the reservoir. PPGs usually

Fig. 5 **a** Dried nanogel. **b** nanogel dispersed in brine [5]

have only one component during injection and display little sensitivity to the physico-chemical conditions in a reservoir, such as pH, salinity, multivalent ions, hydrogen sulfide, and temperature. The swellable particles can be injected into reservoirs to fully or partially plug the fluid flow in fractures or channels [17]. The swelling ratio of PPG depends on its composition when synthesized, as well as the reservoir properties such as temperature, brine salinity and composition [16]. The swelling ratio increases with lower brine salinity and higher temperature. However, PPGs at higher brine salinity tend to be more thermo-stable due to their smaller swelling ratio that increases the crosslinking density; thus resulting in stronger particles [17]. PPGs have been successfully applied in more than 10,000 wells to reduce the permeability of fractures and high permeability channels [16, 85, 86].

PPGs have been widely accepted and are seeing more use by operators due to their unique advantages over traditional in-situ gels, including that:

- PPGs are environmentally friendly, stable in the presence of almost all reservoir minerals and formation water salinities.
- PPGs are both strength- and size-controlled.
- PPGs can preferentially enter into fractures or fracture-feature channels while minimizing gel penetration into unswept zones/matrices. Gel particles with the appropriate size and properties should transport through fractures or fracture-feature channels but should not penetrate into conventional rocks.
- PPGs have only one component during injection. Thus, PPG treatment is a simpler process that does not require many of the injection facilities and instruments that often are needed to dissolve and mix polymers and crosslinkers for conventional in-situ gels.
- PPGs can be prepared with produced water without influencing gel stability. In contrast, traditional in-situ gels are often very sensitive to salinity, multivalent

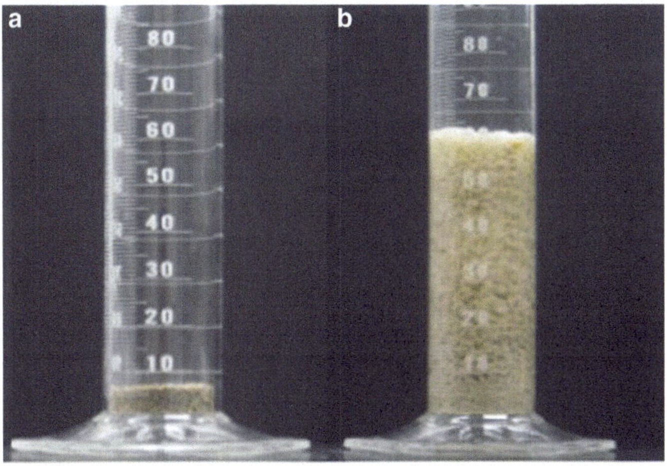

Fig. 6 A picture of a typical preformed particle gel product **a** dried, and **b** swolen in brine

cations, and H_2S in the produced water. This not only saves fresh water but also protects the environment.

4 Conventional Nanoparticles Combined with Surfactants

Previous sections discussed the recovery mechanisms of each kind of particles and surfactants when employed individually. The question that might be raised here is: what will occur if nanoparticles and surfactants are employed together in EOR applications? In fact, nanoparticles have been widely investigated to be combined with surfactants to enhance the strength and stability of nanoparticles, and eliminate some drawbacks related to nanoparticle and other chemicals used in EOR processes, such as the high cost of expensive chemicals. Using nanoparticles in surfactant flooding processes can improve their properties and consequently increase the impact of surfactant solutions on recovery processes. Surfactant-based nanoparticles are functionalized nanoparticles that comprise of a nanoscale part with their surface active groups to perform specific tasks, for example, adsorbing at the oil-water interface to change and alter a portion of their properties including wettability and interfacial tension (IFT).

At the point when surfactants are employed together with nanoparticles, surfactants work as a bridge among nanoparticles and the displacing fluid [157]. The selection of the utilized surfactants chiefly relies on the properties of the nanoparticles and the base-fluid. For example, when there is a need to scatter metal oxide nanoparticles into a non-polar fluid (metal oxide nanoparticles are effortlessly dispersed in polar fluids), the employment of surfactants is needed to improve the stability of the nanoparticles. The level of nanoparticle stability is usually indicated by the zeta potential values. In general, higher absolute zeta potential values refer to higher stability.

Moreover, the relative concentrations of both nanoparticles and surfactants essentially determine the properties of surfactant-based nanoparticles. In case that the concentration ratio of surfactant to nanoparticle is relatively low, only a little portion of the particle surface would be covered with surfactants. However, larger concentration proportions can make the surfactant forms a double layer on the particle's surface, which prompts a hydrophilic nanoparticle surface [45]. The hybrid system of surfactants and nanoparticles has been used to produce stable foams and emulsions [129, 155]. Limage et al. [84] have described that most flocculated nanoparticles are related to hydrophobic particles, containing a mono-layer surfactant on the surface. Cui et al. [37] have indicated that surfactants with single chains are better for foam generation when being combined with nanoparticles since the doable chain surfactants may cause the formation of double layer adsorption on particles at concentrations lower than that of single-chain surfactants.

In general, the interactions among surfactants and nanoparticles can cause a significant change in the surface activity of the surfactant molecules. Essentially, surfactants with higher surface activity (concentrations) are entitled with a higher adsorption into the surface. Therefore, they could greatly decrease the interfacial tension (IFT) and modify the wettability strongly towards a water-wet state. The main mechanisms associated with conventional nanoparticles combined with surfactants are presented as follow.

4.1 Rock Wettability Modification

As nanoparticles are developing in the oil and gas industry, numerous experiments have appeared and demonstrated that the employment of nanoparticles and surfactants is equipped for modifying the wetting property of a reservoir towards a water-wet condition. The extremely small size of nanoparticles allows them to pass through pore throats in typical oil reservoirs and access the residual oil where many other EOR processes are not able to reach into; thereby recovering an amount of the trapped oil [149, 150]. This urged scientists to study and explore the impacts of nanoparticles when employed with surfactants on wettability alteration.

Recently, the impact of surfactant-based nanoparticles on reducing the surface forces has been broadly examined [47, 49, 161]. Table 2 outlines the experimental studies that discuss the wettability alteration of surfactant-based nanofluids.

Binks et al. [25] conducted a thorough experimental study on the behavior of silica nanofluids when combined with cationic surfactants. Their outcomes uncovered that an initially hydrophilic surface without surfactants displayed a contact angle of 8°. The adsorption of surfactants to the air-brine, solid-brine, and in particular, air-solid interfaces resulted in an expansion in the contact angle to a maximum of 63°, followed by a huge reduction at higher concentrations. As a result, silica nanoparticles experienced a change from hydrophilic to hydrophobic, and then back to hydrophilic once again.

Karimi et al. [74] have indicated that the dispersions consisted of ZrO_2 nanoparticles and nonionic surfactants (LA2 and NON-EO4) have the potential to alter the wettability in carbonate reservoirs; thereby can be employed in EOR applications. Their outcomes showed that the combined surfactant-based nanoparticle dispersion is equipped for modifying the wettability from strongly oil-wet to strongly water-wet states. However, the wettability modification resulted from the combination of ZrO_2 nanofluids and nonionic surfactants is a time consuming process that requires a period of at least two days.

Ahmadi and Shadizadeh [3] presented a novel combination between various kinds of nano silica and Zyziphus Spina Christi, a newly developed surfactant, in aqueous solutions for EOR applications. Their procedure was executed to survey the adsorption of surfactant and nano silica in the water phase. The outcomes showed the ability of the novel surfactant to modify the wettability of the surface because of the adsorption on the carbonate surface. Moreover, they performed another experimental study to investigate the impact of nano silica on shale sandstone rock surfaces [2]. Their outcomes demonstrated that hydrophobic nano silica was more effective than hydrophilic nano silica to hinder the absorption losses into the shale sandstone.

Surface modification caused by the combination of $CaCO_3$ nanoparticles and anionic surfactants (linear alkyl benzene sulfonic acid (LABSA) and branched alkyl benzene sulfonic acid (BABSA)) was explored by Song et al. [137]. Their outcomes demonstrated that both surfactant dispersions showed a similar pattern. Essentially, the contact angle increased with surfactant concentrations, reached a maximum value, and after that diminished at elevated concentrations. Their investigation also showed that the earliest increase in contact angle was caused by the arrangement of mono layer of surfactant on particle surface, which resulted in the hydrophobic condition. Nonetheless, when the concentration of surfactant increased above (CMC), bi-layer arrangement of surfactant particles on a surface was formed. This resulted in switching the surface back to a hydrophilic condition, and prompting the decrease in contact angle. In addition, Sharma et al. [132] have demonstrated that dispersions consisted of hydrophilic silica nanoparticles, anionic surfactant (SDS), and polyacrylamide polymer were proficient in altering the wettability from intermediate-wet to strongly water-wet conditions.

Nwidee et al. [108] considered the behavior of surfactant-based nanofluids on the wettability modification of limestone rocks. Various methodologies including contact angle estimations and water imbibition tests were adopted to assess the effectiveness of the combined dispersions. In their experimental study, two distinct surfactants (cationic and nonionic) and metal nanoparticles (zirconium oxide (ZrO_2) and nickel oxide (NiO)) were employed as an function of surfactant type and concentration, nanoparticle-surfactant type and concentration, and imbibition. They have discovered that the contact angle relies upon the nanoparticle-surfactant types and concentration. The nanofluid dispersion composed of the non-ionic surfactant indicated better compatibility with NiO over ZrO_2. However, nanofluids consisted of (ZrO_2 and cationic surfactant) and (NiO and cationic surfactant) consistently diminished

Table 2 Wettability modification of conventional nanoparticle-surfactant flooding experiments [6]

References	NP	Surfactant	Base fluid	Oil type	Core type	Tested parameters	Contact angle (From - To)
[142]	Non-ferrous metal NP	(Anionic) Sulphanole	NA	Heavy	NA	Concentration	40°–19°
[66]	Hydrophobic silica	Non-ionic (TX-100)	NA	NA	NA	Foam stability, concentration	70°–0°
[74]	ZrO_2	Non-EO4 LA2 LA7 Tween-80 Span 20 Span 80 Span 85	Distilled water	Heavy	Carbonate	Concentration, surfactant combinations	180°–30°
[25]	Silica	(Cationic) CTAB CDMAB	NA	NA	NA	PH, foam stability, concentration	35°–63°–8°
[132]	Hydrophilic silica	Anionic SDS	DI	Medium	Sandpack	Concentration, temperature	Intermediate to strongly water wet
[108]	ZrO_2 NiO	TX-100 CTAB	DI	NA	Limestone	Concentration, temperature	155°–18°
[137]	$CaCO_3$	Anionic LABSA, BABSA	NA	NA	NA	PH, concentration	15°–60°
[3]	Aerosil 200 Aerosil R816	(Non-ionic) Zyziphus Spina Christi	DI	NA	Carbonate	Concentration	Wettability change due to adsorption
[54]	Al_2O_3	(Anionic) PRNS	DI	Heavy	Sandstone	Concentration	142°–0°
[20]	Silica	(Non-ionic) Tergitol 15-S-5 Tergitol 15-S-7 Tergitol 15-S-9 Tergitol 15-S-12	Brine	Light	Quartz surface	Temperature, salinity	85°–50°

the contact angle compared with the non-ionic surfactant dispersion. Their outcomes are consistent with the literature that the contact angle is influenced by liquid-rock interactions [20]. Also, they have demonstrated that the employment of nanoparticles can improve the surfactant performance and enhance the steadiness of the formed emulsion. As a result, the inspected nanofluids are practical for in-depth migration in the reservoir and have the capability of settling wettability issues related with limestone formations.

The employment of alumina-based nanofluids that are made of Al_2O_3 nanofluids and anionic PRNS surfactant in modifying the wettability of sandstone rocks was experimentally studied by Giraldo et al. [54]. The outcomes have indicated that the employment of the anionic PRNS surfactant as a surface modifier can be improved when combined with oxide nanoparticles in low concentrations. The contact angle measurements were at first estimated as high as 142° and were decreased to about 0° after the employment of nanofluids. This demonstrated that the wettability of the system was altered from strongly oil-wet to strongly water-wet condition.

Conversely, a laboratory evaluation of a dispersion of metal nanoparticles combined with anionic surfactants indicated that wettability stayed unaltered with nanofluid addition [142]. Nonetheless, flooding tests using cores have indicated that oil recovery has enhanced. Thus, the recovery mechanism of this test study is not to be interpreted by wettability modification.

4.2 *Oil-Water Interfacial Tension Reduction*

Silica Nanoparticles: Silicon dioxide (SiO_2) are the most common compound on earth, as they structure over 90% of the earth's crust. This has made silicates the most regularly utilized, environmentally friendly, and practical nano sized particles. The adjustment in oil-water interfacial tension due to surfactant-based nanoparticles flooding is as yet an open inquiry. Nonetheless, a few investigations are available that examine the viability of nanoparticles when combined with surfactants on interfacial tension of oil-water systems. Table 3 illustrates the laboratory studies that investigate the impact of surfactant-based nanoparticle flooding on IFT.

Ravera et al. [116] investigated the air-aqueous interfacial tension of nano sized silica solutions when combined with cationic surfactants. They have discovered that the employment of the silica nanoparticles reduced the surface and interfacial tension. The impact of silica nanoparticles when combined with ionic and non-ionic surfactant systems on surface and interfacial tension was studied by Ma et al. [87]. They have demonstrated that silica nanoparticles negligibly affect the surface and interfacial tension of nonionic surfactant systems, while expands the surface activity of the anionic surfactant dispersions, and as a result reduces the interfacial and surface tension.

Nonetheless, the interfacial tension of oil-water interface in the presence of silica nanoparticles and cationic surfactants (cetyltrimethyl ammonium bromide) was assessed by Lan et al. [77]. Their investigation demonstrated that silica nanoparti-

Table 3 Interfacial tension reduction of silica nanoparticle-surfactant flooding experiments [6]

References	NP	Surfactant	Base fluid	Oil type	Core type	Tested parameters	IFT mN /m From-To)
[77]	Silica Aerosil 200	CTAB	DI	NA	NA	Concentration, NP size	52–5
[147]	Hydrophilic silica	CTAB	Distilled water	Heptane	NA	Concentration	51–2
[116]	Colloidal silica	CTAB (cationic)	Water	Hexane	NA	Frequency, concentration	72.5–21
[79]	SiO_2	(Anionic) XSA-1416D SS16-47A IAMS-M2-P	DI	Crude oil	Sandstone	Concentration, surfactant type	24–0.2
[67]	Hydrophobic silica	CTAB SDBS CPC	Water	NA	NA	Concentration	73–32
[159]	Hydrophilic fumed silica (A-300) Hydrophobic silica (R-816)	(Anionic) SDS	Distilled water	Medium	Sandpack	Concentration, injection regime	20–2
[114]	Hydrophobic fumed silica	(Non-ionic) Triton X-100	Brine	Heavy	Sandstone	Concentration	29–0.08
[24]	hydrophilic silica	(Non-ionic) alkylpoly-oxyethylene	Water	Medium	NA	Concentration	31–1.7
[132]	Hydrophilic silica	(Anionic) SDS	DI	Medium	Sandstone	NP size, concentration, adding polymer, temperature	NA-2
[143]	Hydrophobic silica	(Anionic) SDS	Brine	Crude	Sandpack	Concentration, temperature	22–6
[87]	Colloidal silica	SDS Triton X-100	Water	Tri-chloro-ethylene	NA	Concentration	40–2
[18]	SiO_2	(Non-ionic) Tween 20	DI	Heavy	Carbonate	Concentration, dye effect	42–10

cles minimally affect the interfacial tension of cationic surfactant dispersions. This insignificant nanoparticle impact is most likely caused by the high level of surfactant adsorption to the rocks' surface. These outcomes are consistent with the behavior of surfactant-polymer dispersions detailed by Bell et al. [19]. Also, Le et al. [79] showed that SiO_2 nanoparticles combined with anionic surfactant dispersions [XSA-1416D, SS16-47A, and IAMS-M2-P] resulted in ultra-low IFT measurements.

Understanding the synergy and interaction between hydrophobic silica nanoparticles and ionic surfactants may reveal an insight into the implementation of such nanoparticles in improved oil recovery applications. The synergy between hydrophobic silica nanoparticles when combined with charged surfactants (CTAB, SDBS, and CPC) was researched by Jiang et al. [67]. In their experimental setup, the concentration of silica nanoparticles was fluctuated, while the surfactant concentration was fixed below CMC levels. It was seen that nanoparticles began to aggregate when the concentration was above 1%. Thus, the interaction between silica nanoparticles and ionic surfactants was tested when the nanoparticle concentration was below 1%. Their results demonstrate that both the surface tension and the zeta potential increased with the employment of nanoparticles to CTAB and CPC surfactants, suggesting that the silica nanoparticles were attracting surfactant particles.

Despite the fact that the addition of particles to surfactant systems does not always improve the stability of emulsions as suggested by Legrand et al. [80], a stability and rheology investigation of oil-in-water emulsion containing a combination of hydrophilic silica nanoparticles and non-ionic surfactant particles was detailed by Binks et al. [24]. The oil-water interfacial tension of the nanofluid combination has significantly reduced from 31 to 1.7 $\frac{\text{mN}}{\text{m}}$. Moreover, Sun et al. [143] examined foam stability by mixing hydrophobic SiO_2 nanoparticles with an anionic surfactant (SDS). The outcomes demonstrated that the decrease in interfacial tension by a factor of 80% after the employment of nanofluids has prompted foam stability, and as expected, their outcomes showed that foam stability reduced with higher temperature.

Qiu et al. [114] reported a laboratory experiment of the potential utilization of hydrophobic fumed silica nanoparticles (CAB-O-SIL TS-530) and a non-ionic surfactant (Triton X-100) to recover heavy oil from the Alaska North Slope oil reservoir. Generally, this kind of nanoparticles was utilized to enhance the stability of the emulsion with the surfactant. Additionally, it is capable of thickening the emulsion and giving a great resistance that decreases the adsorption by the rocks. In this analysis, rheology evaluation on the emulsion has been performed by adding various amounts of nanoparticles. The outcomes have showed that the utilized nanofluids are fit for stabilizing the emulsion due to their large surface area, which can lower the interfacial tension values with immiscible phases. Also, core flooding experiments have demonstrated that the emulsion flooding significantly increased the oil recovery factor after water flooding from 76 to 95% [113]. This improvement showed that the emulsion plugged the water channels, increased the sweep efficiency, and mobilized the residual oil.

Zargartalebi et al. [159] conducted a series of experiments to study the impact of hydrophobic and hydrophilic silica nanoparticles when combined with an anionic surfactant (SDS) on interfacial tension measurements. It was seen that the adsorption level of the utilized surfactant was decreased when combined with nanoparticles. The outcomes of the interfacial tension estimations between surfactant-based nanoparticle dispersions and oil uncovered a strange behavior that began with a fast reduction in low surfactant concentrations, followed with a major increase at higher concentrations.

Furthermore, the impact of hydrophilic SiO_2 nanoparticles mixed with an anionic surfactant (SDS) and a polymer (polyacrylamide, PAM) on wettability modification and interfacial tension reduction was studied by Sharma et al. [132]. Their outcomes demonstrated that super low IFT values were obtained when these nanofluids were utilized, compared to the employment of each solution alone.

Vatanparast et al. [147] conducted multiple experiments for the effect of CTAB surfactant and hydrophilic silica nanoparticle dispersions at low surfactant to nanoparticle proportions, in which only small amount of surfactant was adsorbed at the nanoparticles' surface. They have discovered that hydrophilic silica nanoparticles are not surface modifiers alone, and showed minimal effects on the oil-water interfacial tension. Nonetheless, upon the employment of CTAB surfactant, nanoparticles transform into surface modifiers by adsorbing surfactant particles and firmly influencing the interfacial tension values. Likewise, Bazazi et al. [18] demonstrated that interfacial tension measurements between heavy oil and water diminished 75% after the employment of nonionic surfactant-based silica nanoparticle dispersions.

Metallic Oxide Nanoparticles: Metalic oxide nanoparticles are known as hydrophilic materials. Table 4 lists the experimental studies that investigate the impact of metallic nanoparticles and surfactants on IFT.

Vashisth et al. [146] have revealed that the employment of surfactant displaced nanoparticles from the interface. They have additionally expressed that a total interfacial displacement and oil recovery can only be accomplished when surfactant concentration is above the CMC. Additionally, they showed that the necessary energy to eliminate the trapped nanoparticles from the interface is at least equal to a few thou-

Table 4 Interfacial tension reduction of metallic nanoparticle-surfactant flooding experiments [6]

References	NP	Surfactant	Base fluid	Oil type	Tested parameters	IFT mN/m (From–To)
[46]	ZrO_2	SDS CTAB LA7	Distilled water	Heptane	Concentration, NP size	51–5
[142]	Non-ferrous metal NP	(Anionic) Sulphanole	NA	Heavy	Concentration	18–1
[55]	Al_2O_3, ZrO_2, $Ca_3(PO_4)_2$, TiO_2	(Non-ionic) Triton X-45	DI	NA	NP type, PH, Concentration	72–55
[138]	CuO	Triton X-100	Brine	Heavy	Concentration, temperature	20–0.1

sand times more than the typical "Brownian thermal energy well". Since this energy corresponds to the oil-water interfacial tension [21], the employment of surfactant is equipped for decreasing the desorption energy by reducing the oil-water interfacial tension.

Esmaeilzadeh et al. [46] have detailed the fluid-fluid and fluid-air interfacial tension of zirconium oxide nanoparticles in cationic, anionic, and nonionic surfactant dispersions. They have demonstrated that ZrO_2 nanoparticles adsorbed at the oil-water interface, bringing a reduction in the interfacial tension measurements. Their investigation uncovered that ZrO_2 nanoparticles strongly interacted with all inspected surfactants. It was additionally discovered that the employed nanoparticles had no impact on the interfacial tension when the concentrations of the surfactants are over the critical micelle concentration (CMC). Alternately, ZrO_2 nanoparticles had a lowering impact for all tested surfactants when the concentration was below CMC. It was likewise demonstrated that the minimum interfacial tension value was reported when nanoparticles were blended with the nonionic surfactant (LA7). Moreover, a laboratory investigation of a dispersion consisting of anionic surfactants and metal nanoparticles was studied by Suleimanov et al. [142]. It was indicated that the addition of the nanofluid allowed a 70–90% reduction in interfacial tension on the oil limit, when compared with the employment of each surfactant solution alone.

4.3 Oil Viscosity Reduction and Conformance Control

Adjusting the viscosity of the injected chemicals to match the oil phase is an important procedure to acquire better conformance and mobility control capacities. Multiple laboratory evaluations have demonstrated that the viscosity of fluid dispersions increases with reducing the size of silica nanoparticles [78, 91–94], as delineated in Table 5. Rankin and Nguyen [115] revealed the idea of silica nanoparticle gels for conformance control in heterogeneous and naturally fractured reservoir. They indicated that permeability reduction caused by gelation is achievable at low concentrations of silica nanoparticles.

As clarified earlier, nanoparticles can be utilized to enhance the stability of emulsion and foam due to their surface activities, which lead to the improvement of injection fluids. Furthermore, nanofluids that consist of hydrophilic silica nanoparticles and cationic surfactant (CTAB) were experimentally reported to enhance the stability of emulsions and improve the recovery from heavy oil reservoirs [110]. A potential clarification of this behavior is that the oil-in-water emulsions stabilized by surfactant-based silica nanoparticles may have shown a critical shear thinning behavior with high viscosity at low shear rates, making them possible candidates for conformance control. Ogolo et al. [109] and Ragab et al. [122] demonstrated that aluminum oxide nanofluids had the potential to decrease oil viscosity by breaking carbon-sulfur bonds, which ultimately can be utilized for conformance and mobility control. Moreover, Sharma et al. [131] led a series of experimental studies and demonstrated that the performance of conventional surfactant-polymer flooding was enhanced by an emulsion stabilized with nanoparticle-surfactant-polymer dispersion due to improved water viscosity and oil-water IFT.

Qiu et al. [114] have shown that the viscosity of the utilized emulsion expanded with the number of nanoparticles available in the system, and the emulsion would in general act like a Newtonian fluid. This experiment has also expressed that the utilized nanofluids can improve the mobility of the emulsion by thickening it. Eventually, these surfactant-based nanofluids can diminish the surfactant adsorption by the porous media and limit its utilization in the entire enhanced oil recovery process.

Polyacrylamide nano-spheres have been experimentally evaluated by Wang et al. [148]. These particles were scattered in a combination of emulsions with $NaOH$, which was a critical factor in diminishing IFT by framing in-situ surfactants. The outcomes indicated that oil recovery from a heavy oil reservoir expanded by a factor of 20%. However, the use of this nanofluid in field applications is limited due to economical reasons [105].

The viscosity reduction of heavy oil fields by employing surfactant-based nanoparticle dispersions has been studied by Srinivasan et al. [138]. The outcomes have uncovered that the produced nano-emulsions composed of CuO nanoparticles and Triton X-100 surfactant were effective in diminishing the viscosity of the heavy oil to lower qualities compared with conventional viscosity reducers, particularly at higher shear rates. A huge decrease in interfacial tension was reported after the formation of the oil-in-water emulsion system.

Pei et al. [110] introduced an investigation of silica nanoparticle-surfactant (CTAB) stabilized emulsion to improve heavy oil recovery. They conducted phase behavior and rheology tests to investigate the influence of nanoparticles on the stability and rheological properties of the emulsion system. Afterward, they conducted core flooding experiments to explore the recovery mechanisms for improved heavy oil

Table 5 Oil viscosity reduction of nanoparticle-surfactant flooding experiments [6]

References	NP	Surfactant	Base fluid	Oil type	Core type	Tested parameters	Viscosity, cp (From–To)
[138]	CuO	Triton X-100	Brine	Heavy	NA	Concentration, temperature	600–150 at 70 F° 50–18 at 175 F°
[131]	Hydrophilic SiO_2	SDS	Brine	Light	Sandstone	NP size, injection regime, adding polymer	800–10 at 100 F° 800–7 at 200 F°
[148]	Polyacrylamide microgel nano-spheres	SLPS	DI	Heavy	Sandpack	Concentration, adding polymer	238–5 at 130 F°
[110]	Hydrophilic SiO_2	CTAB	Distilled water	Heavy	NA	Concentration, shear rate	350–2 at 130 F°

recovery by nanofluids. The phase behavior results have uncovered that the expansion of nanoparticles cannot only improve the stability of the emulsion, yet also increase the emulsion's viscosity. The microscopic study proposed that nanoparticles can thicken the emulsion to the needed mobility, prompting a major improvement in sweep efficiency. The heavy oil was emulsified into the water phase to form emulsions with the help of ionic surfactants, which were the recovery mechanisms for enhance oil recovery by silica nanoparticle-surfactant stabilized emulsion.

4.4 Challenges and Limitations of Conventional Nanoparticles Combined with Surfactants

The implementation of various sizes, types, and concentrations of surfactants and nanoparticles prompts different recovery systems. Hence it is important to characterize the objective and target of a surfactant-based nanoparticle dispersion when choosing the size, type and concentration of both surfactants and nanoparticles. In addition, the necessary recovery components could be accomplished by choosing the suitable surfactants and nanoparticles. The choice of the utilized surfactants basically relies upon the properties of nanoparticles and the dispersion. In other words, the concentration of surfactants and nanoparticles characterizes the properties of surfactant-based nanofluids. When the concentration proportion between surfactants and nano-sized particles is moderately low, only a little portion of nanoparticle's surface will be covered with surfactants. Nonetheless, bigger concentration proportions can form a double layer of surfactants on the nanoparticles.

Despite the fact that nanopartices, alone and coupled with surfactants, have shown promising results in the lab, their deployment in field applications is still limited. Generally, a few difficulties must be settled before this innovation is utilized in field applications as a potential EOR process. These difficulties are as per the following:

1. Technical difficulties related with nanoparticles: As stated previously, nanoparticles tend to aggregate and plug pore throats due to their strong interactions, particularly under harsh conditions. Thus, it is urgent to produce homogeneous solutions of nanoparticles and employ strong; yet economical surfactants to improve their stability [43].
2. The accessible experimental studies conducted utilizing metallic oxide nanoparticles demonstrated their capacity as EOR agents. Nonetheless, the quantity of these examinations is limited compared to studies conducted utilizing silica nanoparticles. This is likely because of the accessibility of silicas over metallic oxides. As a result, performing experimental studies on various kinds of nanoparticles, including metallic, magnetic and inorganic, is significant to evaluate their capacity to enhance oil recovery.
3. The absence of investigations utilizing a combination of nanoparticles and surfactants. This absence of experimental investigations on nanofluid blends hinders their potential improvement in EOR applications.

4. The quantity of core flooding experiments is extremely limited. Performing these analyses is altogether significant to picture the oil enhancement level.

As a result, it is critical to perform experimental studies that defeat and overcome the previously mentioned difficulties. Furthermore, building numerical models for different surfactant-based nanoparticles may be useful to comprehend the recovery systems. These models are planned to help in choosing the best and appropriate nanofluid recovery mechanisms for field applications, and diminish the risk related with it.

5 Polymeric Nanogel Combined with Surfactant

Recently, the combination between nanosized particles and surfactants have attracted a great deal of attention by many researchers [74, 87, 95, 152]. Suleimanov et al. [142] have shown that the usage of nanosized particles combined with anionic surfactant permitted a great reduction of surface tension. Moreover, it has been revealed that the usage of nanosized with an anionic surfactant has a major impact on increasing the ultimate oil recovery [54]. Karimi et al. [74] have studied the effect of nanosized particles combined with several surfactants on carbonate reservoir rocks and reported that the combination has a strong capability for oil recovery enhancement. Next, the possible recovery mechanisms of nanogels combined with surfactants are presented.

5.1 *Interfacial Tension Reduction Mechanism*

Nanogels can reduce the interfacial tension by adsorbing at the oil-water interface, which stabilizes oil-in-water emulsions, leading to improvement of the recovered oil from reservoirs [8, 52, 81]. The reduction in interfacial tension mobilizes residual oil, which enhances oil recovery. Nanogels form an irreversible adsorbing layer at the oil-water interface, and tend to deform in a unique fashion. The particles adsorbed in a flattened morphology at the oil-water interface that prevents drops from coalescence [112]. Nanogels are stretched out when the surface coverage is low due to the high free energy gain of covering larger interfacial area as compared to the energy of elastic deformation of the particles [41]. The ability of nanogel dispersions for interfacial tension reduction is strongly linked to its concentration and brine salinity. The dynamic interfacial tension of liquid-liquid interfaces can be divided into four stages: the early stage where both interfacial tension and surface area are decreased rapidly; the second stage where both interfacial tension and surface area are decreased in a slower rate; the third stage where the interfacial tension is decreased in a slower rate and the surface area stays constant; and the late stage where the interfacial tension reaches the equilibrium value [28].

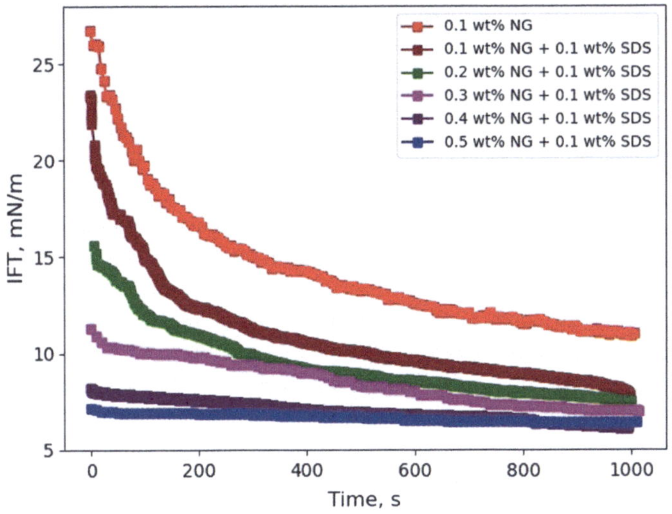

Fig. 7 An illustration of the effect of nanogel combined with surfactant on interfacial tension reduction [7]

Geng et al. [52] have studied the interfacial tension between several nanogels with different surface charges while dispersed in different brine salinities and crude oil/Decane. They have stated that nanogel-decane interfacial tension reduced from 26 to 11, 4, and 8 $\frac{\text{mN}}{\text{m}}$ by cationic, anionic, and neutral charged nanogels, respectively. They observed similar trend between nanogel/crude oil where the interfacial tension reduced from 45 to 16, 5, and 12 $\frac{\text{mN}}{\text{m}}$ by cationic, anionic, and neutral charged nanogels, respectively. They have also shown that the equilibrium interfacial tension between nanogel and oil phases decreased with higher brine salinity. This might be due to the reduction in the electrostatic repulsion among the nanogel particles at the interface, which increases the stability of nanogel layers. Comparing these results with nanoparticle performance reported by Bizmark et al. [27], the decrease of oil-brine interfacial tension in the presence of nanogels was faster due to their high diffusivity.

These observations of nanogels call for theoretical and experimental studies aimed at understanding the synergy of flow type and recovery mechanisms of the combination of nanogels and surfactants as a potential enhanced oil recovery method. Although limited studies have been reported on the ability of the combined technology to reduce the interfacial tension, the few available studies clearly show that surfactants can lower the interfacial tensions of nanogel dispersions [5, 8]. Almahfood et al. [8] have shown that the interfacial tension between anionic nanogel combined with anionic surfactant and oil reduced from 25 to 6 $\frac{\text{mN}}{\text{m}}$. They have indicated that this reduction in the interfacial tension permitted an improvement in the oil recovery from sandstone rocks by a factor of 15% (Fig. 7).

5.2 Emulsion Stability Mechanism

Nanogel particles are capable of stabilizing oil droplets by forming absorbed layers and providing better resistance than emulsions stabilized solely by surfactants. Emulsions stabilized by particles, commonly known as Pickering emulsions, can be used as conformance control agents due to their higher viscosity compared to the displacing fluid [57]. Furthermore, it is well established by multiple studies that the interaction and combination between nanoparticles and surfactants can either stabilize or de-stabilize oil-in-water emulsions [70]. The long term stability is generally a function of surfactant type, composition, and emulsifier concentration [111]. Almohsin et al. [9] have reported that emulsions stabilized by nanosilica and nonionic surfactant could not form a stable emulsified oil droplets. Hence, Pickering emulsions stabilized by conventional nanoparticles and surfactants are not stable for the long-term as coalescence takes place within few hours [38]. Therefore, the use of chemically-proven nano materials along with surfactants seems to be promising to ultimately create long-term stable emulsions.

Microgels and nanogels have been used for many years for multiple applications including pharmaceuticals, medicine and cosmetics [72, 130]. Their application to stabilize oil-in-water emulsions has been reported by Binks et al. [23] and Ngai et al.[106]. In addition, it was observed by few researchers that oil-in-water emulsions stabilized by surfactants and nanosized particles with controlled pH and water concentration can provide significant improvements to emulsion stability by controlling the wettability and degree of flocculation [22]. Almahfood [5] has studied the stability against coalescence and flocculation of oil-in-water Pickering emulsions in the presence of both nanogels and surfactants. In particular, the effect of combin-

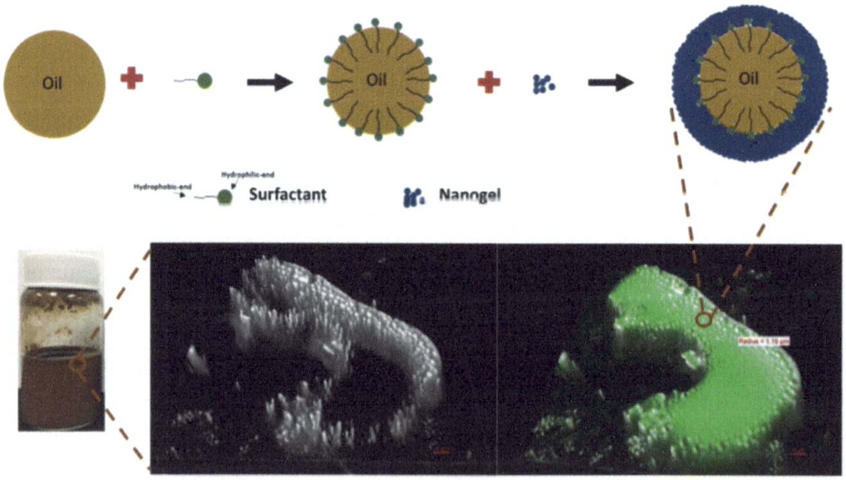

Fig. 8 An illustration of emulsified oil droplet stabilized by nanogel and surfactant. 3-D confocal images show oil droplets in green and nanogel-surfactant clusters in white

ing anionic nanogel with multiple surfactant types in different brine salinities, pH and sonication times on the stability of emulsions was evaluated. The bahavior of emulsions was reported to be dependent on surfactant type and surface charge as emulsions stabilized by anionic nanogel and cationic surfactant produced less stable oil droplets that suffered from flocculation regardless of brine salinity and sonication time. They have also shown that lower brine salinity significantly influenced the stability of the emulsions, especially when nanogels were combined with anionic surfactants. They have also reported that the strong acidic conditions lowered the stability of emulsion systems.

Figure 8 illustrates how nanogels and surfactants might be attached to the oil droplet. Basically, the hydrophobic end of surfactants will be attached to oil molecules. Since nanogels are partially hydrophobic, they will basically fill the spaces between surfactants and oil molecules which leads, at best scenarios, to enhance the stability of the Pickering emulsion. The 3-D confocal images show an oil drop stabilized by nanogel and surfactant. They both clearly illustrate that the combination of nanogel and surfactant will form clusters at the oil-water interface. The 3-D image on the right shows an emulsified oil droplet in green color, and nanogel-surfactant solution, in white color, at the interface. To better visualize the nanogel-surfactant clusters, oil droplet is not shown in the 3-D image on left. Here, the white color represents the solid-like clusters created by the synergy between nanogel and surfactant.

5.3 Plugging Mechanism

Several experimental studies have concluded that nanogels can adsorb and form a blockage in porous media, which reduces the relative permeability of water and improves the oil recovery [10]. Nevertheless, the surface charges of both nanogels and surfactants significantly impacts the aggregation degree of the dispersions, which either reduces or enhances the plugging performance; depending on the surface charge and type of the reservoir rocks. Surface charges also affect the arrangement and adsorption of nanogels and surfactants at rock surfaces [71]. In general, nanogel-surfactant dispersions are attracted to rock surfaces with opposite charges. It is extremely crucial to understand the synergy and charge neutralization between nanogel and surfactant dispersions, and the interactions between these dispersions and rock surfaces to better understand the permeability reduction mechanism and particle transportation in porous media [5, 8].

Almahfood et al. [8] have studied the transportation of an anionic nanogel when combined with anionic surfactant in sandstone reservoirs. They have showed that the combined technology was not suggested as strong plugging agents in sandstone rocks due to their anionic surface charge which limited the plugging efficiency to less than 40%. On the other hand, Almahfood [5] has studied the performance of the combined technology in carbonate reservoirs with low permeability, which are characterized with positive surface charge, and reported their ability to plug pore throats that resulted in improving the oil recovery by a factor of 27%.

5.4 *Wettability Modification Mechanism*

The wettability modification of nanogel dispersions towards water-wet state is caused by their adsorption on rock surfaces. These particles prefer to arrange themselves into a wedge-like structure and begin to force themselves at the oil-water interface. The adsorption basically changes the balance between the capillary and viscous forces [97]. This change creates a wedge-film structure on the solid surface, which helps spreading the nano particles throughout the surface; thus releasing a great portion of the trapped oil [78]. Giraldo et al. [54] have shown that the adsorption of nanogels on sandstone rocks significantly modified the wettability of the oil-water-rock system. Although several studies have reported the ability of nanogel to modify the wettability towards water-wet condition, to the best of our knowledge, their performance has not been tested when combined with surfactants. Thus, more work has to be carried out to investigate the performance of the combined technology on modifying the wettability.

6 PPG Combined with Surfactants

PPG treatments are intended to reduce the volume of water produced with the oil but also can result in improved sweep efficiency. When successful, these gel systems divert a portion of the injected water into areas not previously swept by water. Oil recovery is the product of displacement efficiency (ED) and sweep efficiency (ES). EOR methods focus on increasing either ED by reducing residual oil saturation or ES by correcting reservoir and fluid heterogeneity. Surfactant EOR and gel treatment are two principle EOR methods. Each has limitations that can be largely avoided by combining the two methods.

Surfactant EOR increases displacement efficiency by reducing capillary pressure and altering wettability. Surfactants can be injected either from an injection well (called flooding) or a production well (called huff-puff or soaking). It is well known that the function of surfactant in surfactant flooding is to reduce residual oil in the swept area and to improve displacement efficiency by reducing interfacial tension between oil and water and altering formation wettability. Field tests have shown that surfactant flooding can increase oil recovery by 10–20% after waterflooding. However, early surfactant breakthrough often can occur due to flow short-circuiting. This occurs because surfactant flooding is always performed in mature oil fields where reservoir heterogeneity has been aggravated due to previous oil production and water injection. Early breakthrough wastes surfactants and increases lifting costs.

Surfactant soaking is mainly used to alter reservoir wettability from oil-wet to water-wet for oil recovery improvement, mostly for carbonate and heavy oil reservoirs. During field application, a surfactant solution is first injected into a production well, and then the well need shut off for a few days to allow surfactant enter non-swept regions by spontaneous imbibition; because surfactant solution will preferentially enter the fractures or high permeable channels while little surfactant enters unswept areas during its injection, and lastly the well is put back into production. This method can significantly improve oil recovery in lab scale by spontaneous imbibitions, but it is a slow recovery process because it is limited by the rate of molecular diffusion. Stoll et al. [141] indicated that spontaneous imbibition could not provide an economically interesting opportunity unless external forces enable forced imbibition.

Surfactant EOR can increase displacement efficiency but has little or no effect on sweep efficiency improvement. Gel treatment can improve sweep efficiency but has little or no effect on displacement efficiency improvement. A few studies have been conducted to investigate the idea of coupling PPG treatment and surfactants to improve both sweep and displacement efficiency [36, 40, 103]. Two factors have supported the idea that PPG is a better gel for the combination process. The first factor is that large size of particle gel can only enter high permeability zone/streaks or fractures and will not penetrate into low permeability zones or matrix if particle sizes and strength are properly selected and thus has little damage on low permeable oil zones as shown in Fig. 9. Another factor is the compatibility of PPG and surfactant, and it has been found that: (1) surfactant can greatly reduce PPG strength when mixed together but the strength can be recovered after surfactant is removed; and (2) some surfactants cannot enter the pores in particle gel network and thus there is no surfactant loss to the gel. The two factors make preformed particle gels unique from bulk gels, which was not known before.

Comparing single PPG treatment, the coupled EOR process has the following distinct advantages:

- PPG injectivity can be greatly improved because surfactant can greatly reduce PPG strength.
- The reduced PPG strength can be recovered due to water dilution during water flooding after the treatment, which will increase PPG plugging efficiency.
- The surfactant solution mixed with PPG will be squeezed from gel particles into low permeability zones/areas where crude oils are trapped by capillary force. The surfactant solution can dramatically reduce the capillary force (interfacial tension between oil and water) and release the crude oil.

The combined injection of PPG and surfactant will result in a higher injection pressure gradient in the reservoir because of the high flow resistance resulting from the particle. This increased pressure will produce an additional force to drive surfactant into the matrix or low-permeability areas; thus, forced imbibitions can be practical. It is expected that the successful development of this technology will provide a more cost-effective method for improving oil recovery and reducing water production for mature oilfields. Bai et al. [14] have conducted a series of investigation about the combination technology with the financial support from Research

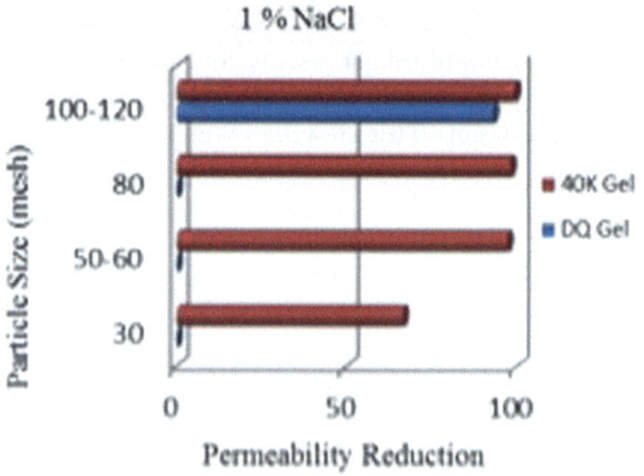

Fig. 9 Effect of Particle size and strength on permeability reduction (the rock permeability is 85 md). 40K is a weak particle gel and DQ is a strong particle gel. As shown in the Figure, DQ has no damage on rock permeability when its sizes are 30, 50–60 and 80 meshes, but it has damage when its size is 100–120. 40K will have damage on rock permeability at all cases

Partnership of Securing Energy for America (RPSEA). In this subsection, we briefly summarize the some research results related to the combination technology.

6.1 *The Compatibility of PPG and Surfactant*

To combine the PPG treatment and surfactant in one process, it is essential to screen both PPG and surfactants that can be compatible with each other. Swelling ratio and swollen particle gel strength are the two main parameter that should be evaluated for the process. Muhammed [101] has extensively studied the impact of anionic, non-ionic and cationic surfactants on the swelling ratio of several PPGs. He found that both the surface charges and concentration of surfactants influenced the swelling ratio of particle gels. Figure 10 shows one example of the effect of surfactant concentration on a commercialized PPG sample. It can be seen that the increase in the surfactant concentrations will lead to low swelling ratio. The influence has the same trend with the ion strength of different concentration brines.

Wu et al. [154] have indicated that a synthesized PPG is highly compatible with some surfactants, and the gel strength in terms of the elastic modulus (G') is reduced by aqueous surfactant solutions. However, after the surfactant has been washed off or diluted to a very low concentration, the gel's strength can be recovered to its initial values in 1.0 wt% NaCl only (Fig. 11). The reduced strength after being mixed with surfactant will improve the PPG's injectivity, but the recovered gel strength after being washed with water (water injection after the coupled treatment) will keep the PPG's blocking efficiency the same as the PPG without surfactant.

Wu et al. [154] also found that the equilibrium surfactant concentration in excess brine increased for most cases after swelling of gel particles as shown in Table 6. This results indicate that the particle gel almost does not absorb most surfactants when it absorbs water from the surfactant solution, especially when the surfactant concentration is above its critical micelle concentration (CMC). Therefore, no surfactant will be lost to the particle gel due to absorption if we can select a proper surfactant.

Muhammed et al. [99] also investigated the interaction between negatively charged PPGs with several surfactants with different surface charges. They have shown that the swelling of PPGs in surfactant solutions could not increase the concentration of anionic and neutral surfactants, but decreased the concentration of cationic surfactants.

6.2 *Oil Recovery Mechanisms of the Combined Method*

Muhammed et al. [100] carried a series of core flooding experiments using both sandstone and carbonate fractured rocks to evaluate the performance of the combined technology. They reported that PPGs could only enter fractures without forming a strong face plugging on the surface of rock with low permeability; thus, they cause little damage to unswept oil zones. This allowed the surfactant to be squeezed from

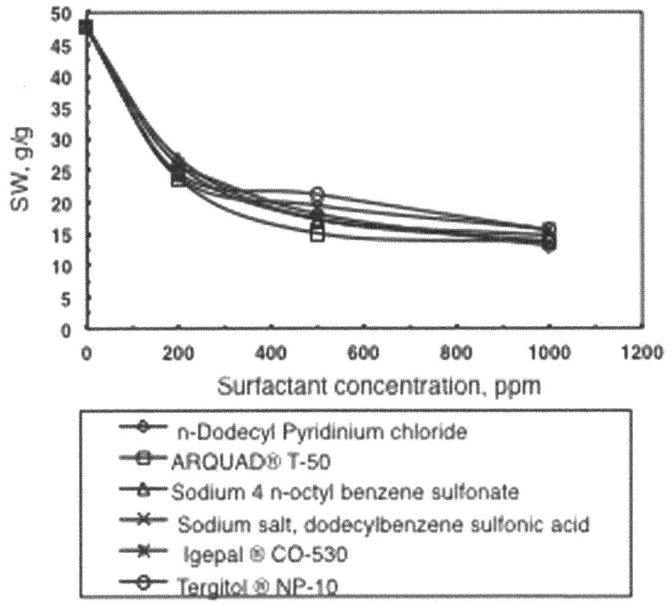

Fig. 10 Equilibrium SW of Cerogel® in different types of surfactants at ambient temperature [101]

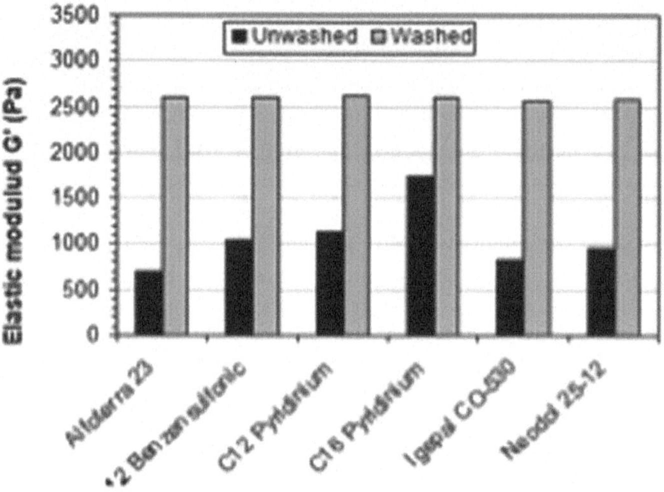

Fig. 11 Elastic modulus, G', of particle gels with and without surfactant. After surfactant molecules have been washed off, the modulus G' increases to the value of the particle gel swollen in 1.0 wt.% NaCl [154]

Table 6 PPG effect on surfactant concentration

Surfactant initial concentration: 200 ppm			At equilibrium of swelling		
Surfactant	λ_{max}	Initial	Net	C_{eq} (ppm)	Conc. Change (%)
n-Dodecylpyridinium chloride	257	2.451	2.326	190	−5.0
(1-Hexadecyl)pyridinium bromide	259	2.106	2.432	231	15.5
Benzalkonium chloride	261	0.179	0.187	209	4.5
Sodium 4-n-octyl benzene sulfonate	260	0.232	0.247	213	6.5
Sodium salt, dodecylbenzene sulfonic acid	260	0.227	0.262	231	15.5
Igepal® CO 530	276	0.806	1.137	282	41.0
Tergitol® NP-10	275	0.429	0.608	283	41.5

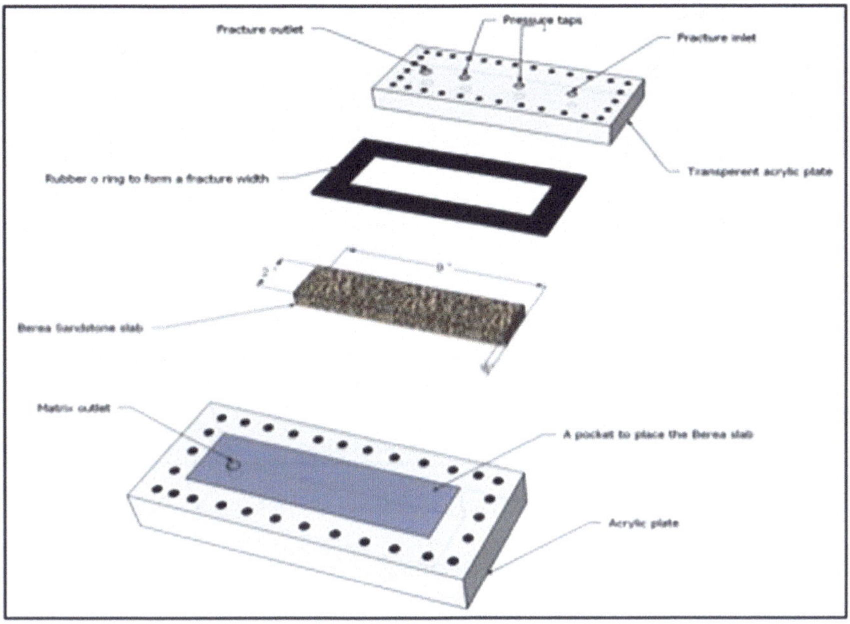

Fig. 12 Schematic diagram of the semi-transparent model [100]

gel particles into unswept zones, where it can modify the wettability and reduce the capillary forces to ultimately reduce residual oil saturation. Figure 12 shows the semi-transparent model that was used to evaluate the combination PPG transport through fracture and the forced imbibition during the process of PPG extrusion through fracture by Muhammed et al. [100]. The semi-transparent model allows one to view the PPG propagation through the fracture, as well as water and/or surfactant penetration through the PPG pack and matrix.

The combination of PPG and surfactant can be practiced through two processes with slightly different mechanisms. The first method to implement the technology is that PPG swollen in brine is injected into the reservoir with channels or fractures, and then surfactant solution is injected into the reservoir, as illustrated in Fig. 13. During PPG injection, the large size particles will preferentially enter the high permeability channels or fractures while they have limited penetration into the low permeability zones/areas or matrix. After PPG injection, surfactant is injected into the reservoir, and it is expected that the surfactant will mainly enter the unswept regions to improve the oil recovery. Figure 14 gives one example of pressure response and oil recovery improvement from lab results. It can be seen significant oil recovery improvement can be reached after PPG injection due to surfactant flooding.

The second method to implement the technology is that PPG swollen in surfactant is injected into the reservoir with channels or fractures, and then water is injected into

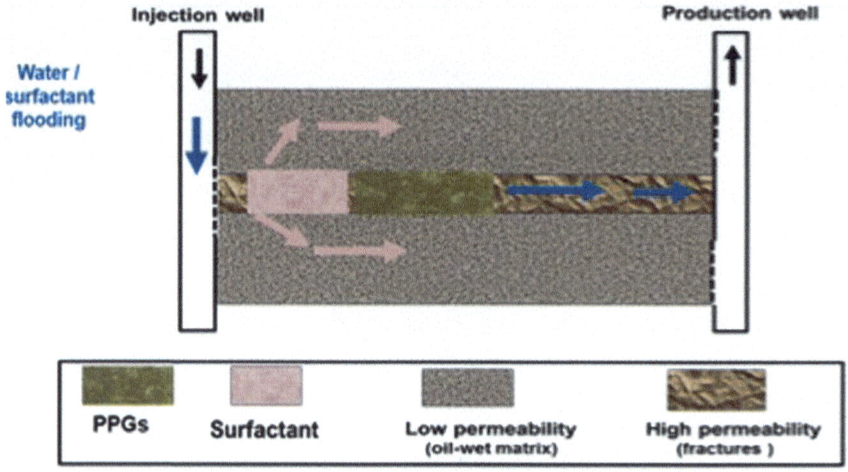

Fig. 13 The combination process of PPG injection followed by surfactant flooding

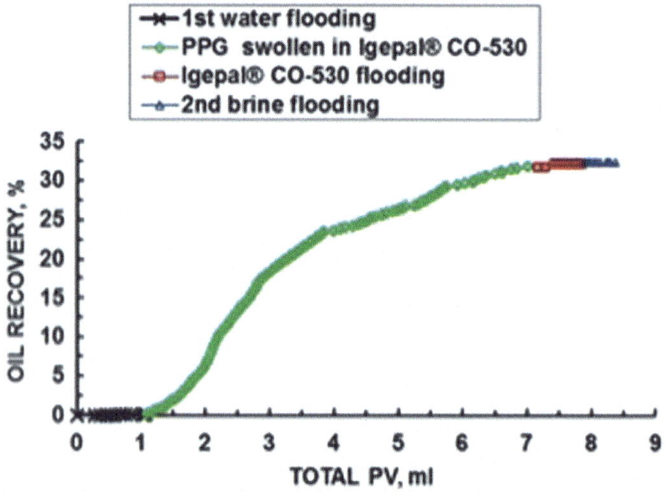

Fig. 14 Oil recovery response from the combination process of PPG followed by surfactant

the reservoir, as illustrated in Fig. 15. Figure 16 shows the oil recovery factor during the process that the PPG swollen in a surfactant was injected into the model. During the initial water flooding, there was no oil being produced from the matrix outlet because the matrix was oil wet and all of the water was directed to the fracture outlet. When the PPG and surfactant mixture was injected, significant oil was produced due to the increased injection pressure resulted from the PPG resistance to the fluid flow in the fracture, which forced the surfactant solution leak off into the matrix through fracture surface to alter the matrix wettability and improve oil recovery. It is expected

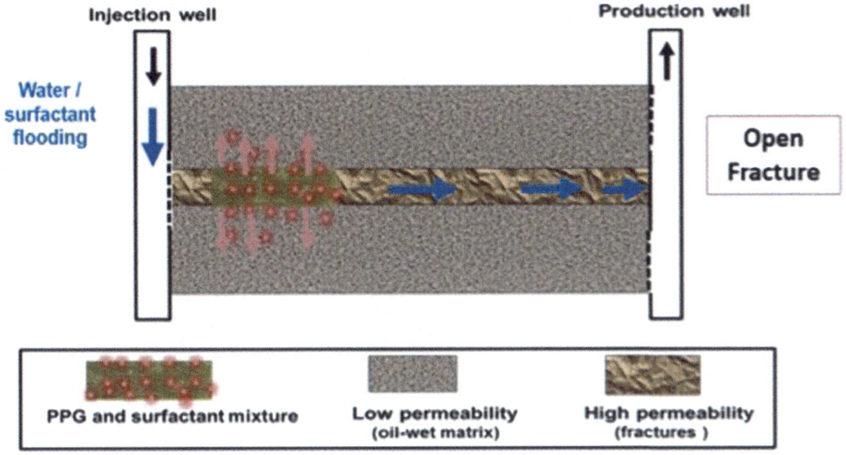

Fig. 15 The combination process implemented by PPG swollen in surfactant solution

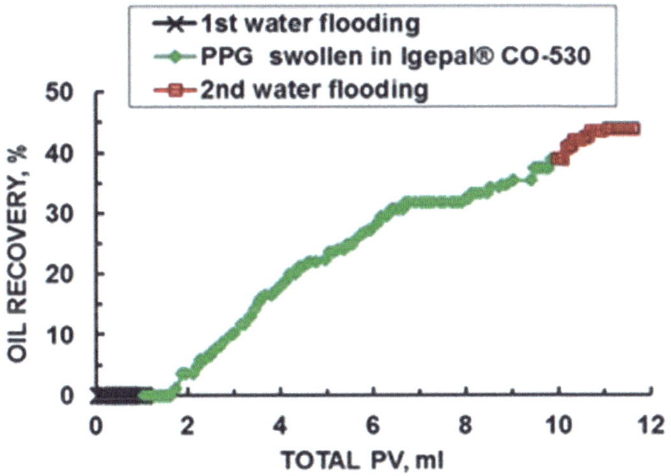

Fig. 16 Oil recovery factor during treatment with PPG swollen in surfactant

that this method could be very effective in carbonate reservoirs because a majority of carbonate reservoirs are featured as a complex fracture and oil-wet matrix system. Surfactant imbibition have been approved to be very effective in the rocks from this type of reservoirs. However, the spontaneous imbibition process relies on the molecular diffusion, which is a pretty slow process and is argued not to be effective in real reservoir conditions. The combination of PPG swollen in surfactant provide a mechanism to force surfactant being imbibed into matrix where the wettability could be modified for enhanced oil recovery.

7 Summary

In this chapter, the EOR mechanisms and performance of different particle systems when combined with surfactants are summarized. Conventional nanoparticle and surface modified nanoparticles have been widely investigated to enhance oil recovery due to their role in EOR interfacial tension reduction, wettability modification, mobility control and sweep efficiency improvement. Synergy has been found when these nanoparticles are combined with surfactants because of their benefit in enhancing the stability of nanoparticles, hybridizing oil recovery improving mechanisms, eliminating some drawbacks related to single method, such as the high cost of chemicals.

Despite their similarities, Nanogels differ from conventional nanoparticles by their ability to deform and swell several times of their original size in diameter when being dispersed in brine with different salinities. Combining polymeric nanogels and surfactant has shown to improve the stability of oil-in-water emulsions and reduce the capillary forces between fluids; thus enhancing the oil recovery. In addition, the synergy and charge neutralization between different nanogel and surfactant dispersions from one side, and their interactions with reservoir rocks from the other side are critically crucial to understand the permeability reduction mechanism and particle transportation in porous media. Combining anionic nanogels and surfactants has shown to reduce the interfacial tension and increase the oil recovery in both sandstone (anionic) and carbonate (cationic) rocks. Nevertheless, combining anionic nanogel and surfactant dispersions in carbonate rocks has a noticeable plugging ability, especially in reservoirs with low permeability, which diverts the flow to unswept areas, compared to a limited impact on sandstone rocks.

Preformed Particle Gel (PPG) treatments have been successfully applied in oilfield to control reservoir conformance. Surfactants have been widely investigated to improve oil recovery from sandstone reservoirs through residual oil reduction process and from carbonate reservoirs through wettability alteration process. It has been approved that surfactant can significantly improve oil recovery from oil-wet carbonate rocks in lab scales; however, the spontaneous imbibition is a slow recovery process unless external force can be added to speed up the imbibition rate. Lab tests have indicated the combination of surfactants and PPG can significantly improve oil recovery. The method could be very effective in carbonate reservoirs because it can provide external forces to push surfactant solution to enter matrix to modify the wettability.

The conclusions reported in this chapter indicate the capability of different preformed particle systems when combined with surfactants on oil recovery enhancement. A major potential advantage of combining particle systems and surfactants is the ability to improve both displacement and sweep efficiency. Different particle gels have different level of contribution towards recovery mechanisms. Combining PPGs and surfactants is particularly promising for highly heterogeneous and naturally fractured reservoirs, while nanosized particles are preferred to be combined with surfactants in reservoirs with low permeability.

References

1. Adibhatia B, Sun X, Mohanty K et al (2005) Numerical studies of oil production from initially oil-wet fracture blocks by surfactant brine imbibition. In: SPE international improved oil recovery conference in Asia Pacific. Society of Petroleum Engineers
2. Ahmadi MA, Shadizadeh SR (2013) Induced effect of adding nano silica on adsorption of a natural surfactant onto sandstone rock: experimental and theoretical study. J Petroleum Sci Eng 112:239–247
3. Ahmadi MA, Shadizadeh SR (2012) Adsorption of novel nonionic surfactant and particles mixture in carbonates: enhanced oil recovery implication. Energy Fuels 26(8):4655–4663
4. Al-Anssari S, Barifcani A, Wang S, Iglauer S (2016) Wettability alteration of oil-wet carbonate by silica nanofluid. J Colloid Interface Sci 461:435–442
5. Almahfood M (2020) Experimental study on a novel EOR method–polymeric nanogel combined with surfactant and low salinity water flooding to enhance oil recovery
6. Almahfood M, Bai B (2018) The synergistic effects of nanoparticle-surfactant nanofluids in EOR applications. J Petrol Sci Eng 171:196–210
7. Almahfood M, B Bai (2020) Characterization and oil recovery enhancement by a polymeric nanogel combined with surfactant for sandstone reservoirs. Petrol Sci, pp 1–13
8. Almahfood MM, Bai B, et al (2020) Potential oil recovery enhancement by a polymeric nanogel combined with surfactant for sandstone reservoirs. In: Offshore technology conference
9. Almohsin A, Alabdulmohsen Z, Bai B, Neogi P et al (2018) Experimental study of crude oil emulsion stability by surfactant and nanoparticles. In: SPE EOR conference at oil and gas West Asia. Society of Petroleum Engineers
10. Almohsin AM, Bai B, Imqam AH, Wei M, Kang W, Delshad M, Sepehrnoori K et al (2014) Transport of nanogel through porous media and its resistance to water flow. In: SPE improved oil recovery symposium. Society of Petroleum Engineers
11. Amanullah Md, Al-Tahini AM et al (2009) Nano-technology-its significance in smart fluid development for oil and gas field application. In: SPE Saudi Arabia section technical symposium. Society of Petroleum Engineers
12. Amott E et al (1959) Observations relating to the wettability of porous rock
13. Anderson WG et al (1986) Wettability literature survey-part 1: rock/oil/brine interactions and the effects of core handling on wettability. J Petrol Technol 38(10):1–125
14. Bai B, Imqam A (2015) Study and pilot test of preformed particle gel conformance control combined with surfactant treatment
15. Bai B, Li L, Liu Y, Liu H, Wang Z, You C et al (2007) Preformed particle gel for conformance control: factors affecting its properties and applications. SPE Reservoir Eval Eng 10(04):415–422
16. Bai B, Liu Y, Coste J-P, Li L et al (2007) Preformed particle gel for conformance control: transport mechanism through porous media. SPE Reservoir Eval Eng 10(02):176–184
17. Bai B, Wei M, Liu Y et al (2013) Field and lab experience with a successful preformed particle gel conformance control technology. In: SPE production and operations symposium. Society of Petroleum Engineers
18. Bazazi P, Gates ID, Nezhad AS, Hossein Hejazi S et al (2017) Silica-based nanofluid heavy oil recovery a microfluidic approach. In: SPE Canada heavy oil technical conference. Society of Petroleum Engineers
19. Bell CG, Breward CJW, Howell PD, Penfold J, Thomas RK (2007) Macroscopic modeling of the surface tension of polymer- surfactant systems. Langmuir 23(11):6042–6052
20. Bera A, Ojha K, Kumar T, Mandal A (2012) Mechanistic study of wettability alteration of quartz surface induced by nonionic surfactants and interaction between crude oil and quartz in the presence of sodium chloride salt. Energy Fuels 26(6):3634–3643
21. Binks BP (2002) Particles as surfactants-similarities and differences. Curr Opin Colloid Interface Sci 7(1):21–41

22. Binks BP, Whitby CP (2005) Nanoparticle silica-stabilised oil-in-water emulsions: improving emulsion stability. Colloids Surfaces A: Physicochem Eng Aspects 253(1–3):105–115
23. Binks BP, Murakami R, Armes SP, Fujii S (2006) Effects of pH and salt concentration on oil-in-water emulsions stabilized solely by nanocomposite microgel particles. Langmuir 22(5):2050–2057
24. Binks BP, Desforges A, Duff DG (2007) Synergistic stabilization of emulsions by a mixture of surface-active nanoparticles and surfactant. Langmuir 23(3):1098–1106
25. Binks BP, Kirkland M, Rodrigues JA (2008) Origin of stabilisation of aqueous foams in nanoparticle-surfactant mixtures. Soft Matter 4(12):2373–2382
26. Binks BP, Lumsdon SO (2000) Influence of particle wettability on the type and stability of surfactant-free emulsions. Langmuir 16(23):8622–8631
27. Bizmark N, Ioannidis MA, Henneke DE (2014) Irreversible adsorption-driven assembly of nanoparticles at fluid interfaces revealed by a dynamic surface tension probe. Langmuir 30(3):710–717
28. Campanelli JR, Wang X (1999) Dynamic interfacial tension of surfactant mixtures at liquid-liquid interfaces. J Colloid Interface Sci 213(2):340–351
29. Chang HL, Sui X, Xiao L, Guo Z, Yao Y, Yiao Y, Chen G, Song K, Mack JC et al (2006) Successful field pilot of in-depth colloidal dispersion gel (CDG) technology in Daqing oilfield. SPE Reservoir Eval Eng 9(06):664–673
30. Chauveteau G, Omari A, Tabary R, Renard M, Veerapen J, Rose J et al (2001) New size-controlled microgels for oil production. In: SPE international symposium on oilfield chemistry. Society of Petroleum Engineers
31. Chauveteau G, Tabary R, Le Bon C, Renard M, Feng Y, Omari A, et al (2003) In-depth permeability control by adsorption of soft size-controlled microgels. In: SPE European formation damage conference. Society of Petroleum Engineers
32. Chengara A, Nikolov AD, Wasan DT, Trokhymchuk A, Henderson D (2004) Spreading of nanofluids driven by the structural disjoining pressure gradient. J Colloid Interface Sci 280(1):192–201
33. Cheraghian G, Hendraningrat L (2016) A review on applications of nanotechnology in the enhanced oil recovery part a: effects of nanoparticles on interfacial tension. Int Nano Lett 6(2):129–138
34. Coste J-P, Liu Y, Bai B, Li Y, Shen P, Wang Z, Zhu G, et al (2000) In-depth fluid diversion by pre-gelled particles. laboratory study and pilot testing. In: SPE/DOE improved oil recovery symposium. Society of Petroleum Engineers
35. Craig FF (1971) The reservoir engineering aspects of waterflooding, vol 3. HL Doherty Memorial Fund of AIME, New York
36. Cui X, Li Z, Cao X, Song X, Chen X, Zhang X, et al (2011) A novel PPG enhanced surfactant-polymer system for EOR. In: SPE enhanced oil recovery conference. Society of Petroleum Engineers
37. Cui Z-G, Cui Y-Z, Cui C-F, Chen Z, Binks BP (2010) Aqueous foams stabilized by in situ surface activation of $CaCO_3$ nanoparticles via adsorption of anionic surfactant. Langmuir 26(15):12567–12574
38. Delmas T, Piraux H, Couffin A-C, Texier I, Vinet F, Poulin P, Cates ME, Bibette J (2011) How to prepare and stabilize very small nanoemulsions. Langmuir 27(5):1683–1692
39. Delshad M, Najafabadi NF, Anderson G, Pope GA, Sepehrnoori K et al (2009) Modeling wettability alteration by surfactants in naturally fractured reservoirs. SPE Reservoir Eval Eng 12(03):361–370
40. Dembo AT, Starodoubtsev SG (2001) Interaction between polyelectrolyte gel- surfactant complexes with oppositely charged polymer and surfactant components. Macromolecules 34(8):2635–2640
41. Deshmukh OS, van den Ende D, Stuart MC, Mugele F, Duits MHG (2015) Hard and soft colloids at fluid interfaces: adsorption, interactions, assembly & rheology. Adv Colloid Interface Sci 222:215–227

42. Diaz D, Somaruga C, Norman C, Romero JL et al (2008) Colloidal dispersion gels improve oil recovery in a heterogeneous Argentina waterflood. In: SPE symposium on improved oil recovery. Society of Petroleum Engineers
43. Ehtesabi H, Ahadian MM, Taghikhani V (2014) Investigation of diffusion and deposition of TiO_2 nanoparticles in sandstone rocks for eor application. In: 76th EAGE conference and exhibition 2014
44. El-Diasty AI, Ragab AMS et al (2013) Applications of nanotechnology in the oil & gas industry: latest trends worldwide & future challenges in Egypt. In: North Africa technical conference and exhibition. Society of Petroleum Engineers
45. Engeset B (2012) The potential of hydrophilic silica nanoparticles for EOR purposes: a literateur review and an experimental study. Master's thesis, Institutt for petroleumsteknologi og anvendt geofysikk
46. Esmaeilzadeh P, Hosseinpour N, A Bahramian, Z Fakhroueian, Arya S (2014) Effect of ZrO_2 nanoparticles on the interfacial behavior of surfactant solutions at air–water and n-heptane–water interfaces. Fluid Phase Equilibria 361:289–295
47. Espinoza DA, Caldelas FM, Johnston KP, Bryant SL, Huh C et al (2010) Nanoparticle-stabilized supercritical CO_2 foams for potential mobility control applications. In: SPE improved oil recovery symposium. Society of Petroleum Engineers
48. Feng ZC (2012) Handbook of zinc oxide and related materials: volume two, devices and nano-engineering, vol 2. CRC press
49. Fletcher A, Davis J et al (2010) How EOR can be transformed by nanotechnology. In: SPE improved oil recovery symposium. Society of Petroleum Engineers
50. Frampton H, Morgan JC, Cheung SK, Munson L, Chang KT, Williams D et al (2004) Development of a novel waterflood conformance control system. In: SPE/DOE symposium on improved oil recovery. Society of Petroleum Engineers
51. Friedheim JE, Young S, De Stefano G, Lee J, Guo Q et al (2012) Nanotechnology for oilfield applications-hype or reality? In: SPE international oilfield nanotechnology conference and exhibition. Society of Petroleum Engineers
52. Geng J, Han P, Bai B et al (2018) Experimental study on charged nanogels for interfacial tension reduction and emulsion stabilization at various salinities and oil types. In: SPE Asia Pacific oil and gas conference and exhibition. Society of Petroleum Engineers
53. Geng J, Pu J, Wang L, Bai B (2018) Surface charge effect of nanogel on emulsification of oil in water for fossil energy recovery. Fuel 223:140–148
54. Giraldo J, Benjumea P, Lopera S, Cortes FB, Ruiz MA (2013) Wettability alteration of sandstone cores by alumina-based nanofluids. Energy Fuels 27(7):3659–3665
55. Gonzenbach UT, Studart AR, Tervoort E, Gauckler LJ (2006) Stabilization of foams with inorganic colloidal particles. Langmuir 22(26):10983–10988
56. Green DW, Paul Willhite G et al (1998) Enhanced oil recovery, vol 6. Doherty Memorial Fund of AIME, Society of Petroleum Engineers Richardson, TX, Henry L
57. Griffith N, Ahmad Y, Daigle H, Huh C et al (2016) Nanoparticle-stabilized natural gas liquid-in-water emulsions for residual oil recovery. In: SPE improved oil recovery conference. Society of Petroleum Engineers
58. Hashemi R, Nassar NN, Almao PP (2014) Nanoparticle technology for heavy oil in-situ upgrading and recovery enhancement: opportunities and challenges. Appl Energy 133:374–387
59. Hendraningrat L, Torsæter O (2015) Metal oxide-based nanoparticles: revealing their potential to enhance oil recovery in different wettability systems. Appl Nanosci 5(2):181–199
60. L Hendraningrat, S Li, O Torsaeter et al (2013) Enhancing oil recovery of low-permeability Berea sandstone through optimised nanofluids concentration. In: SPE enhanced oil recovery conference. Society of Petroleum Engineers
61. Hendraningrat L, Li S, Torsater O et al (2013) Effect of some parameters influencing enhanced oil recovery process using silica nanoparticles: an experimental investigation. In: SPE reservoir characterization and simulation conference and exhibition. Society of Petroleum Engineers

62. Hosseini-Nasab SM, Zitha PLJ, Mirhaj SA, Simjoo M (2016) A new chemical-enhanced oil recovery method? Colloids Surfaces A: Physicochem Eng Aspects 507:89–95
63. Hua Z, Lin M, Dong Z, Li M, Zhang G, Yang J (2014) Study of deep profile control and oil displacement technologies with nanoscale polymer microspheres. J Colloid Interface Sci 424:67–74
64. Huang T, Evans BA, Crews JB, Belcher CK et al (2010) Field case study on formation fines control with nanoparticles in offshore applications. In: SPE annual technical conference and exhibition. Society of Petroleum Engineers
65. Huh C, Nizamidin N, Pope GA, Milner TE, Wang B (2014) Hydrophobic paramagnetic nanoparticles as intelligent crude oil tracers, January 15 2014. US Patent App. 14/765,426
66. Hunter TN, Wanless EJ, Jameson GJ, Pugh RJ (2009) Non-ionic surfactant interactions with hydrophobic nanoparticles: impact on foam stability. Colloids Surfaces A: Physicochem Eng Aspects 347(1):81–89
67. Jiang L, Li S, Yu W, Wang J, Sun Q, Li Z (2016) Interfacial study on the interaction between hydrophobic nanoparticles and ionic surfactants. Colloids Surfaces A: Physicochem Eng Aspects 488:20–27
68. Jiang L, Sun G, Zhou Z, Sun S, Wang Q, Yan S, Li H, Tian J, Guo J, Zhou B et al (2005) Size-controllable synthesis of monodispersed SnO_2 nanoparticles and application in electrocatalysts. J Phys Chem B 109(18):8774–8778
69. Johannessen AM, Spildo K (2013) Enhanced oil recovery (EOR) by combining surfactant with low salinity injection. Energy Fuels 27(10):5738–5749
70. Johansson D, Bergenståhl B, Lundgren E (1995) Wetting of fat crystals by triglyceride oil and water. 1. The effect of additives. J Am Oil Chem Soc 72(8): 921–931
71. Johnson CA, Lenhoff AM (1996) Adsorption of charged latex particles on mica studied by atomic force microscopy. J Colloid Interface Sci 179(2):587–599
72. Kabanov AV, Vinogradov SV (2009) Nanogels as pharmaceutical carriers: finite networks of infinite capabilities. Angewandte Chemie Int Edition 48(30):5418–5429
73. Kanj MY, Rashid Md, Giannelis E et al (2011) Industry first field trial of reservoir nanoagents. In: SPE Middle East oil and gas show and conference. Society of Petroleum Engineers
74. Karimi A, Fakhroueian Z, Bahramian A, Khiabani NP, Darabad JB, Azin R, Arya S (2012) Wettability alteration in carbonates using zirconium oxide nanofluids: EOR implications. Energy Fuels 26(2):1028–1036
75. Kothari N, Raina B, Chandak KB, Iyer V, Mahajan HP et al (2010) Application of ferrofluids for enhanced surfactant flooding in IOR. In: SPE EUROPEC/EAGE annual conference and exhibition. Society of Petroleum Engineers
76. Lake LW (1989) Enhanced oil recovery
77. Lan Q, Yang F, Zhang S, Liu S, Xu J, Sun D (2007) Synergistic effect of silica nanoparticle and cetyltrimethyl ammonium bromide on the stabilization of O/W emulsions. Colloids Surfaces A: Physicochem Eng Aspects 302 (1): 126–135
78. Lau HC, Yu M, Nguyen QP (2017) Nanotechnology for oilfield applications: challenges and impact. J Petrol Sci Eng 157:1160–1169
79. Le NYT, Pham DK, Le KH, Nguyen PT (2011) Design and screening of synergistic blends of SiO_2 nanoparticles and surfactants for enhanced oil recovery in high-temperature reservoirs. Adv Nat Sci: Nanosci Nanotech 2(3)
80. Legrand J, Chamerois M, Placin F, Poirier JE, Bibette J, Leal-Calderon F (2005) Solid colloidal particles inducing coalescence in bitumen-in-water emulsions. Langmuir 21(1):64–70
81. Lenchenkov NS, Slob M, van Dalen E, Glasbergen G, van Kruijsdijk C, et al. Oil recovery from outcrop cores with polymeric nano-spheres. In: SPE improved oil recovery conference. Society of Petroleum Engineers
82. Li J, Niu L, Wu W, Sun M (2020) The reservoir adaptability and oil displacement mechanism of polymer microspheres. Polymers 12(4):885
83. Li S, Hendraningrat L, Torsæter O (2013) A coreflood investigation of nanofluid enhanced oil recovery. J Petrol Sci Eng 111:128–138

84. Limage S, Kragel J, Schmitt M, Dominici C, Miller R, Antoni M (2010) Rheology and structure formation in diluted mixed particle- surfactant systems. Langmuir 26(22):16754–16761
85. Liu Y, Bai B, Shuler PJ et al (2006) Application and development of chemical-based conformance control treatments in china oilfields. In: SPE/DOE symposium on improved oil recovery. Society of Petroleum Engineers
86. Liu Y, Bai B, Wang Y (2010) Applied technologies and prospects of conformance control treatments in china. Oil & Gas Sci Technol-Revue d'IFP Energies nouvelles 65(6):859–878
87. Ma H, Luo M, Dai LL (2008) Influences of surfactant and nanoparticle assembly on effective interfacial tensions. Phys Chem Chem Phys 10(16):2207–2213
88. Mandal A, Bera A, Ojha K, Kumar T et al (2012) Characterization of surfactant stabilized nanoemulsion and its use in enhanced oil recovery. In: SPE international oilfield nanotechnology conference and exhibition. Society of Petroleum Engineers
89. Mcelfresh PM, Holcomb DL, Ector D et al (2012) Application of nanofluid technology to improve recovery in oil and gas wells. In: SPE international oilfield nanotechnology conference and exhibition. Society of Petroleum Engineers
90. Mcelfresh PM, Olguin C, Ector D et al (2012) The application of nanoparticle dispersions to remove paraffin and polymer filter cake damage. In: SPE international symposium and exhibition on formation damage control. Society of Petroleum Engineers
91. Metin C, Bonnecaze R, Nguyen Q et al (2013) The viscosity of silica nanoparticle dispersions in permeable media. SPE Reservoir Eval Eng 16(03):327–332
92. Metin CO, Bonnecaze RT, Nguyen QP (2011) Shear rheology of silica nanoparticle dispersions. Appl Rheol 21(1):13146
93. Metin CO, Lake LW, Miranda CR, Nguyen QP (2011) Stability of aqueous silica nanoparticle dispersions. J Nanoparticle Res 13(2):839–850
94. Metin CO, Baran JR, Nguyen QP (2012) Adsorption of surface functionalized silica nanoparticles onto mineral surfaces and decane/water interface. J Nanoparticle Res 14(11):1246
95. Mohajeri M, Hemmati M, Shekarabi AS (2015) An experimental study on using a nanosurfactant in an EOR process of heavy oil in a fractured micromodel. J Petrol Sci Eng 126:162–173
96. Mohan K, Gupta R, Mohanty KK (2011) Wettability altering secondary oil recovery in carbonate rocks. Energy Fuels 25(9):3966–3973
97. Moradi B, Pourafshary P, Jalali F, Mohammadi M, Emadi MA (2015) Experimental study of water-based nanofluid alternating gas injection as a novel enhanced oil-recovery method in oil-wet carbonate reservoirs. J Nat Gas Sci Eng 27:64–73
98. Moraes RR, Garcia JW, Barros MD, Lewis SH, Pfeifer CS, Liu JC, Stansbury JW (2011) Control of polymerization shrinkage and stress in nanogel-modified monomer and composite materials. Dental Mater 27(6):509–519
99. Muhammed FA, Bai B, Tang T (2012) Experimental study of the interaction between surfactants and super absorbent polymer gel. J Petrol Sci Eng 90:159–164
100. Muhammed FA, Bai B, Imqam A, Almansour AO et al (2014) Preformed particle gel-enhanced surfactant imbibition for improving oil recovery in fractured carbonate reservoirs. In: SPE heavy oil conference Canada. Society of Petroleum Engineers
101. Muhammed FA (2014) Study and pilot test on a novel EOR method-coupling PPG conformance control and surfactant flooding
102. Muller P, Sudre G, Theodoly O (2008) Wetting transition on hydrophobic surfaces covered by polyelectrolyte brushes. Langmuir 24(17):9541–9550
103. Mylonas Y, Staikos G (2001) Study of the complexation of sodium dodecyl sulfate with cationically charged acrylamide and n-isopropylacrylamide based copolymers: the role of the polymer chain hydrophobicity and of its charge density. Langmuir 17(12):3586–3591
104. Naje AN, Norry AS, Suhail AM (2013) Preparation and characterization of SnO_2 nanoparticles. Int J Innov Res Sci Eng Technol 2(12)
105. Negin C, Ali S, Xie Q (2016) Application of nanotechnology for enhancing oil recovery-a review. Petroleum 2(4):324–333

106. Ngai T, Auweter H, Behrens SH (2006) Environmental responsiveness of microgel particles and particle-stabilized emulsions. Macromolecules 39(23):8171–8177
107. Nguyen P-T, Do B-PH, Pham D-K, Nguyen Q-T, Dao D-QP, Nguyen H-A et al (2012) Evaluation on the EOR potential capacity of the synthesized composite silica-core/polymer-shell nanoparticles blended with surfactant systems for the HPHT offshore reservoir conditions. In: SPE international oilfield nanotechnology conference and exhibition. Society of Petroleum Engineers
108. Nwidee LN, Lebedev M, Barifcani A, Sarmadivaleh M, Iglauer S (2017) Wettability alteration of oil-wet limestone using surfactant-nanoparticle formulation. J Colloid Interface Sci
109. Ogolo NA, Olafuyi OA, Onyekonwu MO et al (2012) Enhanced oil recovery using nanoparticles. In: SPE Saudi Arabia section technical symposium and exhibition. Society of Petroleum Engineers
110. Pei HH, Zhang GC, Ge JJ, Zhang J, Zhang Q, Fu LP et al (2015) Investigation of nanoparticle and surfactant stabilized emulsion to enhance oil recovery in waterflooded heavy oil reservoirs. In: SPE Canada heavy oil technical conference. Society of Petroleum Engineers
111. Pichot R, Spyropoulos F, Norton IT (2010) O/w emulsions stabilised by both low molecular weight surfactants and colloidal particles: the effect of surfactant type and concentration. J Colloid Interface Sci 352(1):128–135
112. Pinaud F, Geisel K, Massé P, Catargi B, Isa L, Richtering W, Ravaine V, Schmitt V (2014) Adsorption of microgels at an oil-water interface: correlation between packing and 2d elasticity. Soft Matter 10(36):6963–6974
113. Qiu F, Mamora DD et al (2010) Experimental study of solvent-based emulsion injection to enhance heavy oil recovery in Alaska north slope area. In: Canadian unconventional resources and international petroleum conference. Society of Petroleum Engineers
114. Qiu F et al (2010) The potential applications in heavy oil EOR with the nanoparticle and surfactant stabilized solvent-based emulsion. In: Canadian unconventional resources and international petroleum conference. Society of Petroleum Engineers
115. Rankin KM, Nguyen QP (2014) Conformance control through in-situ gelation of silica nanoparticles. In: Nanotech conference & expo, pp 15–18
116. Ravera F, Ferrari M, Liggieri L, Loglio G, Santini E, Zanobini A (2008) Liquid-liquid interfacial properties of mixed nanoparticle-surfactant systems. Colloids Surfaces A: Physicochem Eng Aspects 323(1):99–108
117. Rousseau D, Chauveteau G, Renard M, Tabary R, Zaitoun A, Mallo P, Braun O, Omari A et al (2005) Rheology and transport in porous media of new water shutoff/conformance control microgels. In: SPE international symposium on oilfield chemistry. Society of Petroleum Engineers
118. Roustaei A, Bagherzadeh H (2015) Experimental investigation of SiO_2 nanoparticles on enhanced oil recovery of carbonate reservoirs. J Petrol Explor Prod Technol 5(1):27–33
119. Roustaei A, Moghadasi J, Bagherzadeh H, Shahrabadi A et al (2012) An experimental investigation of polysilicon nanoparticles' recovery efficiencies through changes in interfacial tension and wettability alteration. In: SPE international oilfield nanotechnology conference and exhibition. Society of Petroleum Engineers
120. Sagir M, Mushtaq M, Suleman Tahir M, Tahir MB, Shaik AR (2020) Surfactant in petroleum industry. Surfactants for enhanced oil recovery applications. Springer, Berlin, pp 13–40
121. Salehi M, Johnson SJ, Liang J-T (2008) Mechanistic study of wettability alteration using surfactants with applications in naturally fractured reservoirs. Langmuir 24(24):14099–14107
122. Salem Ragab AM, Hannora AE et al (2015) A comparative investigation of nano particle effects for improved oil recovery–experimental work. In: SPE Kuwait oil and gas show and conference. Society of Petroleum Engineers
123. Salyer IO (1993) Dry powder mixes comprising phase change materials, May 18 1993. US Patent 5,211,949
124. Schmidt G, Malwitz MM (2003) Properties of polymer-nanoparticle composites. Curr Opin Colloid Interface Sci 8(1):103–108
125. Seright RS et al (2001) Gel propagation through fractures. SPE Prod Facilities 16(04):225–231

126. Seright RS et al (1997) Use of preformed gels for conformance control in fractured systems. SPE Prod Facilities 12(01):59–65
127. Shah RD et al (2009) Application of nanoparticle saturated injectant gases for EOR of heavy oils. In: SPE annual technical conference and exhibition. Society of Petroleum Engineers
128. Shahrabadi A, Bagherzadeh H, Roostaie A, Golghanddashti H et al (2012) Experimental investigation of hlp nanofluid potential to enhance oil recovery: a mechanistic approach. In: SPE international oilfield nanotechnology conference and exhibition. Society of Petroleum Engineers
129. ShamsiJazeyi H, Miller CA, Wong MS, Tour JM, Verduzco R (2014) Polymer-coated nanoparticles for enhanced oil recovery. J Appl Polym Sci 131(15)
130. Sharma S, Sarangdevot K (2012) Nanoemulsions for cosmetics. IJARPB 1(3):408–415
131. Sharma T, Suresh Kumar G, Sangwai JS (2015) Comparative effectiveness of production performance of pickering emulsion stabilized by nanoparticle–surfactant–polymerover surfactant–polymer (sp) flooding for enhanced oil recovery for brownfield reservoir. J Petrol Sci Eng 129:221–232
132. Sharma T, Iglauer S, Sangwai JS (2016) Silica nanofluids in an oilfield polymer polyacrylamide: interfacial properties, wettability alteration, and applications for chemical enhanced oil recovery. Ind Eng Chem Res 55(48):12387–12397
133. Sheng J (2010) Modern chemical enhanced oil recovery: theory and practice. Gulf Professional Publishing
134. Shokrlu YH, Babadagli T et al (2011) Transportation and interaction of nano and micro size metal particles injected to improve thermal recovery of heavy-oil. In: SPE annual technical conference and exhibition. Society of Petroleum Engineers
135. Singh R, Mohanty KK et al (2016) Foams stabilized by in-situ surface-activated nanoparticles in bulk and porous media. SPE J 21(01):121–130
136. Skauge T, Spildo K, Skauge A et al (2010) Nano-sized particles for EOR. In: SPE improved oil recovery symposium. Society of Petroleum Engineers
137. Song E, Kim D, Kim BJ, Lim JC (2014) Surface modification of $CaCO_3$ nanoparticles by alkylbenzene sulfonic acid surfactant. Colloids Surfaces A: Physicochem Eng Aspects 461:1–10
138. Srinivasan A, Shah SN et al (2014) Surfactant-based fluids containing copper-oxide nanoparticles for heavy oil viscosity reduction. In: SPE annual technical conference and exhibition. Society of Petroleum Engineers
139. Standnes DC, Austad T (2000) Wettability alteration in chalk: 2. mechanism for wettability alteration from oil-wet to water-wet using surfactants. J Petrol Sci Eng 28(3):123–143
140. Standnes DC, Austad T (2003) Wettability alteration in carbonates: interaction between cationic surfactant and carboxylates as a key factor in wettability alteration from oil-wet to water-wet conditions. Colloids Surfaces A: Physicochem Eng Aspects 216(1–3):243–259
141. Stoll M, Hofman J, Ligthelm DJ, Faber MJ, van den Hoek P et al (2008) Toward field-scale wettability modification-the limitations of diffusive transport. SPE Reservoir Eval Eng 11(03):633–640
142. Suleimanov BA, Ismailov FS, Veliyev EF (2011) Nanofluid for enhanced oil recovery. J Petrol Sci Eng 78(2):431–437
143. Sun Q, Li Z, Li S, Jiang L, Wang J, Wang P (2014) Utilization of surfactant-stabilized foam for enhanced oil recovery by adding nanoparticles. Energy Fuels 28(4):2384–2394
144. Tarek M et al (2015) Investigating nano-fluid mixture effects to enhance oil recovery. In: SPE annual technical conference and exhibition. Society of Petroleum Engineers
145. Thomas S (2008) Enhanced oil recovery-an overview. Oil Gas Sci Technol-Revue de l'IFP 63(1):9–19
146. Vashisth C, Whitby CP, Fornasiero D, Ralston J (2010) Interfacial displacement of nanoparticles by surfactant molecules in emulsions. J Colloid Interface Sci 349(2):537–543
147. Vatanparast H, Javadi A, Bahramian A (2017) Silica nanoparticles cationic surfactants interaction in water-oil system. Colloids Surfaces A: Physicochem Eng Aspects 521:221–230

148. Wang L, Zhang G, Li G, Zhang JQ, Ding B et al (2010) Preparation of microgel nanospheres and their application in EOR. In: International oil and gas conference and exhibition in China. Society of Petroleum Engineers
149. Wasan D, Nikolov A, Kondiparty K (2011) The wetting and spreading of nanofluids on solids: role of the structural disjoining pressure. Curr Opin Colloid Interface Sci 16(4):344–349
150. Wasan DT, Nikolov AD (2003) Spreading of nanofluids on solids. Nature 423(6936):156–159
151. Wu W, He Q, Jiang C (2008) Magnetic iron oxide nanoparticles: synthesis and surface functionalization strategies. Nanoscale Res Lett 3(11):397
152. Wu Y, Chen W, Dai C, Huang Y, Li H, Zhao M, He L, Jiao B (2017) Reducing surfactant adsorption on rock by silica nanoparticles for enhanced oil recovery. J Petrol Sci Eng 153:283–287
153. Wu Y, Shuler PJ, Blanco M, Tang Y, Goddard WA et al (2008) An experimental study of wetting behavior and surfactant EOR in carbonates with model compounds. SPE J 13(01):26–34
154. Wu Y, Tang T, Bai B, Tang X, Wang J, Liu Y (2011) An experimental study of interaction between surfactant and particle hydrogels. Polymer 52(2):452–460
155. Xu K, Zhu P, Tatiana C, Huh C, Balhoff M et al (2016) A microfluidic investigation of the synergistic effect of nanoparticles and surfactants in macro-emulsion based EOR. In: SPE improved oil recovery conference. Society of Petroleum Engineers
156. Yahya N, Kashif M, Nasir N, Akhtar MN, Yusof NM (2012) Cobalt ferrite nanoparticles: an innovative approach for enhanced oil recovery application. J Nano Res 17:115–126
157. Yu H, Hermann S, Schulz SE, Gessner T, Dong Z, Li WJ (2012) Optimizing sonication parameters for dispersion of single-walled carbon nanotubes. Chem Phys 408:11–16
158. Yu J, Berlin JM, Lu W, Zhang L, Kan AT, Zhang P, Walsh EE, Work S, Chen W, Tour J et al (2010) Transport study of nanoparticles for oilfield application. In: SPE international conference on oilfield scale. Society of Petroleum Engineers
159. Zargartalebi M, Kharrat R, Barati N (2015) Enhancement of surfactant flooding performance by the use of silica nanoparticles. Fuel 143:21–27
160. Zhang P, Tweheyo MT, Austad T (2007) Wettability alteration and improved oil recovery by spontaneous imbibition of seawater into chalk: impact of the potential determining ions Ca^{2+}, Mg^{2+}, and SO_4^{2-}. Colloids Surfaces A: Physicochem Eng Aspects 301(1–3):199–208
161. Zhang T, Davidson D, Bryant SL, Huh C et al (2010) Nanoparticle-stabilized emulsions for applications in enhanced oil recovery. In: SPE improved oil recovery symposium. Society of Petroleum Engineers

Gas Injection

Recovery of Oil Using Surfactant-Based Foams

Kristian Mogensen

Abstract This chapter describes the application of surfactant-based foams for recovery of oil with a focus on subsurface aspects. While the concept of foaming may be qualitatively well understood, the physical behaviour of a foam system comprising gas, brine, and surfactant depends on the type of each of these three constituents and their interaction, in addition to the properties of the porous medium in which the foam is designed to be generated and perhaps propagate. Key physical properties, which must be investigated during a laboratory experimental program, are discussed. A critical review is provided of a number of key applications where foam is utilised for recovery of oil, starting with drilling, completion, and stimulation before moving on to chemical conformance and enhanced oil recovery.

Keywords Film drainage · Bubble coalescence · Foam stability · Foam quality · FGSO · FAWAG · Adsorption · Mobility control

1 Introduction

A stringent definition of foam would be that it is a dispersed medium where gas bubbles are separated by interconnected liquid films called lamellae. Lamellae are thin, on the order of 100 nm. Spanning a 3D network, they connect to one another at so-called Plateau borders. Figure 1, which presents a schematic 2D view of a foam network. Unlike gas hydrates, which trap single gas molecules inside a cage of water molecules, the gas bubbles in a foam contain many molecules.

Soap bubbles are an illustrative everyday example of a foam. Detergents added to the water phase help trap air bubbles and the water jetted from the tap of a hose provides mixing energy for the foam to form. If left untouched, the foam typically lasts only for a few minutes before it breaks due to coalescence of adjacent gas bubbles.

K. Mogensen (✉)
ADNOC, Abu Dhabi, United Arab Emirates
e-mail: Kmogensen@adnoc.ae

T. Solling et al. (eds.), *Surfactants in Upstream E&P*, Petroleum Engineering,
https://doi.org/10.1007/978-3-030-70026-3_10

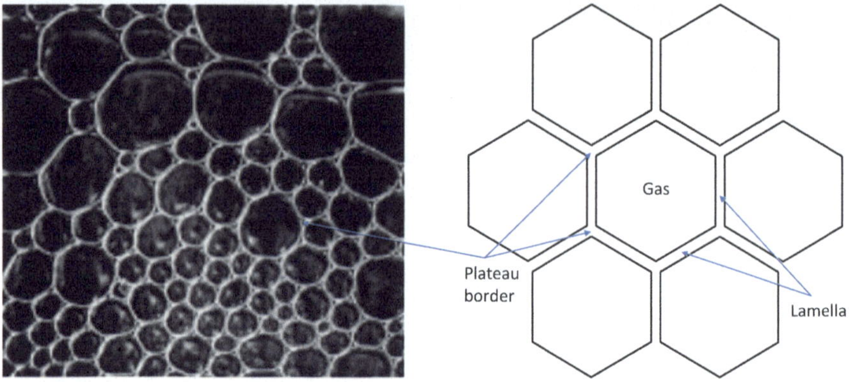

Fig. 1 Left: Actual image of a foam. Right: Schematic of a foam network

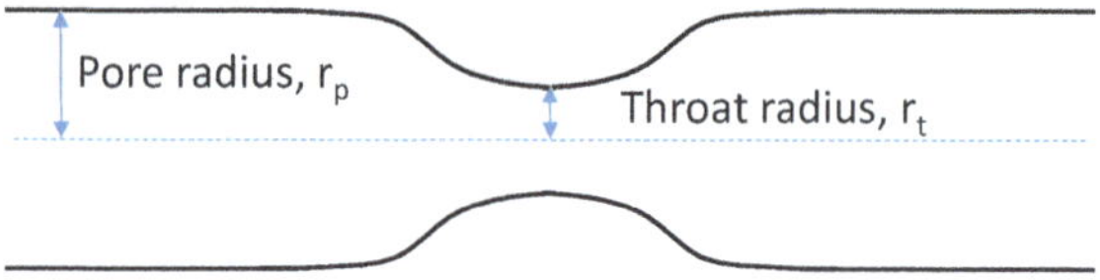

Fig. 2 Schematic of a porous medium consisting of two pores connected by a throat

It is tempting to draw analogies between foams formed in bulk (ex-situ, without a porous medium) and foams which are created inside a pore system (in-situ). However, the two situations turn out to be very different, as will become clear in the next paragraphs.

The topology of the void space of a porous medium is often described as a 3D network of pores connected to each other by restrictions called throats, see Fig. 2. If the fluids filling the void space are water and gas, water preferentially covers the surface of the rock. This molecular adsorption phenomenon is known as wettability and the rock is said to be water-wet. Wettability has profound implications for immiscible displacement of one fluid by another. When two phases are brought into contact with each other, a curved interface will form and the interface curvature is related to the local radius at the position of the interface. The difference in pressure between the two phases is known as the capillary pressure, P_c, which for a circular cross-section is given as a function of interfacial tension and radius of curvature:

$$P_c = \frac{2 \times IFT}{r}.$$

The smaller the radius, the larger the pressure difference. The presence of capillarity during an immiscible displacement leads to a certain amount of trapping of the non-wetting phase, in this case gas. Fundamental flow studies conducted in the

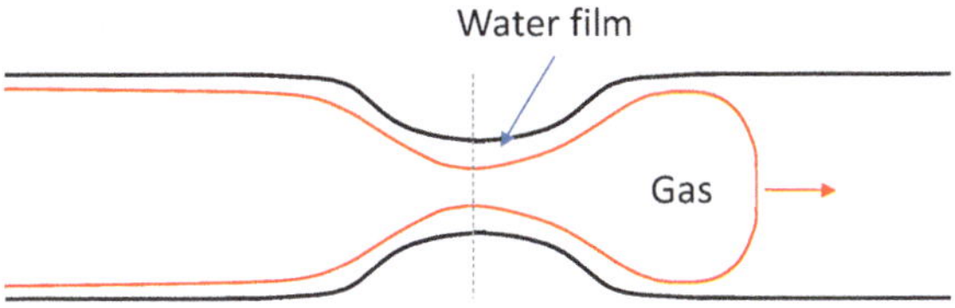

Fig. 3 Advancement of a gas finger in a pore

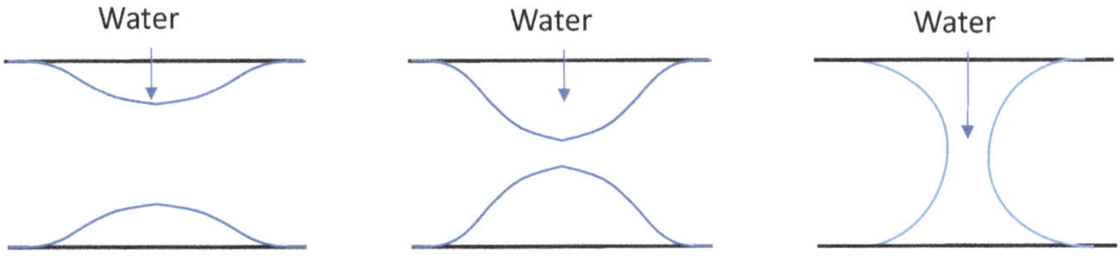

Fig. 4 Liquid film instability leading to snap-off

1980s made use of transparent micro-models where a reproducible 2D pore structure is created by an etching technique. The studies revealed that fluid front advance involves pore-level events which occur within a few milli-seconds. Roof [36] demonstrated that if the aspect ratio, defined as the pore radius divided by the throat radius, exceeds 2 the liquid film wetting the surface is pulled towards the centre of the throat by capillarity. This phenomenon is known as snap-off and is depicted schematically in Figs. 3, 4. Snap-off is the main physical mechanism responsible for trapping of oil and gas by water at the microscopic level, the other mechanism is bypassing due to velocity differences caused by the pore-size distribution.

The advancing gas finger illustrated in Fig. 3 will be snapped off by the water film at the throat resulting in the formation of a trapped gas bubble. The gas trapping process will then repeat itself until the pore on the right-hand side is filled with bubbles. Note that liquid films are thinnest in the pores and thickest at the throats, which is why snap-off occurs at or close to the throats.

Although gas trapping and foam creation are the result of the same underlying capillary-driven mechanism, the two phenomena are clearly different. Trapping of gas bubbles is controlled entirely by the pore topology and does not require surfactants to occur.

The liquid films separating the gas bubbles are very thin and will quickly rupture as a result of film drainage. The role of the surfactant is therefore to stabilize the films by diffusion towards the gas-liquid interface, see Fig. 5. In the presence of surfactants, lamellae division can occur and Plateau borders will emerge.

From a thermodynamic point of view, foam is unstable, because foam destruction leads to a reduced interfacial area. However, according to Chambers and Radke [10], foam can reach a meta-stable configuration which depends on a force balance between the local capillary pressure working towards interfacial area reduction and hence foam destruction on one hand, and then a repulsive contribution to the disjoining pressure, which is affected by the presence of adsorbed surfactant. The net force

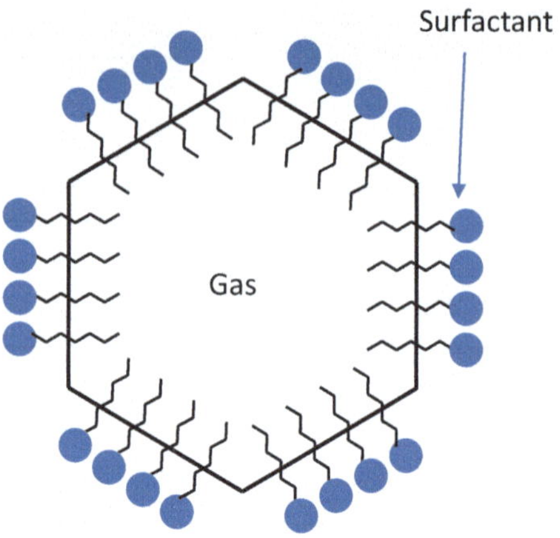

Fig. 5 Film stabilization with a surfactant

depends on the size of the wetting film, h, and a meta-stable situation can arise if the two forces balance each other, see Fig. 6. Note that in the absence of a porous medium, gravity is the main force leading to film drainage and foam break-up, but in a porous medium, gravity does not play an active role. This difference is important to remember when designing foam stability tests. The micro-structure of foam is shaped by the porous medium in which it resides.

Two adjacent gas bubbles with a different curvature will result in a different gas phase pressure inside the bubbles. This pressure gradient will lead to gas diffusion and bubble coalescence and explains why adjacent gas bubbles often have similar curvatures. Gas diffusion rates depend on the curvature difference, the type of gas, and on the solubility of gas in the aqueous phase.

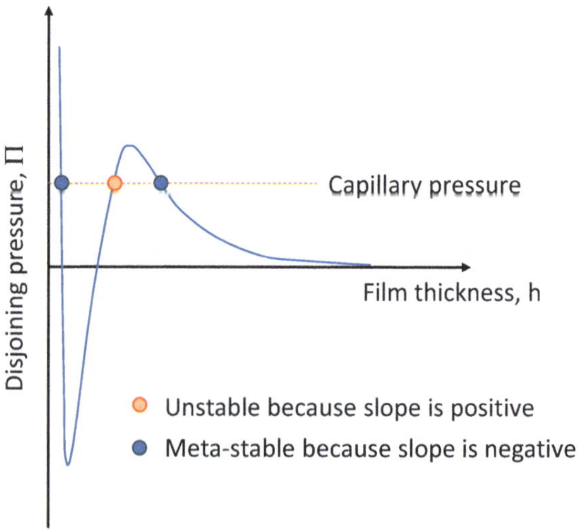

Fig. 6 Force balance, showing disjoining pressure isotherm Adapted from [10]

Figure 6 shows that there exists a maximum capillary pressure, which the disjoining pressure can sustain without breaking the foam. Since gas-water capillary pressure increases with increasing gas saturation, this translates to a maximum gas saturation above which the foam collapses. In such a case, the foam is said to be drying out. A key experimental design parameter is the foam quality defined as the fractional flow of gas. Experience shows that there exists an optimum foam quality, which generates the strongest foam.

A higher capillary pressure would also occur in low-permeability rock because of the smaller pores. Hence, foams will preferentially break down in tight formations. Lake et al. [28] mention that larger aspect ratios lead to quicker bubble coalescence and that larger film thickness variation occurs at higher gas rates.

The presence of oil in the porous medium may destabilize foams. One explanation is that it is predominantly the lighter alkanes which diffuse towards the gas-liquid interface and alter the force balance. This means that light oils are more problematic than heavy oils. Another argument often mentioned is that the surfactant may partially dissolve in the oil phase, depending on the alkane chain length of the surfactant. The dissolution into the oil phase could then lead to oil-water emulsion formation. Wettability also comes into play. Carbonate rocks have a higher affinity towards the polar components in the oil phase and are predominantly mixed-wet or in some cases strongly oil-wet. This means that there will often be a mobile oil phase remaining as an oil layer in each pore after displacement by water or gas. Oil wetness also implies that snap-off of gas by water is hindered. In mixed-wet rocks, foam is therefore expected to form only in pores which are not strongly oil-wet or where the oil saturation has been significantly reduced. As a consequence, foam stands a better chance of surviving in gas caps where there is no oil or in miscible gas injection projects where the gas reduces the residual oil saturation to very low levels. The advantage of foam destruction by oil is of course that foam will never block oil flow; this behavior is exploited in foam gas shut-off treatments of wells coning gas from an overlaying gas cap.

While the concept of foaming may be qualitatively understood, the physical behavior of a foam system comprising gas, brine, and surfactant depends on the type of each of these three constituents and their interaction, in addition to the properties of the porous medium in which the foam is designed to be generated and perhaps propagate. In Sect. 2, we begin with a description of the key properties, which must be investigated during a laboratory experimental program to screen suitable surfactants for foaming potential, not just in bulk but also in-situ. In Sect. 3, we review a number of key applications where foam is applied for recovery of oil, starting with drilling, completion, and stimulation before moving on to chemical conformance and enhanced oil recovery. Finally, in Sect. 4, we present some concluding remarks and share our view on what the future of foam may look like.

2 Laboratory Studies

Reservoirs around the globe vary in terms of oil properties, brine composition, temperature, and rock properties. Chemical systems which seem to work in one environment cannot always be transferred or adapted to different rock and fluid conditions. Therefore, comprehensive laboratory experiments are a mandatory step towards derisking a field trial.

Laboratory experiments themselves are divided into two main parts. The first part involves only bulk fluid tests whereas the second part studies the interplay between fluid and porous medium.

2.1 Fluid-Only Testing

The key goal of the fluid-only testing phase is to identify a surfactant formulation, which

- Is soluble with the carrier fluid, which is most often brine.
- Foams when brought in contact with the selected gas.
- Is stable at reservoir pressure and temperature conditions.
- Maintains foam properties for a prolonged time.
- Is somewhat tolerant towards presence of oil.
- Is commercially available.
- Is environmentally approved for field application.

It is common practice to contact several chemical vendors and test a number of their recommended surfactants. Solubility in brine is very much dictated by alkane chain length although temperature and brine salinity also play a role. Solubility testing is quick and may eliminate some candidates. Surfactants often work within a certain temperature range; for high-temperature applications, only a very limited number of surfactant families are applicable, such as alpha-olefin sulfonates (AOS).

Contrary to surfactants for interfacial tension reduction, no high-throughput screening methods exist for testing a large number of surfactants automatically in terms of foaming capacity. In practice, this makes screening somewhat more tedious without the ability to probe a large number of possible chemical combinations. Therefore, researchers are often restricted to conducting experiments with chemicals pre-screened by the chemical vendors.

In light of the tremendous advances made in computational chemistry over the past few decades, it is this author's belief that foam systems comprising gas mixtures, brines, and surfactants can be modelled with tools such as molecular dynamics or density functional theory because all components have a well-defined chemical structure and because the relevant force fields have been described. Computational screening of surfactants in terms of foaming tendency could then be automated once the agreement between predictions and measurements is demonstrated. Estimation of foam stability, on the other hand, occurs on a length scale which is orders of magnitudes beyond what can be handled in molecular dynamics simulations. Since foam is thermodynamically unstable, the rate of film thinning induced by gravity must be modelled, which requires an estimate of the liquid conductivity as a function of the geometry of the lamellae. In a porous medium, film drainage is a wetting/de-wetting phenomenon, which is treated in some dynamic pore network models, see [3] for details.

The ability to create stable foams is examined next. Ex-situ foam generation requires external energy in the form of mixing to develop. The mixing results in movement of molecules inside the fluid and gives rise to shear forces. A fluid is referred to as sheared when different layers of molecules move past one another within the fluid itself. The relative difference in velocity between molecular layers gives rise to a velocity gradient perpendicular to the main flow direction, which is known as the shear rate. The mixing energy applied in the laboratory should be comparable to the shear rates which can be expected to occur in the field. Shear rates encountered in rocks depend on the fluid rheology as well as on the rock properties. The following expression by Cannella et al. [9] developed for polymer flooding can also be applied to foam:

$$\gamma_{eff} = C \times \left[\frac{3n+1}{4n}\right]^{n/(n-1)} \times \frac{4}{\sqrt{8}} \times \frac{u}{\sqrt{k_{rw} \times k \times S_w \times \varphi}}$$

C and n are fluid rheology properties, u is velocity, k_{rw} denotes the relative permeability to water at water saturation S_w, k is permeability and ϕ is porosity. Berg and van Wunnik [5] provide a detailed review of shear rate determination for pore level calculations and arrive at a simpler expression:

$$\gamma_{eff} = C \times \frac{u}{\sqrt{k \times \varphi}}$$

A typical field-scale velocity is 0.5–1.0 ft/d, but could be an order of magnitude higher close to the wellbore.

A common approach to quantifying foaming capability is to measure the foam height in a capillary tube, either visually or with a light source and a sensor. Neither approach has a good reproducibility. The foam height is tracked versus time and the longer the foam can maintain its structure, the better. Lunkenheimer and Malysa [29] advocated for the use of a foam height ratio defined as the foam height after 5 min relative to the initial foam height. It is debatable whether foam height can be used as a proxy for in-situ foam stability since the pore space provides a 3D geometrical confinement of the foam bubbles which cannot easily be mimicked with other means. Film drainage rates in-situ will be different from the ones obtained by bulk phase experiments because foam destruction ex-situ is caused by gravity whereas capillary pressure is responsible for foam collapse in-situ. Nevertheless, it is argued that a surfactant which fails to foam in a capillary will also not foam inside a porous medium. In other words, foam height is seen as a necessary but insufficient screening criterion.

It is often observed that foam creation, in additional to a threshold shear force, also requires a minimum surfactant concentration. Similar to the critical micelle concentration (CMC), this value increases with the brine salinity, which may pose a logistical and economical challenge in high-salinity formations. In such frontier applications, foam may potentially be piloted in combination with a pre-flush of low-salinity brine to bring down the chemical consumption. A lower salinity also seems to reduce the chemical adsorption.

The choice of gas impacts the foam stability. Since brine can dissolve ten times more CO_2 than methane, CO_2 foam is weakened by diffusion of CO_2 bubbles towards the aqueous phase. CO_2 solubility decreases with increasing brine salinity and temperature, which should then, in theory, lead to better stability. However, few surfactants exhibit the required tolerance towards high-salinity brines and high temperatures. Foams targeting natural gas or nitrogen show better stability than CO_2 foams.

Foam rheology can be assessed by measuring the relationship between shear stress and shear rate, similar to drilling muds or polymers. In general, foam is considered to be visco-elastic, but the power-law stress-strain relation does not have a constant exponent, see [14]. Conventional bulk testing of foams targeting oil recovery applications does not focus on foam rheology, although foam texture can be visually inspected.

2.2 Synergies with Polymers

Foam collapses due to gradual thinning of the liquid films separating the gas bubbles. The observation that the rate of thinning is influenced by the viscosity of the liquid film has led researchers to investigate whether addition of polymer can improve foam stability since polymers are known to increase the viscosity. Friedmann et al. [19] described laboratory studies and preliminary field pilot observations for the Rangely field CO_2 project and referred to the concept as a foam-gel.

Hernando et al. [22] performed both bulk tests and core floods to investigate various combinations of surfactants and polymers and found that associative polymers rather than classical non-ionic polymers were effective for water profile control in both core floods and sandpack experiments. Non-ionic polymers, on the other hand, decreased the foaming tendency as the higher solution viscosity was thought to reduce the surfactant diffusivity towards the gas-water interface. This effect could perhaps have been avoided if the polymer was added after foam was created instead of mixing surfactant and polymer together first.

An important learning point from this study is therefore, that the impact of polymers on foam stability depends on the particular combination of surfactant and polymer. Brine composition, characterized by salinity as well as the amount of divalent ions, will also impact the synergy between surfactant and polymers. Low-salinity brine with salinities less than 1000–2000 ppm are known to increase polymer viscosity which could lead to a stronger foam. Note that if the polymer does improve stability, it should not decrease the foam mobility to a point where the foam cannot be propagated.

2.3 *Synergies with Nano-Particles (NP)*

The past decade has seen a steady increase in the use of nano-technology in various areas such as materials design, biomedicine and electronics, see review by Bennetzen and Mogensen [4]. Cross-disciplinary research has also demonstrated that nano-technology may be applicable in enhanced oil recovery (EOR). Nano-particles are small spherical particles with a diameter in the range of 1–1000 nm with a large area-to-volume ratio. The surface of the particles can be modified by attaching various chemical molecules, a process referred to as conjugation or grafting. The molecular coating of the naked particles can be tailored for a specific application. One such example is described by Espinoza et al. [18], who showed that silica nano-particles coated with poly-ethylene glycol (PEG) could help stabilize CO_2 foams by aggregating at the CO_2-water interface. It is speculated that the synergetic effect would manifest itself in at least three ways; by speeding up the diffusing of surfactant towards the gas-water interface, by reducing the amount of surfactant required to cover the gas-water interface, and via stronger molecular forces preventing the films from draining completely. Espinoza et al. [18] showed that foam remained stable without a surfactant at reservoir conditions. Once pressure was reduced to surface conditions, the foam disintegrated.

The NP-stabilized foam was able to withstand high temperature and remained stable at surfactant concentrations as low as 0.05 wt%, almost two orders of magnitude lower than for conventional applications, although the required surfactant concentration increased with brine salinity. The foam generation itself was brought about by co-injection and required a threshold shear rate to take place. From a field application perspective, co-injection into the wellbore poses some operational challenges. In one scenario, it may cause gas and liquid to segregate preventing the foam from forming; in another scenario, the foam mobility may lead to a significant reduction in injectivity.

2.4 *Fluid-Rock Testing*

Once a subset of surfactants or even just a single surfactant formulation has passed the preliminary screening, the interplay between fluids and rock must be investigated. The following parameters must be assessed:

- Foam generation in the porous medium at realistic shear rates.
- Foam strength as a function of foam quality (gas saturation).
- Optimum foam quality.
- Foam mobility reduction.
- Dynamic adsorption to the rock.
- Pressure gradient needed to mobilize a stagnant foam.
- Tolerance towards oil.

The choice of porous medium varies among investigators and each one has its pros and cons. 2D micro-models, which provided a breakthrough in the understanding of two-flow displacement mechanisms more than thirty years ago have also been used to investigate foam flow. Micro-models have a well-defined pore structure and are ideally suited for imaging which is a major attraction. Unfortunately, the foam creation in real reservoir rocks is impacted by the pore geometry, which is poorly captured by today's 2D micromodels. Other drawbacks to using micro-models is that they do not enable steady-state two-phase flow, they cannot account for wettability variation, effluent analysis is not feasible and flow is dominated by capillary end effects (defined later). It is speculated that the use of 3D printing techniques may pave the way for construction of more realistic micro-models in the future, which may alleviate some of the before-mentioned limitations.

Sandpacks have been used primarily in Academia by researchers who wish to study fundamental properties at larger scale without having access to reservoir core material. Sandpacks are easy to work with, they can be imaged, and may show some similarities to high-permeability sandstone reservoirs but certainly not to carbonate rocks, which have complex pore geometries.

Slimtubes, which can in some way be regarded as sandpacks have also been used for foam testing. With a clear protocol for packing of the sand grains, the slimtube is the only industry-accepted method for evaluating dispersion-free minimum miscibility pressures for gas injection studies. However, confinement of the porous medium inside a steel cylinder does not allow for imaging to take place. The advantage of slimtube testing is that pure 1D flow can be investigated at length scales up to 60 ft.

The best option is to conduct flow experiments with real rocks at realistic flow rates. Experience shows that foam forms within a mixing zone, which can exceed the length of a typical core plug. The solution could therefore be to put several cores in series, a technique known as composite cores. Extreme care must be taken to ensure capillary continuity between consecutive core plugs to avoid introduction of capillary artefacts. Saturation monitoring using CT imaging has proven useful to test for capillary continuity and to study diversion of gas towards unswept parts of the rock. The alternative to composite cores is to select analogue outcrops, such as Indiana limestone. The most popular outcrops can be ordered to possess a certain permeability and with a length suitable for foam flooding (typically 50 cm or more). A key question remains whether the dynamic adsorption in outcrop is similar to that found in the reservoir rock. Note that static adsorption, as measured on a flat polished surface, is much higher than the dynamic adsorption inferred from core floods. The surfactant does not come into contact with the entire pore space, either due to pore geometry restrictions or because of wettability effects.

It is sometimes observed that foam leads to a reduction in residual oil saturation. One reason may be that the surfactant, in addition to creating foam is also capable of lowering the oil-water interfacial tension to a point where residual oil can be mobilized. Another explanation is that the incremental oil is in fact an experimental artefact known as the capillary end effect. Since gas is the non-wetting phase, there will still be connected oil left at the time of gas breakthrough because the gas-oil capillary pressure is positive. Foam creation leads to a higher pressure drop due to the reduced gas mobility. This pressure drop is often sufficient to overcome the gas-oil capillary pressure and push out the mobile oil. To eliminate or reduce the capillary end effect, it is common practice to conduct a bump flood whereby the injection rate is increased tenfold to make sure that all mobile oil is displaced.

The optimum foam quality (gas fractional flow) often lies close to 70% but this can be investigated with a couple of core floods. What defines optimum is the mobility reduction relative to the mobility of gas. Since the gas viscosity is low, as a rule-of-thumb a mobility reduction in the order of 50 or above is often required. However, weaker foams may be preferred in a continuous injection scheme to ensure injectivity remains high whereas strong foam may be required in gas shut-off applications.

Many chemicals tend to adsorb more in carbonates compared to sandstones because adsorption is linked to surface area. For a continuous foam application, a high adsorption will significantly impact the economics, whereas for a near well-bore treatment, the adsorption level is of secondary importance. The adsorption can be inferred based on the breakthrough time of a surfactant-only flood. It is unclear how the adsorption is affected by the foam creation, but it is believed that surfactant is first spent satisfying the adsorption before assisting in foam generation.

Finally, in gas shut-off operations, where the generated foam will remain stagnant after being formed, the pressure drop needed to (re-)mobilize the foam is of interest. In fact, it is worth remembering that most of the foam generated in-situ will remain stagnant and that foam flow occurs in a small portion of the pore network.

3 Foam Applications for Recovery of Oil

The following paragraphs describe various applications of foam in the upstream value chain from drilling and completions to fluid flow diversion in the reservoir. Each of these examples will eventually lead to recovery of more oil, whether directly, such as in enhanced oil recovery, or indirectly, by lowering operational costs.

3.1 Foam for Air Drilling and Corrosion Inhibition

Most drilling operations make use of expensive mud systems to stabilize the wellbore, prevent clay swelling, reduce unwanted fluid influx from the reservoir using appropriate weighing materials, cool the drilling bit and circulate the drill cuttings to surface where they are removed at the shakers. Such mud systems which require several chemical additives are expensive.

In air drilling, compressed air is used to cool the bit and transport drill cuttings to surface. Air is clearly a cheaper option, but suffers from poor heat capacity, hence less cooling effect, and is not suitable for handling influx of reservoir fluids. Generation of air foam increases the cuttings' carrying capacity substantially and enables removal of liquids entering the wellbore. Saline formation water is known to be corrosive towards drilling tools but Meng et al. [31] found that foam, in addition to providing better lifting of fluids and solids, also helped reduce the corrosion rates. The authors conducted laboratory experiments at ambient conditions using a mixture of dodecyl alcohol sulfonate, HPAM and biopolymer as foam stabilizers, in addition to several other additives. With such a large array of chemicals, it is not clear whether it was the foam which provided better corrosion resistance. It is speculated that the two polymers which were added to enhance the foam strength may have diffused towards the metal surface providing a thin coating and thereby shielding the pipe from the corrosive formation water. HPAM is also known for its drag reduction effect whereby a laminar sub-layer is created close to the tubing wall, see [43]. The authors did not address foam destabilization in the presence of hydrocarbons, which could become an issue in situations where a high reservoir pressure causes an unwanted influx of hydrocarbons from the reservoir into the wellbore, a situation known as a kick.

3.2 Additive in Cement Slurry

Cementing operations are critical for providing well integrity and zonal isolation both during and after drilling a well. As for drilling muds, a multitude of chemical additives are required to design a slurry with the appropriate temperature resistance, density, setting time, fluid loss, compressive strength and other important design variables. According to McElfresh and Boncan [30], foam offers the possibility to achieve a low-density, yet high-strength material. The authors mention the use of foam cement in formations which are weak, highly fractured, vuggy, or containing thief zones. From an operational point of view, the drilling mud must be circulated out and replaced with the foamed cement, just like in traditional cementing applications. However, in the case of foamed cement, it is presumably lighter and less viscous than the mud it needs to displace, so care must be taken to avoid viscous instabilities leading to unwanted contamination of the cement with the mud. As a side note, the drilling operations during the Deepwater Horizon incident made use of a foamed cement recipe, which was not sufficiently tested for the particular conditions encountered. Foamed cement is still regarded as a niche application.

3.3 Wellbore Insulation

Well integrity is a major headache for the oil industry. According to Penberthy and Bayless [34] the high heat loss from the wellbore during steam injection operations necessitates a high wellhead temperature to maintain a given steam quality downhole but can lead to thermally-induced stresses causing casing failures. Research was therefore conducted to reduce the heat loss through annulus insulation with a low thermal conductivity fluid. A silicate foam, formed by boiling a sodium silicate solution, turned out to possess excellent insulating properties. Implementing the technique required several steps. The silicate solution was first injected into the annulus and began boiling once steam injection took place in the well. The boiling resulted in foam generation. The excess silicate solution was then displaced by water and lifted out using gas-lift leaving only the foam in the annulus.

3.4 Foam Fracturing Treatments

Hydraulic fracturing is a cornerstone for delivering economic production rates from low-permeability reservoirs. Efficient fracture propagation requires a high-viscosity fracturing fluid with good fluid loss control and with the ability to transport a proppant which serves to keep the fracture open and conductive once created. Aqueous phase fracturing fluids rely on gelling agents to increase the viscosity and control fluid loss. A drawback in low-pressure reservoirs is that back-production of such high-viscosity, high-density fluids requires some sort of artificial lift. According to Gaydos and Harris [20], foam has already been used as a fracturing fluid for several decades due to its excellent fluid loss properties. Foams help minimize water damage to sensitive formations containing clays. Furthermore, when the wellhead pressure is reduced during back-production after the stimulation, the lower hydrostatic pressure in the wellbore helps to lift both gas and liquids. Use of foam therefore speeds up the recovery of fracturing fluids after the stimulation.

3.5 Foam as Additive in Matrix Acidization

The goal of matrix acidization of carbonate formations is to remove drilling-induced reservoir damage and to increase well productivity. The acid, which is typically hydrochloric acid with a concentration in the range of 10–32 wt% reacts with the rock and can, under the right flow conditions, create dissolution patterns referred to as wormholes, which penetrate up to 20 ft into the formation. Wormholing makes more efficient use of the acid which is expensive.

Wormhole growth depends on acid velocity at the tip of the wormhole. In radial flow, the velocity decreases with distance from the well, and furthermore, an increasing amount of acid is spent broadening the stem as well as an increasing number of branches on the wormhole "tree". While some branching is beneficial to the skin reduction, it does limit further wormhole growth away from the wellbore.

Bernardiner et al. [6] investigated the use of foam additives in the acid stimulation treatment with the purpose to promote deeper wormhole penetration by reducing acid leak-off into the side-branches. The authors performed linear core floods and imaged the dissolution patterns during in-situ foam creation. The foam was created by a mixture of nitrogen and dodecyl-benzene sulfonic acid (DBSA) and was able to maintain structure at low pH. One of the stated advantages of the foam was that wormhole propagation was enhanced even at injection rates below the optimum conditions for wormholing. Similar to leak-off control in fracturing applications, foam served to temporarily block unwanted fluid movement while favouring displacement in the main direction of convection (linear in the core flood but radial in a field application). While the experimental evidence speaks for itself, the standard laboratory practice suffered from some limitations. First, linear core floods are now known to artificially enhanced wormhole propagation because the fluid can only exit at the end of the plug. Second, presence of reservoir oil would negatively affect foam stability, especially at reservoir pressure and temperature. Third, the chemical reaction between acid and rock produces carbon dioxide, which is in super-critical state at reservoir conditions and is able to block pore restrictions. Despite these drawbacks, this early attempt to control acid diversion by means of additives has become common industry practice although different additives have been developed since.

3.6 Foam as Additive in Gravel Packs

Weak rocks consisting of loosely held sand grains require screens to prevent solids from entering into the wellbore and reducing flow. Gravel packs are an example of a completion type designed for soft formations. Elson and Anderson [16] proposed to use foam as the carrier fluid instead of polymers in low-pressure reservoirs. The tested foam gravel pack came at half the price of a conventional gravel pack. The authors quoted a number of other advantages but did not comment on the durability of the foam.

3.7 Foam Gas Shut-off

Gas viscosity typically ranges from 0.02 to 0.06 cP whereas liquid viscosity can span several orders of magnitude. This means that gas mobility is often much higher than liquid mobility, which results in an unstable viscous gas-oil displacement.

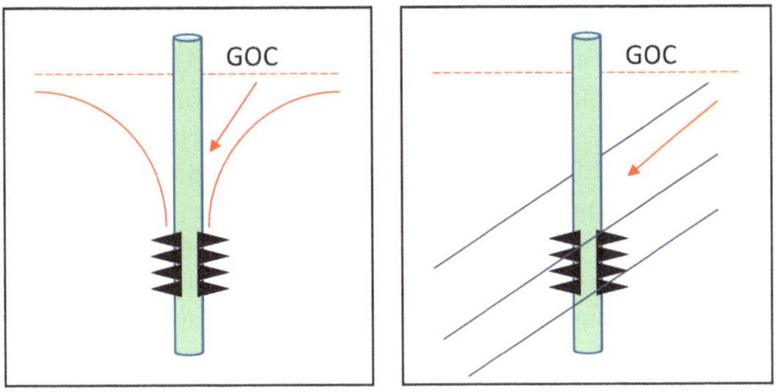

Fig. 7 Situations where FGSO may block unwanted gas Adapted from [38]

In saturated oil reservoirs containing a sizeable gas cap, oil production is often hampered by increasing gas influx from the overlaying gas cap, a phenomenon referred to as gas coning. As the gas saturation around the well increases, so does the gas relative permeability, which leads to higher gas rates thereby choking oil production. The key operational metric is the gas-oil producing ratio (GOR), which must be controlled to maintain stable oil production rates and avoid reaching the maximum gas handling capacity of the surface facilities. Presence of fractures or high-permeability streaks may further accelerate unwanted gas production (Fig. 7).

Need for GOR control was addressed more than eighty years ago by Sullivan [40]. GOR management strategies have been covered by numerous authors, including Kyi et al. [27] and Sarsekov et al. [37]. The key elements is frequent testing, choking back of the worst GOR offenders, well segmentation with zonal control, pressure support by water injection and placement of wells at a safe distance from the gas cap.

However, in mature fields where economics do not allow such major investments, chemical gas shut-off treatments present a low-cost option to temporarily reduce gas production. Such near wellbore treatments require the use of foam, generated by a surfactant tailored to the gas composition, the brine, and the rock, hence the term foam gas shut-off (FGSO). The foam must be strong and should withstand a large pressure drop when stagnant, once placed in the formation. Foam gas shut-off is said to be auto-selective because it is destabilized in the presence of oil; hence, if it is injected close to the gas-oil contact, it will preferentially form in the invaded gas zone. Furthermore, high gas shear rates caused by pressure drawdown around the wellbore will help maintain and regenerate the foam and continue to block or reduce the flow of free gas. In practice, treatments do not have a lasting effect and must be repeated every 6 months or so.

Heuer and Jacobs [23] patented the technology for gas shut off using foam. The first field application, reported by Holm [25], confirmed the laboratory-derived observations of foam as an effective method for decreasing high gas mobility caused by severe gas channeling. Interestingly, the foam also decreased the water production, reflected in a notable decrease in the producing water-oil ratio (WOR).

Since the first reported field trial some fifty years ago, a number of published applications have shown a mix of successes and failures. Aarra and Skauge [1] and Aarra et al. [2] describe the details of an FGSO pilot performed in the Oseberg field, located offshore Norway some 140 km from the coastal city of Bergen. The main objectives were first and foremost to obtain field experience with foam placement and foam generation in a production well, and as a secondary goal to evaluate if foam could be used to reduce gas production caused by gas coning. The pilot well was producing from a 2–3 Darcy homogeneous sandstone through five perforated intervals. After gas breakthrough occurred, only the top perforation interval was opened for foam treatment and back-production. The well was monitored with production logging tools (PLT) during injection and start of back-production after foam placement. Alternating injection of gas and alpha-olefin sulphonate (AOS) surfactant solution was chosen and the surfactant was injected together with seawater in two slugs at 1–2 wt% concentration and then displaced by gas. Production tests prior to the foam treatment were carried out to obtain a baseline GOR level and to calibrate the reservoir simulation model. The foam pilot was deemed operationally successful and showed that foam can be generated by slug injection of gas and surfactant solution.

Following the positive results from Norway, the mid-to-late 1990s saw a number of foam gas shut-off trials around the world. Pilot design for a well in the Rabi field in Gabon was covered by Bouts et al. [8]. The well in question was a vertical producer suffering from severe gas coning originating from an overlaying gas cap. As in the Oseberg field, the reservoir permeability was in the Darcy range, which accelerated gas breakthrough. The authors stressed the importance of proper foam placement as a key success factor. Since the foam is generated in-situ, the injected gas must be able contact the surfactant solution, which in this case was designed to be oil-soluble rather than water-soluble. In the absence of water, one suggestion was to add a solvent to reduce the surfactant mixture density below the oil density and thereby enhance gravity segregation. The recommended surfactant concentration was 1–2 wt%, in line with previous indications that foam generation requires a certain threshold concentration to take place.

The Prudhoe Bay field is one of the largest fields discovered in the United States. Located on the North Slope of Alaska, technologies such as enriched hydrocarbon gas flooding, horizontal drilling, and hydraulic fracturing have been deployed at an early stage to improve recovery. While fractures improve early production of oil, they also accelerate subsequent unwanted production of water and gas. In the case of Prudhoe Bay, the presence of a large gas cap soon resulted in excessive gas production, according to Thach et al. [41]. Prior to the foam pilots, other methods to control GOR included shut-in of high-GOR wells, side-tracking, or cyclic production-injection schemes to modify the sweep patterns. Laboratory studies confirmed that aqueous-phase foams provided larger foam strength than non-aqueous foams and that addition of polymers could further strengthen the foam. A complicating factor was to identify a surfactant which would work at a reservoir temperature of 200 °F. While most commercial products available at the time were found to be unsuitable, several AOS-based system were chosen for further studies. Surfactant chain length was found to play a key role with regards to stability. The shortest chain length generated unstable foam, the largest chain length was more oil-soluble and gave rise to oil-in-water emulsions. In the end, a mix of several surfactants was chosen because it improved the foam stability.

An important aspect when piloting new concepts is to select not one but several wells to evaluate the outcome in a statistically meaningful manner. Some trials may fail whereas others will hopefully be successful. Therefore, large fields with many wells offer better opportunities for testing new technology, including near-wellbore treatments; the scope for field-wide implementation is simply larger. Chukwueke et al. [12] described a field trial with a 50% success rate in Nigeria involving eight wells, two foam systems and two different foam generation techniques. Similar to the previous field trials, the reservoir permeability was above 1 D but instead of relying on sand packs for flow studies, the experimental protocol involved reservoir cores. Increased tolerance towards oil was regarded as desirable, which prompted the use of a combination of water-soluble fluoro-surfactants and the traditional AOS formulations, fortified by addition of a low molecular weight polymer. Fluorinated surfactants have since become subject to import bans in some countries due to environmental concerns.

Mixed results from a number of trials, environmental restrictions, and lack of long-term foam stability have remained an Achilles heel for large-scale foam applications. After a quiet period, interest in foam is picking up again. Noteworthy studies over the past decade include Skoreyko et al. [01], Enick and Olsen [17], and Ocambo et al. [33].

A recent FGSO trial took place in a mixed-wet carbonate reservoir offshore Abu Dhabi with a target reservoir permeability much lower compared to prior applications. A comprehensive laboratory work program was detailed by Skauge et al. [38] involving extended stability tests and bulk rheology experiments followed by core flooding to establish mobility reduction, adsorption, and pressure gradient resistance. Strong foam was eventually obtained with a 5 wt% AOS formulation. Addition of fluorinated surfactants and a new high-temperature resistant polymer did not improve foam stability. Design of the FGSO pilot using the selected surfactant formulations was described by Elhassan et al. [15]. The carbonate reservoir is characterized by a large gas cap overlaying an oil rim, a permeability variation from 5 to 1000 mD, a temperature of 220 °F and a formation brine salinity in excess of 200,000 ppm with more than 20,000 ppm divalent ions. Given the high required surfactant concentration of 5 wt% and the offshore location, the logistics of the operation proved challenging because the footprint had to be limited to a confined space on the barge used for well interventions. Prior to the shut-off, the pilot wells were subjected to production testing and PLT followed by a shut-in period to estimate permeability and skin. The same monitoring campaign was then repeated after the foam treatment to be able to compare changes not only to GOR but also with regards to inflow profile, injectivity/productivity and effective skin. In terms of injection techniques, both surfactant-alternating-gas and co-injection were piloted; the co-injection data showed clear signs of foam generation in-situ.

3.8 *Mobility Control in Gas Floods*

Whereas gels are seen as mobility control agents for aqueous phases, foams represent a mobility control solution for gas-based EOR. Foam stability over prolonged periods still remains a challenge, but foam is in principle well-suited for reduction of fluid mobility in fractures and high-permeability channels. Gland et al. [21] discuss a new development of cationic surfactants for creation of CO_2 foam in carbonates whereas Chevallier et al. [11] discuss foam in naturally-fractured reservoirs in general. The dilemma with foam is that on one hand, foam stability is important. On the other hand, a low gas-oil interfacial tension is highly desirable to force gas into an oil-wet matrix. Creation of a viscous pressure drop due to a strong and stable foam is more important than lowering the gas-oil IFT.

A successful near-wellbore treatment with foam is critically dependent on the correct placement of foam, which should remain strong also at stagnant conditions. In other words, the foam mobility should be low. On the other hand, weaker and more mobile foams are preferable for gas injection aiming at mobility control deeper into the porous formation without impairing injectivity. Therefore, a successful foam system for a near-wellbore treatment cannot be directly transferred to a gas flood requiring mobility control.

In water-alternating-gas (WAG) floods, a water-soluble surfactant slug would be added during the water cycle. The foam would then be (re-)generated during the subsequent gas cycle and the scheme would therefore be referred to as foam-assisted WAG or FAWAG. Turta and Singhal [42] have compiled an extensive list of foam pilots from North America to guide screening and design of foam applications.

3.8.1 Hydrocarbon Gas Foam

The FAWAG concept was piloted in the Snorre field, located 150 km offshore Norway, from 1997 to 2000. Blaker et al. [7] describe how an FGSO treatment in the field was carried out a year earlier to test if foam would block gas movement in-situ. A favorable outcome of the FGSO treatment was regarded as an important step in derisking the larger-scale FAWAG pilot. FAWAG differs from FGSO in a number of ways. FGSO is a near wellbore treatment performed in producing wells involving a limited volume of surfactant. A high surfactant adsorption is of little importance to the economics of FGSO, as long as the stagnant foam is strong enough to significantly reduce gas influx for several months. FAWAG, on the other hand, targets injection wells, and requires a substantial volume of surfactant to propagate foam far into the formation. Both surfactant adsorption and concentration must be as low as possible and the foam strength is a compromise between achieving a much wanted gas mobility reduction and yet maintain ability to inject gas and water to maintain reservoir pressure and sustain production. In terms of injection scheme surfactant-alternating-gas (SAG) appeared to be superior to co-injection of surfactant and gas, which the numerical simulations had been unable to quantify. One important reason is the operational challenges associated with co-injection. Surfactant and gas have to mix at surface in the right proportions at the right pressure and must mix and foam prior to reaching the reservoir to avoid segregation once inside the reservoir. Three WAG cycles were performed and surfactant was added to each of the water cycles. Analysis of injection bottom-hole pressure data showed that the gas injectivity increased during the end of the first gas cycle and this was interpreted as a sign of foam drying out and disintegrating. Data from the second and third cycles indicated that gas and surfactant gradually began following different paths, possibly with the help from natural fractures. Despite operational challenges and difficulty in propagating foam deep into the reservoir, the overall results showed that gas breakthrough was delayed and the gas oil ratio was considerably lowered.

The Cusiana field in Colombia contains a volatile oil in equilibrium with a gas cap. A combination of development strategies had been implemented since the start of production in 1994, such as natural depletion, water injection, and gas recycling into the gas cap for condensate recovery, see [33]. Gas injection also targeted recovery of oil from the oil rim through conversion of old water injection wells. As the gas injection project matured, there was an increasing need for mobility control to improve sweep. Conformance had to tackle not only the unfavorable viscosity ratio between gas and oil but also existence of high-permeability streaks as well as reactivation of fracture corridors. The main treatment involved a surfactant concentration of only 0.2 wt% followed by a non-foaming low-IFT solution to push the foam further into the reservoir. After 2–3 months, the oil rate decline was arrested and the GOR was reduced in a number of wells. One of the drawbacks of the SAG scheme is that the treatment zone for foam is relatively limited. At some distance away from the injection well, gas and surfactant may segregate away from each other just like in a normal WAG situation, and further foam generation is no longer possible.

3.8.2 Carbon Dioxide Foam

Carbon dioxide flooding was initiated in the SACROC unit in the Permian Basin almost fifty years ago, see [13]. Compared to nitrogen and hydrocarbon gas, carbon dioxide benefits from a higher density at reservoir conditions which can match or even exceed that of the reservoir fluid in some cases. Hence, gravity override caused by density differences is not as prominent in carbon dioxide floods. Viscous instabilities caused by unfavorable mobility ratio, in addition to presence of high-permeability channels, on the other hand, are more than enough to cause premature gas breakthrough, even in miscible floods. In mixed-wet reservoirs where injectivity is not impaired by a low water relative permeability end-point, conversion to a tapered water-alternating-gas (WAG) scheme is beneficial for mobility control. In water-wet rocks, conversion to WAG is not an option because it would substantially reduce the injectivity and incremental recovery has to come from continuous gas injection. In both situations (i.e. continuous gas injection and WAG), foam is able to address the need for improved mobility control.

If the target is a continuous CO_2 flood, the absence of an aqueous phase may require thinking outside the box in terms of surfactant selection. One proposal which spurred interest around 2010 involved creation of a nano-particle stabilized foam without the need for a surfactant, see [18]. Laboratory studies showed that the foam would disintegrate at surface following a large pressure reduction. Other studies pointed towards dissolving surfactant in the super-critical CO_2 phase itself [44].

CCUS is an area which may also benefit in the future from stable CO_2 foams.

3.8.3 Nitrogen Foam

Nitrogen foam is relatively well-studied because nitrogen is an inert gas and hence easier to work with in a laboratory. Also, the solubility of nitrogen in brine is much lower than for carbon dioxide, which may help generate and propagate a more stable foam. Nitrogen is often used as a proxy for hydrocarbon gas during laboratory programs.

The most well-described field implementation of nitrogen injection comes from a highly fractured carbonate reservoir in Mexico. Akal, the main field in the large offshore Cantarell complex, has undergone immiscible nitrogen injection since 1997. Rodríguez et al. [35] have summarized the field history and captured some key learnings. The reservoir thickness is close to 4000 ft, which favors a gas-oil gravity drainage (GOGD) scheme. As with other naturally fractured reservoirs, the initial production came from primary depletion where high fracture conductivity contributed to high initial production rates and therefore quicker payback of the investment costs. The recovery factor after primary production was low, which soon prompted the need for a pressure maintenance scheme. Gas injection was identified as the most feasible EOR method, but the choice of gas required detailed studies.

Based on availability, cost, safety and numerous other considerations, nitrogen was selected as the preferred injectant. This is a remarkable project given the fact that although nitrogen makes up almost 80% of the air, it had to be separated from oxygen in an energy-intensive operation onshore and then piped offshore to the field. Also, the breakthrough gas would consist of an increasing amount of nitrogen which would have to be dealt with in the surface facilities since nitrogen has no heating value. Other concerns were mostly reservoir related.

Nitrogen channeling leading to premature breakthrough was seen as the biggest potential drawback to the project but the risk was toned down due to field evidence suggesting very effective gravity segregation was taking place as long as injection was carried out from the top of the reservoir. Not all injection wells were positioned at the top of the structure and nitrogen did break through earlier than expected in some wells. The Cantarell nitrogen project is still unprecedented in terms of scale and must be characterized as a success, regardless of operational issues resulting from early gas breakthrough. Skoreyko et al. [39] refers to three foam pilots being conducted in Cantarell and described the efforts to model the foam process based on laboratory experiments as well as the data recorded during the pilots.

3.9 Mobility Control in Steam Floods

Steam flooding is a thermal EOR technique applicable to shallow reservoirs containing heavy oil. The principle relies on heat transfer from condensation as the super-saturated steam contacts the reservoir fluid. At depths beyond some 2500 ft, steam condenses in the wellbore and becomes hot water, which has a much lower capacity to transfer heat than steam. The steam is most often generated at surface using gas turbines, and in rare instances using solar panels. Steam injection is an energy-intensive operation and it is therefore paramount to make the most efficient use of the steam. As with any injection scheme, mobility control helps the injectant contact the target reservoir fluid. According to the review paper by Hirasaki [24], the use of temperature-resistant foams for steam applications was patented by Needham [32]. The goal was to plug high-permeability channels with foam and hence divert the steam towards unswept zones with lower permeability.

Steam drives are known to reduce the residual oil saturation below the values reported for waterflooding as a result of high-temperature distillation taking place in the reservoir. Since presence of oil can have a detrimental effect on foam stability, much research went into developing a surfactant solution which would not only create a stable foam but also reduce the residual oil saturation. AOS surfactants with longer alkane chain lengths in combination with alkali were found to meet both targets. It must be emphasized that the gas used for foam generation is not water vapour. In most field applications summarized by Hirasaki [24], the gas consisted of nitrogen or air. Foam was either injected continuously or as slugs.

A key metric used to evaluate steam flood performance is the steam-oil ratio, defined as the amount of steam required to yield an incremental barrel of oil relative to a baseline, which is sometimes taken as zero. Observations from various pilots was that even if foam was unable to increase the ultimate recovery factor, it would often accelerate production and hence improve the project economics.

4 Concluding Remarks

The field applications of foam are numerous but the properties of foam are best exploited in situations where long-term stability may not be needed, such as in hydraulic fracturing or in near wellbore treatments where chemical placement can be controlled. Foam for enhanced oil recovery is a topic of active research but has so far failed to gain widespread acceptance as a reliable method for in-depth conformance and mobility control. Such frontier applications require a surfactant which at low concentrations generates a very stable foam that can be propagated from the wellbore and far into the formation, and has low adsorption.

Most EOR processes struggle with high unit technical cost (UTC) and the current oil price environment does nothing to entice operators to initiate multi-year foam pilots. The appetite for risk varies among operators, but the economic upside in terms of improved sweep has to be present to justify continuous injection; i.e. a base case scenario without foam which yields a poor recovery is preferable. Although the concept as such can no longer be regarded as novel, a near wellbore treatment such as FGSO is still regarded as an important stepping stone towards derisking FAWAG. A good start would be to pilot foam in benign conditions such as low salinity and low temperature reservoirs where the foaming agents (i.e., surfactants) are readily available.

The behavior of foam systems comprising gas mixtures, brines, and surfactants can in principle be modelled with tools such as molecular dynamics or density functional theory because all components have a well-defined chemical structure. This author believes that a systematic brute-force computational approach towards screening of surfactants in terms of foaming tendency is needed to develop new chemicals which can maintain longer stability.

Frontier applications of foam involve high-temperature, high-salinity, and low-permeability reservoirs. An earlier paragraph described how small pores lead to a high capillary pressure, which destroy foam. A further complicating factor in low-permeability reservoirs is that some mobility reduction is required without sacrificing injectivity. Katiyar et al. [26] released details about the first hydrocarbon foam pilot in an unconventional reservoir; however, the purpose of the foam was to penetrate the hydraulic fracture network, not the tight matrix. In any case, the operational envelope of chemical EOR to which foam belongs will continue to expand in the coming decade as production of easy oil declines.

References

1. Aarra MG, Skauge A (1994) A foam pilot in a north sea oil reservoir: preparation for a production well treatment. SPE-28599-MS, presented at the SPE annual technical conference and exhibition, New Orleans, Louisiana, USA, 25–28 Sept 1994
2. Aarra MG, Skauge A, Sognesand S, Stenhaug M (1995) A foam pilot test aimed at reducing gas inflow in a production well at the oseberg field. Presented at the 8th European symposium on improved oil recovery, Vienna, Austria, 15–17 May 1995
3. Arash A, Piri M (2015) Direct pore-to-core upscaling of displacement processes: dynamic pore network modeling and experimentation. J Hydrol 522:488–509
4. Bennetzen MV, Mogensen K (2014) Novel applications of nanoparticles for future enhanced oil recovery. IPTC-17857-MS, presented at the international petroleum technology conference, Kuala Lumpur, Malaysia, 10–12 Dec 2014
5. Berg S, van Wunnik J (2017) Shear rate determination from pore-scale flow fields. Transp Porous Media 117:229–246
6. Bernardiner MG, Thompson KE, Fogler HS (1992) Effect of foams used during carbonate acidizing. SPE-21035-PA, SPE production engineering, pp 350–356
7. Blaker T, Aarra MG, Skauge A, Rasmussen L, Celius HK, Martinsen HA, Vassenden F (2002) Foam for gas mobility control in the snorre field: the FAWAG. SPE-78824-PA. SPE Reservoir Eval Eng 5:317–323
8. Bouts MN, de Vries AS, Dalland M, Hanssen JE (1996) Design of near well bore foam treatments for high GOR producers. SPE-35399-MS, presented at the SPE/DOE improved oil recovery symposium, Tulsa, Oklahoma, USA, 21–24 Apr 1996

9. Cannella WJ, Huh C, Seright RS (1988) Prediction of Xanthan Rheology in porous media. SPE-18089-MS, presented at the SPE annual technical conference and exhibition, Houston, Texas, USA, 2–5 Oct 1988
10. Chambers KT, Radke CJ (1991) Capillary phenomena in foam flow through porous media. In: Morrow NR (ed) Interfacial phenomena in petroleum recovery. Marcel Dekker, Inc., New York 10016, New York, USA
11. Chevallier E, Bouquet S, Gland N, Douarche F, Batôt G (2019) Advanced eor foam in naturally fractured carbonates reservoirs: optimal balance between foam and interfacial tension properties SPE-194992-MS, presented at the SPE middle east oil and gas show and conference, 18–21 March 2019. https://doi.org/10.2118/194992-MS
12. Chukwueke VO, Bouts MN, van Dijkum CE (1998) Gas shut-off foam treatments. SPE-39650-MS, presented at the SPE/DOE improved oil recovery symposium, Tulsa, Oklahoma, USA, 19–22 Apr 1998
13. Crameik TD, Plassey JA (1972) Carbon Dioxide injection project sacroc unit, Scurry County, Texas. API-72-D001, annual meeting papers, division of production, Houston, Texas, USA, 6–8 Mar 1972
14. Dollet B, Raufaste C (2014) Rheology of aqueous foams. C R Phys 15:731–747
15. Elhassan M, Keshtta OM, Berrim A, Draoui E, Mogensen K, Levitt D (2019) Design of foam gas shut-off pilot for a giant high-temperature, high-salinity carbonate reservoir. SPE-197322-MS, presented at the Abu Dhabi international petroleum exhibition and conference, Abu Dhabi, UAE, 11–14 Nov 2019
16. Elson TD, Anderson GW (1982) Foam gravel packing. SPE-11013, presented at the annual fall technical conference and exhibition, New Orleans, Louisiana, USA, 26–29 Sept 1982
17. Enick RM, Olsen DK (2012) Mobility and conformance control for carbon dioxide EOR via thickeners, foams, and gels—a detailed literature review of 40 years of research. DOE/NETL 2012/1540; Activity 4003.200.01. Report prepared for the U.S. Department of Energy, 2012
18. Espinoza D, Caldelas FM, Johnston KP, Bryant SL, Huh C (2010) Nanoparticle-stabilized supercritical CO_2 foams for potential mobility control applications. SPE-129925-MS, presented at the SPE improved oil recovery symposium, Tulsa, Oklahoma, USA, 24–28 Apr 2010
19. Friedmann F, Hughes TL, Smith ME, Hild GP, Wilson A, Davies SN (1999) Development and testing of a foam-gel technology to improve conformance of the rangely CO_2 flood. SPE reservoir evaluation & engineering 2(01):4–13. https://doi.org/10.2118/54429-PA
20. Gaydos JS, Harris PC (1980) Foam fracturing: theories, procedures and results. SPE-8961, presented at the SPE-DOE symposium on unconventional recovery, Pittsburgh, Pennsylvania, USA, 18–21 May 1980
21. Gland N, Chevallier E, Cuenca A, Batôt G (2018) New development of cationic surfactant formulations for foam assisted CO_2-EOR in carbonates formations. SPE-193201-MS, presented at the Abu Dhabi international petroleum exhibition & conference, Abu Dhabi, UAE, 12–15 Nov 2018
22. Hernando L, Bertin HJ, Omari A, Dupuis G, Zaitoun A (2016) Polymer-enhanced foams for water profile control. SPE-179581-MS, presented at the SPE improved oil recovery symposium, Tulsa, Oklahoma, USA, 11–13 Apr 2016
23. Heuer GJ, Jacobs CL (1968) Control of gas-oil-ratio in producing wells. U.S. Patent No. 3,368,624
24. Hirasaki GJ (1989) The steam-foam process. SPE-19505-PA, JPT (May 1989), pp 449–456
25. Holm LW (1970) Foam injection test in the Siggins field, Illinois. J Pet Technol 22(12):1499–1506
26. Katiyar A, Patil PD, Rohilla N, Rozowski P, Evans J, Bozeman T, Nguyen Q (2019) Industry-first hydrocarbon-foam EOR pilot in an unconventional reservoir: design, implementation and performance analysis. URTeC 103, presented at the unconventional resources technology conference, Denver, Colorado, USA, 22–24 July 2019
27. Kyi KK, Amin MFBM, Alam BN, Salleh NH, Yaakob AA, Wa WW, Haddad S, Johan Z, Wijoseno DA, Devan T (2016) Production diagnostic analysis and optimisation in complex

high GOR horizontal oil producer. IPTC-18935-MS, presented at the international petroleum technology conference, Bangkok, Thailand, 14–16 Nov 2016
28. Lake LW, Johns RT, Rossen WR, Pope GA (2014) Fundamentals of enhanced oil recovery. Society of Petroleum Engineers, Richardson, Texas, USA
29. Lunkenheimer K, Malysa K (2015) Simple and generally applicable method of determination and evaluation of foam properties. J Surf Deterg 6(1):69–74
30. McElfresh PM, Boncan VCG (1982) Applications of foam cement. SPE-11203, presented at the annual SPE technical conference and exhibition, New Orleans, Louisiana, USA, 26–29 Sept 1982
31. Meng Y, Wan L, Chen X, Chen G, Yang L, Wang J (2005) Discussion of foam corrosion inhibition in air foam drilling. SPE-94469-MS, presented at the SPE international symposium on oilfield corrosion, Aberdeen, United Kingdom, 13 May 2005
32. Needham RB (1968) Plugging of high permeability Earth Strata. US Patent No. 3,412,793
33. Ocambo A, Restrepo A, Cifuentes H, Hester J, Orozco N, Gil C, Castro E, Lopera S, Gonzalez C (2013) Successful foam EOR pilot in a mature volatile oil reservoir under miscible gas injection. IPTC-16984-MS, presented at the international petroleum technology conference, Beijing, China, 26–28 Mar 2013
34. Penberthy WL, Bayless JH (1974) Silicate foam wellbore insulation. SPE-4666-PA, J Pet Technol (June 1974), 583–588
35. Rodríguez F, Ortega G, Sánchez JL, Jiménez O (2001) Reservoir management issues in the cantarell nitrogen injection project OTC-13178-MS, presented at the offshore technology conference, April 30–May 3 2001. https://doi.org/10.4043/13178-MS
36. Roof JG (1970) Snap-off of oil droplets in water-wet pores. SPE-2504-PA, SPE J 10(01):85–90
37. Sarsekov A, Al Neaimi A, Saif OY, Abed A (2017) Integrated workflow of GOR management towards sustainable oil production. SPE-188591-MS, presented at the Abu Dhabi international petroleum exhibition & conference, Abu Dhabi, UAE, 13–16 Nov 2017
38. Skauge A, Aarra MG, Ormehaug PA, Solbakken J, Mogensen K, Masalmeh S (2019) Preparations for foam gas shut off in carbonate reservoirs. SPE-197640-MS, presented at the Abu Dhabi international petroleum exhibition and conference, Abu Dhabi, UAE, 11–14 Nov 2019
01. Skoreyko F, Pino AV, Prada HR, Nguyen QP (2011) Development of a new foam eor model from laboratory and field data of the naturally fractured cantarell field SPE-145718-MS, presented at the SPE reservoir characterisation and simulation conference and exhibition, 9–11 October 2011. https://doi.org/10.2118/145718-MS
39. Skoreyko F, Pino AV, Prada HR, Nguyen QP (2012) Understanding foam flow with a new EOR model developed from laboratory and field data of the naturally-fractured Cantarell field. SPE-153942-MS, presented at the SPE improved oil recovery symposium, Tulsa, Oklahoma, USA, 14–18 Apr 2012
40. Sullivan RJ (1937) Gas-oil ratio control in flowing wells. API-37-103, presented at the 18th annual meeting, Chicago, Illinois, USA, Nov 1937
41. Thach S, Miller KC, Lai QJ, Sanders GS, Styler JW, Lane RH (1996) Matrix gas shut-off in hydraulically fractured wells using polymer-foams. SPE-36616-MS, presented at the SPE annual technical conference and exhibition, Denver, Colorado, USA, 6–9 Oct 1996
42. Turta AT, Singhal AK (1998) Field foam applications in enhanced oil recovery projects: screening and design aspects. SPE-48895-MS, presented at the SPE international oil and gas conference and exhibition, Beijing, China, 2–6 Nov 1998
43. Virk PS (1971) Drag reduction in rough pipes. J Fluid Mech 45:225–246
44. Xing D, Wei B, Trickett K, Mohamed A, Eastoe J, Soong Y, Enick RM (2010) CO_2-soluble surfactants for improved mobility control. SPE-129907-MS, presented at the SPE improved oil recovery symposium, Tulsa, Oklahoma, USA, 24–28 Apr 2010

CO_2-Philic Surfactants: Structure Performance Relationship

M. Sagir, M. B. Tahir, Sayeda Hosna Talebian, Sami Ullah, Reema Ansar, M. Pervaiz, and Khurram Shahazad

1 Surfactants and Foaming Issues

In foam displacement for better oil recovery foamability along with foam, stability is considered among the significant concerns. The generation of foam is not much of a challenge as compared to foam stability. It is affected by many factors. Also, the selection of right surfactants is of more importance [1]. As discussed earlier surfactants are not only needed for bubble formation, but they are necessary for foam stability. A surfactant doesn't need to be an excellent foaming agent as well as good at IFT reduction [2]. Following aspects are considered for evaluation and selection of the surfactants: Foamability, foam stability, multivalent ion resistance, and thermal stability, and compatibility, the effect of the presence of oil, adsorption, salinity and IFT reduction [3]. The other factors affecting foam stability are the quality of foam, type of the polymer/its concentration and surfactant type/surfactant

M. Sagir (✉) · R. Ansar
Department of Chemical Engineering, Faculty of Engineering, Khwaja Fareed University of Engineering and Information Technology, Rahim Yar Khan, Pakistan
e-mail: sagir.utp@gmail.com

M. B. Tahir
Department of Physics, Faculty of Sciences, Khwaja Fareed University of Engineering and Information Technology, Rahim Yar Khan, Pakistan

S. H. Talebian
Oil, Gas and Petroleum Research Centre, Amirkabir University of Technology, Tehran, Iran

S. Ullah
Chemistry Department, Faculty of Science, King Khalid University, Abha, Saudi Arabia

M. Pervaiz
Department of Chemistry, GC University Lahore, Lahore, Pakistan

K. Shahazad
Department of Chemistry, King Abdul Aziz University, Jaddah, Saudi Arabia

T. Solling et al. (eds.), *Surfactants in Upstream E&P*, Petroleum Engineering,
https://doi.org/10.1007/978-3-030-70026-3_11

concentration the in terms of salinity less tolerant surfactants were used to generate foam, and there were also difficulties in foamability in oil-wet carbonate rocks [4].

For foam stability, non-ionic surfactants are of great use where the temperature is not high [5]. Usually, with time there is deterioration of a foam absorbed on the rock matrix with a higher decay at elevated temperatures in the presence of oil. In the case of CO_2 gas, it becomes even more severe. Which makes us focus on the synthesis of new surfactants which have an affinity for CO_2 gas under reservoir conditions, along with their ability to overcome the problems discussed above [6, 7]. These novel surfactants tend to generate foam with an increased degree of stability at higher temperatures and in oil presence where the adsorption issues are reduced to a minimum [8].

2 CO_2-Philic Surfactants

The increase in atmospheric carbon dioxide levels is a significant issue for scientists, engineers, economists and politicians. To tackle this problem, their main focus is on the capture and storage of carbon. The modification in physiochemical properties of carbon dioxide to increase its applications as a solvent will help in maintaining its atmospheric balance by carbon capture and sequestration (CSS) and can be used in EOR processes [9, 10]. For both of these techniques, it is important that we manage the fluid properties [11]. For CO_2-EOR to be a sustainable CCS option, improvements of CO_2 utilization factor is required by implementing conformance/mobilitycontrol techniques. The increase in CO_2 viscosity (that will lead to less viscous fingering in enhanced oil recovery and will give more control in the capture and storage of CO_2) would make CO_2-EOR techniques more economical and feasible by providing conformance control [12]. New chemical techniques have been introduced to CO_2-EOR industry to take advantage of the synergistic combination of chemical additive- and gas-EOR processes. These techniques have been employed as thickening agents, conformance control gels and in-depth mobility control CO_2-foamsurfactants [13, 14].

CO_2-philic surfactants have organic nature and are amphiphilic compounds (possessing hydrophilic as well as lipophilic properties), but the former segments, are replaced by CO_2-philic and CO_2-phobic segments. Usually, the parts that have an attraction for CO_2 are the tails of these surfactants and are regarded as CO_2-philic segments, while the head groups are CO_2-phobic parts. Once you have identified the tail of the surfactant, you can chose the CO_2-phobic segment from known hydrophilic groups [15]. The structure of a CO_2-philic surfactant is shown in Fig. 1.

Although we can increase the viscosity of carbon dioxide by adding self-assembly of polymers; before consideration of these structures, it is necessary to deal with the dilemma of solubility, solvophilicity and CO_2-philicity [16].

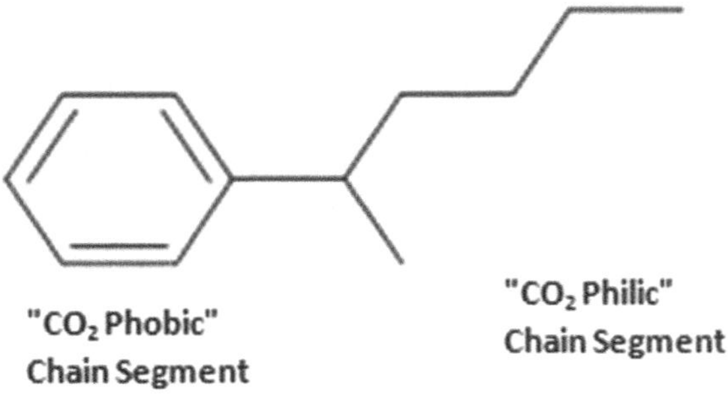

Fig. 1 Structural representation of CO_2-philic surfactant

2.1 Surfactant Tails—Fluorinated Surfactants

Fluorine having higher values of electronegativity and electron affinity causes the fluorocarbons with "n" number of carbon atoms (when $n \geq 4$) have less RI (Refractive indices) and B.P (boiling points) than that of the corresponding HC (For $n < 4$ their behaviour is different) [17]. Also, the molecular volume of fluorocarbons in the liquid state is greater as compared to corresponding hydrocarbons; that's why the value of α/υ, where α denotes polarizability and υ represents volume, and δ is the parameter for Hildebrand solubility parameter are considerably less than corresponding hydrocarbons.

$$\delta = \sqrt{\frac{\Delta H\upsilon - RT}{Vm}} \tag{1}$$

As the value of dielectric constant α/υ and δ for Carbon dioxide is low, the compatibility of fluorocarbons and Carbon dioxide is expected to be more than CO_2. Therefore one can say that the CO_2-philicity of fluorocarbons than the corresponding HC.

The use of FC surfactants has been frequent in the field because of the higher values of FC-chains solubility in liquid and supercritical Carbon dioxide. However, high cost and toxidity concerns have impeded their use in commercial applications.

2.2 Hydrocarbon, Siloxane Based and Oxygenated Surfactants

With the production of hybrid semi-fluorinated surfactants the use of FCs can be considerably reduced, but the main focus is to develop environmental friendly and economic CO_2-soluble hydrocarbon-based, surfactants [18, 19]. These are including silicon-based surfactants, oxygenated HC ionic surfactants, non-ionic surfactants or twin-tailed glycerine surfactants.

2.3 HC-Based Surfactants

As discussed earlier, the majority of hydrocarbons surfactants that are available commercially are not naturally CO_2-philic, but to enhance their philicity, scientists are working on the design and structure of the surfactants [20]. Many factors were identified for affecting CO_2-phlicity of HC-surfactants such as increased branching of tails and methylation lead to more solvophilicity in CO_2 as compared to linear alkanes. That lead to the development of w/c microemulsions having high stability. It was because of weaker interactions b/w surfactant tails and lesser surfactant affinity to H_2O, which leads to an enhanced partition coefficient [21].

3 CO_2-Philic Surfactants for Foam

The use of foam in CO_2-EOR (enhanced oil recovery) is because of its potential to overcome the problems associated with poor Carbon dioxide sweep efficiency [22], as it works to lessen the permeability heterogeneity effect, overcomes instabilities in viscosity, and minimizes the chances of gravity override [23]. The use of previously known surfactants is limited in the presence of oil because of untimely lamellae rupture, requirement of plenty of water for foam generation, adsorption on the rocks causing the surfactant loss and less tolerance in case of salinity, temperature and pressure [8, 24]. The oil droplets could enter and spread into the air-water interface of the foam film according to the accepted theory for the role of oil on foam stability. We can overcome the above-mentioned problems by the surfactant blending along with the addition of CO_2-philic functionalities in the structure of surfactant or the addition of nanoparticles in the foam system [25]. The presence of CO_2-functionalities in surfactant structure can increase the stability of foam, solve mobility issues and speed up foam propagation [22, 26].

CO_2-philic surfactants have resulted in a range of CO_2-based processes [27]. But fluorocarbons being the most effective and expensive CO_2-philes such as poly (perfluoroether), have limited the commercialization of such processes. Recent research has shown the possibility of generation of non-fluorinated, hydrocarbon-based systems having CO_2-philic nature [28]. Many non-fluorous CO_2-philic surfactants from economical raw materials can be made to make these processes economically favourable [29].

3.1 Cooperative C–H···O Hydrogen Bonding

In volatile, low molecular weight solvents CO_2 easily solubilises but solubility decreases for polar materials having higher molecular weights as CO_2 is a poor solvent. In spite of low polarisable properties of carbon dioxide, it is considered

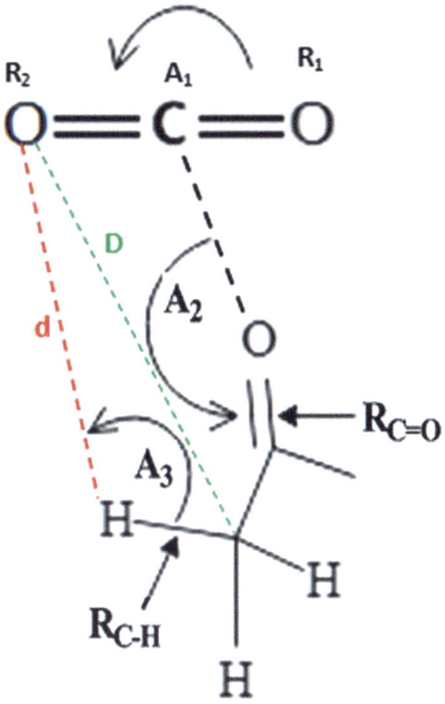

Fig. 2 C–H···O hydrogen bonding in Carbon dioxide Lewis-base interactions

a Lewis acid because of its electrophile nature. Because of this, it takes part in Lewis acid-base interactions. Carbon dioxide gives such interactions with surfactants, polymers etc [30].

Lately, a lot of work has been made on the development of HC-based CO_2-philes having carbonyl groups that make an interaction through a Lewis acid-base interaction with Carbon dioxide molecules, with the provision of required solvation energy for dissolution. On the basis of ab initio calculations that were done on simpler intermolecular complexes of Carbon dioxide with compounds that act as Lewis bases [29]. Results revealed C–H···O hydrogen bonds to be an alleviating interaction that contributes to designing CO_2-philes (Fig. 2).

3.2 *Phase Behaviour of Oxygen-Containing Polymers in CO_2*

To reason, the effect of the functional group of oxygen on the CO_2 and polymer phase behaviour the cloud point curves of many such polymers in carbon dioxide were considered. "CO_2-philicity" can be increased by the addition of an ether oxygen to a HC-polymer. It can be added in the side chain or backbone. It will provide the sites for interaction with carbon dioxide as well as increase the entropy of mixing with the creation of chains having higher free volume. The provision of attractive

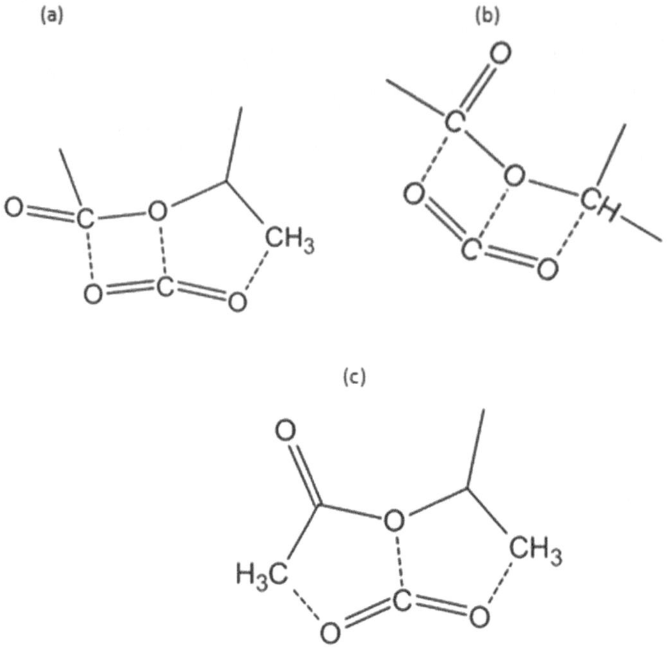

Fig. 3 Schematic of the interactionof Carbon dioxide with H-tail of the molecule

interaction sites for carbon dioxide molecules for ester as well as ether oxygen was revealed with ab initio calculations. However, because of having three binding modes for carbon dioxide interaction instead of one (which is the case for ether functional group) acetate functionalized polymers tend to be Carbon dioxide soluble than ether functionalized polymers. It is proven experimentally that by adding one methylene group b/w acetate-ether group and polymer backbone adversely affect phase behaviour [31].

Fan's work on abinitio molecular simulation studies has revealed that CO_2-philicity is facilitated by the side-chain oxygen (ether or ester) which is regarded as the carbonyl oxygen [32]. The interaction of CO_2 with H-tail of the molecule was observed in three modes which are given in the Fig. 3. Isopropyl acetate molecules were used as an example.

4 CO_2-Philic Surfactants as CO_2-Viscofiers

Besides the benefits of the process of CO_2, EOR by water alternating gas injection method, in a porous media, may restrain the CO_2 mixing with oil; slows down the CO_2's injection and needs water injection facilities [33]. In addition, it accelerates the post production water-oil separation issues; as well, the water production is greater than before [28]. Therefore, to control mobility issues new techniques are under consideration. For this purpose, to introduce dilute CO_2 thickener as a viscofier or the use of foam is a striking option [26]. Some direct thickeners like fluoro thickeners made at University of Pittsburg have been produced, but to achieve the level of the desired CO_2 contact by boosting the CO_2 viscosity no such material is available commercially.

5 Factors Affecting CO_2-Philicity for CO_2-Philic Surfactants

To design a CO_2-philic molecule, it is required to maximise CO_2 attraction by minimising their intermolecular attractions [4]. Some factors affecting CO_2-philicity are branching, hydrophobes, M.W, tail tip and presence of functional groups like propylene, methyl, propylene oxide (PO), carbonyl groups, ter-butyl tip and fewer methylene groups [34–36]. Some of these factors are discussed below.

5.1 Branching

For the CO_2-philicity in case of the hydrophobic segment of these surfactants, the most important factor is branching. With the decrease in chain length, the CMC (Aqueous) enhances; while, there is an increase in CO_2 solubility with increase in branching. According to Ben Tan in case of diacid and diol moiety branching increases the solubility and acyl chains, the increase in solubility because of branching is upto 20 times [4, 5]. Increased solubility into the aqueous phase also will reduce foam sensitivity to water saturation.

The CO_2-philicity for four of the copolymers of vinyl acetate along with comonomers with dipropyl maleate, dibutyl maleate, diisobutyl fumarate and dineopentyl maleate was measured. The cloud point pressure for the polymer having the highest degree of branching was found to be the lowest. The molecular dynamics simulation, along with ab initio calculations, showed that the polymer–CO_2 interactions were reduced due to high branched structure degree. However, the molecular dynamics simulation results for the free volume fraction, cohesive energy density, radial distribution functions and solubility parameter showed that the end groups having highly branched caused the free volume fraction to increase that lead to a lower number of polymer–polymer interactions. Thus, end groups with the highly branched structure degree were an effective way to increase the CO_2-philicity (Fig. 4).

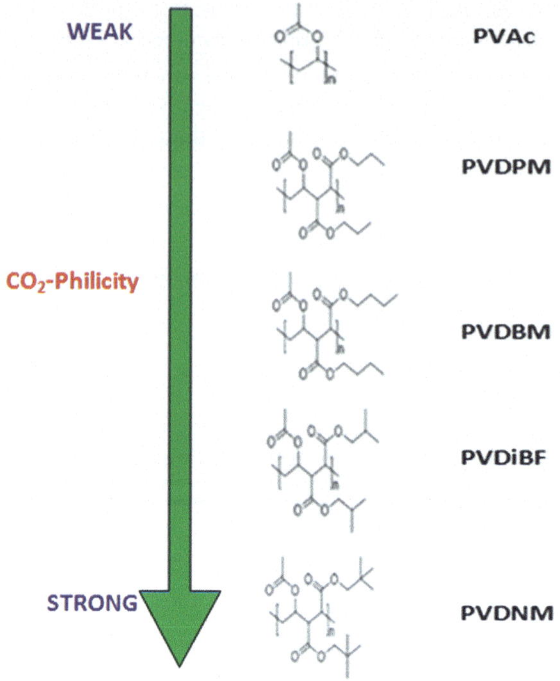

Fig. 4 Trend of CO_2-philicity

5.2 *Number of Tails*

For CO_2-philic compounds, number of tails of the surfuctant affects the solubility significantly. With the addition of tails for such surfactants, the solubility in the CO_2 increases [37]. New studies are focusing on the CO_2-water interface and suggest that for a double-tail surfactant there ought to be increased contact and hence offer more stability for the microemulsion. Similarly, the addition of a third chain increases the solubility even more [38]. It was shown in experimental sandstone coreflooding observations that the limiting factor for increasing the number of tails in surfactant structure is the equilibrium adsorption onto the rock surface.

In supercritical carbon dioxide, the solubility of polar substances can be improved by surfactants having chains of both fluorocarbon and hydrocarbon. An efficient method to carbon dioxide philicity can be variation in the length of hydrocarbon of the hybrid surfactant. The study on the effects of hydrocarbon chain length as well CO_2-philicity of such surfactants (F7Hn, $n = 1, 4, 7$ and 10) in carbon dioxide/water mixtures with the use of molecular dynamics simulations was done. The studies revealed that when the length of hydrocarbon is equal to the length of fluorocarbon (F7H7) the self-assembly time of F7Hn was maximum. This indicates that F7H7 has the highest values of CO_2-philicity due its ability to separate carbon dioxide and water. For the explanation, this behaviour structures of reverse micelle were analyzed. The results show that with a decrease in hydrocarbon chain length, there are two mechanisms. Firstly, separation ability of F7H*n* is reduced due to the reduction of its volume. This leads to the curved conformation of the fluorocarbons. It increases separation ability. But in the case of F7H7 these mechanisms are balanced driving an enhanced separation ability of carbon dioxide and water. Thus curved conformation of the hybrid surfactant tail and increased volume can work for increased CO_2-philicity in F7H*n* surfactants [39].

5.3 *Structural Changes*

The effect of tail-length on CO_2-philicity and CO_2-philic properties has widely been studied in the past. The phase behaviour of different double tailed fluorinated surfactants was studied at different values of temperature and pressure [40]. Those studies led to the optimization of tail length, which was found to be appropriate for greater emulsion formation of CO_2/H_2O at the micro-level. The oligomer's phase behaviour is changed by the variation in end-groups of the oligomer PVAc-OH [41]. Audrey DuPont studied P-phase and T-phase stability, structure aggregation and the effect of chain structure. The chain length effects can be viewed by the surfactant packing and surfactant free volume. Slight structural changes can affect the solubility of CO_2 in esters. Even or odd effects on solubility occur depending upon the number of C atoms [42]. Bray observed the acyl chain length is essential for the solubility of CO_2 for the molecule [43]. As ten carbons enter the chain, they cause the solubility to increase. Small changes in structure affecting the solubility of CO_2 molecule suggest that study of property-structure relationship, can enable us for prediction of parameters affecting CO_2 solubility of molecules [5].

5.4 *No. of Methyl Groups*

As no. of methyl groups increase the CO_2-philicity also increases. Surfactant development also depends upon low M.W, lesser chain length, lower No. of propylene oxide (PO) and methylene groups.

5.5 *Carbonyl Groups*

Eric J. Beckman and Thomas J. Styranec have synthesized compounds that are CO_2-philic from only hydrogen, oxygen, and carbon [44]. They noticed that solubility of Polyether was considerably affected by variation in the side chain or carbonyl group addition in it. Acetate group being added in the side chain enhances solubility to a maximum value then there is decline in the solubility trend. R. Fink et al., observed that with addition of side chains having ester functional group the solubility of CO_2-philic compounds in CO_2 was considerably increased [45]. Bilal Baradie studied the effect of addition of different fluorine and vinyl-acetate groups in the side chain. He noticed a sharp change in polyether solubility with the side chain variation or by adding the carbonyl group [46].

At 298 K PVA (poly vinyl acetate) is miscible with carbon dioxide for a range of molecular weights. Although both poly methyl acrylate and PVA have the same number of C, H and O atoms in their repeat groups but for dissolution of 5 wt% of the former at 298 K cloud-point pressures are considerably greater. This is because of the less accessibility of CO_2 to carbonyl group in poly methyl acrylate [47].

5.6 *Molecular Weight (MW)*

Research has proved that the solubulity of CO_2 is greatly dependant on the MW of the given compound. At lower values of pressure Oligomer PVAc-OH (Molecular Weight <3000 g/mol) tends to be soluble in CO_2, but the decreases in solubility occures with increase in MW. Tan et al determined that the solubility of polymers like PEC and PEE relys on their MW and the chemical structure. As the different solvent, in different amounts, were mixed with Carbon dioxide for the measurement of the compatibility for mixture components, it was noticed that compounds with lower MM (molar mass) having minimum single O-atom in ether, carbonyl and/or acetate groups interact favourably with CO_2 via Lewis base/Lewis acid interactions [48].

5.7 *Interaction Capacity of Carbon Dioxide with Organic Compounds*

Hydrocarbons alone and also with the compounds having a functional group of carbonyl, hydroxyl, carboxyl, thiocarbonyl, sulfonyl have gained attention as CO_2-philic agents. Generally, between carbon dioxide and functionalized organic compounds, the interaction capacity is greater than HC and their derivatives. The addition of carbon dioxide molecules in the former system results in more stable complexes. It has been proved that $\pi \ldots \pi$ linkages b/w aromatic rings and carbon dioxide considerably contribute to the interactions b/w MOF/ZIF materials and carbon dioxide. When dissolved in carbon dioxide, formic acid tends to be the most soluble in comparison with other host molecules. In the case of carbon dioxide interaction, the carbonyl and sulfonyl compounds show more stability than other functionalized groups. Thus the organic compounds are important for the preparation of CO_2-philic materials [49].

5.8 *Stepwise Fluorination*

The fundamental nature of FC and HC interaction of carbon dioxide was explored by studying the outcomes of stepwise fluorination on methane. An optimum value of fluorine density for which CO_2-philicity is maximum was revealed. As we know, carbon dioxide can act as a WLB as well as WLA in intermolecular interactions. It is evident that CO_2-FC and CO_2-HC interaction are, although comparable energetically, yet their fundamentals are different. The interaction of fluorocarbons is via C-atom of carbon dioxide while that of hydrocarbons is from O-atoms. Also in case of partly fluorinated HC there is an explicit interaction of carbon dioxide with a fluorine atom. In such cases, the C−H bonds tend to contribute to CO_2-philicity via weak C−H···O interactions [50].

6 Conclusion

Despite the advantages of the CO_2 such as being non-toxic, inexpensive, and non-flammable gas, overall process sweep efficiency is restricted by its low density, low viscosity, and increased mobility. The low density results in the poor process sweeps efficiency because of increase in velocity of CO_2 gas and leads to viscous fingering and a breach of the inserted CO_2 gas at early stages. This undesired mobility ratio brings the process to reduced sweep efficiency, and consequently, a low oil recovery is expected [51].

To reach the maximum value of oil recovery, concerns regarding CO_2 mobility have to be resolved. By the applications of foam, CO_2 mobility can be efficiently controlled [52]. Foam can reduce the velocity of Carbon dioxide, and it lowers the breakthrough pointof the inserted gas along with lessening the cap gas production. There also exist some disadvantages of foams such as instability under reservoir conditions, i.e., increased salinity, elevated temperature, mainly in the presence of oil. Foam stability is enhanced because of CO_2-philic surfactants. The surfactants have a head as an H_2O soluble part and tail as oil-soluble part [22]. Surfuctants have many applications in Enhanced Oil Recovery (EOR) [53]. CO_2-philic surfactants are used to overcome CO_2 mobility issues. Like other surfactants, these also have a head and a tail being two distinct parts, but in this case, the tail has an affinity for CO_2 to attain a certain degree of stability for the foam.

References

1. Sagir M et al (2020) Foams in EOR. In: Surfactants for enhanced oil recovery applications. Springer, Berlin, pp 41–63
2. Sagir M et al (2020) Surfactants as emulsification agents for IFT reduction in EOR applications. In: Surfactants for enhanced oil recovery applications. Springer, Berlin, pp 65–87

3. Belhaj AF et al (2020) The effect of surfactant concentration, salinity, temperature, and pH on surfactant adsorption for chemical enhanced oil recovery: a review. J Pet Explor Prod Technol 10(1):125–137
4. Sagir M et al (2020) CO_2 philic surfactants, switchable amine-based surfactants and wettability alteration for EOR applications. In: Surfactants for enhanced oil recovery applications. Springer, Berlin, pp 89–102
5. Mushtaq M, Tan IM, Sagir M (2014) New surfactants for EOR applications: effect of chain length on performance. In: AIP conference proceedings. American Institute of Physics, College Park
6. Sagir M et al (2014) Novel surfactant for the reduction of CO_2/brine interfacial tension. J Dispersion Sci Technol 35(3):463–470
7. Sagir M, Talebian SH (2020) Screening of CO_2-philic surfactants morphology for high temperature-pressure sandstone reservoir conditions. J Petrol Sci Eng 186:106789
8. Azam MR et al (2014) Kinetics and equilibria of synthesized anionic surfactant onto berea sandstone. J Dispersion Sci Technol 35(2):223–230
9. Orr FM, JJ Taber (1984) Use of carbon dioxide in enhanced oil recovery. Science 224(4649):563–569
10. Plasynski S et al (2009) Progress and new developments in carbon capture and storage. Crit Rev Plant Sci 28(3):123–138
11. Talebian SH et al (2015) Static and dynamic foam/oil interactions: potential of CO_2-philic surfactants as mobility control agents. J Pet Sci Eng 135:118–126
12. Sagir M et al (2014) CO_2 mobility and CO_2/brine interfacial tension reduction by using a new surfactant for EOR applications. J Dispersion Sci Technol 35(11):1512–1519
13. Rindfleisch F, DiNoia TP, McHugh MA (1996) Solubility of polymers and copolymers in supercritical CO_2. J Phys Chem 100(38):15581–15587
14. Cummings S et al (2012) Amphiphiles for supercritical CO_2. Biochimie 94(1):94–100
15. Sagir M et al (2014) FAWAG using CO_2 philic surfactants for CO_2 mobility control for enhanced oil recovery applications. In: SPE Saudi Arabia section technical symposium and exhibition. Society of Petroleum Engineers
16. Peach J, Eastoe J (2014) Supercritical carbon dioxide: a solvent like no other. Beilstein J Org Chem 10:1878–1895
17. Kissa E (2001) Fluorinated surfactants and repellents, vol 97. CRC Press, Boca Raton
18. Sarbu T, Styranec T, Beckman EJ (2000) Non-fluorous polymers with very high solubility in supercritical CO_2 down to low pressures. Nature 405(6783):165–168
19. Hollamby MJ et al (2009) Tri-Chain hydrocarbon surfactants as designed micellar modifiers for supercritical CO_2. Angew Chem Int Ed 48(27):4993–4995
20. Consan KA, Smith RD (1990) Observations on the solubility of surfactants and related molecules in carbon dioxide at 50 C. J Supercrit Fluids 3(2):51–65
21. Trickett K et al (2010) Rod-like micelles thicken CO_2. Langmuir 26(1):83–88
22. Sagir M et al (2018) CO_2 foam for enhanced oil recovery (EOR) applications using low adsorption surfactant structure. Arab J Geosci 11(24):789
23. Sagir M et al (2018) CO_2 capture, storage, and enhanced oil recovery applications
24. Bhatti MP, Sagir M (2018) CO_2 capture, storage and EOR application in reference module in materials science and materials engineering
25. Mushtaq M et al (2015) Effect of pH on the static adsorption of foaming surfactants on Malaysian sandstone. Arab J Geosci 8(10):8539–8548
26. Sagir M, Tan IM, Mushtaq M (2014) CO_2 philic surfactant as possible mobility control agent in EOR applications. In: AIP conference proceedings. American Institute of Physics, College Park
27. Talebian SH, Sagir M, Mumtaz M (2018) An integrated property-performance analysis for CO_2-philic foam-assisted CO_2-enhanced oil recovery. Energy Fuels 32(7):7773–7785
28. Sagir M et al (2015) Synthesis and CO_2 EOR applications of foaming surfactants. J Pet Environ Biotechnol 7(257):2

29. Wang Y et al (2009) Design and evaluation of nonfluorous CO_2-soluble oligomers and polymers. J Phys Chem B 113(45):14971–14980
30. Baradie B et al (2004) Synthesis and solubility of linear poly(tetrafluoroethylene-co-vinyl acetate) in dense CO_2: experimental and molecular modeling results. Macromolecules 37(20):7799–7807
31. Kilic S et al (2007) Phase behavior of oxygen-containing polymers in CO_2. Macromolecules 40
32. Fan X et al (2005) Oxygenated hydrocarbon ionic surfactants exhibit CO_2 solubility. J Am Chem Soc 127(33):11754–11762
33. Sagir M et al (2020) Surfactants for enhanced oil recovery applications. Springer, Berlin
34. Hoefling T, Enick R, Beckman E (1991) Microemulsions in near-critical and supercritical carbon dioxide. J Phys Chem 95(19):7127–7129
35. Enick R et al (1998) Phase behavior of CO_2–perfluoropolyether oil mixtures and CO_2–perfluoropolyether chelating agent mixtures. J Supercrit Fluids 13(1–3):121–126
36. Kilic S et al (2009) Influence of tert-amine groups on the solubility of polymers in CO_2. Polymer 50(11):2436–2444
37. Sagir M et al (2020) CO_2-philic surfactants structural morphology prerequests for CO_2 philicity for foam durability for EOR applications. In: Analytical chemistry-advancement, perspectives and applications. IntechOpen
38. Sagisaka M et al (2007) Optimum tail length of fluorinated double-tail anionic surfactant for water/supercritical CO_2 microemulsion formation. Langmuir 23(17):8784–8788
39. Wang M et al (2016) The self-assembly structure and the CO_2-philicity of a hybrid surfactant in supercritical CO_2: effects of hydrocarbon chain length. Soft Matter 12(39):8177–8185
40. Eastoe J et al (2003) Design and performance of surfactants for carbon dioxide. ACS Publications, Washington DC, USA
41. Tan B, Bray CL, Cooper AI (2009) Fractionation of poly (vinyl acetate) and the phase behavior of end-group modified oligo (vinyl acetate) s in CO_2. Macromolecules 42(20):7945–7952
42. Dupont A et al (2004) Hybrid fluorocarbon– hydrocarbon CO_2-philic surfactants. 1. Synthesis and properties of aqueous solutions. Langmuir 20(23):9953–9959
43. Bray CL et al (2010) Polymer CO_2 solubility. Structure/Property relationships in polyester libraries. Macromolecules 43(22):9426–9433
44. Sarbu T, Styranec TJ, Beckman EJ (2000) Design and synthesis of low cost, sustainable CO_2-philes. Ind Eng Chem Res 39(12):4678–4683
45. Fink R et al (1999) Toward the development of "CO_2-philic" hydrocarbons. 1. Use of side-chain functionalization to lower the miscibility pressure of polydimethylsiloxanes in CO_2. J Phys Chem B 103(31):6441–6444
46. Baradie B, Shoichet MS (2002) Synthesis of fluorocarbon−vinyl acetate copolymers in supercritical carbon dioxide: insight into bulk properties. Macromolecules 35(9):3569–3575
47. Bao L et al (2017) Enhancement of the CO_2-philicity of poly (vinyl ester) s by end-group modification with branched chains. J Supercrit Fluids 127:129–136
48. Miller MB et al (2012) Solid CO_2-philes as potential phase-change physical solvents for CO_2. J Supercrit Fluids 61:212–220
49. Khanh PN, Trung NT (2018) Understanding interaction capacity of CO_2 with organic compounds at molecular level: a theoretical approach. In: Carbon dioxide chemistry, capture and oil recovery, p 105
50. Raveendran P, Wallen SL (2003) Exploring CO_2-philicity: effects of stepwise fluorination. J Phys Chem B 107(6):1473–1477
51. Sagir M, Mushtaq M, Hashment MR (2015) CO_2 enhanced oil recovery by using surfactants as possible mobility control agents. In: SPE Middle East oil & gas show and conference. Society of Petroleum Engineers
52. Khalil F, Asghari K (2006) Application of CO-foam as a means of reducing carbon dioxide mobility. J Can Pet Technol 45(05)
53. Sagir M et al (2020) Surfactant in petroleum industry. In: Surfactants for enhanced oil recovery applications. Springer International Publishing, Cham, pp 13–40

Stimulation

Applications of Surfactants as Fracturing Fluids: Chemical Design, Practice, and Future Prospects in Oilfield Stimulation Operations

Nilanjan Pal and Amit Verma

Abstract Oil and gas recovery from subsurface reservoir formations requires the application of appropriate stimulation and production techniques, aimed at restoring sufficient pressure difference within drilled formations. Proper implementation of surfactants aids in enhanced fluid connectivity of the reservoir at initial stages of well stimulation, as well as maintain long-term hydrocarbon production. Nowadays, it is being considered as an effective alternative to conventional fracturing fluids such as polymers, gels, etc. due to low cost of application, alteration of inter-molecular interactions, and prevention of insoluble residues' formation. It is evident that the physicochemical attributes of surfactant-based fracturing fluids can be suitably modified through the use of combination of additives such as friction reducers, clay stabilizers, acids, iron-control agents, cross-linking agents, non-emulsifiers, buffers, inhibition agents, gels, and associated gel breakers. The primary objective of this method lies in minimizing the extent of oil-water block near the wellbore matric and develop pore-connectivity in hydrocarbon pay-zones to attain good recovery characteristics. Surfactant fracturing fluids, if injected properly, are capable of reducing flowback, improving fluid stability and effective clean-up. Therefore, it is a possible route for petroleum engineers and fracture design professionals to produce oil and gas from low permeability reservoir zones via hydraulic fracturing technique, whilst attaining maximum recovery efficacy, production rate and economical operation. This chapter provides a detailed description of design and methodology of surfactants as fracturing fluids in the petroleum industry.

Keywords Surfactants · Fracturing fluid · Geotechnical & Process considerations · Chemical design · Functional evaluation · Fracture pore connectivity

---

N. Pal (✉)
Department of Petroleum Engineering, Indian Institute of Technology (ISM) Dhanbad, Dhanbad, Jharkhand, India
e-mail: nilanjanpaul01@gmail.com

A. Verma
School of Petroleum Technology, Pandit Deendayal Energy University, Gandhinagar, Gujarat, India
e-mail: amit.verma@spt.pdpu.ac.in

T. Solling et al. (eds.), *Surfactants in Upstream E&P*, Petroleum Engineering,
https://doi.org/10.1007/978-3-030-70026-3_12

1 Introduction

Oil and gas recovery from subsurface reservoir formations requires the application of appropriate stimulation and production techniques, aimed at restoring sufficient pressure difference within drilled formations. These permit the lifting of in-situ hydrocarbon reserves trapped due to capillary forces previously existing within rock pore-spaces [63, 62]. All resources are not conveniently located in accessible regions and may be present in heterogeneous formation layers with complex, impermeable oil windows (pay zones). Therefore, the use of favorable well stimulation techniques such as fracturing, acidization etc. are essential to access a larger area of the reservoir [21, 27]. In the past few years, hydraulic fracturing technologies have been increasingly used on a commercial scale that eventually became crucial to production operations in the mature oilfields. Proper planning and accountability of hydraulic fracturing projects are useful to devise facilities, implement the solution and attain cost-profitability for the industry.

Hydraulically fracturing is a widely employed field of hydrocarbon production in the petroleum industry with significant potential in terms of technical as well as beneficial economic ties. It involves the introduction of a proppant-laden fluid, which effectively perforates the otherwise hydrocarbon-containing tight formations. The fractures, so formed, retain their connectivity due to the presence of proppant. After that, the fracturing fluids allowed to flow-back completely, and thereafter, the oil zones are produced via pressure drive and subsequent methods [5]. The associated loss of hydrocarbon conductivity is one of the major problems affecting fracturing results to achieve an efficient stimulation plan [18, 59]. To maximize recovery from the porous network, the fracturing process creates an open pathway for hydrocarbon flow.

Fracturing fluid normally consist of high viscosity components, which can not only create an effective fracture but also transport the proppant (sand) to the fracture zones. The fracturing fluid must contain sufficient gelation property to support the proppant under dynamic shear conditions. It must be designed, whilst keeping in mind, the reservoir characteristics, in-situ fluid properties and geotechnological conditions.

The use of conventional polymers is facing operational difficulties owing to the formation of insoluble residues, high cost and improper planning on the part of oilfield professionals [77]. Insoluble polymer fragments form large flocs with inter-molecular interactions, which plug the fractured regions within reservoir formations and reduce the conducive property of proppant-packed network of interconnected (fracture) zones. However, these drawbacks can be controlled by addition or replacement of different types of "surface-active agents" or surfactants in fracturing fluid compositions [34, 38, 49]. Surfactants function by altering the extent of inter-molecular polymer interactions to inhibit the build-up of oligomeric aggregates. Burman and Hall [13] showed that better fluid loss control and flow-back efficiency could be achieved with surfactant-based fracturing fluids. Other works by Paterniti [48], Xu and Fu [83] and Xu [82] further corroborate this observation, and stress on the employment of appropriate surfactant type/dosage in hydraulic fracturing applications. In this chapter, the implications of surfactant-assisted hydraulic fracturing processes have been studied in detail. The geological considerations, formation evaluation, fluid characterization, and optimization have been presented herein to assess the functionality of surfactants in well stimulation operations. This constitutes an essential application of surfactant-based fluid systems in the area of enhanced hydrocarbon production within the purview of the petroleum sector.

2 Reservoir Evaluation and Geotechnical Considerations

Before the design and approval of any fracturing project, it is pivotal to consider reservoir characteristics, as well as that of present fluid phases and surrounding formations [53, 61, 71]. Permeability zones within conventional reservoirs exhibit values of the order of Darcys, whereas that of tight formations have permeability levels in terms of milliDarcies. However, significant variations may exist among different oil pay sections in a reservoir, which may further complicate the planning process. Therefore, no true relationship exists between porosity and permeability, especially when developing strategies to enhanced oil flow through the production wells. Evaluation of fracture length is also an intricate factor in understanding fracture pattern/geometry and determining well spacing from a development standpoint [32]. During selection, the reservoir parameters/variables that need considerations with critical information as follows:

(a) Formation permeability;
(b) In-situ stress distribution;
(c) In-situ fluid saturation and viscosity properties;
(d) Vertical depth of the reservoir;
(e) Reservoir pressure and temperature;
(f) Skin factor effect, which identifies if the reservoir is stimulated or damaged;
(g) Well-bore condition and extent of completion.

The typical skin factor values range from 100 for poorly consolidated gravel pack, to −6 for an extensive hydraulic fracture with infinite-conductivity. These values differ with nature of the reservoir, type of drilling and completion operations performed, reservoir fluid properties and location of oil pay zones. As a result, no single technique of hydraulic fracturing has ever worked universally. Each method has unique methodologies and benefits that cater to reservoir requirements through specified fracture treatment and fluid design. For example, ductile formations require greater proppant placement ability of fracking fluids as compared to that of brittle formations. This variation in porosity and consequently permeability is because of two reasons; gas desorption from shale surface (unlike conventional reservoirs) and increasing significant stresses by pressure depletion. The critical life-time of a reservoir is influenced by compaction and available pore volume. Reservoir porosity may decrease with the domination of the compaction effect over porosity change against desorption, and vice-versa. Seismic technologies serve as a useful tool to map conventional as well as unconventional formations, to plan a beneficial fracturing treatment operation. In onshore formations, these include exploding dynamite and vibroseis, or measuring vibrations produced by purpose-built trucks. Since marine seismic survey technologies are much better as compared to land seismic tools, they are now being customized for use in onshore reservoirs around the world. Vertical wells are not very conducive in case of tight gas formations, which results in the need to establish directional drilling procedures with an effective hydraulic fracturing plan for the greatest possible amount of hydrocarbon extraction. This also lessens the drilling footprint and lowers the cost of drilling as compared to multiple well developments.

Reservoir evaluation before, during and after the fracturing treatment is necessary to interpret rock-fluid characteristics, develop a combination of datasets measured inside the wellbore to detect the amount of oil/gas reserves and alter stimulation strategies, if necessary. Formation evaluation not only provides information about the properties such as thickness, permeability, fluid saturation, porosity, in-situ stress and conductivity; but also assesses the ability of a wellbore to produce hydrocarbons [33]. The feasibility of a fracturing process is dependent on the knowledge of geological properties of the reservoir, rock-fluid interactions, porosity–permeability analyses in different crude oil-containing sections, contaminants present if any, location of aquifers and groundwater sources, and depth of the formation. Hence, the geotechnical considerations for a reservoir for predicting its suitability in the hydraulic fracturing process can be summarized by determination of the lower limits for porosity, permeability, and upper limits for water saturation. A baseline estimate of permeability range, rock nature, the volume of oil/gas reserves, and process cost must be available for analyses by oilfield managers, engineers and project analysts prior to application of a proposed hydraulic fracturing process. Techniques which favour improvement of fluid conductivity and permeability characteristics of reservoir formations at the minimum coat and longer efficacy is amenable and likely to be more successful for the industry.

3 Fluid Design and Characterization

The composition and properties of the fracking fluid candidate directly control the economic productivity of the hydraulic fracturing treatment process. Fracturing involves pumping high viscosity fluids under high pressure to segregate the rocks in reservoir formation. Sometimes, acidization is performed in conjunction with fracturing technologies to re-develop natural fissures, which were present in the reservoir formation before compaction and cementation. A fracturing fluid must possess sufficient stability, viscosity, proppant carrying capacity and fluid loss control ability. The introduction of suitable proppants, attainment of fluid characteristics, and ideal pumpability rates are vital from the technical viewpoint. While fracking fluids are being pumped into the system, the formed fractures are held open by fluid pressure. However, once this process is stopped, and the injection pressure disrupts, the minimum principal stress will act to close the created fracture pores. However, sufficient optimization and design will lead to retention of proppant materials within fracture zones, even after the removal of fracturing fluids. This results in the establishment of an open, conductive fracture zone within hydrocarbon-bearing reservoirs.

Fluid viscosity and pump rate are essential parameters that simultaneously control the net pressure differential required to attain the favorable fracture depth/width. This must ensure sufficient conductivity to allow displacing fluid to transport deep into the formation and proppant (such as sand) to enter the fracture. Adequate viscosity is helpful in decreasing frictional pressure losses during pumping. Stability of fracking fluids in high-temperature conditions is necessary to attain the desired flowability, minimize adsorption losses and decrease the formation of in-situ insoluble aggregates, which lessen the efficacy of fractures. A combination of additives such as acids, friction reducers, clay stabilizers, biocides, scale inhibitors, gels, buffers, gel breakers, scale inhibition agents, cross-linkers, non-emulsifiers, and iron control agents in surfactant-based fracturing fluids is incorporated in order to achieve optimal formulation for use in fracturing.

4 Physicochemical Attributes of Surfactant-Based Fracturing Fluids

4.1 Friction Reduction (FR) Capacity

Friction reduction is an important property of hydraulic fracturing fluids. Generally, conventional polymers and novel surfactants are employed to reduce pipe friction and allow the job to occur successfully under desired pressure [7]. In the absence of friction reducers (FR), frictional pressure inside the pipe reaches very high values in the presence of high flow rates. The concentration of FR varies in the range 0.5–1.0 gpt (gallons per thousand gallons of water), depending on the quality of concentrate and solvent fluids. The type of source water (freshwater/reused water), salinity and quality of FR affects the cost and efficacy of fracturing operation. Common FR materials are used in dry powder as well as liquid (with mineral oil base). At the end of fracturing, oxidizer or enzymatic compounds are added as FR breakers to degrade filter cakes/fluids and prevent damage to fracture conductivity. Polymer/surfactant in low concentrations serve primarily in reducing friction loss along the flow-lines, whilst employing slickwater and hybrid fracturing jobs.

4.2 Low Pipe Frictional Pressure

Low frictional pressures are considered beneficial during fracturing operations. During fluid injection, the friction pressure is a function of fluid viscosity, fluid density, flow rate, and diameter of pipe/pore spaces within the formation. For example, usage of a smaller diameter pipe generally causes friction pressure to increase. In addition, proper selection and composition of fracturing fluid is necessary to derive improved flow conductivity and decreased friction pressure. As the rate of fluid pumping rate increases, the frictional pressure increases as well. However, insignificant flow rates reduce the operational capacity of the fracturing process, thereby creating a need to formulate an optimal fluid flow-rate. Rabaa [50] found that the stress field altered after the creation of fracture; the subsequent created fracture would be affected by the new stress field and would not be parallel to the first fracture. In another work, Zhou et al. [86] reported that hydraulic fracture was a dominating fracture with multiple random branches within high horizontal stress difference, while the hydraulic fracture was partly vertical (planar fracture with branches) within the scope of low horizontal stress difference. Weijers and co-workers [78] observed that the formation of transverse fractures with low flow-rate, low viscosity and high-stress contrast, whilst axial fractures were initiated during fracturing application for horizontal well-bores. Numerical simulation tools are effective for elucidating the mechanisms responsible for friction pressure reduction and scale inhibition during hydraulic fracturing.

4.3 *Tortuosity*

Tortuosity is defined as a measure of the restricted, convoluted pathways between the perforations and fracture zones. This phenomenon is severe in horizontal wells, moderate-to-high inclined wells, hard rock reservoirs, perforated wells; not a problem in vertical wells. The addition of surfactant-polymer based fluids with viscoelastic properties can solve the tortuosity issue, and successfully carry the proppant sand particles in-between the formations [10]. This leads to a decline in surface-treating pressures as soon as the fracking fluid reaches the perforations. However, if the sand slug causes an increase in pressure with a considerably sharp or smooth break-in pressure, it indicates the absence of tortuosity problem. Finally, if sand hits the perforations and no impact is obtained, problems with tortuosity are very unlikely to exist. Furthermore, the differential (>100 psi) between the closure pressure and instantaneous shut-in pressure (ISIP) also suggests the possibility of tortuosity. It can be mitigated by pumping low-concentration proppant slugs, loading of strong gelling agents (>15 lb system), and flow-rate increase. Tortuosity is dependent on the formation factor; as well as the ratio of a fluid's diffusion coefficient (when not confined) to effective diffusion coefficient (confined in porous medium).

4.4 Stability

Formulation of optimal fracturing fluids is a complex matter due to stabilization and compatibility requirements on the part of different compositions within a single system. A fracturing fluid must exhibit stability at elevated temperatures, high pumping rates, and dynamic shear conditions. Failure to comply with these parameters may cause the fluids to degrade, and settle out of the dispersed proppant(s) prematurely. Commercially, fracturing fluids are aqueous-based liquids with the ability to be either gelled or foamed. Addition of surfactants aid in retaining the proppant carrying ability of fluids, creating a conductive flow path from the formation to the wellbore, and reduce the quantity of insoluble aggregates formed within the formation during operation [58]. An important measure of stability for fracking fluids is viscosity enhancement. Fracturing fluids, when designed for a certain reservoir, are specifically tested to confirm suitable rheological attributes under dynamic flow conditions. Gelled based fracking fluids are favourable for high-temperature applications, whereas foam fluids are employed in sensitive operations with environmental concerns [64]. With respect to time, stability to a few hours or days is generally preferred. Despite the presence of dissolved solids and contaminants, it is necessary to achieve desired viscosity without flocculation/coagulation tendencies. Reusability, viscosity and temperature resistivity of fluids translate to the ability of a relatively low volume of stable fracturing fluids to displace and propagate a large quantity of proppant [12, 79].

4.5 Flow Pumpability and Flow Loop Testing

Significantly high pumping rates establish beneficial oil and gas production rates during fracturing stimulation into conventional/unconventional reservoir formations. However, many technical and environmental constraints associated with this approach need to be resolved. The main factor whilst increasing flow pumpability is the tubular friction pressure. The frictional pressure limit must be exceeded by injection pressure, to reduce the hydraulic power demand by 80% of the initial pumping energy requirement [10, 75]. Addition of appropriate surfactant fluids to fracturing fluid compositions improves the pumpability of fluids, as well as the flowback post-fracturing. Flow loop experiments identify the optimal flowability of different fracturing fluids and constitute a significant element of hydraulic fracturing fluid design. Herein, the pressure is allowed to drop while maintaining a constant rate of flow rate, as friction reducer and other components are added to the frac fluid system. It is pivotal to evaluate the relationship between different fracturing constraints such as FR such as polymer/surfactant, proppant, and additional components; and assess the specific function and influence of each chemical on hydraulic fracturing process.

5 Surfactant as Fracturing Fluid

Hydraulic fracturing, informally known as "fracking," is a reservoir development technique, which involves the injection of water, sand, and chemicals under high pressure. This process is primarily intended to create and establish new fracture zones within the rock and increase the fluid connectivity of existing fractures. Though it is generally used in low-permeability formations such as tight oil, shale, and some coal beds, it can also be effectively employed to improve the producing life of a mature conventional well. The first field application of hydraulic fracturing comprised "slickwater multistage horizontal stimulation" or "slickwater frac" in the year 1947. However, modern fracturing practices have been continuously developed since then, keeping in mind the efficacy and cost-effectiveness of the project. For example, an advanced form of multistage fracturing was employed in 1998 in the Barnett Shale reserve, Texas, the United States, with the injection of more water and higher pump rate. A stage in the well life-cycle occurs when no additional oil can be produced, despite large reserves due to high capillary forces, reservoir heterogeneities and gravity drainage. Hydraulic fracturing is a promising way to alleviate this problem. In contrast, unconventional formations would be economically feasible in the presence of hydraulic fracturing techniques.

Surfactant injection decreases the oil-aqueous interfacial tension and 'wets' oil-saturated rock surfaces. This is useful in effectively removing oil–water blocks during hydraulic fracturing processes in and near the wellbore matrix [6, 85]. It is undeniable that surfactant fracturing is necessary to develop hydrocarbon pay-zones with good recovery results [65]. The various functional advantages of surfactant-induced hydraulic fracturing include:

- Connect fracture zones with existing natural fractures
- Reduce the formation of insoluble residues within the reservoir formation
- Increase the oil window or the degree of formation contact with the wellbore
- Reduce the drilling of infill wells with horizontal fracturing strategy
- Enhance the oil displacement ability of polymer-based fracturing fluids
- Suppress the formation of detrimental in-situ emulsion droplets
- Reduce sand production by reducing the pressure drop around the well
- Increase flow connectivity within low-permeability reservoirs with geological complexities
- Increase flow-rates from damaged wells (and reduce skin damage).

Surfactant addition, in proper formulations, can suppress the formation of in-situ emulsion phases and mobilize a greater amount of oil. Emulsion fluids stabilized by surfactant are characterized by smaller oil dispersions, which could effectively squeeze through tiny fractures [17]; and additional benefits are attained through improved mobility ratio and oil sweep efficiencies. In unconventional and impermeable formations, the formation of large-sized emulsion droplets and polymeric chains/aggregates must be avoided to prevent effective plugging of tiny pores [55]. Hydraulic fracture fluid systems are unique for each reservoir, and depends on the geology, reservoir fluid characteristics and degree of pacing/heterogeneity of the reservoir. Table 1 shows some recent research works for the utilization of various surfactants for the hydraulic fracturing application.

6 Components of Surfactant-Based Hydraulic Fracturing Fluids

For hydraulic fracturing, different components/additives are selected by the industry, depending on the properties of the reservoir and fluids. Surfactants are gaining rapid interest as the primary constituents of conventional as well as novel fracking fluids. Table 2 shows a list of additives employed to avoid problems associated with oil production, rock permeabilities and environmental contamination, respectively.

7 Different Kinds of Fracturing Fluids

The base fluid can be categorized mainly into the water-based and oil-based fracturing fluid. Water-based fluids have been the primary fluids over the oil-based fluids since the introduction of thick water base gels, and the research and developments in the last 50 years are mainly devoted to water-base fluids. However, there has also been a need for oil-based fluids to treat the water-sensitive formations, [3, 26, 36, 60]. The other categories include acid–base, foam-base, emulsion-based, and alcohol-based fracturing fluids.

7.1 Water-Based Fracturing Fluids

Water fracturing fluids, for example, slick-water, linear & cross-linked polymer-based and viscoelastic based fluids have been used in many reservoirs fields as the conventional fracturing fluids. The slick-water type of fracturing fluids are widespread and frequently used for most of the oil and gas fields. The slick-water fracturing fluids mainly consist of a high percentage of water (greater than 90%) and supported with a minimum concentration of polymers (guar gum, xanthan gum, etc.) so that the viscosity of the water-based fluids is enhanced which consequently helps in proppant carrying capacity and transportation into the fractures points [43]. Slick-water can improve fracture length by creating very long skinny fracture, whereas the fracture width is mainly increased by gelled fluids [11, 40, 35]. Slick-water type of the fracturing fluids approach is simple to tackle and has been observed to generate small fractures [24, 23, 70].

Table 1 Earlier literature showing research works in the field of surfactant-assisted fracturing applications in the pilot and field operations

References	Surfactant used for the fracturing fluid characterization	Investigation parameters and results/outcomes
[39]	A novel anionic VES fracturing fluid "D3F-AS05"	D3F-AS05 fracturing fluid controls fracture geometry without compromising proppant transport. Real-time application of the devolved fracturing fluids in various oilfields in China
[28]	Bioterge AS-40; Viscoelastic surfactant (VES)	The temperature has an indirect effect on the foam rheology, viscosity of foam decreases with increase in temperature. Pressure effect on the foam viscosity is insignificant
[56]	Cocamidopropyl betaine (CAPB) ($C_{19}H_{38}N_2O_3$), sodium dodecylbenzene sulfonate (SDBS) ($C_{18}H_{29}NaO_3S$)	VES gels are stable in the temperature range (10–35 °C). The wormlike micellar network characterizes it. The improved rheological behaviour enhanced the fracturing application
[9]	Zwitterionic surfactant Cocamidopropyl betaine (CAPB), anionic sodium dodecyl Sulphate (SDS), sodium oleate (NaOA)	Pseudo-plastic and shear thinning nature of the VES fluids results in low frictional losses during pumping of the fluids downhole in an oil well. Enhanced viscoelasticity, good miscibility and better static proppant suspension capacity are obtained
[84]	0.5% VES (BET/SLP) fluid mixed with 0.25% HMP	The dynamic rheological properties of the VES fluid shows high viscoelasticity, in which the elastic moduli are higher than the loss moduli. The fluid has 50% lower formation damage than conventional guar
[24, 23]	Sodium Dodecyl Sulfonate (SDS), Sodium dodecylbenzene sulfonate (SDBS), Cocamidopropyl betaine (CAPB)	Proppant settling and foam stability were significantly affected by variation in the time of fracture closure. Proppant settling was enhanced with an increase in the fracture closure time
[2]	Alfa olefin sulfonate (AOS), Sodium chloride salt	Pressurized foam rheometer model 8500. Power-law model was modified, and the effect of shear rate and surfactant concentration was incorporated. Power-law model indexes (n, K) were depended on the surfactant and salinity effect

(continued)

Table 1 (continued)

References	Surfactant used for the fracturing fluid characterization	Investigation parameters and results/outcomes
[15]	Sodium dodecyl sulfate (SDS), ammonium lauryl sulfate, Isoamyl alcohol (3-methylbutan-1-ol), used as co-surfactant	Rheology of the gels shows shear thinning behaviour with good viscoelasticity. Elasticity is dominant over the viscous nature of the gel fluid, which helps to suspend and transport the proppant carrying capacity. The mixture of ALS and SLS shows a better gel system with higher viscosity compared to individual surfactants
[69]	Alfa Olefin Sulfonate (AOS) at 0.5 wt. %) HPAM at 100 ppm and NaCl at 1.0 wt. %	80% of foam quality fluids carry and transport the proppant very efficiently within the lamellas with the significantly less vertical setting. 70% of foam quality fluid was not so efficient due to liquid drainage and less viscosity. Proppant bed forms near the injection well
[81]	Viscoelastic surfactant (0.4% VES + 0.15% SSN)	Core displacement analysis reveals that the high compatibility between the gel, core and formation water. The field application in Qinshui Basin of Shanxi Province shows that the production of the well, which is fractured by the developed VES clean fluid. It has a vital application in the coalbed gas
[45]	VES containing both unsaturated carbon–carbon double bond and amide group	Novel Gemini VES fracturing fluid has good heat resistance Gemini VES was improvised VES fracturing fluid, whose viscosity could be maintained about 40 mPa·s at 160 °C
[1]	Alfa olefin sulfonate (AOS), betaine Sodium chloride salt	The modified power-law models for polymer-free supercritical CO2 foam (AOS and betaine) is a function of temperature, pressure, and shear rate. Empirical correlations were found to be significant for the all tested temperature and pressure
[74]	Alfa olefin sulfonate, sodium dodecylbenzene sulfonate, Cocamido- propyl betaine	Thermally stable foam enhanced the viscosity and elastic properties of the fluids, and capable of carrying proppants by reducing formation damage

(continued)

Table 1 (continued)

References	Surfactant used for the fracturing fluid characterization	Investigation parameters and results/outcomes
Chaudhary et al. [14]	Sodium Lauryl Ethyl Sulphate (SLES) + Palmitic Acid, Silica Sand, Propylene Glycol Potassium Chloride	Stability of the grafted copolymer foam is higher than conventional foam fluid system. Improvement in the Proppant carrying capacity was reported with an increase in the foam quality. The reduced permeability value of up to 82% was reported
[16]	CTAB, citric acid (CA), and maleic acid (MA)	VES fluids were showed the shear thickening behaviour through the formulation of mixing long chain cationic surfactant with organic acids

Polymer-based fluid systems, consisting of high-molecular-weight components, are conventionally employed for well stimulation and other production operations. For example, linear polymer fluids are thermally unstable under high-pressure, high temperature (HPHT) conditions. However, in the presence of nanoparticles, these fluids were cross-linked to attain thermal stability. A detrimental effect of polymer fracturing fluids is related to formation damage issues due to pore-plugging, and the existence of insoluble residues. Polymer fluids are unfavourable to control the growth of fracture height, fracture length, and to improve fluid permeability. However, such fluids show good proppant carrying ability to producing zones of interest. Therefore, the use of surfactants can help mitigate these problems. Viscoelastic surfactant (VES) based fracturing fluid has been used since 1997, which is an alternative to conventional polymer and can develop sufficient viscosity to create fractures and transport proppants. VES fluids are effective agents to fracture low and high permeability regions within the reservoir. These fluids exhibit excellent stabilization, rheological attributes and low formation damage characteristics as compared to cross-linked polymers. Surfactant based fracturing fluids are associated with easy preparation technique, low cost/complexities, and a lesser number of chemicals required. Conventional polymer fluids, on the other hand, are much more complex in the presence of other phenomena (such as polymer hydration, cross-linkers, beakers etc. Surfactant-stabilized fracturing systems achieve high fracture conductivity, stability and proppant suspension ability [39].

Table 2 Different components of the hydraulic fracturing fluid(s)

Composition	Examples	Behaviour	Functional application
The aqueous phase and enhancers/proppants (approximately 98% v/v composition)			
Water	Seawater, formation water, deionized/treated water	A part of water/aqueous returns with formation water phase as produced water, whilst the remaining stays within the reservoir. This depends on the type of reservoir and chemical fluid used	Expands the fracture and delivers proppant (sand) deep into the formation
Proppant	Sand, ceramics, resin-coated sand	Remains within the formation zones to hold the fractures in-place, post-stimulation operation	Improves oil and gas productivity by establishing fracture zones in low-permeability reservoirs
Polymer/gels	Polyacrylamides, copolymers and gelation agents	Enter into the formations, and improve the rheological characteristics of fracturing fluid	Improve viscosity, thermal stability and prevents emulsion formation
Surfactant	Ionic, Nonionic, Zwitterionic species	Forms stabilized aggregates/micelles in bulk solution phase and improve network structure in polymer-based fluid systems to attain favourable interfacial, stabilization and rheological characteristics	Reduce IFT, alter wettability, reduce/prevent the formation of insoluble residues with "clean-up" after fracturing
Other additives (approximately 2% v/v composition)			
Friction reducers (FR)	Surfactant, foam, polymer, gel, nanoparticles	Remains in the formation to allow effective propagation/transport of fracturing fluids	Reduces frictional pressure during a fracking operation
FR breakers	Hydrogen peroxide, oxalate	Reacts with FR to contribute to their breakdown and degradation; consumed by natural microbes	Permits breakdown of friction reducer (FR) in fluid; to cause easier fluid flow back to the wellbore

(continued)

Table 2 (continued)

The aqueous phase and enhancers/proppants (approximately 98% v/v composition)			
Composition	Examples	Behaviour	Functional application
Crosslinkers	Borate (high pH & low-to-moderate temperatures); zirconate (low pH & elevated temperatures)	Interacts with frac components to form ions/salts information, which is returned with produced water	Helps in maintaining fluid viscosity at varying shear and temperature conditions
Acids	Hydrochloric acid	Reacts with the formation minerals to result in the creation of salts, water and neutralized carbon dioxide	Dissolves minerals to initiate fractures/cracks within the rock
Clay stabilizers	Potassium chloride, sodium chloride, calcium chloride	Interacts with clays through sodium–potassium ion exchange	Restricts swelling behaviour of clays within the formation
Gelling agents	Guar, polyacrylamide, hydroxyethylcellulose, other polymers	Enhances fluid viscosity and thermal stability	Improves proppant suspension and propagation ability of fluid
Gel breakers	Acids, bleach, hydrogen peroxide, oxalate	Reacts with cross-linker and gel information; decrease fluid viscosity to improve flow back	Allows delayed breakdown of the gel
Corrosion/scale inhibitors	Ethylene glycol	Forms bonds with metal surfaces such as fluid pipe; designed to be bio-degradable by microbes	Prevents scaling and corrosion of the pipe
Anti-bacterial agents & other biocides	Oxidizing biocide: chlorine, bromine, ozone, chlorine dioxide Non-oxidizing biocide: aldehydes, bronopol, DPNPA, acrolein	Reacts with bacteria and other micro-organisms existing in the treatment fluid and formation	Kill bacteria to control fluid rheology
Non-emulsifiers	Polymer; NE-1225, NE-43R, NE-43X (ChemEOR) NE-200, NE-300, NE-400 (Tetra Co.)	Influences molecular arrangement to prevent the formation of agglomerates; returns to the surface with produced water/produced oil and natural gas streams	Prevents the formation of undesirable emulsions with the formation during operation by separating in-situ oil/water mixtures

(continued)

Table 2 (continued)

The aqueous phase and enhancers/proppants (approximately 98% v/v composition)			
Composition	Examples	Behaviour	Functional application
pH adjusters/Buffers	Acidic/basic types: Potassium carbonate, acetic acid	Reacts with in-situ or existent acidic. Basic agents in stimulation fluid to attain close-to-neutral pH	Retains efficacy of FR, breakers, gelling agents and cross-linkers
Iron control agents	Ammonium chloride, Ethylene, Citric and other weak acids, Glycol	Reacts with minerals in formation to generate salts, water and carbon dioxide, while reducing the percentage of dissolved iron	Prevents precipitation of metal, minimise the formation of insoluble residues and prevent plugging-off of formations

7.2 Oil-Based Fracturing Fluids

Oil-based fracturing fluids were implemented in fracturing treatments at the beginning stage, and the reason was their compatibility with almost all kinds of formation. However, the higher cost, safety, and environment concerns limited their usage and led to the initiation of a water-based fracturing fluid system. Gelled crude oil, diesel, and kerosene had found its application in the past as an oil-based fracturing fluid. Though LPG has been used for stimulating conventional reservoirs for the last 50 years, now it is being adapted for unconventional reservoirs like shale gas and tight sands as they eliminate phase trapping by exhibiting high capillary pressure thus improving the recovery. They demonstrate various advantages like reduced water usage, fewer chemical additives, increased productivity, no fluid loss, rapid clean up, and full fluid compatibility with shale reservoirs, which are sometimes water-sensitive [26]. However, its massive application has been limited due to the higher investment cost, and it requires manipulation of large amounts of flammable proppant, [3, 26, 36, 60].

7.3 Alcohol-Based Fracturing Fluids

Methanol has been infrequently used as an alcohol-based fracturing fluid in Argentina and Canada (from the 1990s to 2001) for the reservoir with irreducible high water or hydrocarbon saturation (minimal fluid recovery), high clay content-low permeability reservoirs, and low bottom hole pressures due to its properties like a low freezing point, high water solubility, low surface tension and high compatibility with the formation. Methanol (less viscous than water) has been gelled using foaming with guar or synthetic polymer and CO_2 and has also been metal crosslinked. However, the three to four times higher cost than water-based fluid and issues related to safe handling (low flash point makes it highly ignitable) have made a shift from using methanol as a base fluid to methanol as just an additive [3, 26], 36, 60].

7.4 Acid-Based Fracturing Fluids

Acid fracturing is generally used in carbonate/limestone reservoirs to "etch" the channels in the rock. For these types of fluids, the formation should be slightly soluble in acid to etch 'artificial' channels within the fractured wells. Its usage is limited to only carbonate reservoirs and cannot be applicable to fracture the coal bed methane, sandstone, and shales reservoirs [26].

7.5 Emulsion-Based Fracturing Fluids

An emulsion-based fluid is a mixture of two or more immiscible liquids mainly developed to reduce or eliminate the usage of water in water-sensitive reservoirs. One such fluid is an emulsion of CO_2 in the aqueous alcohol-based gel applied in the western Canadian sedimentary basin in 1981, and such fluids have been significantly used in tight gas and low-pressure applications. The fluid provides advantages similar to the conventional high-quality CO_2 foam but with higher water loading [26].

7.6 Foam Fracturing System

Foam based fracturing fluids have been used in the petroleum industry mainly for the unconventional low permeable reservoirs, water-sensitive formation generally for undersaturated gas reservoirs, and areas having water scarcity. Foam based fracturing fluid are considered the best for unconventional reservoirs since it causes less damaging in water-sensitive formations with easy cleanup and less water to recover post-fracturing [52, 72]. Foams are produced on-site by a mixture of two phases, i.e. liquid and gas. Moreover, surfactants are used to reduce the interfacial tension between the two phases, which consequently enhance foaming capacity and the stability of foam [47, 73].

The foam quality is an important property for effective fracturing. In preparation of foam-based fluid, it is required to maintain the desired quality (percentage of gas volume) of the generated foam as given by Eq. (1) [25, 74]. During foam production, the internal phase- gas and external phase, the mixture of surfactant and water are mixed. Initially, the surfactant is combined with an external phase (mainly water) [25]. Then after, the prepared mixture of surfactant solution and gas (mainly N_2/CO_2)

are together pumped into the formation through the wellbore [22, 67]. Proppants are combined with the foam fluid before pumped into the wellbore so that clogging inside pipeline and foam generator can be avoided.

$$Q = \frac{V_{gas}}{V_{gas} + V_{liquid}} \quad (1)$$

where Q is the foam quality, V_{gas} is the total volume in the foam, and V_{liquid} is the volume of liquid in the foam.

Foam fracturing fluid can be classified based on gas usage during foam preparation [26, 54] as given below:

1. Water-Based Foam (a combination of water, foaming agent and CO_2/N_2 gas)
2. Acid-Based Foam (the combination of acid, foaming agent and N_2 gas)
3. Alcohol-Based Foam (the combination of methanol, foaming agent and N_2 gas)
4. CO_2-Based Foam (the combination of liquid CO_2 and N_2 gas).

Carbon dioxide (CO_2) and nitrogen (N_2) are the most used gases for the generation of the foam fracking fluids. CO_2 based foams have wider application, and it has a higher hydrostatic pressure as compared to N_2, and more suitable for reservoirs having higher breakdown pressure. N_2 foam injected at low hydrostatic pressure requires high surface treating pressures in contrast to the CO_2 foam fluids [44]. Therefore, N_2 foam fracturing fluid is mostly affected by high surface injecting requirements [24, 23, 76]. The comparison of the various fracturing fluids in terms of advantages and disadvantages as mentioned in Table 3, and followed with the brief discussion are presented.

8 Hydraulic Fracturing Process Considerations

The selection of fracturing fluid is a critical decision. It encompasses a number of factors such as reservoir temperature, reservoir pressure, the expected value of fracture half-length, and any water sensitivity. The following list shows the industrial and professional standards to be maintained whilst designing a fracture process:

- Fluid must create a fracture wide enough, and pump proppants at concentrations high enough, to achieve the flow conductivity.
- The model should account for compromise fracture length, and conductivity in situations, wherein substantial damage to the formation may occur around the fracture.
- Transverse fractures are tough to achieve and require a greater degree of planning as compared to longitudinal fractures, but more favourable for production viewpoint.
- Fracture size must be controlled during the process.

Table 3 Different types of fracturing fluids employed in the petroleum sector

Fluid type	Properties of fluids (ambient conditions)	Advantages	Limitations	Remark	References
Water-based fluid	Density = 1 g/cm3 Apparent viscosity = 2 cP	Formation damage reduced Economical reservoir volume is higher during fracturing Better fracture confining	Proppant suspension capacity is poor Freshwater consumption is high Environmental problems	Water usage is 99.5%, and other chemicals are 0.5%, i.e., friction reducers, pH-adjusting agents, etc Not suggested for water-sensitive reservoirs	[67]
Oil-based fluid	Density = 0.85 g/cm3 Apparent viscosity = 100 cP	Water requirement is less Logistic cost reduced Rate of recovery is high Fast clean-up of well	Not Economical Initial set up cost is high flammability issues are very high	Usage of Crude oil, kerosene, and diesel oil Recommended for water-sensitive reservoir	[26, 47]
Acid-Based Fluid	Density = 1.2 g/cm3 Apparent viscosity = 2 cP	Proppant usage is a decline Water requirement is reduced	Not economical Never recommended for carbonate formations Very fast and frequent acid interaction with the reservoir	Usage of Hydrochloric acid, acetic and formic acids Recommended for limestone reservoir	[20, 26]
Alcohol-based fluid	Density = 0.8 g/cm3 Apparent viscosity = 0.5 cP	Fast clean up of well Corrosion or scale inhibition Friction reduces	Flammability creates the problem of safety Proppant suspension capacity is not good	Methyl and isopropyl alcohol are used Recommended for low-permeable and dry gas reservoir	[19, 57]

(continued)

Table 3 (continued)

Fluid type	Properties of fluids (ambient conditions)	Advantages	Limitations	Remark	References
Emulsion-based fluid	Density = 0.75 g/cm3 Apparent viscosity = 750 cP	Water requirement is a decline The additive is very less required Enhanced productivity	Not Economical Logistic needs are high	Non-mixable fluid are used such as oil Recommended for low-pressure reservoir	[37, 42]
Foam-based fluid	Density = 0.25 g/cm3 Apparent viscoity = 150 cP	Water requirment is very less (only 5–30%) Foramtion damage is less Proppant suspension capacity improved	Initial running cost is very high Logistic usage is high High temperature is highly unstable	Foam is a combination of liquid and gas(nitrogen or carbon dioxide) Recommended for water-sensitive and unconventional reservoirs	[30] , [31, 75–74]

- Parameters such as geometry, fluid characteristics, reservoir heterogeneities, permeability, and formation thickness play an essential role in project feasibility.
- Information regarding in-situ stresses is necessary to predict fracture half-length, width, height and complexities prior to production testing.
- Rock properties such as ductility and depth also provide an impetus to the formation of an appropriate fracture.

The extent of a created fracture and its resulting propagation is controlled by the in-situ fluid characteristics, upper confining zone, injected fluid parameters and reservoir heterogeneities. Previous studies have proved the influence of fractures on reservoir characteristics. The significant phenomena affecting rock behaviour are porosity, rock-volume shrinkage due to dolomitization, porosity increase due to solution, and other geological factors. Fracturing plan should also involve effective flowback model after the completion of fracture treatment during diagnostics and monitoring.

9 Applications of Surfactants as Fracturing Fluids

The different surfactant based fracturing fluids have been developed with time. Mathis et al. [46] presented that the proppant suspension capacities of viscoelastic surfactant fracturing fluids were not due to the drastic enhancement in the viscosity. They concluded that the proppant carrying capacity of viscoelastic surfactant is mainly because of the elastic nature and structure of fluid rather than viscosity [66] reported the rheology and phase behaviour of sodium oleate surfactant. The increase in the concentration of sodium oleate surfactant (above CMC) leads to self assembles into worm-like micelles either in the presence of inorganic or binding salt that screens the inter-micellar electrostatic interactions and reduces the micellar surface charge. In 2006, Sullivan et al. were utilized the zwitterionic surfactant fluid in high permeability reservoir that leads to low friction pressure, effective proppant transport and high proppant pack conductivity. Welton et al. [80] developed an anionic surfactant based fracturing fluid with improved fluid loss and de-emulsification characteristics, which do not adversely change rock-wettability. Nonionic Tween and Brij surfactants form worm-like micelles in solution phase, which can be employed a fracturing fluid.

Similarly, an anionic surfactant with an easy method of synthesis, favourable viscosity, low frictional resistance, and enhanced stability at 30–100 °C was reported as a potential fracking agent by Khair and others [39]. The fluid exhibited good suspension and proppant transportation attributes at lower viscosities than conventional systems. Thampi and co-workers [68] compared the effect of co-solvents and branched alcohols on phase behaviour and physicochemical properties of viscoelastic surfactant-based gel fracturing fluids. Gel-stabilized systems have viscoelasticity much greater than the minimum requirement for the fracturing application [68]. Rao et al. [51] discussed ionic liquid-based microemulsions as fracturing fluids over a wide range of temperature, i.e. 278–423 K. Viscoelastic surfactant-stabilized fluids characterized by wormlike micellar structures are considered as potent, functional alternatives to hydraulic fracturing applications [4, 41].

Surfactant molecules form micelles in the bulk phase, with the polar head pointing towards aqueous phase and non-polar tail oriented toward oil (hydrocarbon) phase. Baruah et al. [7] worked on the effect of concentration on the micellar arrangement and physicochemical properties of sodium oleate (NaOA) based fracturing fluids, and identified the existence of a liquid crystal phase from loosely packed surfactant molecule patterns to form lamellar hexagonal structures. These properties contribute to extraordinary proppant suspension characteristics at low viscosities than polymer-containing fluids. Another work on mixed ionic-ionic surfactant confirmed the sensitivity of formulation characteristics to the quantity of surfactant, cosurfactant, hydrocarbon and aqueous phases involved [8]. The developed lamellar crystals are characterized by pseudoplastic attributes, which is desirable for pumping under high shear conditions and transporting proppant (sand) effectively to the fracture zone. Additionally, the rheology of polymer fluids is completely reversible with no permanent degradation properties, even under high shear. VES fluids easily segregate into low viscosity components via wormlike-to-spherical micelle transition at the end of fracturing jobs, which allows them to recover from sub-surface formations.

10 Summary and Outlook

The impact of hydraulic fracturing on the petroleum production industry is becoming an increasingly sought topic of interest globally. This method is a promising area with the capability to create a large fracture network in low permeability formations and achieve economical production results. However, groundwater can enter into the oil-producing zones during this process, which can have adverse repercussions on both production and environmental aspects. Conventional polymer-induced fracturing routes are associated with drawbacks in the current industry owing to issues of less fluid loss control, the formation of insoluble residues, and flow back. The field of surfactant has generated a marked improvement in optimizing and attaining beneficial hydraulic fracturing solutions. Systems consisting of a single surfactant, mixed surfactant, hybrid formulations with polymer/nanoparticles, have proved to be effective fracturing fluids based on a documented history of experimental and numerical simulation investigations. Surfactant based fluids help in mitigating stability, flow back and water-blockage issues faced by the oilfield managers during operation. Surfactants are effective clean-up additives, which reduce the amount of residues or precipitates remaining within the reservoir formation post-application.

Furthermore, surfactant-based fluids reduce interfacial tension, alter wettability to a water-wet state, reduce flow friction and provide good proppant suspension for fracturing jobs in difficult, complicated formations. Polymer-surfactant aggregate structures show better rheological attributes in comparison to (only) polymer systems. Earlier reports by researchers and academicians have corroborated to the favorability of surfactant-assisted hydraulic fracturing in the petroleum industry. Hence, the introduction of surfactants in fracturing operations provides a sustainable fracturing technique to meet the needs of on-site fracturing considerations from the industry viewpoint. This chapter provides a detailed, systematic description of the concepts, function and prospects of surfactant-based hydraulic fracturing in the oil & gas sector.

References

1. Ahmed S, Elraies KA, Hashmet MR, Alnarabiji MS (2018) Empirical modeling of the viscosity of supercritical carbon dioxide foam fracturing fluid under different downhole conditions. Energies 11:782. https://doi.org/10.3390/en11040782
2. Ahmed S, Elraies KA, Hashmet MR, Hanamertani AS (2017) Viscosity models for polymer free CO_2 foam fracturing fluid with the effect of surfactant concentration, salinity and shear rate. Energies 10:1970. https://doi.org/10.3390/en10121970
3. Al-Muntasheri GA (2014) A critical review of hydraulic fracturing fluids over the last decade. Soc Pet Eng Product Oper 29:243–260
4. Angelescu D, Khan A, Caldararu H (2003) Viscoelastic properties of sodium dodecyl sulfate with aluminum salt in aqueous solution. Langmuir 19(22):9155–9161. https://doi.org/10.1021/la034770a
5. Balhoff M, Miller M (2005) An analytical model for cleanup of yield-stress fluids in hydraulic fractures. SPE J 10(01):5–12. https://doi.org/10.2118/77596-pa

6. Bang V, Yuan C, Pope GA, Sharma MM (2008) Improving productivity of hydraulically fractured gas condensate wells by chemical treatment. OTC19599, Presented at the SPE offshore technology conference held in Houston, Texas, USA, 5–8 May
7. Barati R, Liang J (2014) A review of fracturing fluid systems used for hydraulic fracturing of oil and gas wells. J Appl Polym Sci 131(16), n/a-n/a. https://doi.org/10.1002/app.40735
8. Baruah A, Pathak A, Ojha K (2015) Study on the thermal stability of viscoelastic surfactant-based fluids bearing lamellar structures. Ind Eng Chem Res 54(31):7640–7649. https://doi.org/10.1021/acs.iecr.5b01612
9. Baruah A, Pathak AK, Ojha K (2016) Study on rheology and thermal stability of mixed surfactant (non-ionic–anionic) surfactant based fracturing fluids. AIChE J 62(6):2177–2187
10. Belyadi H, Fathi E, Belyadi F (2019) Chapter five—hydraulic fracturing fluid systems. In: Belyadi H, Fathi E, Belyadi F (eds) Hydraulic fracturing in unconventional reservoirs, 2nd edn. Gulf Professional Publishing, pp 47–69. ISBN 9780128176658. https://doi.org/10.1016/B978-0-12-817665-8.00005-9
11. Bolt M, Medlin W (1985) SPE-14468-MS, Presented at SPE annual technical conference and exhibition. Society of Petroleum Engineers, Las Vegas, Nevada, 22–26 Sept, pp 1.https://doi.org/10.2118/14468-MS
12. Bonekamp JE, Rose GD, Schmidt DL, Teot AS, Watkins EK (1993) Viscoelastic surfactant based foam fluids, US patent 5 258 137, assigned to Dow Chemical Co. URL: https://www.freepatentsonline.com/5258137.html
13. Burman JW, Hall BE (1986) Foam as a diverting technique for matrix sandstone stimulation. Paper SPE 15575 presented at the SPE annual technical conference and exhibition, New Orleans, Louisiana, USA, 5–8 Oct. https://doi.org/10.2118/15575-MS)
14. Chaudhary S, Singh S, Singh VK (2019) A novel approach for formulating CO_2 foam based fracturing fluid by synthesized grafting copolymerization to enhance its stability for HPHT shale reservoirs. Pet Chem Indus Int 2(2):1–6
15. Chauhan G, Baruah A, Ojha K (2017) Effects of nanoparticles and surfactant charge groups on the properties of ves gel. Braz J Chem Eng 34(1). São Paulo (Jan/Mar). https://doi.org/10.1590/0104-6632.20170341s20150094
16. Chieng ZH, Mohyaldinn ME, Hassan AM, Bruining H (2020) Experimental investigation and performance evaluation of modified viscoelastic surfactant (VES) as a new thickening fracturing fluid. Polymers 12(7):1470
17. Cobos S, Carvalho MS, Alvarado V (2009) Flow of oil–water emulsions through a constricted capillary. Int J Multiph Flow 35:507–515
18. Cooke CE Jr (1975) Effect of fracturing fluids on fracture conductivity. J Pet Tech 27(10):1273–1282. https://doi.org/10.2118/5114-PA
19. Donaldson EC, Alam W, Begum N (2014) Hydraulic fracturing explained: evaluation, implementation, and challenges. Elsevier, Houston, Texas
20. EPA (Environmental Protection Agency) (2004) Appendix A-department of energy-hydraulic fracturing white paper. Evaluation of impacts to underground sources of drinking water by hydraulic fracturing of coalbed methane reservoirs study. Publication No. EPA 816-R-04–003
21. Ely JW (1985) Handbook of stimulation engineering. PennWell Publishing, Tulsa, Oklahoma
22. Eren T (2004) Foam characterization: bubble size and texture effects, MSc Thesis, Middle East Technical University
23. Fei Y, Pokalai K, Johnson R, Gonzalez M, Haghighi M (2017) Experimental and simulation study of foam stability and the effects on hydraulic fracture proppant placement. J Nat Gas Sci Eng. https://doi.org/10.1016/j.jngse.2017.08.020
24. Fei Y, Zhu J, Xu B, Li X, Gonzalez M, Haghighi M (2017) Experimental investigation of nanotechnology on wormlike micelles for high-temperature foam stimulation. J Ind Eng Chem 50:190–198
25. Gajbhiye R, Kam S (2011) Chem Eng Sci 66(08):1536. https://doi.org/10.1016/j.ces.2010.12.012
26. Gandossi L (2013) An overview of hydraulic fracturing and other formation stimulation technologies for shale gas production. Jrc Technical Reports, Report Eur 26347 En

27. Gidley JL, Holditch SA, Nierode DE, Veatch RW (1990) Hydraulic fracturing to improve production. In: Monograph series SPE 12. Society of Petroleum Engineers, Richardson, Texas
28. Gu M, Mohantay K (2015) Rheology of polymer-free foam fracturing fluids. J Petrol Sci Eng 134:87–96
29. Gu M, Fan M, Chen C (2017) SPE-185071-MS presented at SPE unconventional resources conference. Society of Petroleum Engineers, Calgary, Alberta, Canada, 15–16 Feb
30. Gu M, Mohanty K (2014) Int J Rock Mech Min Sci 70:273. https://doi.org/10.1016/j.ijrmms.2014.05.013
31. Gupta D, Leshchyshyn TT, Hlidek BT (2005) SPE-97211-MS, presented SPE annual technical conference and exhibition, 9–12 Oct. Society of Petroleum Engineers, Dallas, Texas, p 1.https://doi.org/10.2118/97211-MS
32. Holditch SA, Jennings JW, Neuse SH (1978) The optimization of well spacing and fracture length in low permeability gas reservoirs. Paper No. SPE-7496. In: Proceedings. SPE annual fall technical conference and exhibition, Houston, Texas, October 1e3. Society of Petroleum Engineers, Richardson, Texas
33. Holditch SA, Robinson BM, Whitehead WS (1987) Prefracture and postfracture formation evaluation necessary to characterize the three dimensional shape of the hydraulic fracture. In: Proceedings. SPE formation evaluation. Society of Petroleum Engineers, Richardson, Texas
34. Hornof V, Neale GH, Chaaraoui A (1983) Viscosity of surfactant-polymer solutions. Presented at the international symposium of oilfield and geothermal chemistry, Denver, Colorado, USA, 1–3 June. SPE-11775-MS. https://doi.org/10.2118/11775-MS.
35. Hu X, Wu K, Li G, Tang J, Shen Z (2018) Pet Sci Eng 167:110. https://doi.org/10.1016/j.petrol.2018.03.081
36. Jennings AR (1996) Fracturing fluids—then and now. J Petrol Technol 47:604–610
37. Jha PK, Mahto V, Saxena V (2014) Emulsion based drilling fluids: an overview. Int J ChemTech Res 6(4):2306–2315
38. Kalpakci B, Klaus EE, Duda JL (1981) Flow characteristics of surfactant solutions in porous media and their role in permeability modification. SPE J. 21(06):709–720. https://doi.org/10.2118/9930-PA
39. Khair EMM, Shicheng Z, Shanbo M, Mei Z (2011) Performance and application of new anionic D3F-AS05 viscoelastic fracturing fluid. J Petrol Sci Eng 78(1):131–138
40. Kincaid KP, Snider PM, Herring M, Mahoney RP, Soane D (2013) SPE-163818-MS, Presented at SPE hydraulic fracturing technology conference, 4–6 Feb. The Woodlands, Texas, USA. Society of Petroleum Engineers, pp 1
41. Koehler RD, Raghavan SR, Kaler EW (2000) Microstructure and dynamics of wormlike micellar solutions formed by mixing cationic and anionic surfactants. J Phys Chem B 104:11035–11044
42. Laurain A (2014) Analysis of fracturing fluid system, effect of rock mechanical properties on fluid selection. AGH Drill Oil Gas 31(1):167–178
43. Liang F, Sayed M, Al-Muntasheri GA, Chang FF, Li L (2016) Petroleum 2(2016):26. https://doi.org/10.1016/j.petlm.2015.11.001
44. Luo X, Wang S, Wang Z, Jing Z, Lv M (2014) Experimental research On rheological properties and proppant transport performance of Grf–CO_2 fracturing fluid. J Petrol Sci Eng 20:154–162
45. Mao J et al (2018) Viscosity reduction mechanism in high temperature of a Gemini viscoelastic surfactant (VES) fracturing fluid and effect of counter-ion salt (KCl) on its heat resistance. J Petrol Sci Eng 164:189–195. https://doi.org/10.1016/j.petrol.2018.01.052
46. Mathis SP, Pitoni E, Ripa G, Ferrara G, Conte A, Ruzic M (2002) VES fluid allows minimized pad volumes and viscosity to optimize Frac-Pac geometry: completion type evolution in Barbara field, Central Adriatic Sea. SPE 78317:1–7
47. Montgomery C (2013) Chapter 2 fracturing fluid components. In: Effective and sustainable hydraulic fracturing. https://doi.org/10.5772/56422
48. Paterniti ML (2009) Microemulsion surfactant increases production in the codell formation of the DJ basin. Presented at the SPE rocky mountain petroleum technology conference, Denver, Colorado, USA, 14–16 Apr. SPE-116237-MS. https://doi.org/10.2118/116237-MS

49. Piculell L, Egermayer M, Sjostrom J (2003) Rheology of mixed solutions of an associating polymer with a surfactant. Why are different surfactants different? Langmuir 19(9):3643–3649. https://doi.org/10.1021/la020912.
50. Rabaa WE (1989) Experimental study of hydraulic fracture geometry initiated from horizontal wells. In: SPE annual technical conference and exhibition. Society of Petroleum Engineers, Inc, San Antonio, Texas, Copyright 1989
51. Rao VG, Banerjee C, Ghosh S, Mandal S, Kuchlyan J, Sarkar N (2013) A step toward the development of high-temperature stable ionic liquid-in-oil microemulsions containing double-chain anionic surface active ionic liquid. J Phys Chem B 117:7472–7480
52. Reidenbach V, Harris P, Lee Y, Lord D (1986) SPE-12026-PA. SPE Prod Eng 1(01):31. https://doi.org/10.2118/12026-PA
53. Reinicke A, Rybacki E, Stanchits S, Huenges E, Dresen G (2010) Hydraulic fracturing stimulation techniques and formation damage mechanisms: implications from laboratory testing of tight sandstoneeproppant systems. Chem Erde 70(S3):107e117
54. Reynolds M, Bachman R, Peters W (2014) SPE-168632-MS SPE hydraulic fracturing technology conference, 4–6 Feb, The Woodlands, Texas, USA. Society of Petroleum Engineers, pp 1. https://doi.org/10.2118/168632-MS
55. Rickman RD, Jaripatke O (2010) Micro-emulsion/surfactant packages for shale and tight gas reservoirs. Paper SPE 131107 presented at the deep gas conference and exhibition, Manama, Bahrain, 24–26 Jan. https://doi.org/10.2118/131107-MS
56. Różańska S (2015) Rheology of wormlike micelles in mixed solutions of cocoamidopropyl betaine and sodium dodecylbenzenesulfonate. Colloid Surface Physicochem Eng Aspect 482:394–402. https://doi.org/10.1016/j.colsurfa.2015.06.045
57. Saba T, Mohsen F, Murphy B, Garry M, Hilbert B (2012) White paper-methanol use in hydraulic fracturing fluids. Document No. 1103844.000 0101 0711 TS26. Exponent, Maynard, MA
58. Shah SN, Fakoya MF (2013) Rheological properties of surfactant-based and polymeric nanofluids. SPE/ICoTA coiled tubing & well intervention conference & exhibition. https://doi.org/10.2118/163921-ms
59. Sherman JB, Holditch SA (1991) Effect of injected fracture fluids and operating procedures on ultimate gas recovery. Presented at the SPE gas technology symposium, Houston, Texas, USA, 23–25 Jan. SPE-21496-MS. https://doi.org/10.2118/21496-MS
60. Smith MB, Montgomery C (2014) Hydraulic fracturing. CRC Press, Taylor And Francis, New York
61. Smith MB, Hannah RR (1996) High-permeability fracturing: the evolution of a technology. SPE J Pet Technol 48(7):628e633
62. Speight JG (2016) Handbook of hydraulic fracturing. Wiley, Hoboken, New Jersey
63. Speight (2014) The chemistry and technology of petroleum, 5th edn. CRC Press, Taylor & Francis Group, Boca Raton, Florida
64. Stacy AL, Weber RB (1995) Method for reducing deleterious environmental impact of subterranean fracturing processes. US patent 5 424 285, assigned to Western Co. North America. URL: https://www.freepatentsonline.com/5424285.html
65. Stoll M et al (2008) Toward field-scale wettability modification—the limitations of diffusive transport. SPE Reserv Eval Eng 11(03):633–640
66. Sullivan PF, Gadiyar B, Morales RH, Hollicek R, Sorrelles D, Lee J (2006) Optimization of a viscoelastc surfactant (VES) fracturing fluid for application in high-permeability formations. SPE 98338:1–8
67. Sun X, Liang X, Wang S, Lu Y (2014) J Pet Sci Eng 119:104. https://doi.org/10.1016/j.petrol.2014.04.017
68. Thampi NV, Ojha K, Nair UG (2013) Effect of branched alcohols on phase behaviour and physicochemical properties of Winsor IV microemulsions. J Surfs Detergents 17(2):371–381
69. Tong S, Singh R, Mohanty KK (2018) A visualization study of proppant transport in foam fracturing fluids. J Nat Gas Sci Eng 52:235–247. https://doi.org/10.1016/j.jngse.2018.01.030
70. Tong S, Singh R, Mohanty KK (2017) SPE-187376-MS, Presented at SPE annual technical conference and exhibition, 9–11 Oct. Society of Petroleum Engineers, San Antonio, Texas, USA.https://doi.org/10.2118/187376-MS

71. Veatch Jr RW (1983) Overview of current hydraulic fracturing design and treatment technology, part 1. J Pet Technol 35(4):677e687. Paper SPE-10039-PA. Society of Petroleum Engineers, Richardson, Texas
72. Verma A, Chauhan G, Baruah PP, Ojha K (2018) Morphology, rheology and kinetics of nano silica stabilized gelled foam fluid for hydraulic fracturing application. J Ind Eng Chem 57:13449–13462
73. Verma A, Chauhan G, Ojha K (2017) Synergistic effects of polymer and bentonite clay on rheology and thermal stability of foam fluid developed for hydraulic fracturing. Asia-Pac J Chem Eng 12(2):872–883
74. Verma A, Chauhan G, Ojha K, Padmanabhan E (2019) Characterization of Nano iron stabilized foam fracturing fluid for unconventional gas reservoirs. Energy Fuels 33:10570–10582
75. Virk PS (1975) Drag reduction fundamentals. AICHE J 21(4):625–656
76. Wanniarachchi W, Ranjith P, Perera M, Lashin A, Al Arifi N, Li J Geophys. Geo-Energy Geo-Resour 1(1):121. https://doi.org/10.1007/s40948-015-0015-x
77. Weaver J, Gdanski R, Karcher A (2003) Guar gum degradation: a kinetic study. Presented at the international symposium on oilfield chemistry, Houston, Texas, USA, 5–7 Feb. SPE-80226-MS. https://doi.org/10.2118/80226-MS
78. Weijers L, Pater CJD, Owens KA, Kogsbøll HH (1994) Geometry of hydraulic fractures induced from horizontal wellbores. SPE Prod Oper 9(2):87e92
79. Welton TD, Todd BL, McMechan D (2010) Methods for effecting controlled break in pH dependent foamed fracturing fluid. US patent 7 662 756, assigned to Halliburton Energy Services, Inc. (Duncan, OK). https://www.freepatentsonline.com/7662756.html
80. Welton TD, Bryant J, Funkhouser GP (2007) Anionic surfactant gel treatment fluid. SPE 105815:1–8
81. Xu J, Ka J, Wen S, Liu B, Tang J (2018) The performance evaluation and field application of low temperature coalbed gas clean fracturing fluid. Open J Yangtze Oil Gas 3:231. https://doi.org/10.4236/ojogas.2018.34020,04
82. Xu L (2013) Optimizing surfactant additives for low-permeability wet gas formations: a laboratory approach for correlating capillary pressure and emulsion tendency with surfactant performance. Presented at the SPE eastern regional meeting, Pittsburgh, Pennsylvania, USA, and 20–22 Aug. SPE 165697-MS. https://doi.org/10.2118/165697-MS
83. Xu L, Fu Q (2012) Ensuring better well stimulation in unconventional oil and gas formations by optimizing surfactant additives. Presented at the SPE western regional meeting in Bakersfield, California, USA, 19–23 Mar. SPE 154242-MS. https://doi.org/10.2118/154242-MS
84. Yang J, Cui W, Guan B, Lu Y, Qiu X, Yang Z, Qin W (2016) Supramolecular fluid of associative polymer and viscoelastic surfactant for hydraulic fracturing. SPE Prod Oper 31(4):318–324. https://doi.org/10.2118/175762-PA
85. Yuan C (2012) The chemical treatment to remove liquid block in hydraulic fractured well—a simulation study with leak-off. In: Proceedings of the 2012 energy conference-developing resources for sustainability, June 11–13, Port-of Spain, and Trinidad. SPE 158190-PP
86. Zhou J, Jin Y, Chen M (2010) Experimental investigation of hydraulic fracturing in random naturally fractured blocks. Int J Rock Mech Min Sci 47 (7):1193e1199

Application of Surfactants in Well Stimulation

Pooya Khodaparast and Kristian Jessen

Abstract Development and application of technology to achieve and maintain economic production rates from oil and gas wells has always been a central challenge in the oil and gas industry. The challenge can, in simple terms, be defined as how one can maximize the surface area that dictates the communication between the subsurface and the well, while at the same time, improve the flow of hydrocarbons from the stimulated drainage area/volume into the well. In this Chapter, we provide a review of the use of surfactants in well stimulation. We start with a brief overview of the various mechanisms by which surfactants can be used to achieve specific goals. We then provide a summary of laboratory-scale efforts and observations related to the application of surfactants. Reports and findings from field-scale application/tests are then summarized, and we complete the Chapter with a review of novel developments/applications that are currently subject to research.

Keywords Well stimulation · Surfactants · Fundamentals · Laboratory Studies · Field Applications

1 Introduction

Development and application of technology to achieve and maintain economic production rates from oil and gas wells has always been a central challenge in the oil and gas industry. The challenge can, in simple terms, be defined as how one can maximize the surface area that dictates the communication between the subsurface and the well, while at the same time, improve the flow of hydrocarbons from the stimulated drainage area/volume into the well. To address this challenge, any technology that aims to increase the stimulated area/volume of a well and/or enhances the effective permeability of the drainage area/volume, is commonly referred to as well stimulation.

P. Khodaparast · K. Jessen (✉)
University of Southern California, Los Angeles, USA
e-mail: jessen@usc.edu

T. Solling et al. (eds.), *Surfactants in Upstream E&P*, Petroleum Engineering,
https://doi.org/10.1007/978-3-030-70026-3_13

While different opinions exist on how to classify well stimulation, one can define main categories based on the general approach:

1. Use of mechanical force, delivered by different means, to fracture the subsurface and increase the contact area and promote higher production rates. Hydraulic fracturing is one method, where fluids are pumped at high pressure into the well to achieve this objective in the subsurface, combined with the application of proppants (sized particles) to keep the fractures open. The goal here is to increase the surface area (communication between well and subsurface) and the hydraulic conductivity of the stimulated volume. Hydraulic fracturing (or fracking) is commonly performed on low-permeability (tight, unconventional) reservoirs.
2. Treatment of formation damage (e.g. blockage of pathways for fluid flow) due to (i) drilling and completion (fracking) of the well, or (ii) subsequent operating at thermodynamic conditions that can trigger such damage (e.g. asphaltene precipitation). Acidization (injection of an acidic solution e.g. hydrochloric acid) is a well-known process of this category.

From a more fundamental point of view, altering the chemistry of the well surroundings is a key ingredient in many well-stimulation approaches: Acidization is essentially a chemical treatment procedure that often includes the use of corrosion inhibitors to protect well pipes [16]. Other processes based on the use of chemicals include application of precipitation inhibitors in old, heavy-oil wells, iron control chemicals in acidization processes, and wettability-altering chemicals like surfactants that are used to increase the effective mobility of oil (and/or gas) by modifying the state of rock-fluid or fluid–fluid interfaces. While hydraulic fracturing aims to increase the flow rate by mechanical means, several methods to further enhance this process incorporate chemicals: Proppants (resin-coated sands and multifunctional proppants), polymers, and surfactants (including viscoelastic surfactants as gelling agents) are examples of chemical enhancement during fracturing that all fit under the umbrella of well stimulation.

In the following sections, we provide a review of the use of surfactants in well stimulation. We start with a brief overview of the various mechanisms by which surfactants can be used to achieve specific goals. We then provide a summary of laboratory-scale efforts and observations related to the application of surfactants. Reports and findings from field-scale application and tests are then summarized, and we complete the chapter with a review of novel developments and applications that are currently subject to research.

2 Mechanisms

Surfactants can serve in several roles during well stimulation processes including: (i) agents that can alter the wetting characteristics/state of the rock in favor of a more mobile hydrocarbon phase, (ii) agents to decrease interfacial tension and improve flowback and cleanup of fracturing fluid, and (iii) fracturing fluid agents that, when used in conjunction with internal breakers, allow for improved flowback and cleanup through viscosity reductions. In the following subsections, we provide a brief review of the multifaceted application of surfactants in well stimulation.

2.1 Alteration of Wettability

The effect of wettability on relative fluid mobility during multiphase flow in porous media is well established [33]. Carbonate reservoirs contain more than half of the oil discovered to date [57, 67]. One important characteristic of this class of rocks is their preferential wetting state—they are usually oil-wet as they carry a positive charge, and commonly present negative-charged petroleum acids increase the affinity of the hydrocarbon phase towards the rock surfaces. A similar condition is present in unconventional formations, where the source and reservoir rock are one and the same [6, 46]. It is thus no surprise that such formations are often preferentially oil-wet. Consequently, a significant portion of the oil in place cannot be displaced by traditional techniques such as water injection, as the oil is simply "stuck" onto the rock surfaces. If the wetting characteristics are to shift towards water, the oil would be released more readily from the rock surface and either displaced in globules or in emulsified form by the injected water. A cationic surfactant can effectively serve this purpose by changing the charge on the hydrocarbon interface as demonstrated in Fig. 1.

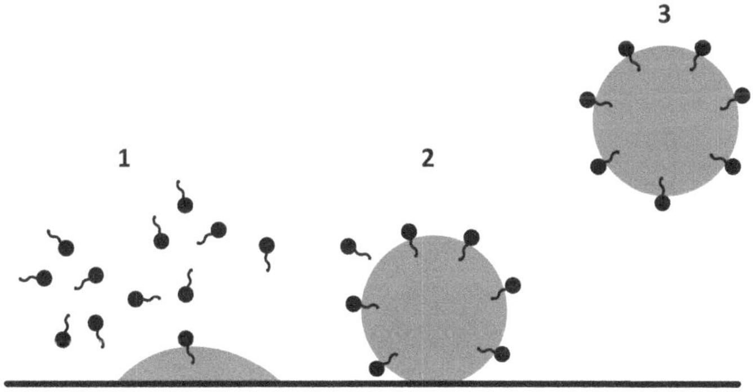

Fig. 1 Displacement of oil from oil-wet surface by surfactant solution—roll-up (1), detachment (2), and solubilization (3) of oil droplets or globules [100]

2.2 Reduction of Interfacial Tension

In any oil and gas operation, the coexistence of multiple immiscible or partially miscible phases results in capillary forces that dictate, in part, the fluid distribution within a porous material. The magnitude of the interfacial tension (IFT) between the fluid phases is proportional to the capillary pressure, and any reduction in the IFT via e.g. use of surfactants can accordingly modify fluid distributions within a porous material. The main principle of this idea is conveniently illustrated by the Capillary Desaturation Curve [50]: A good portion of the hydrocarbon in place is never produced from a porous medium due to the competition between capillary and viscous forces. The capillary number, N_{Ca}, is commonly used to describe the ratio of viscous to capillary forces:

$$N_{Ca} = \frac{\mu U}{\sigma},$$

where μ is viscosity of the displacing phase, U is the fluid velocity, and σ the interfacial tension between the displacing phase and the displaced phase. A reduction of the interfacial tension therefore increases the capillary number and, in turn, decreases the residual saturation as illustrated in Fig. 2.

From Fig. 2, we observe that the residual saturations of both wetting and nonwetting phases are reduced when the IFT is reduced (capillary number is increased). Accordingly, the use of surfactants to reduce the IFT, can be applied to achieve an efficient displacement of both hydrocarbon (oil and gas) and aqueous phases. The latter case is extremely important for (i) cleanup after drilling where water-based mud is commonly used to remove rock cuttings, and (ii) for hydraulic fracturing where water-based mixtures are the primary choice of frac slurries and proppant

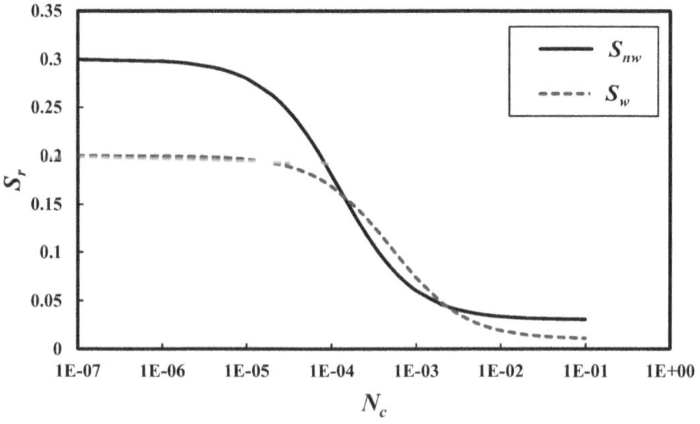

Fig. 2 Effect of capillary number on the residual saturation of both the wetting (S_w) and the nonwetting (S_{nw}) phase. Adapted from [50]

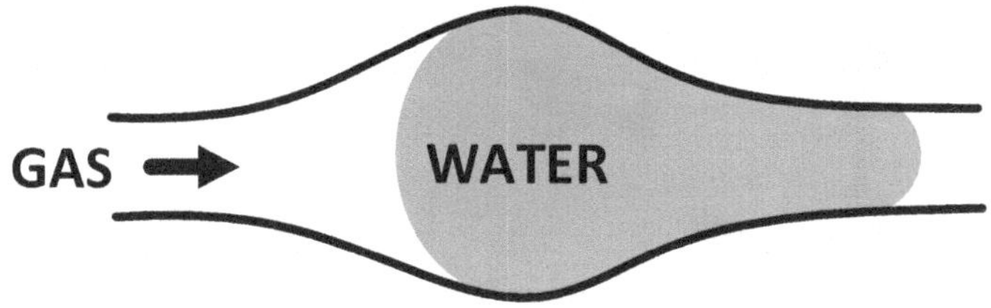

Fig. 3 Illustration of water blocking a pore throats in a gas reservoir

carrier: A high water saturation at the well-rock interface will block subsequent flow of oil or gas (see example of water blocking gas in Fig. 3) and the use of surfactant to facilitate displacement of this water is highly desirable.

In early tests of this concept, to facilitate recovery of fracturing fluid, fluorochemical surfactants were added to the injected stream to decrease the interfacial tension between aqueous and hydrocarbon phases [23, 24]. During flowback (production) from the formation, the reduced IFT provides for more efficient removal of residual fracturing fluids that will otherwise decrease the mobility of the hydrocarbon phase(s).

In addition to well clean-up, after stimulation, a reduction of the IFT via application of surfactants can help remove any condensate (liquid) that has buildup around a gas production well due to the substantially lower pressures near the well. Condensates are valuable components of wet gas, and leaving them stranded in the subsurface is not only an economic loss of its own, but can severely impact the near-well fluid mobility and hence the flow rate during wet-gas production operations [18, 32, 36, 52].

2.3 *Viscoelastic Surfactants (VES) as Gelling Agents*

An ideal fracturing fluid should provide a sufficiently high viscosity to be able to suspend and transport proppant particles, while also break into a lower viscosity fluid during flowback and cleanup, so that a well can deliver optimal hydrocarbon production rates. Polymer has been used to achieve the desired high viscosity, but complications can arise when attempting to break the mixture viscosity during the cleanup phase [13, 104]. Figure 4 provides an example of residual gel due to incomplete cleanup. Polymers such as Guar may result in lowered proppant-pack mobility and a less effective fracture treatment [73].

A more recent method suggests to completely replace polymer, and use viscoelastic surfactants instead. Nehmer [64] introduced the idea of using VES as a gelling agent to replace the use of polymer and proposed using certain water-soluble surfactants (quaternary ammonium compounds) that more readily form long cylindrical micelles, as illustrated in Fig. 5.

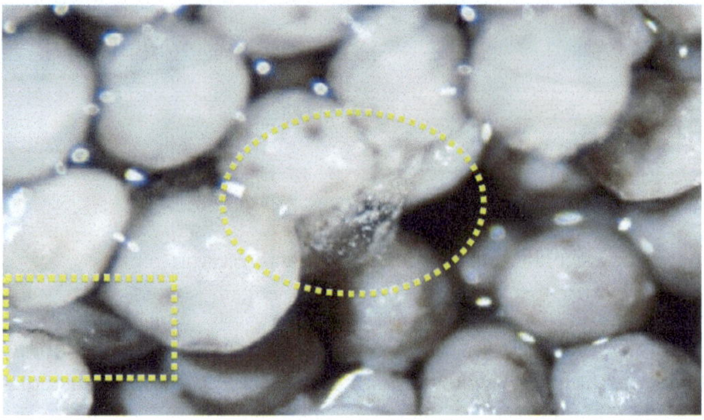

Fig. 4 An example of residual gel damage after cleanup process [68]

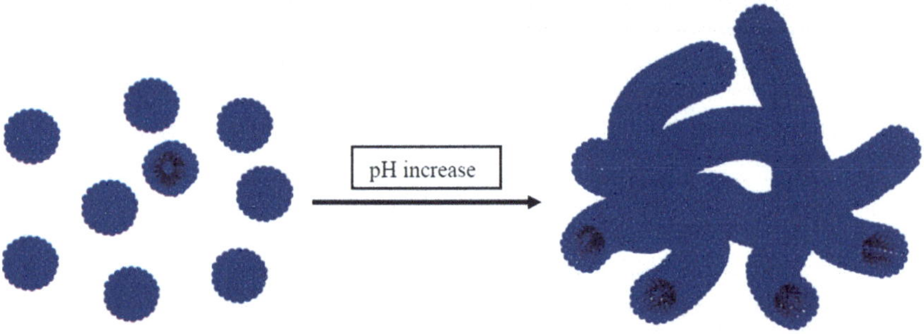

Fig. 5 Spherical versus cylindrical micelles and how the latter can entangle and increase viscosity [107]

These rodlike inclusions mimic long-chained polymer molecules and result in the desired increase in water viscosity. The preferred interfacial curvature depends on the ionic strength and temperature of the brine among other thermodynamic variables. Thus, tuning water salinity or pH may be necessary for a successful gelation. Upon reaching the hydrocarbon components in the reservoir, the rod-shaped micelles swell, and the preferred curvature shifts towards favoring spherical micelles. This natural "breaking" of micelles eliminates the need for an additional "breaker" component as required for more traditional polymer solutions. Kang et al. [49] provides a detailed review of VES and their application in hydraulic fracturing.

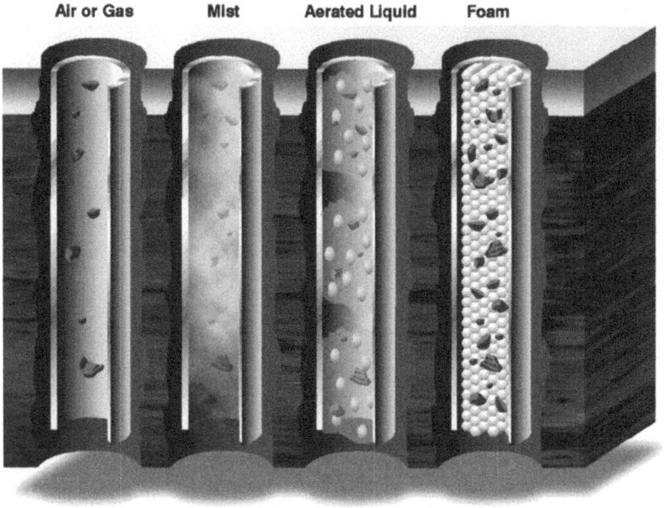

Fig. 6 Various lightweight drilling fluids to suspend and transport cuttings; these methods decrease near-well formation damage by drilling muds [85]

2.4 Foaming Agents

Water-based frac slurries impose post-treatment complications that can push project economics to the limit of feasibility. Moreover, formation damage due to mud-cake invasion can severely affect near-well permeabilities. To this end, researchers have proposed using foams that are stable and can provide sufficient transport properties to carry the proppants during well stimulation [15, 39–41, 56, 88, 92, 93], or carry the drill cuttings when used as a drilling fluid (see Fig. 6).

The latter application in drilling does not exactly fall into the category of well stimulation, and the reader is referred to Thomas and Wilkes [85] for further detail.

2.5 Induced Microcracks and Acceleration of Crack Growth

Sharing a similar concept with the wettability alteration mechanism, Xu et al. [101, 98] proposed that an accurately tailored fracturing fluid can spontaneously spread inside the matrix, thereby increasing the contacted area (see Fig. 7). Tomography images of core plugs that were "soaked" with surfactant containing fluids penetrated almost twice as deep as those without surfactants.

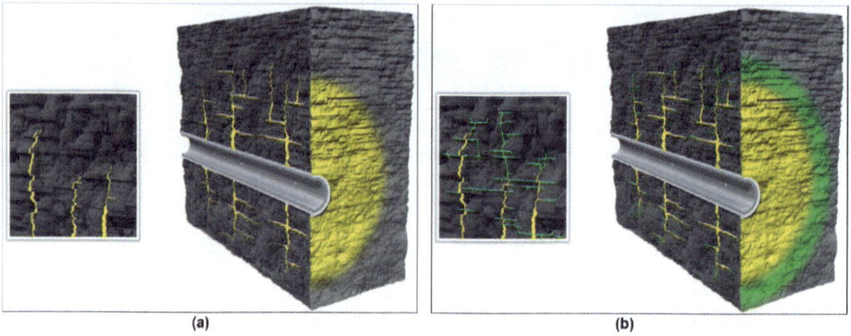

Fig. 7 Schematic of the effects of surfactant in crack initiation and propagation; **a** frac fluid without surfactant and **b** with surfactant [98]

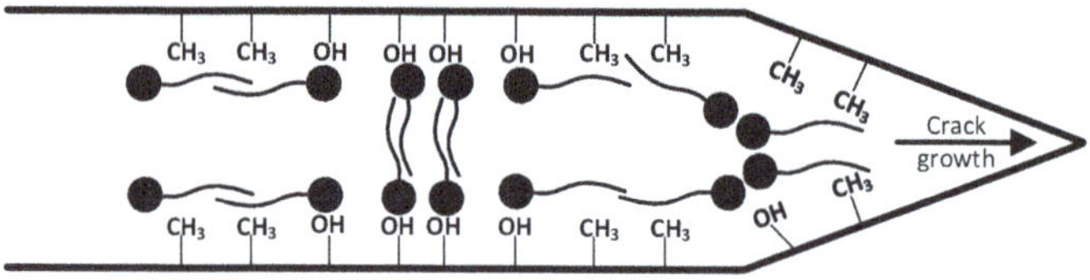

Fig. 8 Crack propagation resulting from surfactant interaction with methyl and hydroxyl groups in organosilicate [101]

More specifically, according to Xu et al. [101], the hydrophilic heads of surfactant molecules strongly interact with the hydroxyl groups in organosilicate rock surfaces, while the hydrophobic tails interact with the methyl groups. Therefore, a local pressure difference builds up and results in liquid migration in the cracks, inducing further growth of cracks (see Fig. 8).

2.6 *Anti-sludge and Acid Retarding Agents*

Application of acid treatment in wells is very common in the petroleum industry. However, the process can introduce complications and damage the formation and its essential flow properties: Acid-in-oil emulsions and asphaltene precipitates can form at certain operational conditions. These particles can also cause challenges in the operation of surface facilities. The naturally present resins in hydrocarbon micellize the aromatic ring structures known as asphaltenes. These particles can aggregate to form deposits that can block pore throats (see Fig. 9). Also, upon contact with the protons (H^+) released from the acids, the commonly negative-charged particles are neutralized and form large aggregates. Surfactants can disperse the asphaltene particles and therefore act as asphaltene inhibitors during acid treatments.

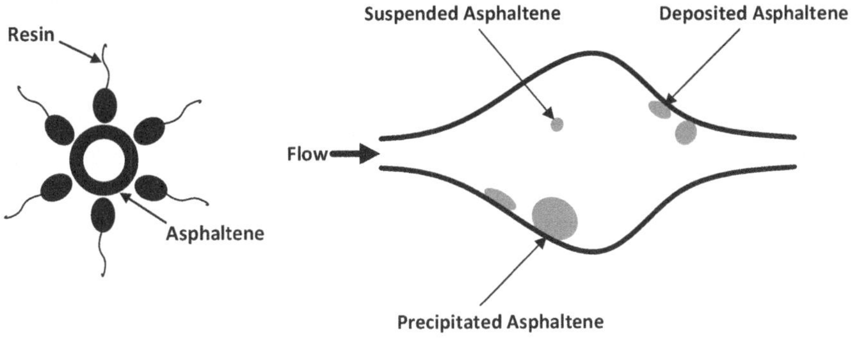

Fig. 9 Deposit formation due to contact with acid and related blockage of pore throats [59]

Acid treatments aim to clean formation damage, but if the implemented acid reacts with the formation, it may be lost before probing and treating the damaged zone and introduce additional damage. Carbonates (calcites in particular) are extremely reactive with commonly used acids such as HCl. One proposed preventative measure is to use acid-in-oil (e.g. diesel) emulsions, with an emulsifier (surfactant) to stabilize the mixture (see Fig. 10). The diesel acts as a barrier to slow down the rock-acid reaction, allowing the acid to penetrate deeper into the formation and improve well productivity.

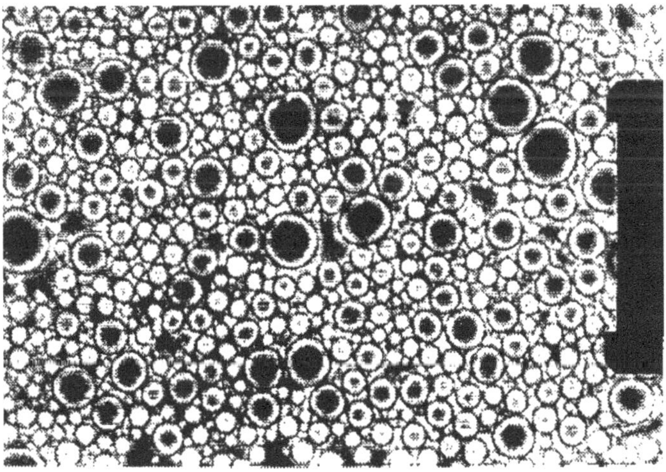

Fig. 10 Micrograph of emulsified acid (x250) [7]

3 Laboratory-Scale Studies

A large fraction of the literature on the application of surfactants in well stimulation describes results and observations from laboratory-scale studies. The literature has traditionally reported on experimental observations from various applications, while a great body of research has been dedicated, more recently, to the design and modeling of new surfactant systems via molecular simulation. In this section, we focus primarily on laboratory experiments and the reader is referred to the reviews by Ahmadi et al. [5], Padding et al. [66], Theodorakis et al. [84] for additional overview of advances in molecular design and simulation of surfactant systems.

Gogarty et al. [37] studied the treatment of injection wells using micellar methods based on the idea of an IFT reduction. Their laboratory-scale results on core floods in sandstone showed an eightfold increase in water relative mobility after the treatment, which improves water injectivity and subsequently the sweep efficiency: Water injection rate, at constant pressure, was found to increase by 370 percent. Based on the IFT reducing property, surfactants were almost always used in hydraulic fracturing operations as well [47].

IFT reduction is an important mechanism to facilitate cleanup after the stimulation of production wells, such as hydraulic fracturing [17, 76]. Fluorochemical surfactants have been reported to achieve low IFTs at very small concentrations and within a wide range of temperatures: Clark et al. [24] reported a stable range of temperatures for fluorochemical surfactants and little to negligible formation damage due to the possible adsorption onto sandstone rocks. They reported, however, higher adsorption rates for cationic fluorochemicals. Much lower surface tensions (~50%) were achieved by this class of surfactants than conventional hydrocarbon surfactants. Stable emulsions were also less likely to form when exposed to hydrocarbons in the formation.

Fluorinated surfactants are also known to be successful wettability altering chemicals (see Sect. 2.1). Detailed reviews of application of various wettability altering chemicals (especially fluorinated surfactants) for condensate removal can be found in the literature [35, 48], though other more conventional surfactants are also used. Dees et al. [31] report results of using different surfactants on Austin Chalk (limestone) and conclude that cationic surfactants perform much better than nonionic counterparts in releasing the oil from the originally oil-wet rock. While Conway et al. [25] reported insignificant improvement in gas permeability through surfactant additives, Li and Firoozabadi [51] were first to report several fluorochemical surfactants that could successfully alter the wetting state from strongly water-wet to intermediate gas-wet and increase gas permeability in chalk and sandstone at room temperature. Several studies followed their promising results to characterize the alteration at higher, more practical temperatures [80, 79]. Fahes et al. [34] and later Wu and Firoozabadi [94] showed that alteration of wettability is due to chemical reaction and are therefore permanent. Liu et al. [54] showed that both water and gas permeabilities improve upon alteration using fluorosurfactants (see Fig. 11), a finding later corroborated by Wu and Firoozabadi [95].

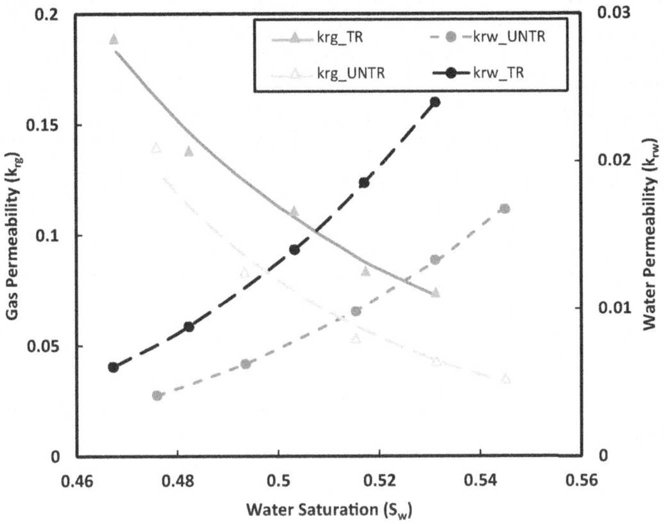

Fig. 11 Fluorosurfactants increase both water and gas relative permeabilities [95]

Chang et al. [21] were possibly the first to propose a self-diverting-acid (SDA) process using viscoelastic surfactant in acid treatment of highly permeable and heterogeneous sandstone reservoirs. They reported superior cleanup results even before VES contacts hydrocarbon. Chang et al. [20] later reported observations from field-case studies using VES as SDA on high permeability, heterogeneous wells in the Gulf of Mexico with effective and successful results. Daniel et al. [30] studied effects of brine density, pH, surfactant concentration, and various additives on the apparent fluid viscosity and elasticity, parameters that are of essence in design of fracturing fluids. Numerous researchers continued to investigate the performance of VES, and the combined mechanisms it offers, and report considerably less rig time and friction losses for VES compared to polymer, as well as a reduction in the required number of frac-stages relative to other fluid systems [28, 26, 63, 80–83], Nasr-El-Din and Samuel [61]. Nasr-El-Din et al. [62] reported on laboratory experiments with enzyme-viscoelastic surfactant mixtures in cleanup cocktails to replace polymers. Though results document success in cleaning and IFT reduction, they also find lower VES viscosities in the presence of enzymes, which highlights the importance of further research in design of frac-fluid mixtures. They also observed mixture phase separation at high temperatures (250 °F). Crews et al. [27] reported on new stabilizer formulations that prevent phase separation and help VES maintain high viscosities at high temperatures (up to 300 °F), and introduced new internal breakers that can help with viscosity reduction in gas wells for better cleanup.

Accurate measurements of the VES concentration in effluent streams (produced fluids) provide important information related to retention due to various interactions in the subsurface. Yu et al. [107] proposed a two-phase titration method to measure VES concentration (errors less than 1.5%) and were able to investigate the presence of acid additives on the accuracy of the measurements. They did not observe any interference between acids and calcium chloride on measurement accuracy. However, they do report high retention rates of VES for their experiments and concluded that longer soak periods and better design of the breaker are needed to further improve the cleanup by this approach.

Cetin and Nasr-El-Din [19] studied the effects of various additives and contaminants present in the fracturing process. They reported that e.g. methanol (additive) concentration reduces apparent viscosity of VES systems. Demulsifier additives were reported to cause an initial increase in the apparent viscosity, while temperatures higher than 100 F can cause the opposite effect.

Regardless of the method used, design and sequencing of the stimulation fluid mixtures can be extremely sensitive to the conditions in the subsurface, and there is no silver bullet. Any procedure should take into consideration the rock minerology as well as operating conditions, such as temperature and prevailing brine chemistry. For VES systems used in hydraulic fracturing, surfactant chain length and structure, concentration, nature and charge of the hydrophilic head, temperature and pressure are among the many factors affecting the degree of worm-like micellization and related rheological properties (see Fig. 12) that are important in the design [14, 53, 72, 103].

Considering wettability alteration and IFT reduction mechanisms in tight rocks, Wang et al. [89] studied several surfactants via spontaneous imbibition in Bakken shale. They document internal olefin sulfonates, one ethoxylate nonionic and a cationic surfactant as being able to withstand the high temperatures of the subsurface (105–120 °C). The former two surfactants demonstrated a larger tolerance to high salinity conditions. Highly saline conditions proved to be essential for imbibition since low salinity fluids failed to imbibe and produce any oil. Another study on the same formation proposed an improved surfactant formulation that is more tolerant of extremely high salinity brines [110]. Xu and Fu [99] reported a better performance for weakly emulsifying surfactants compared to their non-emulsifying counterparts in the Eagle Ford shale formation.

Morsy and Sheng [58] proposed spontaneous imbibition of surfactant mixtures as preflood schemes for higher recovery in the Bakken shale formation. They attributed higher recovery rates after the imbibition process to wettability alteration, but an argument can be made for crack initiation and propagation as later described by Xu et al. [101, 98]. Wettability alteration indeed contributes to higher production, as has been shown both in tight formations and conventional reservoirs such as carbonates [1],[3, 8, 11, 12, 74]. Neog and Schechter [65] studied the effect of several proprietary and traditional nonionic surfactants on the Wolfcamp shale wetting state through imbibition experiments. The intermediate-wet shale shows preference towards water after imbibition for certain surfactant mixtures that deliver a moderate IFT reduction. The inverse correlation of wettability and IFT reduction can be justified by the preferential partitioning of surfactant molecules on the rock surfaces versus the oil-water interface. As such, if IFT decreases drastically, bulk concentration of surfactant would be less, which indicates wettability may not be substantially altered.

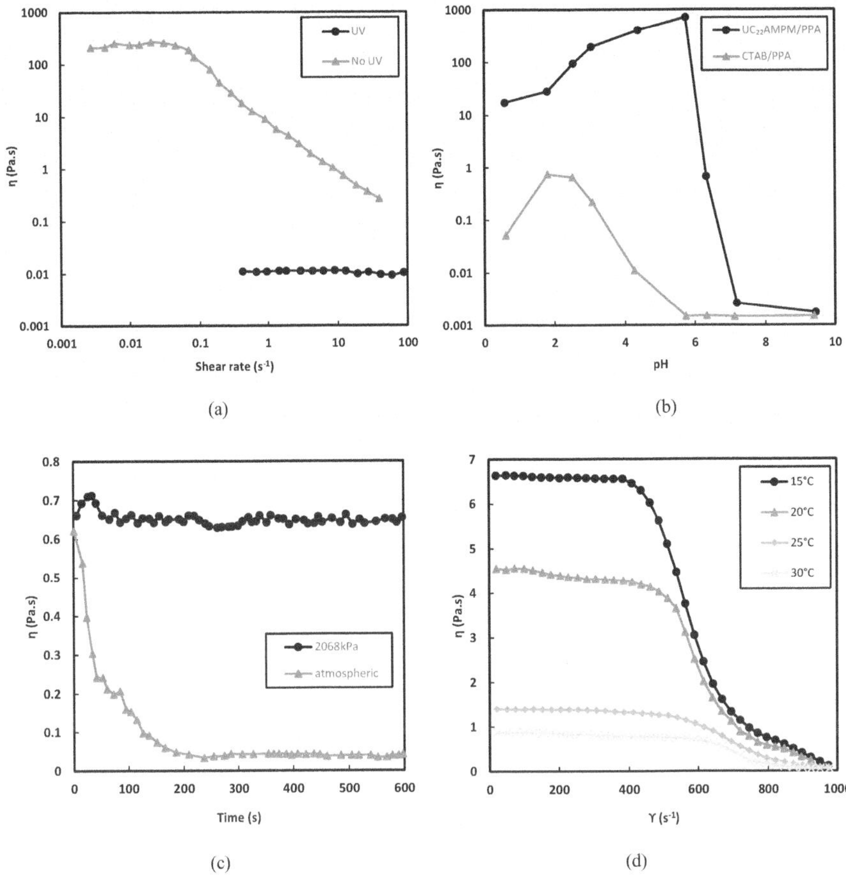

Fig. 12 Effects of (**a**) Ultraviolet light, (**b**) pH, (**c**) pressure, and (**d**) temperature on rheological properties of various VES systems. Adapted from: [55, 90, 14, 103]

As discussed above, surfactants can affect a stimulation process through an array of mechanisms. Whether spontaneous imbibition of surfactant mixtures increases recovery through (a) wettability alteration or by (b) inducing and propagating micro-cracks, both processes require significant interaction times. Therefore, it is natural to propose shut-in or well-resting periods, after the well stimulation, for the processes to take place and achieve the desired outcome of higher production rates. In repeated cycles, this periodic process is known in the industry as huff-n-puff [75], while Wang et al. [91] suggest that a more accurate name may be huff-n-soak-n-puff to indicate the shut-in period. Huff-n-puff operations are well-known for injection of CO_2 (or other gases) in tight formations, but surfactants and chemical blends have also been proposed. More recently, Zeng et al. [108] and Zhang et al. [109] have proposed a combination of gas and surfactant chemical blends.

4 Field Tests

While a large body of literature details observations and analysis from laboratory-scale experiments, less detailed literature is available for well- or field-scale applications of surfactant systems. We interpret this a result of the proprietary nature of both the service companies implementing well stimulations and the oil and gas industry. Here we summarize some of the literature reporting on findings from larger-scale implementation of surfactant systems in well stimulation.

Smith and Murphy [76] report excellent cleanup results from their surfactant-containing "emulsion frac" fluid: The average swabbing (cleanup) times were reduced to 1–1.5 days using emulsion frac, relative to 5–7 days when conventional water-frac methods were implemented. The authors hint on a possible wettability alteration near the well as one mechanism responsible of achieving the recorded increase in production.

Clark et al. [24] report on three field tests that implemented 3M's fluorochemical surfactants (FC-C's), all of which were demonstrated successful in improving well productivity. Pendergrass et al. [70] incorporated surfactants as IFT reducing agents to address water blockage in their target high-clay-content formation. They report on four field cases implementing their non-acid reactive solution (NARS), all of which are demonstrated to improve well productivity.

Patemiti [69] report on successful implementation of a surfactant system as an IFT reducing emulsifying agent for the Codell formation production wells in Colorado. They report an improved productivity in stimulated wells, including a lower producing gas-to-oil ratio (GOR).

Nasr-El-Din et al. [60] report on six different field stimulation treatments using viscoelastic surfactants in acid systems. They found improvements in friction inhibition and higher injection rates (see Fig. 13), a reduced number of stages required to meet production targets, shorter rig time, and a smaller volume of chemicals needed relative to polymer-based acid systems. They do report, however, that VES-based acid systems interact unfavorably with ferric ions and hydrocarbons, an observation

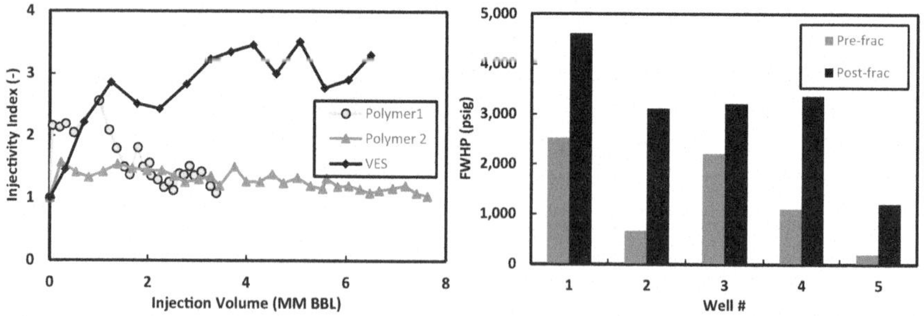

Fig. 13 (Left) Stimulation of a "dead" well with VES result in higher injectivities. (Right) flowing wellhead pressure before and after surfactant-based acid treatment for five different gas production wells [60]

that was later corroborated in other studies [19]. As mentioned previously, rod-shape to spherical micellar transitions occur when hydrocarbons are emulsified and swell the cylinders, resulting in a sharp drop in viscosity. Presence and strength of different ions (such as ferric) can also change the preferred curvature and hence affect cylinder to sphere transition [87], [86].

Al Mutawa et al. [10] used VES as a diversion agent in heavily water-producing parts of a well, so that the acid treatments are more focused on the oil-producing zones and water cut is decreased through a local permeability reduction. They report a 360% increase in oil production, and a 40% decrease in water production following the treatment approach. Stanislav [77] also reports on the use of a viscoelastic diverting agent (VDA) in successful field cases of acid treatment in Eastern Siberia.

Stewart et al. [78] reviewed performance of twenty completions in oil and gas wells over the span of 21 years from 1973 and 1994. The report laboratory results on field samples that indicate "deep bed" filtration and plugging of the pore throats, fracture packs, and perforations where polymer fluids were used. By contrast, the VES systems allow for more efficient cleanup of the injected fluids. They reported a 590% improvement of productivity when using VES frac-pack as compared to conventional gravel-pack in a shallow gas reservoir, indicating that VES systems are economically feasible for gas reservoirs. Also, in a shallow oil reservoir case, incorporating VES was demonstrated to eliminate the need for any additional artificial lift system.

Samuel et al. [73] report that field implementations of VES systems facilitate operational simplicity and notably improve productivity, both onshore and offshore. For example, one shallow gas well in Kansas obtained a productivity that was 52% higher than the average of 12 adjacent wells after application of VES. Pitoni et al. [71] report on a successful screenless completion method that was implemented in the Giovanna field, offshore Italy. VES systems were used, and no sand proppants were produced following the stimulation.

5 New Directions for Surfactant Systems

Viscoelastic gels, using novel and proprietary surfactants, have recently attracted researchers' attention [43, 105]. This structure can be formed by a combination of cationic and anionic surfactants. These new structures were developed in a quest for more effective VES systems that can produce higher viscosities at lower concentrations. VES systems still need improvements in certain aspects including cost, operational temperature, and leak-off characteristics [29, 45]. Nanoparticles such as ZnO and MgO have recently been demonstrated to enhance VES-based frac fluids in terms of thermal stability, leak-off, and viscosity [9]. He et al. [44] suggest the use of a sacrificial agent in the fracturing fluid to avoid premature adsorption of surfactant on rock. They show that a properly selected agent does not affect fracturing fluid stability and compatibility. Yekeen et al. [106] provides a detailed review of application of nanoparticles in hydraulic fracturing.

While, conventionally, laboratory research of VES systems has been focused on rheological characterization of these systems, more advanced techniques (NMR, Cryo-TEM, SANS) can provide valuable information about their microstructural evolution and their compatibility with other additives [4].

Despite the successful application of hydraulic fracturing and related development of numerous assets, there are several drawbacks associated with this technique. Post-processing and reuse of frac fluids, due to their potential negative impact on the environment, has been a source of contention as well as active research: Intermediates resulting from the breakdown of certain surfactants can be carcinogenic and toxic (e.g. alkylphenol ethoxylate intermediates) [22]. Therefore, it is important to account for waste treatment and reuse of frac fluids in the design process. Gupta and Hlidek [38] estimate that more than 50% of the recovered (produced) water from frac jobs they studied in Canada could be reused. Yan et al. [102] propose a reusable VES fracturing fluid that can regain high values of viscosity after flowback by altering the pH value, while Wu et al. [96] report on a novel surfactant (erucamidopropyl dimethylamine, or EA) that exhibits very high viscosities upon contact with supercritical CO_2. Wu et al. [97] studied effects of pressure, temperature, and salinity on performance of the novel VES system and reported high endurance relative to other options in the industry (see Fig. 14). Reuse of the proposed VES system is achieved by removing CO_2 from the mixture.

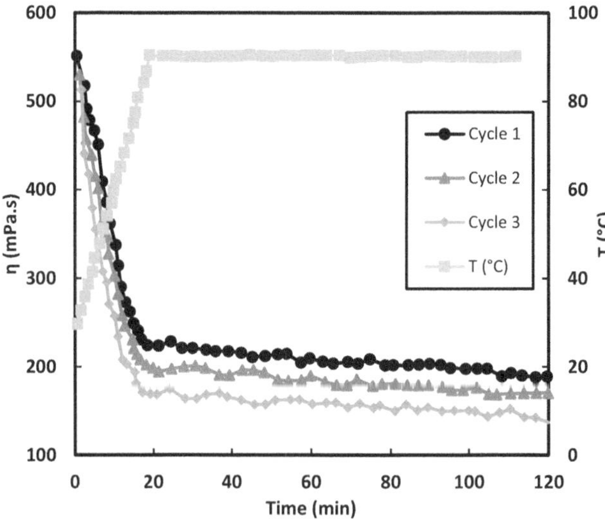

Fig. 14 Change of viscosity upon reuse of pH-responsive surfactant system in fracturing cycles [102]

6 Summary

In this chapter, we have attempted to cover a plethora of important subtopics that all pertain to the application of surfactants in well stimulation. Their amphiphilic nature makes surfactants an ideal additive in environments where aqueous and oleic phases are present. To optimize production, researchers aim to engineer these molecules and all the parameters that affect their interactions with present components and minerals.

Admittedly, covering all aspects of the applications, and the ongoing research in this field, would require several volumes. Our approach, therefore, was focused on drawing an introductory image of these applications, the mechanisms behind them, the relevant challenges faced by the industry, and to refer the reader to more comprehensive review papers for further investigation.

References

1. Adejare OO, Nasralla RA, Nasr-El-Din HA (2012) The effect of viscoelastic surfactants and a mutual solvent on the wettability of a carbonate rock. In: SPE production and operations symposium, proceedings, vol 2, pp 1133–1143. https://doi.org/10.2118/157323-ms
2. Adibhatla B, Sun X, Mohanty KK, Houston U (n.d.) Brine imbibition
3. Adibhatla B, Mohanty KK (2008) Oil recovery from fractured carbonates by surfactant-aided gravity drainage: laboratory experiments and mechanistic simulations. SPE-99773. SPE Reservoir Eval Eng 11(1):119–130. https://doi.org/10.2118/99773-PA
4. Afra S, Samouei H, Nasr-El-Din H (2020) NMR investigation of viscoelastic surfactants compatibility with corrosion inhibitors. In: International petroleum technology conference 2020, IPTC 2020. https://doi.org/10.2523/iptc-19601-ms
5. Ahmadi M, Hou Q, Wang Y, Chen Z (2020) Interfacial and molecular interactions between fractions of heavy oil and surfactants in porous media: comprehensive review. Adv Colloid Interface Sci 283. https://doi.org/10.1016/j.cis.2020.102242
6. Ahmed U, Meehan DN (2016) Unconventional oil and gas resources: exploitation and development. CRC Press
7. Al-Anazi HA, Nasr-El-Din HA, Mohamed SK (1998) Stimulation of tight carbonate reservoirs using acid-in-diesel emulsions: Field Appl 1:9–17. https://doi.org/10.2118/39418-ms
8. Al-Muntasheri GA, Sierra L, Garzon FO, Lynn JD (2010) Water shut-off with polymer gels in a high temperature horizontal gas well: a success story. Saudi Aramco J Technol, SUMMER
9. Al-muntasheri GA, Advanced E, Aramco S (2017) Nanoparticle-enhanced hydraulic-fracturing fluids : a review. May
10. Al Mutawa M, Al Matar B, Dashti AA, Kelkouli R, Liu H, Razouqi M (2008) High-water-cut wells stimulation combined viscoelastic surfactant. In: Proceedings—SPE international symposium on formation damage control, vol 1, pp 170–174. https://doi.org/10.2118/112176-ms
11. Alshehri J, Kovscek AR (2012) An X-Ray CT study of multidimensional imbibition in dual porosity carbonates
12. Babadagli T (2003) Evaluation of EOR methods for heavy—oil recovery in naturally fractured reservoirs. J Pet Sci Eng 37(1–2):25–37. https://doi.org/10.1016/S0920-4105(02)00309-1
13. Barati R, Liang JT (2014) A review of fracturing fluid systems used for hydraulic fracturing of oil and gas wells. J Appl Polym Sci 131(16):1–11. https://doi.org/10.1002/app.40735

14. Baruah A, Shekhawat DS, Pathak AK, Ojha K (2016) Experimental investigation of rheological properties in zwitterionic-anionic mixed-surfactant based fracturing fluids. J PeT Sci Eng 146:340–349 https://doi.org/10.1016/j.petrol.2016.05.001
15. Bentsen NW, Veny JN (1976) Preformed stable foam performance in drilling and evaluating shallow gas wells in Alberta. JPT J Pet Technol 28:1237–1240. https://doi.org/10.2118/5712-PA
16. Bhardwaj A, Hartland S (1993) Applications of surfactants in petroleum industry. J Dispersion Sci Technol 14(1):87–116. https://doi.org/10.1080/01932699308943389
17. Briaud JL, Riner KB, Ohya S (1984) Cyclic pressuremeter tests for cyclic lateral loads. In: Proceedings of the annual offshore technology conference May(2):275–282. https://doi.org/10.4043/4678-ms
18. Castelijns JHP, Hagoort J (1984) Recovery of retrograde condensate from naturally fractured gas-condensate reservoirs. Soc Petrol Eng J 24(6):707–717. https://doi.org/10.2118/11199-PA
19. Cetin S, Nasr-El-Din HA (2017) Rheological study of a novel sulfobetaine surfactant-based acid system. In: SPE middle east oil and gas show and conference, MEOS, proceedings, pp 2363–2377. https://doi.org/10.2118/183652-ms
20. Chang F, Acock AM, Geoghagan A, Huckabee PT (2001) Experience in acid diversion in high permeability deep water formations using visco-elastic-surfactant. In: SPE—European formation damage control conference, proceedings, pp 77–81.https://doi.org/10.2118/68919-ms
21. Chang F, Qu Q, Frenier W (2001) A novel self-diverting-acid developed for matrix stimulation of carbonate reservoirs. In: Proceedings—SPE international symposium on oilfield chemistry, pp 477–482.https://doi.org/10.2523/65033-ms
22. Chen H, Carter KE (2017) Characterization of the chemicals used in hydraulic fracturing fluids for wells located in the Marcellus Shale Play. J Environ Manage 200:312–324. https://doi.org/10.1016/j.jenvman.2017.05.069
23. Clark HB (1980) Use of fluorochemical surfactants in nonaqueous stimulation fluids. JPT J Pet Technol 32(10):1695–1697. https://doi.org/10.2118/7894-PA
24. Clark HB, Pike MT, Rengel GL (1982) Water-soluble fluorochemical surfactant well stimulation additives. JPT J Pet Technol 34(7):1565–1569. https://doi.org/10.2118/9008-PA
25. Conway M, Smith K, Thomas T, Schraufnagel R (1995) The effect of surface active agents on the relative permeability of brine and gas in porous media. In: SPE international symposium on oilfield chemistry, pp 395–401.https://doi.org/10.2523/28982-ms
26. Crews JB, Huang T (2008) Internal breakers for viscoelastic-surfactant fracturing fluids. JPT J Pet Technol 60(3):70–71. https://doi.org/10.2523/106216-ms
27. Crews JB, Huang T, Wood WR (2008) New technology improves performance of viscoelastic surfactant fluids. SPE Drill Complet 23(1):41–47. https://doi.org/10.2118/103118-pa
28. Crews JB (2005) Internal phase breaker technology for viscoelastic surfactant gelled fluids. In: SPE International symposium on oilfield chemistry proceedings. https://doi.org/10.2523/93449-ms
29. Daeffler C, Perroni D, Makarychev-Mikhailov S, Mirakyan A (2019) Internal viscoelastic surfactant breakers from in-situ oligomerization. In: Proceedings—SPE international symposium on oilfield chemistry, pp 8 9.https://doi.org/10.2118/193563-ms
30. Daniel S, Morris L, Chen Y, Brady ME, Lungwitz BR, George L, Kranenburg A, Ali SA, Twynam A, Parlar M (2002) New visco-elastic surfactant formulations extend simultaneous gravel-packing and cake-cleanup technique to higher-pressure and higher-temperature horizontal open-hole completions: laboratory development and a field case history from the North Sea.https://doi.org/10.2523/73770-ms
31. Dees JM, Freet TG, Hollabaugh GS (1990) Horizontal well stimulation results in the Austin Chalk Formation, Pearsall Field, Texas. In: Proceedings—SPE annual technical conference and exhibition, Pi(4), pp 665–676. https://doi.org/10.2523/20683-ms
32. Denney D (2007) Laboratory and field-trial results of condensate-bank removal in retrograde-gas reservoirs. JPT J Pet Technol 59(4). https://doi.org/10.2118/0407-0072-jpt

33. Donaldson EC, Thomas RD, Lorenz PB (1969). Nettability determination and its effect on mecovery mllclency. SPE J 9(1):13–20
34. Fahes M, Imperial C, Firoozabadi A, Engineering R (2005) SPE 96184 wettability alteration to intermediate gas-wetting in gas/condensate reservoirs at high temperatures
35. Ganie K, Idris AK, Mohshim DF, Wan Sulaiman WR, Mohd Saaid I, Abdul Malik A (2019) A review on the wettability alteration mechanism in condensate banking removal. J Pet Sci Eng 183(May):106431.https://doi.org/10.1016/j.petrol.2019.106431
36. Garzon FO, Al-Anazi HA, Leal JA, Al-Faifi MG (2006) Laboratory and field trial results of condensate banking removal in retrograde gas reservoirs: case history. In: Proceedings—SPE annual technical conference and exhibition, vol 4, pp 2386–2395.https://doi.org/10.2523/102558-ms
37. Gogarty WB, Kinney WL, Kirk WB (1970) Injection well stimulation with micellar solutions. J Pet Technol 22(12):1577–1584. https://doi.org/10.2118/2412-pa
38. Gupta DVS, Hlidek BT (2010) Frac-fluid recycling and water conservation: a case history. SPE Prod Oper 25(1):65–69. https://doi.org/10.2118/119478-PA
39. Harris PC (1995) Comparison of mixed-gas foams with N2 and CO2 foam fracturing fluids on a flow-loop viscometer. SPE Prod Facil 10(3):197–203. https://doi.org/10.2118/20642-pa
40. Harris PC, Heath SJ (1996) Rheology of crosslinked foams. SPE Prod Facil 11(2):113–116. https://doi.org/10.2118/28512-PA
41. Harris PC, Reidenbach VG (1987) High-temperature rheological study of foam fracturing fluids. JPT, J Pet Technol 39(5):613–619. https://doi.org/10.2118/13177-PA
42. Harris PC (1992) Application of foam fluids to minimize damage during fracturing, pp 673–678. https://doi.org/10.2523/22394-ms
43. He K, Xu L (2017) Unique mixtures of anionic/cationic surfactants: a new approach to enhance surfactant performance in liquids-rich shale reservoirs. In: Proceedings—SPE international symposium on oilfield chemistry, pp 969–978. https://doi.org/10.2118/184515-ms
44. He K, Yue Z, Xu L (2016) Using sacrificial agents to enhance surfactant performance in the Eagle Ford shale. In: 78th EAGE conference and exhibition 2016: efficient use of technology—unlocking potential. https://doi.org/10.2118/180156-ms
45. Hull Katherine L, Mohammed S, Al-Muntasheri Ghaithan A (2016) Recent advances in viscoelastic surfactants for improved production from hydrocarbon reservoirs. SPE J 21(4):1340–1357. https://doi.org/10.2118/173776-pa
46. Hyne NJ (2012) Nontechnical guide to petroleum geology, exploration, drilling, and production. PennWell Books
47. Jennings AR (1975) The effect of surfactant-bearing fluids on permeability behavior in oil—producing format ions. In: Society of petroleum engineers—fall meeting of the society of petroleum engineers of AIME, FM 1975. https://doi.org/10.2118/5635-ms
48. Jin J, Sun J, Rong K, Lv K, Nguyen TAH, Wang R, Huang X, Bai Y, Liu J, Wang J (2020) Gas-wetting alteration by fluorochemicals and its application for enhancing gas recovery in gas-condensate reservoirs: a review. Energies 13(18). https://doi.org/10.3390/en13184591
49. Kang W, Mushi SJ, Yang H, Wang P, Hou X (2020) Development of smart viscoelastic surfactants and its applications in fracturing fluid: a review. J Petrol Sci Eng 190(February):107107. https://doi.org/10.1016/j.petrol.2020.107107
50. Lake LW, Johns RT, Rossen WR, Pope GA.(2014) Fundamentals of enhanced oil recovery
51. Li K, Firoozabadi A (2000) Experimental study of wettability alteration to preferential gas-wetting in porous media and its effects. SPE Reservoir Eval Eng 3(2):139–149. https://doi.org/10.2118/62515-pa
52. Li S, Zheng X, Dai Z, Luo K, Chen G, Liu N (2001) Investigation of revaporization of retrograde condensate. In: Proceedings of the middle east oil show, pp 597–601. https://doi.org/10.2118/68170-ms
53. Liu S, Wang X, Chen L, Hou L, Zhou T (2014) Aggregation morphologies of a series of heterogemini surfactants with a hydroxyl head group in aqueous solution. Soft Matter 10(45):9177–9186. https://doi.org/10.1039/c4sm01524j

54. Liu Y, Zheng H, Huang G, Li G, Li K (2006) Improving production in gas/condensate reservoirs by wettability alteration to gas wetness. In: Proceedings—SPE symposium on improved oil recovery, vol 2, pp 670–675.https://doi.org/10.2118/99739-ms
55. Lu Y, Zhou T, Fan Q, Dong J, Li X (2013) Light-responsive viscoelastic fluids based on anionic wormlike micelles. J Colloid Interface Sci 412:107–111. https://doi.org/10.1016/j.jcis.2013.09.014
56. Lv Q, Li Z, Zheng R (2019) Study of ultra-dry CO_2 foam fracturing fluid enhanced by graphene oxide. In: International petroleum technology conference 2019, IPTC 2019. https://doi.orghttps://doi.org/10.2523/iptc-19295-ms
57. Masalmeh SK, Wei L, Blom C, Jing XD (2014) EOR options for heterogeneous carbonate reservoirs currently under waterflooding. In: Society of petroleum engineers—30th Abu Dhabi international petroleum exhibition and conference, ADIPEC 2014: challenges and opportunities for the next 30 years, vol 3, pp 2374–2402. https://doi.org/10.2118/171900-ms
58. Morsy S, Sheng JJ (2014) Surfactant preflood to improve waterflooding performance in shale formations. In: Society of petroleum engineers—SPE Western North American and Rocky Mountain joint meeting
59. Nasr-El-Din HA (2010) Surfactant use in acid stimulation. Surfactants. https://doi.org/10.1017/cbo9780511524844.010
60. Nasr-El-Din HA, Samuel M, Tibbles R (2004) Lessons learned from using viscoelastic surfactants in well stimulation. SPE Prod Oper 22(1):112–120. https://doi.org/10.2118/90383-pa
61. Nasr-El-Din HA, Samuel M (2007) Lessons learned from using viscoelastic surfactants in well stimulation. SPE Prod Oper 22(1), February. https://doi.org/10.2118/90383-MS
62. Nasr-El-Din HA, Al-Otaibi MB, Al-Qahtani AA, Al-Fuwaires OA (2006) Filter-Cake cleanup in MRC wells using enzyme/surfactant solutions
63. Nasr-El-Din HA, Samuel E, Samuel M (2003) Application of a new class of surfactants in stimulation treatments. In: Proceedings—SPE international improved oil recovery conference in asia pacific, pp 373–382. https://doi.org/10.2118/84898-ms
64. Nehmer WL (1988) SPE 17168 viscoelastic gravel-pack carrier fluid. SPE J 217–226
65. Neog A, Schechter DS (2016) Investigation of surfactant induced wettability alteration in Wolfcamp shale for hydraulic fracturing and EOR applications. In: SPE—DOE improved oil recovery symposium proceedings. https://doi.org/10.2118/179600-ms
66. Padding JT, Boek ES, Briels WJ (2005) Rheology of wormlike micellar fluids from Brownian and molecular dynamics simulations. J Phys Condens Matter 17(45). https://doi.org/10.1088/0953-8984/17/45/021
67. Pal S, Mushtaq M, Banat F, Al Sumaiti AM (2018) Review of surfactant-assisted chemical enhanced oil recovery for carbonate reservoirs: challenges and future perspectives. Pet Sci 15(1):77–102.https://doi.org/10.1007/s12182-017-0198-6
68. Palisch T, Duenckel R, Bazan L, Heidt HJ, Turk G (2007) Determining realistic fracture conductivity and understanding its impact on well performance—theory and field examples. In: SPE—hydraulic fracturing technology conference, pp 462–474.https://doi.org/10.2118/106301-ms
69. Patemiti M (2009) SPE 116237 ME surfactant increases production in the Codell formation of the DJ Basin. In: SPE Rocky mountain petroleum technology conference, pp 20–27
70. Pendergrass J, Campbell P, Bell D, Bishop M, Copeland J (1984) Nonacid reactive solution for deep well stimulation. Society of petroleum engineers of AIME, (Paper) SPE, pp 153–160. https://doi.org/10.2118/12625-ms
71. Pitoni E, Devia F, James SG, Heitmann N (2000) Screenless completions: cost-effective sand control in the Adriatic Sea. SPE Drill Complet 15(4):293–297. https://doi.org/10.2118/67836-PA
72. Różańska S (2015) Rheology of wormlike micelles in mixed solutions of cocoamidopropyl betaine and sodium dodecylbenzenesulfonate. In: Colloids and surfaces a: physicochemical and engineering aspects, vol 482, pp 394–402. https://doi.org/10.1016/j.colsurfa.2015.06.045

73. Samuel M, Card RJ, Nelson EB, Brown JE, Vinod PS, Temple HL, Qu Q, Fu DK (1997) Spe 38622:553–559
74. Sharma G, Mohanty KK (2013) Wettability alteration in high-temperature and high-salinity carbonate reservoirs. SPE J 18(4):646–655. https://doi.org/10.2118/147306-PA
75. Shuler PJ, Lu Z, Ma Q, Tang Y (2016) Surfactant huff-n-puff application potentials for unconventional reservoirs. In: SPE—DOE improved oil recovery symposium proceedings. https://doi.org/10.2118/179667-ms
76. Smith, CE, Murphy DL (1974) Field evaluation of gas well stimulation methods—Sutton County, Texas, pp 83–90.https://doi.org/10.2118/4800-ms
77. Stanislav M (2016) SPE-182122-MS well stimulation techniques on carbonate reservoirs of Eastern Siberia : case history relevance of Eastern Siberia oil-and-gas province development characteristics of carbonate reservoirs geology. Structural-Textural Features of Interstitia
78. Stewart BR, Mullen ME, Howard WJ, Norman WD (1995) Use of a solids-free viscous carrying fluid in fracturing applications: an economic and productivity comparison in shallow completions. In: SPE—European formation damage conference, proceedings, EFDC, pp 379–392. https://doi.org/10.2118/30114-ms
79. Tang GQ, Firoozabadi A (2003) Wettability alteration to intermediate gas-wetting in porous media at elevated temperatures. Transp Porous Media 52(2):185–211. https://doi.org/10.1023/A:1023579706686
80. Tang GQ, Firoozabadi A (2000) Relative permeability modification in gas-liquid systems through wettability alteration to intermediate gas-wetting. SPE Reservoir Engineering (Society of Petroleum Engineers), B, pp 203–217. https://doi.org/10.2523/62934-ms
81. Taylor D, Santhana Kumar P, Fu D, Jemmali M, Helou H, Chang F, Davies S, Al-Mutawa M (2003) Viscoelastic surfactant based self-diverting acid for enhanced stimulation in carbonate reservoirs. In: SPE European formation damage control conference proceedings. https://doi.org/10.2523/82263-ms
82. Taylor KC, Nasr-El-Din HA (2002) Coreflood evaluation of in-situ gelled acids. In: SPE international formation damage control symposium proceedings. https://doi.org/10.2118/73707-ms
83. Taylor KC, Nasr-El-Din HA (2003) Laboratory evaluation of in-situ gelled acids for carbonate reservoirs. SPE J 3291–3300. https://doi.org/10.2118/71694-ms
84. Theodorakis PE, Smith ER, Craster RV, Müller EA, Matar OK (2019) Molecular dynamics simulation of the superspreading of surfactant-laden droplets. A review. Fluids 4(4):7–10. https://doi.org/10.3390/fluids4040176
85. Thomas TR, Wilkes TM (2010) The use of surfactants in lightweight drilling fluids. Surfactants. https://doi.org/10.1017/cbo9780511524844.009
86. Tlusty T, Safran SA, Menes R, Strey R (1997) Scaling laws for microemulsions governed by spontaneous curvature. Phys Rev Lett 78(13):2616–2619. https://doi.org/10.1103/PhysRevLett.78.2616
87. Tlusty T (2000) Self-assembling networks in microemulsions and dipolar fluids
88. Valko P, Economides MJ, Baumgartner SA, McElfresh PM (1992) The rheological properties of carbon dioxide and nitrogen foams. 1. https://doi.org/10.2118/23778-ms
89. Wang D, Butler R, Liu H, Ahmed S (2011) Surfactant formulation study for Bakken shale imbibition. In: Proceedings—SPE annual technical conference and exhibition, vol 1, pp 545–558. https://doi.org/10.2118/145510-ms
90. Wang P, Kang W, Yang H, Yin X, Zhao Y, Zhu Z, Zhang X (2017) PH-responsive worm-like micelles based on microstructural transition in a C22-tailed cationic surfactant-aromatic dibasic acid system. RSC Adv 7(60):37699–37705. https://doi.org/10.1039/c7ra07000d
91. Wang Y, Bai B, Gao H, He L, Zhao F (2008) Enhanced oil production through a combined application of gel treatment and surfactant huff'n'puff technology. In: Proceedings—SPE international symposium on formation damage control, vol 2, pp 896–909
92. Ward VL (1986) N_2 and Co_2 in the oil field: stimulation and completion applications. SPE Prod Eng 1(4):275–278. https://doi.org/10.2118/12594-pa

93. Watkins EK, Wendorff CL, Ainley BR (1983) A new crosslinked foamed fracturing fluid. In: Proceedings—SPE annual technical conference and exhibition. https://doi.org/10.2118/12027-ms
94. Wu S, Firoozabadi A (2010) Permanent alteration of porous media wettability from liquid-wetting to intermediate gas-wetting. Transp Porous Media 85(1):189–213. https://doi.org/10.1007/s11242-010-9554-3
95. Wu S, Firoozabadi A (2011) Simultaneous increase in gas and liquid relative permeabilities and reduction of high-velocity coefficient from wettability alteration. J Can Pet Technol 50(2):17–23. https://doi.org/10.2118/144637-PA
96. Wu X, Zhang Y, Sun X, Huang Y, Dai C, Zhao M (2018) A novel CO_2 and pressure responsive viscoelastic surfactant fluid for fracturing. Fuel 229(April):79–87. https://doi.org/10.1016/j.fuel.2018.04.081
97. Wu X, Song Z, Zhen J, Wang H, Yao L, Zhao M, Dai C (2020) A smart recyclable VES fluid for high temperature and high pressure fracturing. J Pet Sci Eng 190. https://doi.org/10.1016/j.petrol.2020.107097
98. Xu L, He K, Rane JP, Yin X, Neeves K (2015) Spontaneously imbibed fluids for increasing contact area between hydraulic fracturing fluids and formation matrix in liquids-rich shale plays. In: Society of petroleum engineers—SPE liquids-rich basins conferenceNorth America, LRBC. https://doi.org/10.2118/175536-ms
99. Xu L, Fu Q (2012a) Ensuring better well stimulation in unconventional oil and gas formations by optimizing surfactant additives. Soc Pet Eng Western Reg Meet 949–955.https://doi.org/10.2118/154242-ms
100. Xu L, Fu Q (2012b) Proper selection of surfactant additives ensures better well stimulation in the unconventional oil and gas formations. In: Society of petroleum engineers—SPE middle east unconventional gas conference and exhibition 2012, UGAS—unlocking unconventional gas: new energy in the middle east, pp 590–595
101. Xu L, He K, Nguyen C (2015) Insights into surfactant containing fracturing fluids inducing microcracks and spontaneously imbibing in shale rocks. In: Society of petroleum engineers—SPE/CSUR unconventional resources conference. https://doi.org/10.2118/175959-ms
102. Yan Z, Dai C, Zhao M, Sun Y, Zhao G (2016) Development, formation mechanism and performance evaluation of a reusable viscoelastic surfactant fracturing fluid. J Ind Eng Chem 37:115–122. https://doi.org/10.1016/j.jiec.2016.03.012
103. Yan Z, Dai C, Zhao M, Zhao G, Li Y, Wu X, Liu Y, Du M (2015) Multi-responsive wormlike micelles based on N-alkyl-N-Methylpiperidinium bromide cationic surfactant. J Surfactants Deterg 18(5):739–746. https://doi.org/10.1007/s11743-015-1718-2
104. Yang J, Cui W, Guan B, Lu Y, Qiu X, Yang Z, Qin W (2016) Supramolecular fluid of associative polymer and viscoelastic surfactant for hydraulic fracturing. SPE Prod Oper 31(4):318–324. https://doi.org/10.2118/175762-PA
105. Yang J, Guan B, Lu Y, Cui W, Qiu X, Yang Z, Qin W (2013) Viscoelastic evaluation of gemini surfactant gel for hydraulic fracturing. In: SPE—European formation damage conference, proceedings, EFDC, vol 2, pp 942–946. https://doi.org/10.2118/165177-ms
106. Yekeen N, Padmanabhan E, Idris AK, Chauhan PS (2019) Nanoparticles applications for hydraulic fracturing of unconventional reservoirs: a comprehensive review of recent advances and prospects. J Pet Sci Eng 178(February):41–73. https://doi.org/10.1016/j.petrol.2019.02.067
107. Yu M, Mahmoud MA, Nasr-El-Din HA (2009) Quantitative analysis of an amphoteric surfactant in acidizing fluids and coreflood effluent. In: Proceedings–SPE international symposium on oilfield chemistry, vol 2(1), pp 865–878
108. Zeng T, Miller CS, Mohanty KK (2020) Chemical blend-CO_2 huff-n-puff for enhanced oil recovery in shales. SPE/AAPG/SEG unconventional resources technology conference 2020, URTeC 2020, pp 1–10. https://doi.org/10.15530/urtec-2019-362
109. Zhang F, Adel IA, Park KH, Saputra IWR, Schechter DS (2018) Enhanced oil recovery in unconventional liquid reservoir using a combination of CO_2 Huff-n-Puff and surfactant-assisted spontaneous imbibition. In: Proceedings—SPE annual technical conference and exhibition, 2018-Sept. https://doi.org/10.2118/191502-ms

110. Zhou J, Cutler J, Morsy S, Morse A, Sun H, Qu Q (2014) Enhancing well stimulation with improved salt tolerant surfactant for Bakken formation. In: SPE—DOE improved oil recovery symposium proceedings, vol 3, pp 1542–1553. https://doi.org/10.2118/169141-ms

Corrosion Inhibition

Fundamental and Application of Surface Active Agents in Petroleum Industry as Corrosion Inhibitors

M. A. Hegazy and A. S. El-Tabei

Abstract Surface active agents (surfactants) are abundant in nature, manufacturing, and daily life. They find applications in various petroleum production operations. This chapter discusses many applications of surfactants as corrosion inhibitors in the petroleum industry like as oil and gas production, refinery processes, water flooding, acid retarders (acid corrosion inhibitor) in both acidization of oil and gas wells and chemical cleaning processes for heat exchangers, separator vessels, drain of oil and gas production stations, water flooding stations in addition to cooling tower treatment. The higher inhibition efficiency of surfactants as corrosion inhibitors is due to adequate solubility and rate of migration of the inhibitor from the bulk solution to the metal surface. In addition, strong binding of the surfactant head groups to the metal surface and self-assembly of hydrocarbon tails to form a hydrophobic barrier.

1 Introduction

Surface active agents (surfactants) are abundant in nature, manufacturing and daily life [1–5]. In various technological applications, such as lubrication, dispersion, corrosion inhibitors, stabilization, and enhanced oil recovery, they perform a valuable role. It can be argued that surface active agents are the world's most predominant chemicals.

M. A. Hegazy (✉) · A. S. El-Tabei
Egyptian Petroleum Research Institute (EPRI), Nasr City 11727, Cairo, Egypt

T. Solling et al. (eds.), *Surfactants in Upstream E&P*, Petroleum Engineering,
https://doi.org/10.1007/978-3-030-70026-3_14

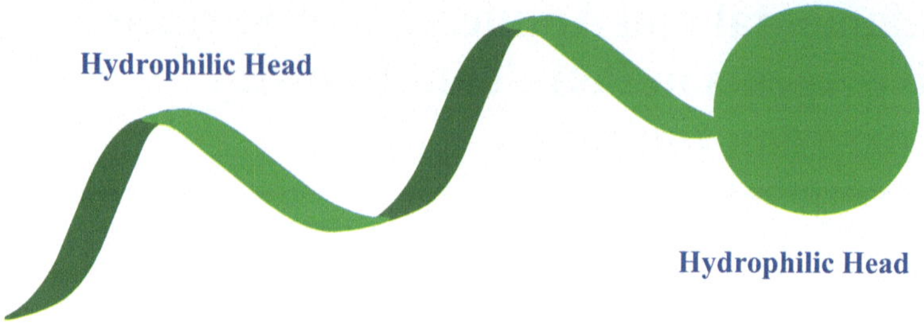

Fig. 1 Schematic illustration of a surface active agent structure

2 Surface Active Agent Fundamental

2.1 Surface Active Agents' Definition

A surface active agent is a material that has the property of adsorbing onto interfaces when present in a system at low concentration and of altering the surface or interfacial free energies of those surfaces (or interfaces) to a marked degree. The word "interface" implies a boundary among any two immiscible phases while the word "surface" refers to an interface where one phase is a gas, typically air [6].

Surface active agents are generally organic substances that are amphiphilic, i.e. they comprise both hydrophobic (tails or non-polar) and hydrophilic groups (heads or polar groups). The hydrophobic group is typically a long chain hydrocarbon radical, composed of 8–22 carbon atoms, straight or branched saturated or not, probably connected with paraffin, isoparaffin, naphthenic, aromatic, alkyl aromatic or alkyl naphtha aromatic hydrocarbons with varying compositions and molecular weights. The hydrophilic group may be ionic (cationic/anionic), nonionic or zwitterionic based upon the existence of their head group; Fig. 1 illustrates the surface active agent structure. Therefore, a surface active agent comprises both a part which is insoluble in water (oil soluble) and the other part which is soluble in water. It diffuses in water and adsorbs at air-water interfaces or the oil-water interface in case of combining water with oils. The hydrophobic water-insoluble group may extend into the air or oil phase from the bulk water phase, while the water-soluble head group stays in the water phase [7–15].

2.2 Classification of Surface Active Agents

Typically, surface active agents are categorized in accordance with the nature of the head group which found in the molecule. They are commonly categorized into five groups (anionic, cationic, amphoteric, nonionic and gemini surface active agents) [6, 15] as shown in Fig. 2.

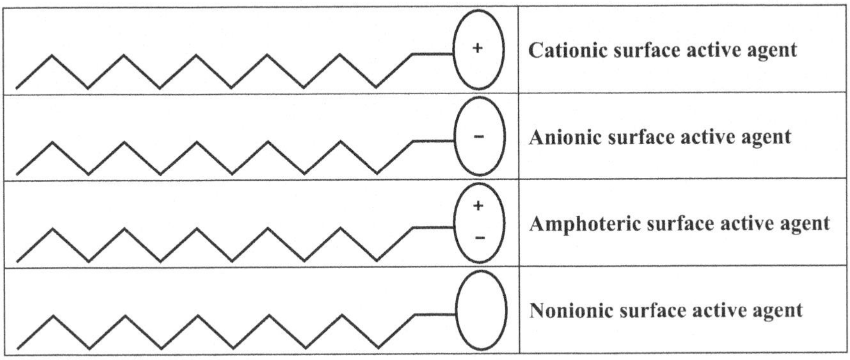

Fig. 2 Schematic illustration of the surface active agents classification

2.2.1 Anionic Surface Active Agents

The active surface portion of the molecules is negatively charged in an aqueous solution. This type of surface active agents accounts for around 50 percent of the world's production and is the most widely used. The most significant explanation for their popularity is the simplicity and low cost of production. Anionics are primarily utilized in detergent formulations and the strongest detergents are derived from alkyl and alkyl aryl chains in the range C12–C18. The most commonly used counter ions are sodium, potassium, calcium, ammonium, and different protonated alkyl amines. Sodium and potassium provide water solubility; while calcium and magnesium contribute to oil solubility. Amine/alkanol amine imparts both water and oil solubility [16, 17]. The example of anionic surfactants is illustrated in Fig. 3:

Fig. 3 Structure of some representative anionic surface active agents

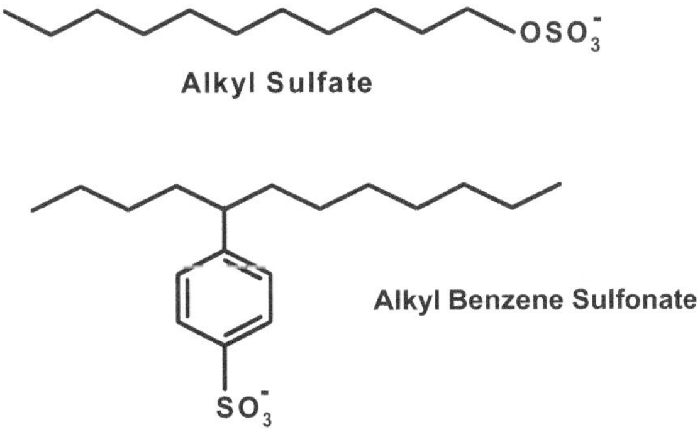

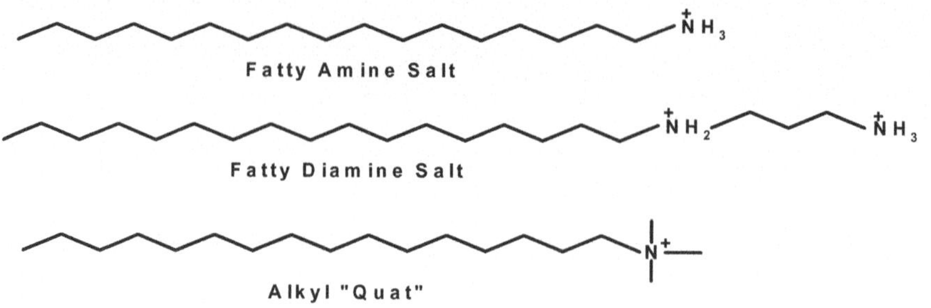

Fig. 4 Structure of some representative cationic surface active agents

2.2.2 Cationic Surface Active Agents

The hydrophilic moiety of a cationic surface active agent carries a positive charge when dissolved in aqueous media. As dissolved in water, cationic surface active agents are dissociated into an amphiphilic cation and an anion, often halogen-type. A very significant proportion of this molecules class is focused on quaternary nitrogen compounds such as alkylammonium halides. They are mostly utilized for specialized applications, such as, in disinfectant formulations, corrosion inhibitors, and fabric softeners, etc. [18]. The typical examples are given in Fig. 4.

2.2.3 Amphoteric or Zwitterionic Surface Active Agents

If both anionic and cationic dissociation occurs on a single surface active agent molecule, this is known as an amphoteric or zwitterionic. Some amphoteric surface active agents are pH-insensitive, while others are cationic at low pH and anionic at high pH, with an amphoteric behavior at moderate pH. They are usually very costly and so their use is limited to very special uses, such as cosmetics, which are mainly highly biologically compatible and low-toxic. Betaines, aminoacids, sulfobetaines and phospholipids are among the amphoteric examples. The structure of the amphoteric surface active agent is seen in Fig. 5.

Fig. 5 Structure of amphoteric surface active agent

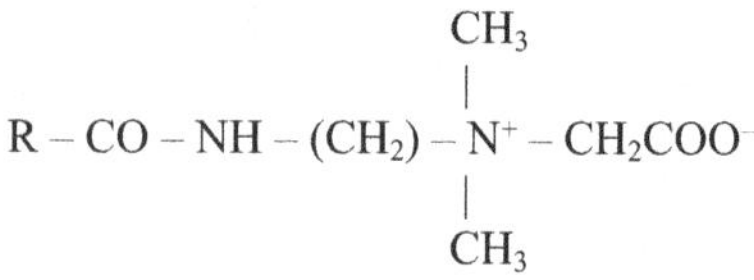

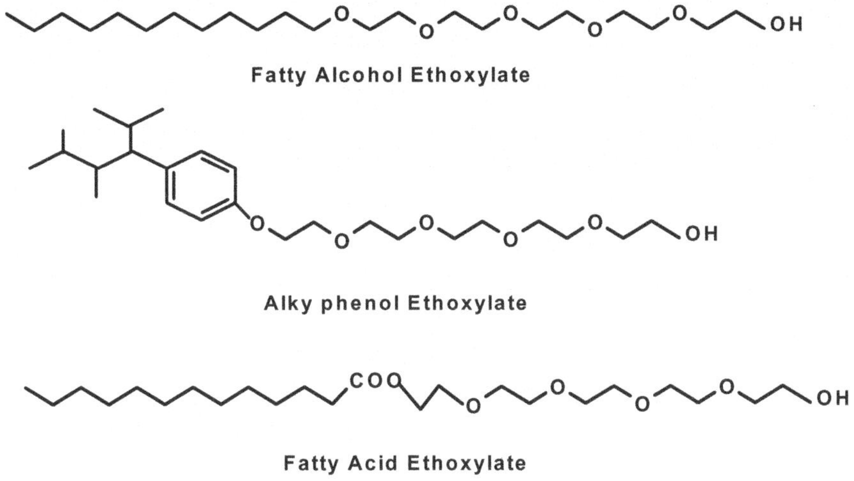

Fig. 6 Structure of some representative nonionic surface active agents

2.2.4 Nonionic Surface Active Agents

Nonionic surface active agents have a head which is polar, but uncharged. Ethoxylated polyethylenes are common examples. This is a surfactant with hydrophilic radicals which, when dissolved in water, do not ionize. Since this type is less affected by water hardness or electrolyte, it can be utilized with the other surfactant types. With the advantages of easy application such in penetration, emulsification, dispersion, detergency, etc., the use of nonionic surface active agents has recently increased considerably and has become a significant surface active agent parallel to anionic surface active agents. They can be categorized as an ether type, an ester type, an ether/ester type and other forms based on the nature of major bonding in the molecule [19, 20]. The common examples are illustrated in Fig. 6.

2.2.5 Gemini Surface Active Agents

A new surface active agent class recently created, named gemini or dimeric, has two hydrophilic groups and two tails per surface active agent molecule. Via a rigid or flexible spacer group of differing lengths (most commonly a methylene spacer or an oxyethylene spacer), the twin parts of the surface active agents are linked as shown in Fig. 7. The nature of spacer can vary greatly: short (2 groups of methylene) or long (until 12 groups of methylene), rigid (stilbene) or flexible (methylene chain) and polar (polyether) or nonpolar (aliphatic, aromatic) variants are possible [21]. The head part can be negative (sulfonate, phosphate, carboxylate & sulfate), positive (ammonium), or nonionic (polyether sugar). They consist of three structural elements, a hydrophobic, a hydrophilic group, and their spacer, which can

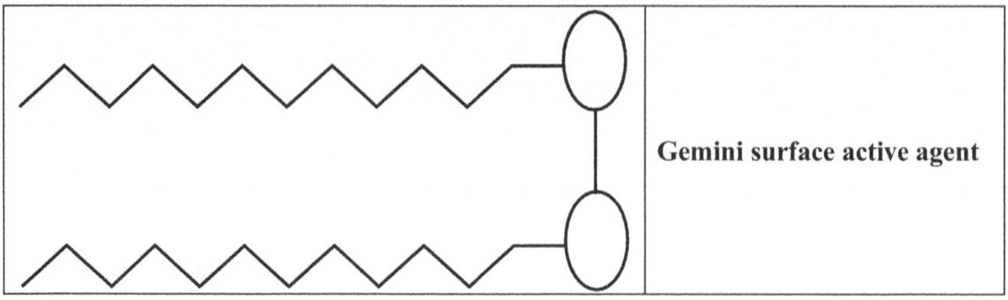

Fig. 7 Schematic illustration of a gemini surface active agent structure

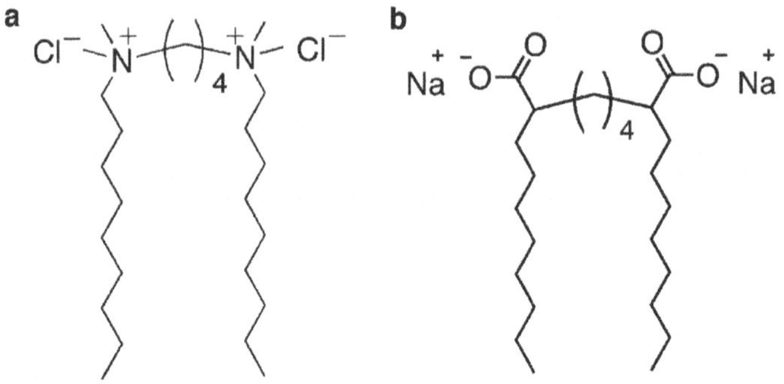

Fig. 8 Structure of some representative ionic gemini surface active agents

differ in order to alter the surfactant's properties. Although there are two equivalent head groups and two equivalent chains for the vast majority of geminis, unsymmetrical geminis are also identified. Geminis with three or more polar groups or tails have recently been synthesized. Compared to traditional single-headed and single-tailed surfactants, gemini surfactants show extraordinary many superior properties. They have lower critical micelle concentrations (CMC) values, higher surface properties and lower surface tension at the CMC, better solution properties, including hard water resistance, higher wetting times and lower Krafft points [22–26]. Gemini surface active agents are also classified into three categories: (i) Ionic gemini, (ii) Nonionic gemini and (iii) Heterogemini surfactants. Ionic gemini surfactants are also categorized as anionic gemini and cationic gemini surfactants as shown in Fig. 8.

2.3 *Micelle Formation*

Surface active substances are distinguished by a vital and efficient phenomenon called micellization. It is the tendency of the surfactant molecules to create a unified aggregative structure made up of well-arranged molecules at a characteristic concentration. That aggregate is known as the micelle as seen in Fig. 9. Micelles are thermo-

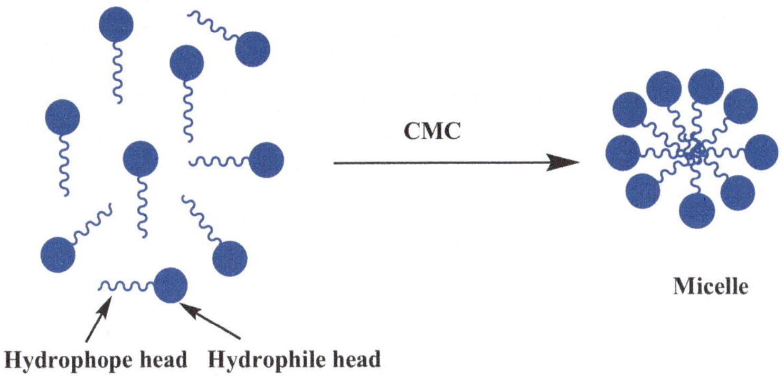

Fig. 9 Schematic illustration of the micelle formation

dynamically and geometrically stable phase, taking place in the bulk of the solution and created to minimize the polar /non-polar interaction between the surfactant and aqueous phase molecules [27–29].

At low concentrations, the surfactant molecules are ordered on the surface. The surface tension of the solution quickly reduces as more surfactants are applied, since more and more surfactant molecules are on the surface. The adding of surfactant molecules contributes to the creation of micelles as the surface becomes saturated. This concentration point is considered the critical micelle concentration. The concentration at which surfactants start to form micelle is known as the critical micelle concentration (CMC). When micelles shape in water, the tails create a core capable of encapsulating the oil droplet, and their (ionic /polar) heads provide an exterior shell which maintains good contact with water. As the surfactants are assembled in oil, the aggregate is pointed to as the reverse micelle. In the reverse micelle, the heads are in the core and the tails maintain contact with the oil. Figure 10 indicates the relationship between surface tension versus the surface active agent concentration.

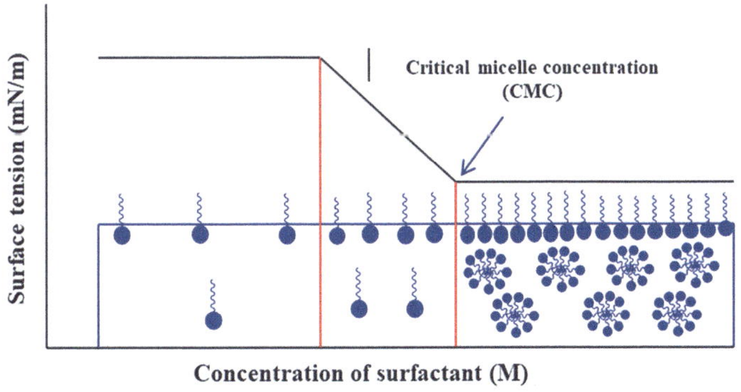

Fig. 10 The relation between surface tension versus the surface active agent concentration

2.4 *Factors Influencing the Critical Micelle Concentration*

A variety of factors e.g., hydrocarbon tail length, the nature of the counter ions, temperature, and the existence of salts and organic additives, influence the CMC values and thus surfactants have characteristic CMC values under given conditions [30, 31].

2.4.1 Hydrophobic Group

The increase of the surfactant molecules' hydrophobic chain length reduces their critical micelle concentration to lower values. The phenyl groups, unsaturation sites and branching have a diminishing impact on CMC. Substituents of other functional groups, particularly fluorine atom, through the hydrophobic chain significantly reduce CMC.

2.4.2 Hydrophilic Group

The hydrophilic group has a significant impact on the surfactant solutions' CMC. Ionic surfactants have much higher CMC's than those corresponded for the nonionics at identical hydrophobes. Zwitterionics tends to have a slightly smaller CMC than ionics with the same hydrophobic group. In addition, the hydrophilic groups' position affects the surfactants' CMCs. It has been observed that while the charge on the ionic hydrophilic group is near to the α-carbon, the CMC is higher. The CMC values declined with a reduction in the number of ethylene oxide units for nonionic surfactants.

2.4.3 Counter Ions

In the case of a typical cationic surfactant, the micellar size increases as the counter ion is altered in accordance with the sequence $Cl^- < Br^- < I^-$ and for a common anionic surfactant in accordance with $Na^+ < K^+ < Cs^+$. Organic counter-ion ionic surfactants, e.g. maleates have fewer CMCs and higher aggregation numbers than inorganic counter-ion.

2.4.4 Temperature Influence

At a characteristic temperature called the cloud point, aqueous solutions of many non-ionic surfactants become turbid. There is a rise in the size of the micellar at temperatures up to the cloud point and a resulting decline in CMC. The micellar properties of ionic surfactants are influenced a relatively minor by the temperature.

2.4.5 Electrolyte Addition

The addition of electrolyte to ionic surfactant solutions reduces the CMC and increases the micellar size. The electrolyte actually decreases the repulsive forces between the charged head groups at the micelle surface, allowing the micelle to form. The micelles of ionic surfactants may become non-spherical with high electrolyte concentrations.

3 Application of Surface Active Agents in Petroleum Industry

Surfactants find applications in various petroleum production operations. They perform a significant role, such as in acidization process, foam, transportation, corrosion inhibition, cement slurries, fracturing fluids, drilling fluids, chemical cleaning, demulsification, water flooding, chemicals, metal working fluids and steam flooding. They are also critical for improving the economics of production and for the complete petroleum recovery. This chapter discusses their application points, the logic behind their use and more general formulations.

This chapter discusses many applications of surfactants as corrosion inhibitors in in petroleum industry like:

1. Oil and gas production.
2. Refinery processes.
3. Water flooding.
4. Acid retarders (acid corrosion inhibitor) in both:
   a. Acidization of oil and gas wells
   b. Chemical cleaning processes for heat exchangers, separator vessels, drain of oil and gas production stations, water flooding system, ….. etc.).

3.1 Surface Active Agents as Corrosion Inhibitors in Oil and Gas Production

The active surface agents have extraordinary ability to impact the surface and interface properties. This capability has been widely used to prevent corrosion of metals and alloys. Adsorption is essential for inhibition of corrosion and the main action of the functional group of surfactants is to be adsorbed on the metal surface. It was found that the adsorption of the surfactant molecules on the metal surface is responsible for inhibiting the metal corrosion. The adsorption degree depends on the inhibitor's chemical structure, the metal surface condition, the corrosion media type, the nature of metal, and the mode of adsorption [32]. Corrosion inhibition of surfactant molecules may be due to either physical or chemical adsorption on the metal surface. Electrostatic interaction between the charged hydrophilic groups and the charging active centres on the metal surface contributes to physisorption. Chemisorption creates bonds with the metal surface. Bonding takes place by electron sharing or transferring between metal and inhibitor. The adsorbed surfactant molecules create a monolayer or bilayer and protect metallic surfaces from being corrodent attacked. The surfactant inhibitors possess many benefits compared with traditional corrosion inhibitors i.e. they have high inhibition performance, low cost, low toxicity and easy manufacturing [33, 34]. The surfactant molecules adsorption to a metal surface dramatically alters the corrosion-resistant properties and, as a result, studies of the relationship between adsorption and corrosion inhibition are of great significance [35–38]. The surfactants' ability to adsorb onto a metallic surface is typically closely related to their ability to aggregate and form micelles [39–43]. Consequently, when assessing the surfactant efficiency as a corrosion inhibitor, the CMC is an essential parameter. Surfactants have been shown to have a strong inhibition performance near their CMC value [44–48]. When the surfactant concentration on the solid surface is high enough, ordered structures are formed that lowers the corrosion reaction by blocking the metallic surface [49]. Below the CMC, individual surfactant molecules or monomers tend to adsorb on exposed interfaces, so interfacial aggregation reduces surface tension; which is associated with corrosion inhibition. Above the CMC, more than one monolayer covers the surface and forms a protective layer on the metal surface. Thus, any additional surfactant applied to the solution above the CMC will result in the micelles formation or several adsorbed layers on the surface. As a result, the surface tension and even the corrosion current density are not greatly altered above the CMC. Therefore, an efficient surfactant inhibitor is one that aggregates or adsorbs at low concentrations. In general, lower is the CMC of the surfactant; greater is its tendency to adsorb at the solid surface [50–57].

3.2 Surface Active Agents as Corrosion Inhibitors in Acidization of Oil and Gas Wells

Acidizing is a method used in the oil and gas production industry to prolong the usable life of an oil and gas well. The acidization process includes injecting acid into the well to dissolve the rocks that cover the well contours.

Acidizing promotes production rates via forming channels through the rock from which oil and gas will pass into the reservoir. Another advantage of acidization of a well is that it will dissolve some loose debris in the well.

The treatment of a reservoir formation with a stimulation fluid is containing a reactive acid. In sandstone formations, the acid interacts with the soluble substances in the formation matrix to expand the pore space. In carbonate formations, the acid dissolves the whole forming matrix. In either case, the matrix acidifying procedure increases the formation permeability to allow the increased production of reservoir fluids. Matrix acidization processes are preferably conducted at a high rate, but at treatment pressures below the fracture pressure of the formation. This causes the acid to enter the formation and prolong the depth of treatment while preventing disruption to the reservoir formation.

Progress in oil well acidizing research was primarily related to the use of additives to boost the efficiency of acid and to resolve a wide variety of problems. Great strides have been made in the development of:

1. Surfactants to reduce emulsion formation, improve wettability of rock by acid, to speed cleanup, and prevent sludge formation.
2. Inhibitors to reduce corrosion.
3. Buffering agents to control pH.
4. Retarders to retard reaction rates.
5. Friction reducers.

Many additives are applied to the acid to reduce the adverse effects of the acid and maximize the overall potency of the acidization treatment. Surfactants are usually used in the acid formulation in order to execute one or more essential functions. Surfactants encounter different chemical species that can influence their efficiency.

3.3 Surface Active Agents as Corrosion Inhibitors in Chemical Cleaning Process

While the hydrochloric acid solution (HCl) is an aggressive solution, it is widely used in large industrial applications such as cleaning, pickling, descaling, and etching of mild steel [58, 59]. Therefore, the use of corrosion inhibitors is a very important issue for saving machinery, particularly those made mostly of mild steel. The efficiency of the corrosion inhibitors depends upon their chemical composition, but a rich functional group is favored, which interacts with vacant d-orbital including phosphorous, oxygen, sulphur, nitrogen, double bond, aromatic ring and triple bond, for enhanced adsorption of corrosion inhibitors [55, 59]. Therefore, by recognizing the unusual adsorption affinity of the surfactant on the surface, the researchers focused on enriching the surfactant structure with some active core for use as corrosion inhibitors in addition to the hydrophobic surfactant tail showing an amazing role in shielding the surface from corrosion by creating a protective film from the aggressive medium.

4 Surface Active Agents as Corrosion Inhibitors

4.1 Cationic Surface Active Agents as Corrosion Inhibitors

Many studies have assessed the adsorption and corrosion effects of cationic surfactants on ferrous and nonferrous metals [60–64]. Quaternary ammonium salts have exhibited high inhibition efficiency in acidic media up to 97% for carbon steel corrosion [65]. Temperature effects on corrosion inhibition and the inhibitor efficacy at higher acid intensity have also been investigated. The newly Tri-cationic surfactant's inhibition effect on carbon steel corrosion was examined at various temperatures in 0.5 M H_2SO_4 [66]. The metallic corrosion inhibition of the mono-, di- and tetra-cationic surfactants in 1 M HCl solution on carbon steel were studied [67].

The cationic surfactants are preferably used as corrosion inhibitor in:

I. Oil & gas production.
II. Acidization processes.
III. Water injection system.
IV. Acid cleaning process.
V. Cooling tower treatment.

4.2 Anionic Surface Active Agents as Corrosion Inhibitors

Anionic surfactants were also utilized for a variety of metals and alloys in a wide range of corroding medium as effective corrosion inhibitors [68–70]. In the presence of anionic surfacant p-myristyloxy carbonyl methoxy-p′-sodium carboxylate-azobenzene, the corrosion inhibition for mild steel was examined by chemical and electrochemical techniques in 1 M H_2SO_4 [71].

The anionic surfactants are preferably used as corrosion inhibitor in:

I. Water injection system.
II. Acid cleaning process especially when H_2SO_4 is used.

The anionic surfactants can be used with nonionic and not cationic surfactants. The addition of anionic surfactants to cationic surfactants contributes to forming a colloidal solution. This isn't desirable.

4.3 Nonionic Surface Active Agents as Corrosion Inhibitors

Recently, one of the furthermost vital presentations of surfactants is as effective corrosion inhibitors in the oil industry [72]. The effective surfactants are which containing heteroatoms such as O, N, P, or S, π-electrons, and aromatic rings [73, 74]. The efficiency of surfactants as inhibitors has mainly depended on the ability of head group to get adsorbed onto the metal surface, forming sequential layers and preventing the metal surface from exposing to the corrosive medium [75]. The surfactant dose and contact time with metal surface have an effect on the efficiency of adsorbed film [76]. The nonionic surfactants have various features such as high inhibition efficiency, low toxicity, low cost, and ease of production [77]. The nonionic surfactants are preferably used in:

I. Acidization processes.
II. Water injection system.
III. Acid cleaning process.

In addition, the nonionic surfactants give better performance when added to cationic surfactants. The nonionic surfactants can be used with both nonionic and cationic surfactants. The performance of many nonionic surfactants as corrosion inhibitors has also been documented [78–80].

4.4 Gemini Surface Active Agents as Corrosion Inhibitors

Gemini or dimeric surface active agents are a new surfactant type, which has gained significant interest in corrosion inhibition in recent years [81–84]. Recent studies have focused on the gemini surfactants. The comparison between dimeric and monomeric surfactants used so far. The dimeric surfactants are better than monomeric surfactants in some points [85–87]:

1. Less effect the environment.
2. Less toxic.
3. More thermally stable.
4. More favorable surface properties.
5. Antimicrobial activity.

5 Mechanism of the Corrosion Inhibition by Surface Active Agent Inhibitors

Corrosion inhibition has complex mechanism and depends on the formation of mono or multidimensional protective layers on the metal surface. The protective nature of the surface layer depends on many factors: interaction between inhibitors and substrate, incorporation of the inhibitor in the surface layer, chemical reactions, electrode potentials, concentration of the inhibitor, temperature and properties of the corresponding surface, etc. The first stage in the action mechanism of the surfactants as corrosion inhibitors in aggressive media is adsorption of the surfactant molecules onto the metal surface. The adsorption process is influenced by the nature and the surface charge of the metal, the chemical structure of the surfactant, and the nature of the aggressive electrolyte. Adsorption of the surfactant molecules on the metal surface can be expressed according to the following equation:

$$\text{Surfactant}_{(\text{sol.})} + \text{nH}_2\text{O}_{(\text{ads.})} \rightarrow \text{Surfactant}_{(\text{ads.})} + \text{nH}_2\text{O}_{(\text{sol.})}$$

where n is the number of water molecules removed from the metal surface for each molecule of surfactant adsorbed. Adsorption of the surfactant molecules occurs

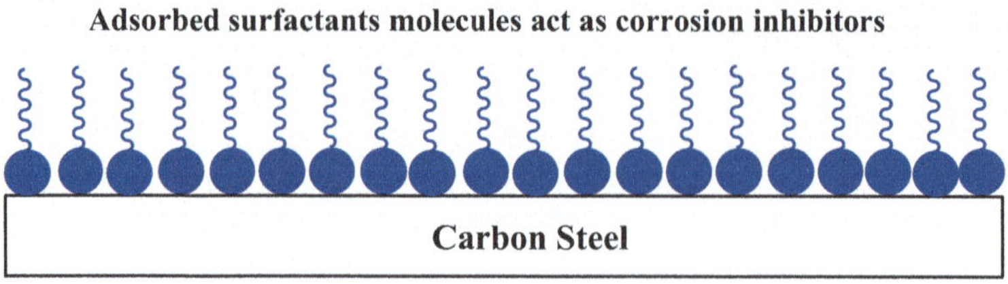

Fig. 11 Illustrate the adsorption of surface active agent corrosion inhibitor on pipeline

because the interaction energy between the surfactant molecules and the metal surface is higher than that between water molecules and the metal surface. So the inhibition effect by surfactants is attributed to the adsorption of the surfactant molecules via their functional groups onto the metal surface. The adsorption rate is usually rapid and hence, the reactive metal is shielded from the aggressive environment. At low concentrations, the inhibitor adsorbs parallel or tilted onto the steel surface. At the critical micelle concentration (CMC), monolayer coverage is achieved and the tail groups are parallel to each other and perpendicular to the metal surface which becomes hydrophobic [88]. The efficiency of the surfactants is higher than the other organic inhibitors because of the existing surfactant compounds at the interface between corrosive media and steel surface by more concentration.

The model suggests for an efficient corrosion inhibitor as follow:

1. Adequate solubility and rate of transport of the inhibitor from solution to the surface.
2. Strong binding of the surfactant head groups to the metal surface.
3. Self-assembly of head groups to form a dense and ordered layer.
4. Self-assembly of hydrocarbon tails to form a hydrophobic barrier as shown in Fig. 11.

Cationic [41, 89], anionic [90, 91], non-ionic [92, 93] and gemini surfactants [94–96] have been used as corrosion inhibitors for iron and steels in both HCl and H_2SO_4 solutions.

References

1. Evans DF, Wennerstrom H (1994) The colloidal domain where physics, chemistry, biology and technology meet. VCH, New York
2. Myers D (1992) Surfactant science & technology, 2nd edn. VCH, New York
3. Karsa DR (15–17th Apr 1986) Industrial applications of surfactants, the proceedings of a symposium organized by the North West Region of the Industrial Division of the Royal Society of Chemistry, University of Salford
4. Karsa DR (19–20th Apr 1989) Industrial applications of surfactants II, the proceedings of a symposium organized by the North West Region of the Industrial Division of the Royal Society of Chemistry, University of Salford

5. Karsa DR (16–18th Sept 1991) Industrial applications of surfactants III, the proceedings of a symposium organized by the North West Region of the Industrial Division of the Royal Society of Chemistry. University of Salford
6. Rosen MJ (2004) Surfactants and interfacial phenomena, 3rd edn. Wiley, New York
7. Rosen MJ (1989) Surfactants and interfacial phenomena, 2nd edn. Wiley, New York
8. Myers D (1988) Surfactant science and technology. VCH, New York
9. Mittal KL (1979) Solution chemistry of surfactants. Plenum, New York
10. Mittal KL, Fendler EJ (1982) Solution behavior of surfactants. Plenum, New York
11. Tadros ThF (1984) Surfactants. Academic Press, London
12. Mittal KL, Lindman B (1984) Surfactants in solution. Plenum, New York
13. Mittal KL, Bothorel P (1987) Surfactants in solution. Plenum, New York
14. Rosen MJ (1987) Surfactants in emerging technologies. Dekker, New York
15. Rosen MJ (1984) Structure/performance relationships in surfactants. American Chemical Society, Washington
16. Cross J (1977) Anionic surfactants, chemical analysis. Dekker, New York
17. Lucassen EH (1981) Anionic surfactants, physical chemistry of surfactant action. Dekker, New York
18. Jungermann E (1970) Cationic Surfactants. Dekker, New York
19. Cross J (1986) Nonionic surfactants, chemical analysis. Dekker, New York
20. Schick MJ (1987) Nonionic surfactants, physical chemistry. Dekker, New York
21. Hait SK, Moulik SP (2002) Curr Sci 82:1101
22. Tsujii K, Mino J (1978) J Phys Chem 82:1610
23. Zana R (2002) Adv Colloid Interface Sci 97:205
24. Keiper M, Angew JS (2002) Chem Int Ed 39:203
25. Zana R, Xia J (2002) J Colloid Interface Sci 248:203
26. Alargova RG, Kochijashky II, Sierra ML, Zana R (1998) Langmuir 14:5412
27. Shinoda K, Nakagawa T, Tamamushi I, Isemura T (1963) Colloidal surfactants, some physicochemical properties. Academic Press, New York
28. Friberg SE, Lindman B (1992) Surfactants in science and technology. Dekker, New York
29. Mittal KL (1977) Micellization, solubilization, and microemulsions. Plenum, New York
30. Mandal B, Nair BU (1991) J Phys Chem 95:9008
31. Sucheta A, Haque IU, Rusling JF (1992) Langmuir 8:1633
32. Mazhar A, Badaway WA, Abou-Romia MM (1986) Surf Coat Techol 29:335
33. El-Tabei AS, Elsharaky EA, El-Tabey AE (2016) Int J Electrochem Sci 11:10978
34. Kumar S (2008) E-J Chem 5:275
35. Bolzan E, Wakenge IB, Piatti RCV, Salvarezza RC, Arria AJ (2001) J Electroanal Chem 501:241
36. Stipnisek-Lisac E, Gazivoda A, Madzarac M (2002) Electrochim Acta 47:4189
37. Sahin M, Bilgic S, Yilmaz H (2002) Appl Surf Sci 195:1
38. Bastidas JM, Pinilla P, Cano E, Polo JL, Miguel S (2003) Corros Sci 45:427
39. Free ML (2002) Corros Sci 44:2865
40. Asefi D, Mahmoodi NM, Arami M (2010) Colloids Surf A 355:183
41. Qiu LG, Xie AJ, Shen YH (2005) Appl Surf Sci 246:1
42. El Achouri M, Hajji MS, Kertit S (1995) Corros Sci 37:381
43. Wang WL, Free ML (2004) Corros Sci 46:2601
44. Malik MA, Hashim MA, Nabi F, AL-Thabaiti SA, Khan Z (2011) J Electrochem Sci 6:1927
45. Frignani A, Tassinari M, Meszaro L, Trbanelli G (1991) Corros Sci 32:903
46. El Achouri M, Kertit S, Gouttaya HM, Neiri B, Bensouda Y, Perez L, Infate MR, El Kacemi K (2001) Prog Org Coat 43:267
47. Algaber S, El Nemma EM, Saleh MM (2004) Mater Chem Phys 86:26
48. Chebabe D, Chikh ZA, Dermaj A, Rhattas K, Jazouli T, Hajjaji N, El Mdari E, Srhiri A (2004) Corros Sci 46:2701

49. Houyi M, Shenhao C, Bingsheng Y, Shiyong Z, Xiangqian L (2003) Corros Sci 45:867
50. Hegazy MA, Azzam EMS, Kandil NG, Badawi AM, Sami RM (2016) J Surfact Deterg 19:861–871
51. Bentiss F, Traisnel M, Lagrenée M (2000) Corros Sci 42:127
52. Bentiss F, Lebrini M, Lagrenée M (2005) Corros Sci 47:2915
53. Azzam EMS, Hegazy MA, Kandil NG, Badawi AM, Sami RM (2015) Egypt. J Petrol 24:493–503
54. Paria S, Khilar KC (2004) Adv Colloid Interface Sci 110:75
55. Branzoi V, Branzoi F, Baibarac M (2000) Mater Chem Phys 65:288
56. Atkin R, Craig VSJ, Wanless EJ, Biggs S (2003) J Colloid Interface Sci 266:236
57. Popova A, Christov M, Raicheva S, Sokolova E (2004) Corros Sci 46:1333
58. Aslam R, Mobin M, Aslam J, Lgaz H, Chung I-M (2019) J. Mat. Res. Tech. 8:4521–4533
59. Hegazy MA, Abd El Rehim SS, Badawi AM, Ahmed MY (2015) RSC Adv 5:49070–49079
60. Hegazy MA, Atlam FM (2016) J Mol Liq 218:649–662
61. Rajendran S, Apparao BV, Palaniswamy N (2000) Anti-Corros Methods Mater 47:11
62. Wang L, Yin GJ, Yin JG (2001) Corros Sci 43:1197
63. Al-Rawashdeh NAF, Maayta AK (2005) Anti-Corros Methods Mater 52:160
64. El Maghraby A, Soror TY (2010) Adv Appl Sci Res 1:156
65. Soror TY, El-Ziady MA (2003) Mater Chem Phys 77:697
66. Hegazy MA, Ahmed HM, El-Tabei AS (2001) Corros Sci 53:671
67. El-Tabei AS, Hegazy MA, Bedair AH, Sadeq MA (2014) J Surfact Deterg 17:341
68. Hegazy MA, Abd El-Rehim SS, Badr EA, Kamel WM, Youssif AH (2015) J Surfact Deterg 18:1033
69. Kellou-Kerkouche F, Benchettara A, Amara S (2008) Mater Chem Phys 110:26
70. Villamil RFV, Corio P, Agostinho SML, Rubim JC (1999) J Electroanal Chem 472:112
71. Luo H, Guan YC, Han KN (1998) Corro. Sci. 54:619
72. Khadom AA, Abd AN, Ahmed NA (2018) S Afr J Chem Eng 25:13–21
73. Bedaira MA, El-Sabbah MMB, Fouda AS, Elaryian HM (2017) Corros Sci 128:54–72
74. Adewuyi A, Göpfert A, Wolff T (2014) Ind Crops Prod 52:439–449
75. Abd-Elaal AA, Shaban SM, Tawfik SM (2017) J Assoc Arab Univ Basic Appl Sci 24:54–65
76. Bin-Hudayb NS, Badr EE, Hegazy MA (2020) Materials 13:2790–2817
77. Hegazy MA, Sami RM, Labena A, Wadaan MAM, Hozzein WN (2020) Mater Sci Eng, C 110:110673
78. Quej-Aké L, Contreras A, Aburto J (2015) Int J Electrochem Sci 10:1809
79. Hegazy MA, El-Tabei AS, Bedair AH, Sadeq MA (2012) Corro Sci 54:219
80. Hegazy MA, El-Tabei AS, Ahmed HM (2012) Corros Sci 64:115
81. Sharma HK, Quraishi MA (2007) Indian J. Chem Tech 14:494
82. Hegazy MA, El-Tabei AS, Bedair AH, Sadeq MA (2015) RSC Adv 5:633
83. El-Tabei AS, Hegazy MA (2015) Chem Eng Commun 202:851
84. Hegazy MA, El-Tabei AS (2013) J Surfact Deterg 16:221
85. Menger FM, Littau CA (1991) Gemini-surfactants: synthesis and properties. J Am Chem Soc 113:1451–1452
86. Zana R (1996) Gemini (dimeric) surfactants. Curr Opin Colloid Interface Sci 1:566–571
87. Kunieda H, Masuda N, Tsubone K (2000) Langmuir 16:6438–6444
88. Stoyanova E, Sokolova EI, Raicheva SN (1997) Corros Sci 39:1595
89. Driver R, Meakins RJ (1977) Br Corr J 12:46
90. Hosseini M, Mertens SFL, Arhadi MR (2003) Corr Sci 45:1473
91. Mu GN, Zhao TP, Liu M, Gu T (1996) Corrosion 52:853
92. Elachouri M, Hajji MS, Salem M, Kertit S, Aride J, Coudert R, Essassi E (1996) Corrosion 52:103
93. Osman MM, Shalaby MN (1997) Anti-Corros Methods Mater 44:318

94. El-Tabei AS, Hegazy MA (2013) J Surfact Deterg 16:757
95. El Achouri M, Infante MR, Izquierdo F, Kertit S (2001) Corros Sci 43:1935
96. Tammam RH, Fekry AM, Saleh MM (2016) Int J Electrochem Sci 11:1310

Hydrate Inhibition

The Role of Surfactants in Gas Hydrate Management

Jyoti Shanker Pandey, Adam Paul Karcz, and Nicolas von Solms

Abstract This chapter provides an introductory understanding of the role of surfactants in the formation of gas hydrates. The main theories that have been developed over the past decades are discussed with support from computational aspects that have become increasingly useful in this regard. Particularly for surfactants, the structure-property relations are key in the full understanding of their behavior in the context of hydrate formation kinetics and equilibria, which are presented with evidence from various studies. Furthermore, surfactants can benefit from co-promoters that may be utilized in hydrate formation, so we present some details to highlight the importance of their interactions. More recently, bio-based surfactants have gained interest out of environmental concerns, and we showcase some of the most interesting cases of their implementation. Although there have been many examples of how gas hydrates can be used for cold storage, hydrogen storage, and other industrial applications, the usage of surfactants or other additives has not been well supported with clear fundamental understandings. Thus, there have been endeavors to gain these insights via computational tools that span different scales, like quantum mechanics and molecular dynamic simulations. The use of these tools is explained with examples. Combining all these different aspects, we hope to provide some understanding of the role of surfactants in current and emerging hydrate management technologies.

J. S. Pandey · N. von Solms (✉)
Center for Energy Resource Engineering (CERE), Department of Chemical Engineering, Technical University of Denmark, 2800 Kgs. Lyngby, Denmark
e-mail: nvs@kt.dtu.dk

A. P. Karcz
Department of Energy Conversion and Storage, Solid State Chemistry, Technical University of Denmark, 2800 Kgs. Lyngby, Denmark

T. Solling et al. (eds.), *Surfactants in Upstream E&P*, Petroleum Engineering,
https://doi.org/10.1007/978-3-030-70026-3_15

1 Introduction

Gas hydrates are ice-like crystalline compounds having different guest molecules surrounded by water cages formed at high pressure but low temperature conditions. These hydrates look like ice; however, they have different physical and chemical properties [1]. The guest molecules could be gases, such as small molecules (e.g., CH_4 or CO_2) or large molecules like propane. Guest molecules are entrapped and stabilized within cages due to intermolecular forces, and cages are formed by water molecules connected to each other via hydrogen bonding. Based on three-dimensional orientation, the hydrate structure can be divided into three types: s-I, s-II, and s-H. Gas hydrate formation is a crystallization process which occurs in different stages, mainly nucleation, growth and agglomeration [2]. Gas hydrates can store a large volume of gas such that 1 m^3 of hydrate volume can store up to 163 m^3 of gas. Gas hydrates also offer additional advantages over conventional gas storage technologies, such as being environmentally friendly, non-explosive, and low mantainence. Gas hydrates applications can be divided into four categories with respect to their origin and application, represented in Fig. 1.

Gas hydrates are found in nature within sediments and are considered a source of gas supply for future generation. These untouched gas hydrate reservoirs are in the continental shelf in the marine environment as well as cold regions on land, such as permafrost regions in Alaska, Canada, China, and Russia. It is estimated that these reservoirs contain 1.5×10^{16} m^3 of gas [3, 4]. Hydrates are also believed to present on other planets, including Mars and Saturn [5]. Gas hydrates formation in oil and gas pipeline is considered an engineering challenge in the petroleum industry as hydrate formation could block and damage the pipelines [6, 7]. Studies in this context are focused on preventing hydrate formation by injecting chemicals known has hydrate inhibitors. Based on the mechanism, these inhibitors are categorized as high dosage thermodynamic hydrate inhibitors (THI, 20–40 wt%) and low dosage hydrate inhibitors (LDHI, 0.1–5 wt%). Due to environmental impact consideration,

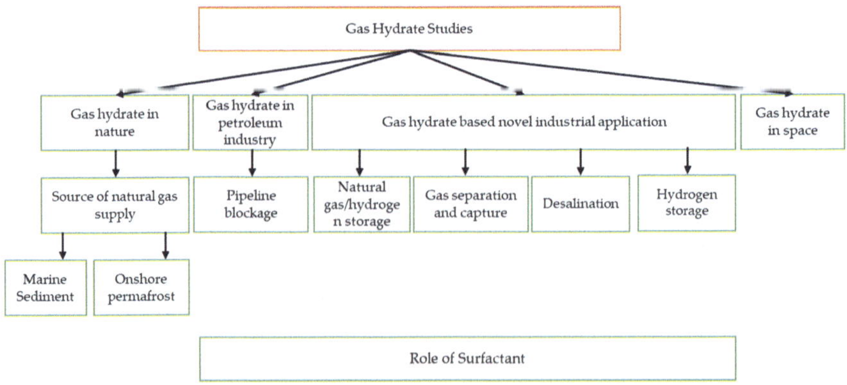

Fig. 1 Role of surfactant in different gas hydrate based applications

LDHI is used more frequently than THI [8]. The most common surfactants used under LDHI type are quaternary ammonium surfactants characterized having n-butyl or n-pentyl functional groups and long alkyl chains [9].

Opposite to hydrate inhibitors, a different class of chemicals, known as promoters, improve formation kinetics needed in different industrial applications [10], such as natural gas storage [11], hydrogen storage [12], hydrate-based pre- and post-combustion CO_2 separation, capture, storage, transport [13–15], and hydrate-based desalination [16].

During laboratory-based hydrate studies, gas hydrate is formed at gas-water interface as gas solubility in water is slow and gas saturated water is crystallization process is very slow to happen. A thin hydrate layer forms first at gas-liquid interface which later grows towards the gas phase; however, growth is limited by mass transfer of gas molecule through the thin hydrate layer [17]. To move toward commercialization, the kinetics of gas hydrate formation require drastic improvement. This could be achieved through mechanical techniques as well as adding surface-active chemicals that could reduce the mass transfer barrier and enhance the gas-liquid contact area without changing the hydrate phase equilibrium [18]. These surface-active agents are used in small concentrations and known as hydrate promoters. Surfactants [19] and hydrophobic amino acids [20] are known to be hydrate promoters. Recently low dosage methanol is also considered as hydrate promoter due to its near similar behaviour as a surfactant at low concentration [21].

This chapter is focused on discussing the role of surfactants as surface-active agents during gas hydrate formation and dissociation. Formation kinetics depends on guest molecule, pressure, temperature, and reactor design. Surfactant performance is also system-dependent, including the difference in reactor design, pressure, and temperature conditions, the difference in hydrate forming gas mixture as well as the role of supporting material due to difference in thermal conductivity and surface-to-volume ratio [22–24]. In this chapter, we discuss the available mechanisms and current status of surfactant application in gas hydrate management briefly.

2 Role of Surfactant Molecular Structure on Hydrate Promotion

Surfactants are known to affect the kinetics of the hydrate formation. Kinetics of hydrate formation can be divided into different stages, starting from dissolution, nucleation, growth, and agglomeration [1]. Surfactants facilitate faster nucleation by reducing the surface free energy by absorbing into the aqueous-hydrocarbon (gas/liquid) interface [25]. Surfactants also enhance the mass transfer by improving the hydrocarbon solubility into the water. Surfactants play an essential role at the gas-liquid and liquid-hydrate interface. A surfactant could occupy the area at the interface and could also hinder hydrate formation. In this context, sodium dodecyl sulfate is considered to be the most effective surface-active agent to enhance nucleation

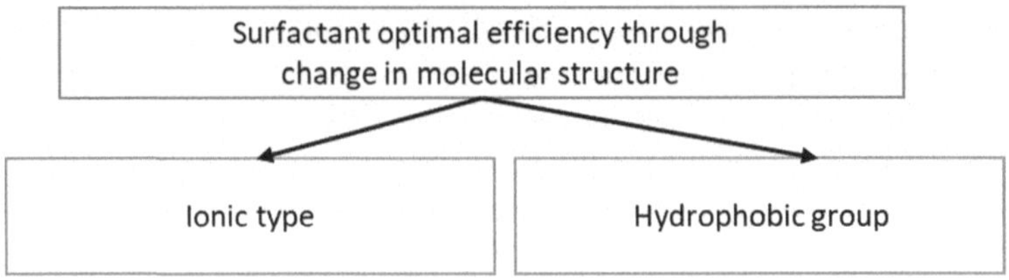

Fig. 2 Role of molecular structure of surfactant optimal efficiency during gas hydrate formation

and growth phases [26–28]. Surfactants are composed of hydrophobic "tails" and hydrophilic headgroups within a single molecule, which determine the surfactant properties. Surfactant molecules could diffuse from the bulk phase to the liquid-gas interface, such that the hydrophilic end stays in the liquid phase while the hydrophobic end stays in the gas phase. This would lead to a change in surface tension, modification in contact angle, and change in surface charge and surface viscosity [29]. At a given concentration, surfactant molecules bind together in different shapes and orientations, called micelles [30]. Above the critical micelle concentration (CMC), the hydrophilic part covers the hydrophobic group and supports the organic compound (methane or CO_2) solubility [31]. Surfactants are well used chemical substances to enhance surface activity that control spreadability, wetting, foaming, etc. [19]; however, it is still unclear about the key mechanism responsible and the role of concentration during formation mechanism. Insights into the molecular structure of surfactant can enhance our understanding of its role in hydrate formation kinetics. The key component of the surfactant structure is given in Fig. 2.

2.1 *Ionic Type*

Surfactants tested for hydrate formation are from anionic, cationic, and nonionic natures. Studies confirmed that anionic surfactants produced better promotion compared to cationic and nonionic ones at lower concentration (100–500 ppm). At higher concentration (>800 ppm), the difference in promotion ability decreased [32, 33]. Among three homologuous anionic surfactants, Sodium dodecyl sulfate (SDS), Sodium tetradecyl sulfate (STS) and Sodium hexadecyl sulfate (SHS), it was found that SDS is most effective above 1000 ppm concentration for methane hydrate formation while STS has shown same promotion behaviour at 100 ppm [34]. SHS was not effective compared to SDS and SHS. In another study on the comparative performance of different surfactants (anionic/cationic/nonionic) during CO_2 hydrate formation, anionic surfactant SDS was the most effective among all three. Nonionic surfactants are more effective compared to cationic surfactant [35]. In another study, when *n*-dodecyltrimethylammonium chloride (DTAC) and Tween 20 were used for the Tetrahydrofuran (THF/H_2) system and THF/Methane system, both surfactants showed different behavior. This led to the conclusion that the role of surfactant during

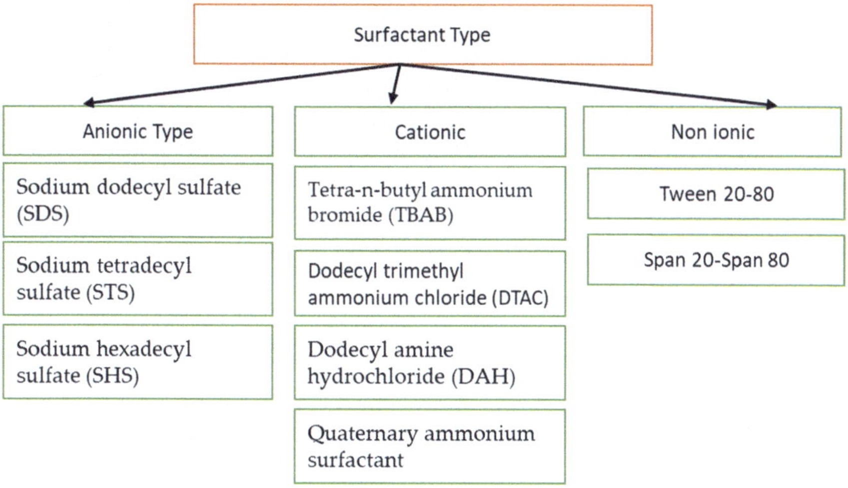

Fig. 3 Popular surfactans used in gas hydrate studies

hydrate formation is dependent on guest molecules as well as the system (single guest molecule vs. mixed hydrate) [36]. Figure 3 provides the details of the key surfactant in each category. Apart from traditional classification, novel surfactants, including biosurfactants and Gemini surfactant, have also been used in hydrate-based studies [37–39]. Figure below describe some popular surfactants used frequently in gas hydrate studies.

2.2 *Properties of Hydrophobic Group*

Size of the hydrophobic and hydrophilic groups controls the surfactant properties, such as interfacial tension. Large groups have lower interfacial tension than smaller groups [40, 41] Properties of the hydrophobic group affect the surfactant properties and are of greater research interest [42]. Kumar et al. [19] have provided a summary of the properties of the hydrophobic group and their effect on the surfactant properties. Three key elements include the change in the length of the hydrophobic group, branching, and unsaturation in the hydrophobic group and the presence of an aromatic nucleus in the hydrophobic group. The hydrophobic group-controlled solubility of surfactant in water and organic solvent, biodegradability and packing of surfactant at the interface. With an increase in the length of the hydrophobic group, the solubility of surfactant in water decreases, but in organic solvent it increases. Apart from that, biodegradability and surfactant absorption at the interface also increases as the length of the hydrophobic part increases [19]. Okutani et al. [34] studied the effect of alkyl chain length, using three surfactants (SDS, STS, and SHS) having the same headgroup ($-OSO_3^-Na^+$) but different carbon chains (12, 14, or 16, respectively).

They concluded that surfactant with larger carbon number could be useful even at lower concentrations. On the other hand, Dicharry et al. [43] have tested the effect of carbon chain for sulphate-based surfactants and found out that higher carbon chain-based surfactants could readily absorb on hydrate surfaces by forming hemimicelles, hence promoting hydrate formation. The difference in carbon chain length creates different solubility and packing area. By looking at the available research, it can be concluded that sulphate-/sulfonate-based surfactants have shown the best promotion capabilities with 12–14 carbon chain as an optimal solution. As per adsorption and mass transfer theory, an increase in chain length could decrease the surfactant hydrate promotion efficiencies.

3 General Theories Behind the Surfactant-Based Promotion

Kalogerakis et al. [18] were one of the first to study the role of surfactants during hydrate formation. During nucleation, hydrate film formed at the gas-liquid interface, which further isolates gas phase from the liquid phase and allows only gas molecule reaching to liquid phase through diffusion. The exact mechanism behind the role of surfactant during hydrate formation is not yet agreed upon. Many theories have been proposed. In the following section, we have discussed some well-known theories available in the literature and describe in the Fig. 4 [31].

3.1 Micelles Formation Theory

In one notable research proposing micelles theory, critical micelle concentration (CMC) was reported as 242 ppm using SDS during natural gas hydrate formation

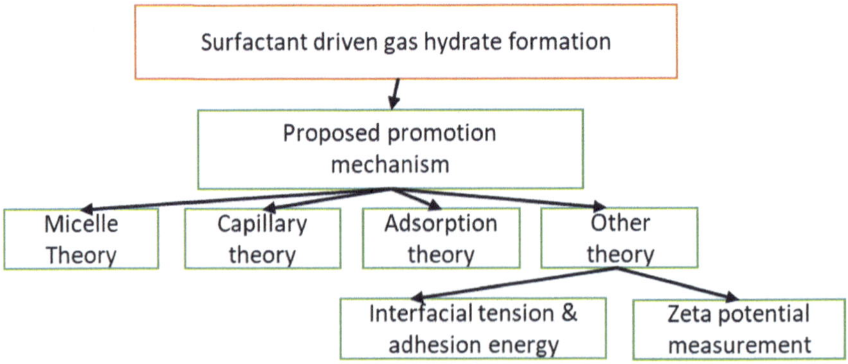

Fig. 4 Surfactant based key hydrate formation theories

[28]. Experiments also suggest that the CMC value of natural gas-water solution decreases as the pressure decreases [28]. Other studies suggested that when surfactant SDS concentration was above the CMC, a 700-fold increase in the rate of hydrate formation was observed due to enhanced guest molecule solubilities [44]. Change in gas solubility due to change in surfactant concentration was also measured for the ethylene and SDS system, both at ambient pressure and under hydrate formation conditions [45]. Solubility behavior for ethylene and methane was similar under hydrate formation conditions in the presence of SDS [46]. Presence of CMC for SDS during methane hydrate formation was also confirmed [47]. The CMC-based hydrate formation mechanism is explained in the figure below. However, some researchers have disputed this theory, and citing that a decrease in the rate of methane hydrate formation above CMC was also observed for cationic and anionic surfactants [48, 49]. Few studies also suggested that SDS at even very low concentration (10 ppm) could promote methane hydrate formation [17] or a single molecule itself can assist in hydrate formation [50]. In another study, it was observed that the CMC value of surfactant does not depend on the hydrate formation conditions and remains unchanged [51].

3.2 *Capillary Driven Growth*

Presence of capillary action during hydrate formation was demonstrated by Watanabe et al. using SDS and Difluoromethane (HFC-32) gas [52]. Visual observation suggested that crystals were initially formed both at the liquid-gas interface as well as reactor sidewall at different spots. Thereafter, when crystals grew in size, coalesence took place with each other, and crystals started to grow at the reactor sidewall. After that, hydrate grew in a downward direction in the solution phase to maintain contact and cause a decrease in the solution surface. Additionally, rippling motion of liquid at the reactor wall also confirmed the presence of capillary action during the formation. Other studies also confirmed moving of gas-liquid interface in upward direction along the reactor wall in the presence of SDS and did not change the hydrate thermodynamics [17, 53]. This behaviour was not observed in the absence of surfactant for the pure water-gas system.

The capillary mechanism is said to be caused by either change in hydrate morphology [54–56] or due to change in the wettability [57–60]. Wang et al. have used anionic surfactants SDS, Sodium dodecyl benzene sulfonate (SDBS) and Sodium dodecyl sulfonate (SDSN) having different wettability on the reactor side and found that SDBS performed poorly compared to other surfactants due to weaker wettability on the reactor wall [60]. Wang et al. [59] have also achieved directional hydrate growth by controlling and varying the wettability of the solid surface inside the glass tube. Wettability of surfactant solution is also controlled by surfactant concentration. SDS wettability also changed due to the difference in its concentration [57]. NMR and Raman studies also confirmed that SDS has two different growth mechanisms at 25 ppm and 500 ppm. At higher concentration, water converted into an intermediate solid-state and then combined with methane gas. This was not observed at the lower concentration [61].

The key limitation of this theory is that most of the observations regarding the capillary mechanism are for anionic surfactants, including SDS, STS, and SHS [34, 53] and for specific guest molecules, methane and ethane. In the presence of the CO_2 molecule and an anionic surfactant such as SDS, no capillary mechanism was observed at high driving force [62]. Other studies involving various surfactants, such as lithium dodecyl sulfate (LDS), dodecyl benzene sulfonic acid (DBSA), sodim oleate (SO), dodecyl alcohol ethoxylates (AEO), cetyltrimethylammonium bromide (CTAB) from cationic and non-ionic categories, did not confirm the capillary mechanism responsible for hydrate growth [63–65]. Therefore, it can be concluded that capillary-based hydrate growth is dependent on surfactant type, surfactant concentration, and guest molecules. For example, Molokitina et al [62] performed microscopic investigation of the CO_2 hydrate formation mechanism in bulk water phase in the presence of SDS under different mass transfer barrier and visualized the hydrate growth pattern as shown in Fig. 5. It can be seen that driving force changes the mechanism of gas hydrate formation at the gas/liquid interface, such that capillary-driven movement is observed at low driving force. Further research is required to understand the factors influencing the capillary mechanism and how this mechanism could be improved to achieve enhanced growth.

3.3 Adsorption Theory

Theory of surfactant adsorption on the hydrate surface was first proposed based on zeta potential measurements [66, 67] and is proposed as a cause behind improved hydrate formation kinetics [26, 66, 68]. Different types of surfactants were absorbed through a different mechanism and due to difference in surfactant concentration. For example, an ionic surfactant is considered to absorb on the hydrate surface under the influence of electrostatic forces while a non-ionic surfactant adsorbs through hydrogen bonding [69]. At concentrations below the CMC, adsorption behavior follows Henry's law, confirmed by Scamerhorn et al. [70]. Above the CMC, surfactants form hemimicelles which are an aggregate form of surfactant due to tail-tail interactions between surfactant molecules. Also above the CMC, adsorption is independent of the concentration [71].

3.4 Interfacial Tension and Adhesion Energy

This theory is based on the few studies that suggest that during the gas hydrate formation, the contact angle between the liquid-gas interface and solid-state change in the presence of the surfactant [53, 72]. Addition of surfactant decreases the surface tension of the aqueous phase, thus decreasing the contact angle. This causes creation

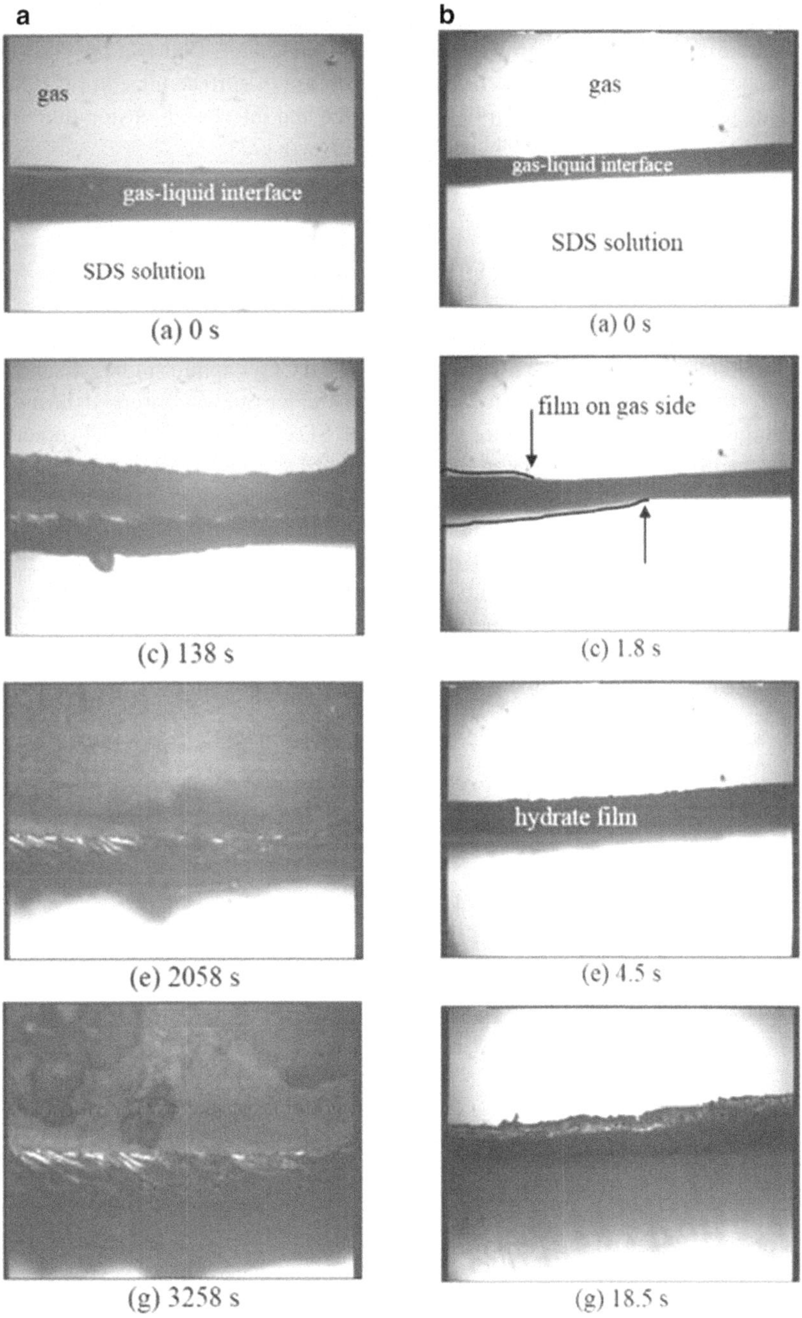

Fig. 5 **a** CO_2 hydrate film propagation along the gas-liquid interface (capillary-driven) in the presence of SDS (1000 ppm) and low mass transfer driving force. **b** CO_2 hydrate film propagation along the gas-liquid interface on both gas and liquid sides under high mass transfer driving force (not capillary-driven) in the presence of SDS (1000 ppm) [62]

of a film-like interface along the reactor wall and becomes a preferred location for hydrate nucleation and growth [53]. In the presence of surfactant, the solid surface becomes more water-wet due to the decrease in surface tension [72]. Song et al. suggested that the reduction in contact force and interfacial tension due to the presence of surfactant caused enhance hydrate growth [73].

3.5 *Zeta Potential Measurement*

Zeta potential measurement was used to explain the synergy between THF and SDS as reported by Torre et al. [74] for gas hydrate based CO_2 capture studies. In another study, Torre er al. [75] has suggested no mass transfer barrier was observed during the gas to liquid mass transfer in the presence of both SDS and SDS+THF. Zeta potential measurement has suggested that in the mixture of SDS and THF, THF hydrate stays dispersed due to electrostatic repulsion caused by adsorption of DS- anions on the hydrate surface hence porous texture allow CO_2 diffusion into the liquid phase [67].

During our recent study focused on kinetics of methane hydrate formation in the presence of SDS and effect of SDS concentration on the formation kinetics. We observed change in formation kinetic behavior around 2000–3000 ppm concentration. To suggested that behavior can be explained in terms of dual effect of absorption and surface tension. Trend in key kinetic properties as a function of SDS concentration is illustrated in Figs. 6 and 7 [47].

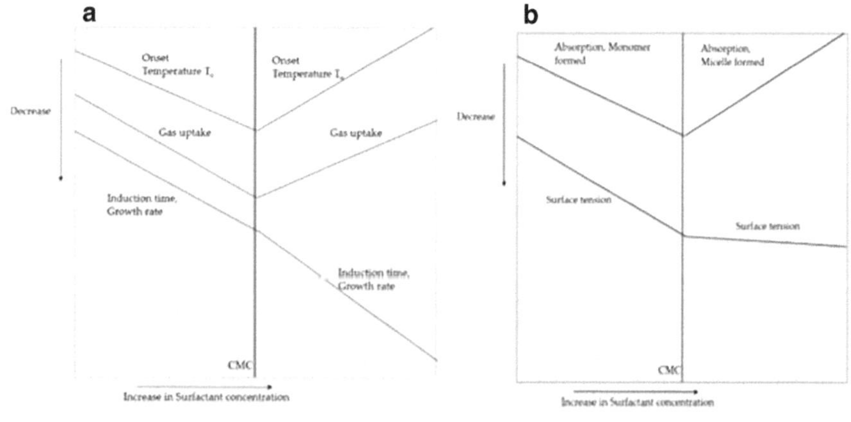

Fig. 6 **a** Key kinetic properties trend during methane hydrate formation as a function of SDS concentration (500-3000 ppm). Experimental observation confirm the presence of CMC between 2000–3000 ppm [47]. **b** Effect of change in SDS concentration on the absorption and surface tension (500–3000 ppm) [47]

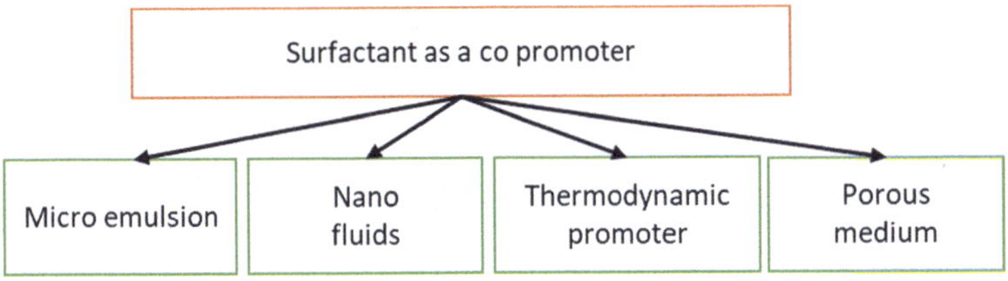

Fig. 7 Different association of surfactants as copromoter

4 Role of Surfactant with Co-promoting Gas Hydrate Formation

Surfactants have also played a key role as co promoter along with different agents and medium to enhance and stablise hydrate formation. Figure below describe their key associations.

4.1 Microemulsions

Surfactants also serve as supports for other systems that could achieve rapid hydrate formation, including microemulsions [76, 77], dry water, and nanofluids. In water-in-oil emulsions, methane molecules disperse into the oil phase and later diffuse and reach the water droplet surface to react. Each water droplet serves as an isolated nucleation front, thus able to create highly efficient water-to-hydrate conversion. Many factors, such as pressure, temperature, stirring, and initial water volume, control the stability and droplet size of the emulsion. Water cut controls not only the gas-liquid contact interface but also controls the water droplet size [78–81]. There are some challenges, such as hydrate film formation at the water droplet surface could retard gas transfer into water droplet, as well as formation and dissociation cycles could potentially destabilize the hydrate [82, 83].

4.2 Nano-Fluids

Nano-fluids are seen as the potential alternative to accelerate gas hydrate formation due to their superior thermal conductivity that addresses the exothermic and thermal inhibition during formation [84, 85]. Among different nanofluids, metal nanofluids such as silver nanoparticles have been tested repeatedly. Silver nanoparticles have been used to study methane and ethane hydrate nucleation and improvement in gas uptake, and the induction time was recorded. It is suggested that nanoparticles help during formation by providing nucleation sites, reducing effective interfacial tension and wetting angle between hydrate and particle [86–88]. A key challenge during the application of nanoparticles for gas hydrate formation and dissociation is their stability during cycling. This could destroy nanoparticle stability and, thus, reusability. To stabilized nanoparticles, surfactants such as SDS are also used. SDS in low concentration has been used along with Al_2O_3-, ZnO-, CuO-based nanoparticles [89–91].

4.3 Thermodynamic Promoters

During the last decade, much attention has been placed on non-mechanical techniques to improve gas uptake and reduce nucleation time and the stochastic nature of hydrate nucleation and formation. These non-mechanical techniques consist of using chemicals categorized into kinetic and thermodynamic promoters. Thermodynamic promoters allow hydrate formation at moderate pressure and temperature condition by reducing the required formation pressure and increasing the temperature, making them sought for novel hydrate-based industrial applications. Frequently used thermodynamic promoters include tetrahydrofuran (THF), Tetra n-butyl ammonium bromide (TBAB), and cyclopentane (CP). The key disadvantage of using thermodynamic promoters includes loss of gas uptake due to the occupancy of the cages by the promoter molecule and slower formation kinetics. To overcome these challenges, surfactants such as SDS have been used along with thermodynamic promoters, including THF and CP, and have received greater attention due to more efficient performance compared to thermodynamic promoters.

Kumar et al. [92] have studied the role of SDS and THF on formation and dissociation kinetics of the methane hydrate in an unstirred reactor configuration. Their results concluded that SDS and THF could provide optimal configuration for methane hydrate storage and transportation at atmospheric pressure. Veluswamy et al. [93] used SDS (100 ppm) with THF (5.6 mol%) to achieve methane hydrate formation within 1 h at ambient temperature 293.2 K and 7.2 MPa. They also highlighted the synergetic effect between SDS and THF was visible only at ambient temperature. When the temperature was reduced to 283.2 K, the addition of SDS decreased the gas uptake by 20% and 60% at 72 bar and 30 bar, respectively. Mech et al. [94] also studied SDS along with THF and TBAB at 276.15 K and different pressures from 30 bar to 75 bar in a stirred tank reactor. They concluded that SDS at 600 ppm concentration, THF plus SDS had higher uptake at 75 bar while TBAB plus SDS had higher gas uptake at 30 bar. Kakati et al. [95] had tested the mixture of SDS and THF on the mixture of methane, ethane and propane and found that a THF and SDS combined system can be used to improve formation and thermodynamics of the natural gas storage in the form of the hydrates.

Additionally, SDS is used with another thermodynamic promoter in applications such as gas separation, CO_2 capture, desalination, and hydrogen energy storage, and it is further discussed in §5.

4.4 *Porous Medium*

Pan et al. [96] have discussed the presence of a porous medium and SDS on formation kinetics. Presence of porous medium improve the heat transfer as well as greater higher surface area leads to quicker 2D nucleation. It is usually challenging to form methane hydrate in the porous medium as a porous medium act as a thermodynamic inhibitor and controlled by the pore size and salinity [97]. It is general practice to add surfactant to enhance the kinetics of hydrate formation [17] however, the surfactant may influence wettability as well as can create an air bubble during gas injection; therefore, the surfactants have been ignored traditionally during the modelling of the hydrate formation process. Presence of surfactant leads to enhanced formation rate and much higher gas uptake. Results suggest that both particle size and water saturation play an important role in the hydrate formation kinetics. Particle size control interface-specific area, pore volume and pore size distribution whereas water saturation control water migration as well as hydrate distribution which intern controls formation and dissociation kinetics [96]. Addition of the surfactant in the liquid phase lowers the mass transfer resistance at the gas-liquid interface and reduce the surface tension. SDS, in particular, found effective because of hydrophobic active groups in the ionized SDS which would accumulate around the surface of the silica sand to avoid any contact with water in the solution. Many surfactant micelles are formed that further help methane dissolves more due to micelle solubilization [98]. In another study for SDS and water system, the presence of SDS increase the ethane hydrate dissociation rate and weaken the self-preservation tendency of the gas hydrate [99]. Some of the key research results are summarized in Table 1.

Amino acids are seen as eco friendly replacement of surfactant for gas hydrate based research and hydrophobic amino acids are seen as good alternative of SDS. Figures 8 and 9 compare the kinetics of methane hydrate formation between SDS and Amino acids at same concentration (3000 ppm) in different sands with four different sand particle sizes.

5 Application of Surfactant

Surfactants have been used as both as inhibitors as well as promoter in different industries and application and their usage is summarize in Fig. 10.

5.1 *Surfactant-Based Hydrate Inhibition*

Quaternary ammonium surfactants in low dosage amount are used as anti-agglomeration (AA) compounds for hydrate inhibition [9] to avoid hydrate plug formation in oil and gas pipelines. AA compounds form a well-structured thin layer

Table 1 Key experimental work focused on the kinetics of hydrate formation in the presence of porous media and surfactant

Gases	Porous medium	Promoter	Key observation	Ref.
CO_2	Silica gels (mesh size: 60–120, 100–200 and 230–400)	Tween-80, SDS (50, 2000 and 4000 ppm), DTAC	Dispersed liquid phase in pore space. Enhanced mass transfer	[35]
	Multi-walled carbon nanotubes (MWCNT), hydroxylated MWCNT, carboxylated MWCNT (COOH-MWCNT) (0.005–0.1 wt%)	SDS (0.03 wt%)	No effect on CO_2 hydrate phase equillbrium in the presence of nano fluids. 0.01 and 0.05 wt% COOH-MWCNT in the presence of 0.03 wt% SDS achieved maximum hydrate formation rate	[100]
	Nanoparticles of Al_2O_3 (0.1–0.6 wt%), cerium oxide (CeO_2), silicon dioxide (SiO_2) (0.1 wt%)	THF (7.8, 10 and 20 wt%), SDS (0.05–0.8 wt%)	CO_2 hydrate formation rate increased by 3.74 times in the presence of 0.6 wt% and 0.2 wt% Al_2O_3 Presence of 10 wt% THF into 0.6 wt% and 0.2 wt% Al_2O_3 results into optimum performance	[101]
CO_2 (80.6%) + N_2 (19.4%)	Soda glass BZ-01 (0.105–0.125 mm), BZ-02 (0.177–0.250 mm), BZ-04 (0.350–0.500 mm)	THF (3 mol%), SDS (1000 mg L^{-1})	Induction time τind and equllibirum (Peq) and were reduced by 3-mol% THF and 1000-mg L–1 SDS	[102]
CH_4	Fixed bed Alumina & Silica particles (2 mm–6 mm)	SDS (300 ppm)	Smaller particle size lead to larger gas uptake and lower induction time. Gas uptake is larger when alumnia particles are present. Presence of SDS increased the storage capacity in porous media 2-4 times compare to pure water case	[103]

(continued)

Table 1 (continued)

Gases	Porous medium	Promoter	Key observation	Ref.
CH_4	Four type silica sand Sand 1(46.4–245 μm) Sand 2 (160–630 μm) Sand 3 (480–1800 μm) Sand 4 (1400–5000 μm)	SDS (500-3000 ppm) Amino acids (3000 ppm)	Increase in particle size lead to lower induction time and lower gas uptake when the initial water saturation is 35% or above due to pore filling hydrate morphology	[104]

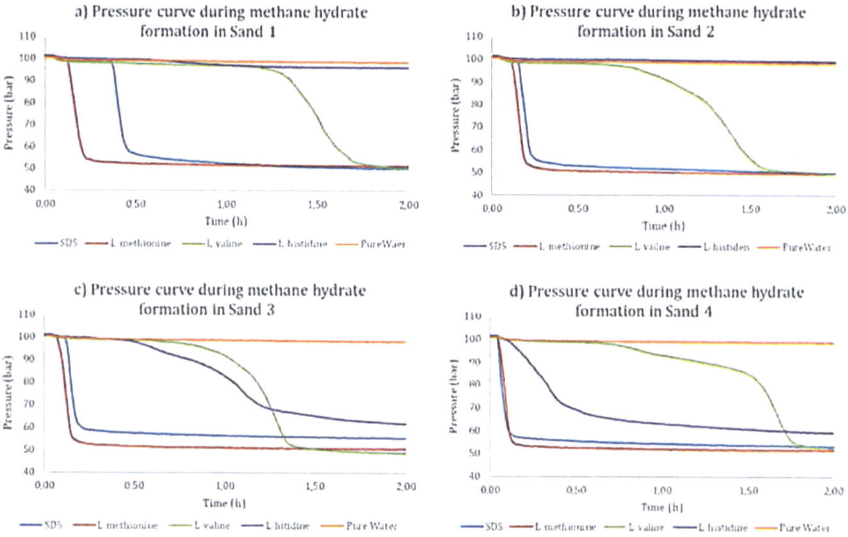

Fig. 8 Pressure variation during isothermal experiments at 100 bar and 1°C starting condition comparing the methane hydrate formation kinetics for SDS and four selected amino acids (L-valine, L-methinonine, L- histidine, L-arginine). Results suggest that SDS and L-methionine hydrate promotion capabilities are near similar in porous medium with different physical properties [104]

that further slows down hydrate aggregation and stops the crystal growth process due to mass transfer barrier [105, 106]. The thin layer could be formed either between water and oil or oil and hydrates and containssurfactants and alkanes from the oil phase [107–110]. Apart from ionic surfactants, Sorbitan type Span-20 to Span-80 are also used as AA compounds [111]. When used along with thermodynamic inhibitor (MeOH) or salts in the aqueous phase, they improve the inhibition efficiency [112, 113]. Increase in salinity increases ionic AA inhibition efficiency without disturbing emulsion stability [114]

Molecular simulation shows that AA could also promote hydrate growth [115] which could be used in hydrate based novel application such as natural gas storage and transportation, desalination, and other emerging applications.

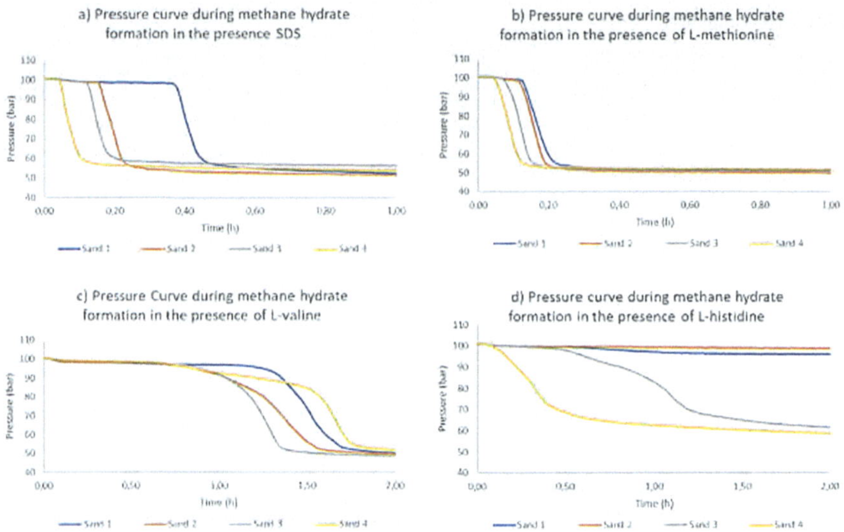

Fig. 9 Pressure variation during the isothermal experiments at 100 bar and 1°C comparing the methane hydrate formation kinetics for SDS and four selected amino acids (L-valine, L-methinonine, L- histidine, L-arginine) at given type of sand. Results indicate the hydrate formation rate increase as sand particle size increases in the presence of SDS and hydrophobic amino acids [104]

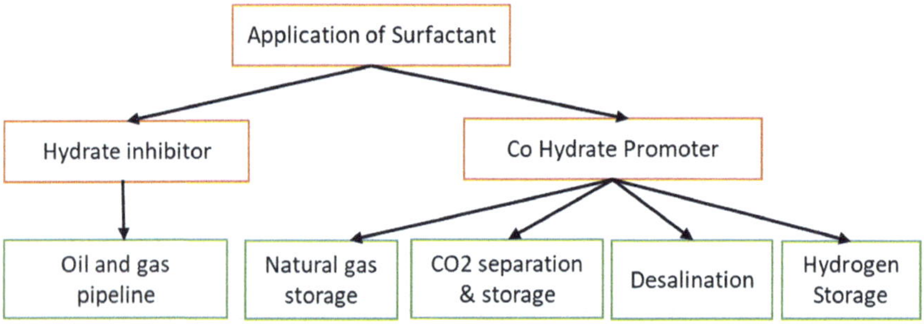

Fig. 10 Usage of Surfactant in different industries and applications

5.2 *Natural Gas Storage and Transportation*

Most crucial industrial development is in the field of natural gas (NG) storage and transportation, as gas hydrates offer additional benefits compared to traditional technologies like liquidified natural gas (LNG) and compressed natural gas (CNG). Methane gas hydrates offer high gas storage capacity in a solid-state, such that 170× methane gas per volume of hydrates (V/V) can be stored at moderate low-temperature (260–270 K) at atmospheric pressure due to self-preservation tendency shown by pure methane hydrates [116]. The storage and transport of NG in solid form could be alternative to CNG [11]. In this regard, hydrate pelletization technology has been evaluated for the storage and transportation of NG [117, 118]. Stability was further improved in the presence of SDS [119] and maintained for 256 h with decomposition measured to be only 0.04% at 1 atm and 268.2 K. This ultrastability was caused due to different hydrate morphology in the presence of surfactant. Therefore, gas hydrates at subzero temperature show reduced gas leakage and offer added advantages compared to CNG and LNG transportation.

5.3 Hydrate Based Desalination and Produced Water Treatment

The feasibility of produced water and seawater treatment via gas hydrate formation was first demonstrated during the 1960 s [120]. This technique is based on the fundamental understanding of gas hydrates, that the chemical structure of gas hydrates includes only water and host molecules and excludes all salts and other impurities in unreacted water [121, 122]. Hydrate-based desalination has shown better efficiency compared to traditional desalination techniques, such as reverse osmosis and multistage distillation, at higher salinity levels up to 25%. Such high salinity in produced water has been reported in shale gas projects and CO_2 injection-based projects. Due to several reasons, including energy consumption, technology immaturity, low efficiency, hydrate-based technology was not used in desalination plants [123]. Recent studies using cyclopentane (CP) have shown promising results to be considered for desalination studies as CP forms hydrates with pure water under atmospheric pressure at 7 °C and is immiscible with pure water; therefore, it can be recycled after hydrate dissociation [124–126]. When water and CP come together, an emulsion is formed, and the use of CP is advantageous as it can be recovered at the end of dissociation. In the presence of promoter the hydrate formation rate, gas uptake, water recovery, and salt rejection improve. To make desalination more attractive, it is also suggested to combine cyclopentane with another guest molecule-based hydrate application, such as gas separation or gas capture to optimize energy consumption and improve the salt removal efficiency [127, 128]. Using surfactant with CP can bring more drawbacks than an advantages because its presence makes the hydrate former difficult to separate after dissociation [129]. Erfani et al. [130] studied the effect of 14 nonionic surfactants on the formation kinetics of CP hydrate and found that presence of surfactant decreased induction time and enhanced the hydrate formation rate. The surfactant, which generates an oil-in-water emulsion, performed better than water-in-oil emulsion. Lim et al. [24] found that SDS changes the CP hydrate morphology, which includes rectangular tree-like or fiber-like crystals, and no change in CP hydrate shell thickness was observed in the presence of surfactant [131]. To summarise based on results, key surfactants tested along with CP hydrate for desalination include LAE8EO, TritonX-100, NPE6EO, SDS, Dodecyl-trimethylammonium bromide (DTB), Span-20, DDBSA (Dodecyl Benzene Sulfonic

Acid) and Tween 20 when used at a suitable concentration. Many surfactants, when added, modify CP hydrate morphology and physical properties. In the presence of surfactant CP solubility also enhances in water, and, hence, the removal of surfactant from the water at the end of dissociation is required as surfactants traditionally are toxic and not environmentally friendly. Use of bio-surfactants could be recommended in this application as they are biodegradable [37, 38, 132]

5.4 Hydrate Based CO_2 Separation, Capture and Storage

CO_2 separation, capture, and storage are important technology considerations to reduce greenhouse gas emission from industrial plants, including chemical, power, cement, etc. This technology includes pre and post-combustion CO2 capture. A pre-combustion gas mixture contains a CO_2/H_2 gas mixture, also known as fuel gas [14], while the post-combustion gas mixture includes CO_2/N_2 mixture known as flue gas. Hydrate-based CO_2 capture is proposed as a novel technique for CO_2 separation from fuel and flue gas mixtures [133–136]. The difference in CO_2 concentration in hydrate and in vapor phase acts as the main driver to separate CO_2 from the gas mixture during hydrate-based separation [15]. The key thermodynamic promoter used to achieve moderate operating conditions includes tetrahydrofuran (THF), tetrabutylammonium bromide (TBAB), tetra-n-butylammonium nitrate ($TBANO_3$), tetrabutylammonium fluoride (TBAF), and dodecyl trimethyl ammonium chloride (DTAC). Among these, THF is the most extensively used thermodynamic promoter for CO_2 separation and capture from the gas mixture.

Thermodynamic promoters are able to lower operational pressure or increase the temperature but do not have the influence of kinetics of the hydrate formation, which is essential for commercialization. Therefore, kinetic promoters including surfactants like SDS and SDBS (sodium dodecylbenzene sulfonate) have been used extensively along with thermodynamic promoters [137]. Among all tested surfactants, SDS was the most efficient. The CO_2–water system in the presence of SDS has shown lower induction time for CO_2 hydrate formation. CO_2 solubility increases in the presence of SDS due to a decrease in surface tension at the liquid-gas interface. Higher solubility causes faster nucleation and a further decrease in induction time [19, 62, 138] The growth rate is highest at 500–1000 ppm concentration, and higher concentration does not improve the growth rate and gas uptake [20]. In another study, it is suggested that SDS concentration has no effect on gas separation efficiency and only affects the rate of hydrate formation [139, 140]. When SDS is used with cyclopentane, no improvement in the kinetics of CO_2 hydrate formation are observed [136]. Presence of SDS during CH_4-CO_2 hydrate swapping can enhance CO_2 storage into methane hydrate reservoirs without disturbing geological formation [141] (Figure 11).

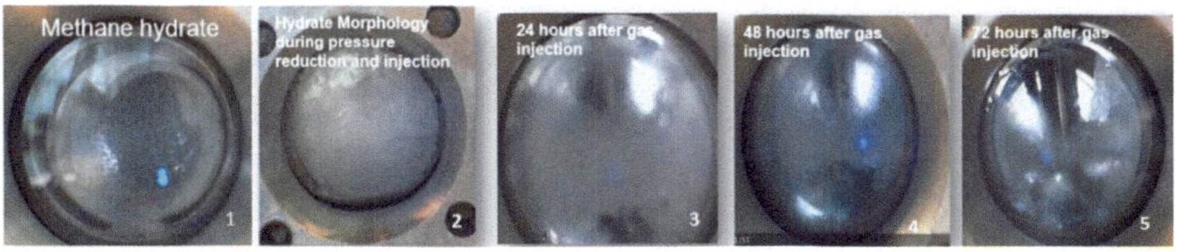

Fig. 11 Change in hydrate morphology in the presence of SDS 500 ppm in bulk water. Change in morphology is recorded before and after CO_2 injection into methane hydrate. Pictures also show the porous methane hydrate morphology. Methane hydate formed are porous in nature (Fig. 1). Figures 2–5 shows the morphgology change after CO_2 injection and 72 h after CO_2 injection [141]

5.5 Hydrate Based Hydrogen Storage

Application of hydrogen hydrates for stationary hydrogen storage has not taken off as hydrogen hydrates are formed at very high pressure (at the scale of GPa) at given ambient temperature. Research is focused on the use of kinetic and thermodynamic promoters and co-guest molecules to achieve moderate operating condition as well as faster formation kinetics [12, 142–144]. Some of the thermodynamic chemicals being tested repeatedly include TBAB, $TBANO_3$, THF, and CP [145–148]. Thermodynamic promoters occupy cages and reduces the hydrogen storage volume within hydrate. Apart from low hydrogen storage, low formation kinetics and risk of hydrogen diffusion through cages hinder adopting hydrate-based hydrogen storage methods at commercial scale [149–151]. Some researchers have tried to improve the hydrogen storage efficiency through different techniques [152–154]; however, there less attention is given to kinetics improvement, and very few studies have discussed the role of surfactant during hydrogen hydrate formation. SDS was found to be effective at the small concentration (5–500 ppm) during mixed hydrogen/propane hydrate formation studies, and two-stage hydrate growth was observed [155], a significant finding showing that micelles are unnecessary to impact hydrate formation (Fig. 12). Profile et al. [156] have invented a new technology with the help of aerosol OT surfactant(AOT), THF, and water and with the use of nanotechnology and isooctane.

5.6 Drawbacks of Surfactants

Surfactants have been studied extensively as kinetic promoters for hydrate formation; however, few studies have discussed the disadvantages of surfactants. The key disadvantage is that surfactants create foam even at low concentration (100–1000 ppm) during the degassing operation [157–159]. Due to foam formation, gas production rates can be very slow,which could be undesirable for industrial-scale applications. There are also concerns about surfactant biodegradability and their effect on environment [160]; therefore, current research is focused on environmentally friendly

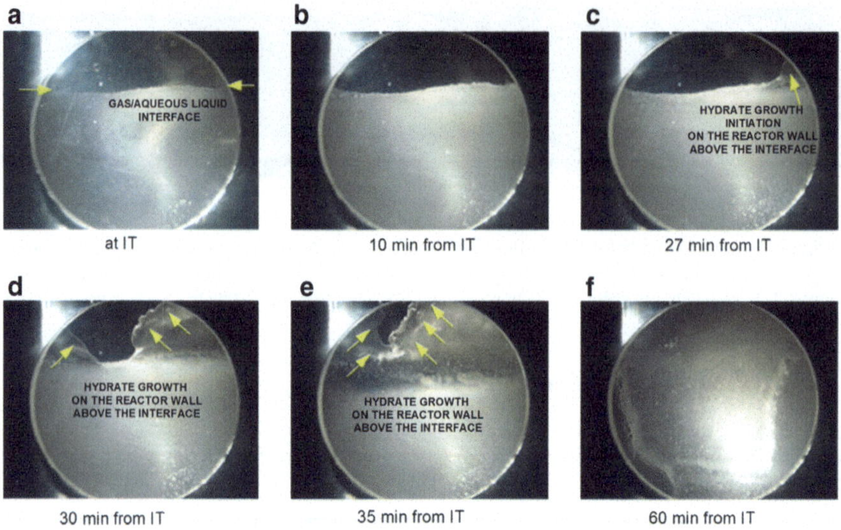

Fig. 12 Evidence of hydrate nucleation and growth at low concentration of SDS (25 ppm) within a mixed hydrogen–propane hydrate system performed at 274.2 K and 8.5 MPa in a stirred tank reactor. Reprinted from [156]

substitutes, such as amino acids [20] or new classes of biosurfactants [38] having similar kinetic promotion capabilities as anionic surfactants.

6 Computational Studies on the Role of Surfactant During Hydrate Formation

The first attempt to understand the blockage of pipelines due to gas hydrates has been by Hammerschmidt [6] in 1934. Among the first additives used to prevent the formation of gas hydrate included anti-freezing agents, such as methanol or ethylene glycol (EG). [161] Their effect in the solution resulted in a shift toward lower temperatures and higher pressures, due to a leftward change in the equilibrium phase boundary conditions. This came about because the hydrogen bonding between water and additive molecules affected the activity of water and the propensity to form hydrate cages [162], giving rise to the class of chemicals that are known as thermodynamic inhibitors. Such inhibitors have become commonplace in the oil and gas industry as preventative measures for gas hydrate formation and consequent pipeline blockage [163]. Conversely, there are chemicals that instead improve the formation behavior of hydrate by causing a shift of the phase boundary to the right, and they are thus known as thermodynamic promoters. These promoters are typically trapped in the hydrate cage along with the gas molecules, aiding in the stabilization of the hydrate crystalline structure at higher temperatures and/or lower pressures.

Thus, it is very valuable for oil and gas industries to be able to predict which additives and the correct amount that would be necessary to remove the risk of pipeline blockage resulting from gas hydrates. Hammerschmidt developed an early formula, considering the temperature that would suppress hydrate formation because of the inclusion of inhibiting chemicals, ΔT_H, as described in the work [6].

$$\Delta T_H = \frac{k_H w_{\text{add}}}{M_{\text{add}}(1 - w_{\text{add}})}$$

Here, k_H is a dimensionless constant that depends on the type of inhibitor, w_{add} is the mass fraction additive in aqueous solution, and M_{add} is its molecular weight. The equation is fairly simple and has relatively good accuracy, which makes it popular even to this day, despite more advanced models that have been developed since the 1950 s. However, because the accuracy of suppression temperature depends on estimation of the hydrate equilibrium temperature considering purity of samples, there is a higher chance of errors in calculation. In comparison, thermodynamic models primarily depend on the chemical potentials of each chemical in every phase being equal. Thus, thermodynamic models make it possible to include additives into calculations, insofar as parameters are provided for predicting their chemical potentials.

In the last decades, there has been a larger prevalence of first principles-based modeling since computational power has greatly increased over that time. This has made it possible to predict the properties of long chain molecules, typical of surfactants, ionic liquids, or amino acids, through methods like density functional theory (DFT). DFT typically has N_e^{2-3} scaling, where N_e is the number of electrons, which is why it has been a more recent endeavor for such molecules. More specifically, much research has been performed to understand the mechanisms of thermodynamic and kinetic hydrate inhibition (THI and KHI, respectively) arising from gas hydrate additives, both from experimental and computational aspects [164–167]. Very recently, Lee et al. were able to find synergistic effects when more than one inhibitor is utilized through a combination of both methods [168]. Statistical thermodynamics has gained much traction in the past years, since it can incorporate DFT-calculated properties into its calculation routine (e.g. the conductor-like screening model with real solvation, COSMO-RS) [169] to give information like reaction constant, activities, and Henry constants, to name a few [170–172].

Further implementation of first principles methods has been seen in the realm of artificial intelligence, as machine learning methods have been able to make predictions about properties using training sets with both experimental and computational data. The advantage of machine learning is that it can provide savings in computational time, given a large enough training set to give accurate results. Going into the future, as large scale operations will begin to implement gas hydrate production for various applications, computational fluid dynamics (CFD) and plant-scale simulations will benefit from incorporating these into a multi-scale approach, since each of these techniques span different time- and length-scales (see Fig. 13). Already,

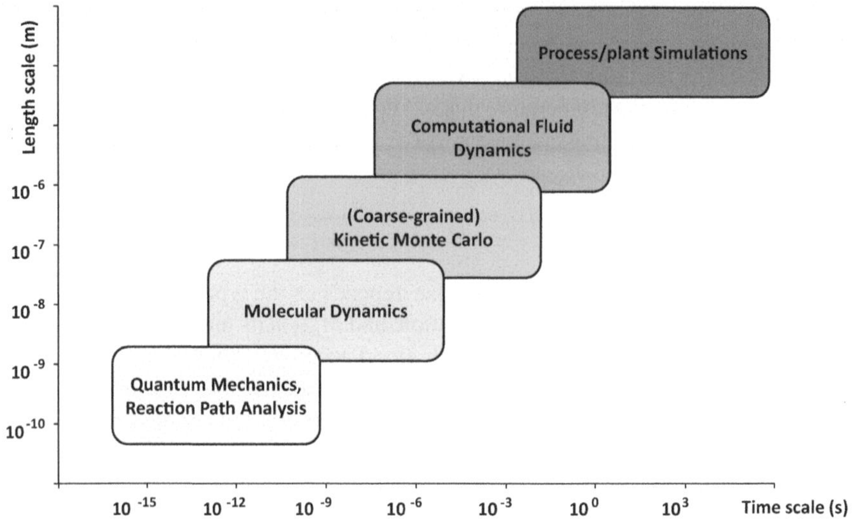

Fig. 13 How multi-scale modeling could link methods of different length- and time-scales

there has been some CFD simulations studying hydrate formation, and inclusion of additives will become increasingly prevalent.

Some key features of each of the methods above can be summarized in Table 2. In the following sections, the atomistic to micro-scale methods described above are presented with examples.

Table 2 Key usages and applications of the different computational methods

Method	Information gathered
Quantum mechanics	Relaxed structures of molecules/materials, interactions of small molecules, transition state searches
Molecular dynamics	Search for conformers, energy landscape of reactions, predict spectroscopic data
Kinetic monte carlo	Surface diffusion and growth, movement of defects and dislocations, viscoelasticity of crosslinking
Computational fluid dynamics	Movement of machinery, simulation of laminar/turbulent flows, heat transfer, aerodynamics, reacting flows and combustion
Process/plant simulations	Thermophysical properties, unit operation properties, chemical reactions/kinetics, environmenta/safety factors

6.1 Empirical Modelling—The Electrolyte Model

One particularly important application of models and property prediction, prior to the use of computationally-assisted models, pertains to equilibrium phase diagrams for gas hydrates and their additives, and the most common method of doing so—with good accuracy—has included estimated values for the activity of water as a function of additive. Dickens and Quinby-Hunt developed an electrolyte-based model [173] that could generate the equilibrium points for dissolved salts as additives, although it could be extended to ionic liquids and surfactants that can be viewed as separated charges [174], as has been proposed by Nashed et al. [175]. regarding ionic liquids as additives for methane hydrates. The electrolyte-based model is an adaptation of Pieroen's model [176], and has been utilized by many other works [177–180]. These models are based on classical thermodynamics, so assumptions, like negligible amount of gas in the hydrate and hydrate dissociation enthalpy (ΔH_{dis}) is constant over a small temperature range, are made, showing that additives decrease water activity (a_w). Thus, Nashed et al. made the following relationship between hydrate formation temperatures of pure water and the additive, T_w and T_{add}, respectively [175]:

$$\ln a_w = \frac{\Delta H_{\text{dis}}}{nR}\left(\frac{1}{T_w} - \frac{1}{T_{\text{add}}}\right)$$

where n is methane hydrate hydration number [181], and R is the universal gas constant. CSMGem software was used to calculate the water- methane hydrate dissociation temperature. The activity is also related to the change in freezing point due to additive, T_f and $T_{f,\text{ice}}$, for with and without additive, respectively, and the heat of fusion of ice:

$$\ln a_w = \frac{\Delta H_{\text{fus,ice}}}{R}\left(\frac{1}{T_{f,\text{ice}}} - \frac{1}{T_f}\right)$$

Combining the two equations above, one would be able to determine the effect of additives on the hydrate formation temperature.

In the work by Nashed et al. they were able to determine the phase boundaries of methane hydrates in the presence of 10 wt% ionic liquid solution experimentally, and found good agreement with the models above [175]. It was found that their impact caused a shift toward lower temperature and higher pressure, and a mean temperature reduction of 0.37–1.52 K was observed within their pressure range (5.1–11.1 MPa), depending on the type of ionic liquid used. More importantly, they drew relationships between the inhibition effect with chemical structure (e.g. cation/anion combinations and alkyl chain length), leading them to conclude that thermodynamic inhibitors are not involved in the formation of hydrate cages.

6.2 Quantum Mechanics

The presently most popular implementation of computational quantum mechanics is density functional theory (DFT), although others are used to lesser degrees, like configuration interaction, coupled cluster, and increasingly Møller-Plesset perturbation theory. One of the basic types of calculation that can be performed with DFT is the determination of the relaxed structure of a molecule and its corresponding energy. Using this information, one could, for example, use DFT (and, similarly, moleculdar dynamics) to determine the interaction energy, E_{int}, of a gas hydrate molecule with an inhibitor, given the gas hydrate energy, E_{GH}, the additive energy, E_{add}, and the energy of the interacting species, via:

$$E_{\text{int}} = E_{\text{GH+add}} - (E_{\text{GH}} + E_{\text{add}})$$

As an example, Lee et al. calculated the interaction energy between a cage and inhibitor molecule (amino acids and ionic liquids), and they found that the ionic liquid 1-butyl,3-methylimidazolium tertrafluoroborate had a greater probability of hydrate inhibition due to a stronger interaction energy compared to amino acid glycine [168]. This and many other binding energies can be compared in Table 3. Such relationships could be used in the reserve, to also determine improved hydrate formers, such as some surfactants (e.g. sodium dodecyl sulfate, SDS). Furthermore, these DFT calculations are used as a database to calculate larger systems, such as molecular dynamics simulations or statistical thermodynamics. Transition state theory is a method to determine the free energy barrier of reactions, of which the formation of gas hydrates could be applied, and DFT is commonly applied to it. However, molecular dynamics can be utilized the same way, as has been performed by Sicard et al. to understand how anti-agglomerants control methane transport with hydrates [106].

6.3 Molecular Dynamics

Molecular dynamics simulation is a very diverse and expanding field that can be utilized to calculate properties like surface tension, surfactant (reverse-) micellization in water/oil systems, bilayer and thin film formation, and looking just for such properties of SDS, there are already several sources [189–198]. Other studies of surfactant molecules have found a difference in the self-assembled structures due to changing conditions [199, 200].

More recently, a study by Choudhary et al. looked into the role of SDS (1 wt%) on methane hydrate formation in comparison with pure water [201], and it is one example of how simulations can be used to predict the influence of surfactants on gas hydrate growth behavior (Fig. 14). It was found that SDS tended to adsorb onto the hydrate surface because of its hydrophobic tail binding to openings in the cages

Table 3 Binding free energies calculated in literature from DFT or molecular dynamics simulations

Compound	sI Binding Free Energy [KJ/mol]	sII Binding Free Energy [KJ/mol]	Ref.	Note
CH_4	−32.51	−28.44	[182]	
CH_4	−30.42	−26.86	[183]	
C_2H_6	–	−39.45	[182]	
CO_2	–	−40.21	[182]	
N_2	−19.02	−20.05	[182]	
PVP 1-mer	6	–	[184]	
PVP 8-mer	−9	–	[184]	
PVP 16-mer	−21	–	[184]	
ChCl	−59.82, −118.21	−162.64, −155.37	[183]	
$ChTfn_2$	−12 to 132	−20 to 80	[185]	
ChOAc	28 to 88	8 to 118	[185]	
DB^3ACl	−1.2 to 68.6		[186]	*
PheAcA		−27.92	[187]	
NapAcA		−34.15	[187]	
PyrAcA		−53.75	[187]	
L-histidine	−47.20		[188]	**
Bicine	−48.91		[188]	**
L-serine	−44.73		[188]	**
Tricine	−29.47		[188]	**
Glycine	−47.88		[188]	**
Glycine	−52.46		[168]	
L-tyrosine	−63.68		[188]	**
L-threonine	−38.81		[188]	**
CAPB	−63.45		[188]	**
Betaine	−59.79		[188]	**
Proline	−48.41		[188]	**
Tryptophan	−44.43		[188]	**
$[BMIM][BF_4]$	−393.46		[168]	

PVP = polyvinylpyrrolidone, ChCl = choline chloride, $ChTfn_2$ = choline bistriflamide, ChOAc = choline acetate, DB^3ACl = n-dodecyl-tri(n-butyl)-ammonium chloride, PheAcA = 1-phenylacetic acid, NapAcA = 2-napthylacetic acid, PyrAcA = 1-pyreneacetic acid, CAPB = cocamidopropyl betaine, $[BMIM][BF_4]$ = 1-butyl-3-methylimidazolium tetrafluoroborate. *Molecular dynamics simulation with mixed hydrate sizes. **Averaged over interaction sites, binding with water molecule

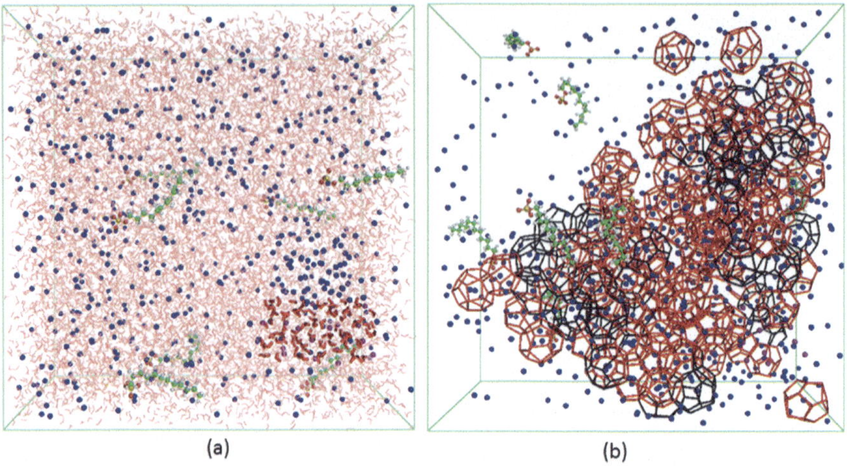

Fig. 14 Molecular dynamics simulation of **a** a mixture of methane, water, and 1% SDS and **b** a snapshot after 1 μs at 270 K and 100 bar, showing significant 5^{12} (black cage) and $5^{12}6^2$ (red cage) hydrate formation (water molecules hidden in **b**). Key for dots: liquid water (light red), seed water (dark red), methane (blue), methane of seed (magenta), carbon of hydrophobic SDS tail (green), sulfur (yellow), hydrogen (white), and oxygen of SDS (red). Reprinted from [201]

of growing hydrates, which they report caused it to stabilize the nascent nuclei. This was expected to reduce the surface energy and thus also the nucleation barrier and induction time. They believe that these adsorbed surfactant molecules might change the morphology of the hydrate as it grows of larger length and time scales. The porosity could increase the mass transfer of guest molecule, leading to improved growth kinetics of the gas hydrate.

Another molecular dynamics study of SDS with methane hydrates found that micellization of the molecule was not required during the formation of gas hydrates [50].

Great interest has been invested in understanding formation kinetics using molecular simulations [202]. One such study by Walsh et al. [203] found that during the nucleation process the interaction of guest molecules with the faces and surfaces of partial hydrate cages led to the formation of the full gas hydrate. This and the many other studies have been able to improve the understanding of how additives may affect hydrate formation, such the simulations of CO_2 and CH_4 gas hydrates [204] leading to the further study on the impact of tetrahydrofuran on their formation [205].

In another important study, Carver et al. [206], using Monte Carlo simulations, found that the effect of PVP as an inhibitor depended on pendant hydrogens on the hydrate surface being available as adsorption sites, since PVP would lie along the surface and block these sites. Furthermore, Bui et al. studied how anti-agglomerants could either enhance or impair hydrate formation [107]. Similarly, other groups studied how sodium chloride might influence the adsorption behavior of anti-agglomerants [207, 208], surfactants like SDS, and hydrocarbons [209]. In another study by Bui et al. [105], they were able to reproduce micromechanical force experiments using equilibrium molecular dynamics.

Large systems have also been utilized with stochastic models like Kinetic Monte Carlo simulations. Understandings of agglomeration [210] and surface diffusion [108] behaviors of hydrates due to surfactacts have been particularly impacted by such Big Picture approaches.

6.4 Continuum Solvation Model

As mentioned above, conductor-like screening model with real solvation (COSMO-RS) can be employed to calculate many thermodynamic properties, in an approach different from molecular dynamics [169]. Taking advantage of the large database and promising nature (low vapor pressure, tunability, and bifunctionality) of ionic liquids, Bavoh et al. screened potential candidates for gas hydrate inhibition, rather than relying on trial and error, which are time intensive [211].

Athough group contribution (GC) methods, such as UNIQUAC Functional-group Activity Coefficients (UNIFAC), are commonplace and reliable for thermodynamic property prediction, they are lacking in accuracy when the groups are less studied experimentally or have long chains, such as in the case of ionic liquids. Thus, using a tool which is based on first principles becomes more attractive, considering its high accuracy [212]. It is already commonly used in pharmaceutical research and chemical engineering [213–215].

Bavoh et al. presented COSMO-RS as a novel prescreening tool for ionic liquid-based hydrate mitigation by correlating their calculated hydrogen bonding energies (E_{HB}) with suppression temperature, ΔT_H, in comparison with induction time studies. Their work was able to describe the factors that impact the E_{HB} of ionic liquids in terms of hydrogen bond donors and acceptors, through the use of sigma profiles and potentials that are generated with COSMO-RS.

Similar studies could be performed with a focus, for example, on screening of surfactant molecules that can aid in gas hydrate formation, since the work by Bavoh et al. is one of the first to apply COSMO-RS calculations for gas hydrate applications, and it was limited to ionic liquids. There are still many ionic liquids that are not included in the database of commercially available software, which would also expand the scope of ionic liquids screening.

6.5 Machine Learning

An example of the application of machine learning for studying gas hydrates can be found in the work done by Xia et al. [216]. They incorporated a fusion modeling method that could be used to predict CO_2 solubility in hydrates as related to nine ionic liquids. Using data collected from literature, they divided them into three sets, some for training, some for validation, and the rest for testing. With the training set, they were able to incorporate the back propagation neural network, support vector machine, and extreme learning machine. From these, three sub-models with the best performance were chosen according to the validation set. Afterwards, linear fusion models were included via the minimum square error and information entropy methods. Lastly, the prediction performance of these last sets of models was evaluated with the test set, which found that their linear fusion model was the best performing, with the information entropy method having better predictions. However, they do clarify that although their predictions worked well, they were not guaranteed to work on ionic liquids outside of the nine they tested, but it is something that is amenable to future expansion and has tremendous time savings compared to the alternatives, and it could rival COSMO-RS in the prediction of thermodynamic properties.

7 Closing Remarks and Future Prospects

In this chapter, an introduction into surfactant-based hydrate promotion studies was provided. Key areas that were touched include a summary of the theories proposed and how they have been implemented computationally, as well as discussions on surfactant-supported promoters. Based on the discussion, the following remarks can be added as the conclusion and future prospects

- Many theories are suggested to explain the role of surfactants during hydrate promotion. Micellar and capillary-based theories are the two most prominent ones. The capillary effect in particular plays a critical role.
- The molecular structure of a surfactant controls the hydrate formation efficiency, and surfactants with optimal structures would outperform the rest of the surfactants. Key factors that play a part in enhancing the efficiency include the ionic strength and chain length.
- Surfactants also play a key role along with other co-promoters since surfactants affect formation in different ways, including stabilization, kinetic improvements, etc., at a suitable concentration.
- The success of hydrate-based industrial applications depends on optimal use of surfactants along with other promoters and mechanical techniques.
- Use of biosurfactants is gaining attention due to their biodegradability. More research is required regarding the use of biosurfactant in desalination.
- Use of surfactants to improve hydrate formation kinetics is not well discussed in emerging technologies, including hydrogen storage, cold storage, etc.
- Computational modeling has given a large number of insights into the understanding experimental observations, and they have assisted in smart design selection of surfactants and other additives.

- Multi-scale modeling from quantum-scale to plant-scale will prove to be a major breakthrough in hydrate formation and storage, since it would be able to predict the appropriate surfactants and physical conditions that make the applications feasible and efficient.

References

1. Sloan ED, Koh CA, Koh C (2007) *Clathrate hydrates of natural gases*, 3rd edn, CRC Press
2. Koh CA (2002) Towards a fundamental understanding of natural gas hydrates. Chem Soc Rev 31:157–167
3. Makogon YF (2010) Natural gas hydrates—a promising source of energy. J Nat Gas Sci Eng
4. Boswell R, Collett TS (2011) Current perspectives on gas hydrate resources. Energy Environ Sci 4:1206–1215
5. Shina K, Kumarb R, Udachina KA, Alavia S, Ripmeester JA (2012) Ammonia clathrate hydrates as new solid phases for Titan, Enceladus, andother planetary systems. Proc Natl Acad Sci USA 109:14785–14790
6. Hammerschmidt EG (1934) Formation of gas hydrates in natural gas transmission lines. Ind Eng Chem 26:851–855
7. Jamaluddin AKM, Kalogerakis N, Bishnoi PR (1991) Hydrate plugging problems in undersea natural gas pipelines under shutdown conditions. J Pet Sci Eng 5:323–335
8. Kelland MA (2006) History of the development of low dosage hydrate inhibitors. Energy Fuels 20:825–847
9. Chua PC, Kelland MA (2018) Study of the gas hydrate antiagglomerant performance of a series of mono-and bis-amine oxides: dual antiagglomerant and kinetic hydrate inhibition behavior. Energy Fuels 32:1674–1684
10. Englezos P, Lee JD (2005) Gas hydrates: a cleaner source of energy and opportunity for innovative technologies. Korean J Chem Eng 22:671–681
11. Veluswamy HP, Kumar A, Seo Y, Lee JD, Linga P (2018) A review of solidified natural gas (SNG) technology for gas storage via clathrate hydrates. Appl Energy 216:262–282
12. Veluswamy HP, Kumar R, Linga P (2014) Hydrogen storage in clathrate hydrates: current state of the art and future directions. Appl Energy 122:112–132
13. Dashti H, Yew LZ, Lou X Recent advances in gas hydrate-based CO2 capture.pdf
14. Babu P, Linga P, Kumar R, Englezos P (2015) A review of the hydrate based gas separation (HBGS) process for carbon dioxide pre-combustion capture. Energy 85:261–279
15. Li A, Wang J, Bao B (2019) High-efficiency CO_2 capture and separation based on hydrate technology: a review. Greenh Gases Sci Technol 9:175–193
16. Babu P, Nambiar A, He T, Karimi IA, Lee JD, Englezos P, Linga P (2018) A review of clathrate hydrate based desalination to strengthen energy-water nexus. ACS Sustain Chem Eng 6:8093–8107
17. Gayet P, Dicharry C, Marion G, Graciaa A, Lachaise J, Nesterov A (2005) Experimental determination of methane hydrate dissociation curve up to 55 MPa by using a small amount of surfactant as hydrate promoter 60:5751–5758
18. Kalogerakis N, Jamaluddin AKM, Dholabhai PD, Bishnoi PR (1993) Effect of surfactants on hydrate formation kinetics. In: Proceedings of the proceedings of the 1993 spe international symposium on oilfield chemistry, Society of Petroleum Engineers 1:375–383
19. Kumar A, Bhattacharjee G, Kulkarni BD, Kumar R (2015) Role of surfactants in promoting gas hydrate formation. Ind Eng Chem Res 54:12217–12232
20. Pandey JS, Daas YJ, Solms N (2020) Von screening of amino acids and surfactant as hydrate promoter for CO_2 capture from flue gas

21. Kvamme B, Selvåg J, Saeidi N, Kuznetsova T (2018) Methanol as a hydrate inhibitor and hydrate activator. Phys Chem Chem Phys 20:21968–21987
22. Linga P, Clarke MA (2017) A review of reactor designs and materials employed for increasing the rate of gas hydrate formation. Energy Fuels 31:1–13
23. Kumar A, Sakpal T, Linga P, Kumar R (2015) Enhanced carbon dioxide hydrate formation kinetics in a fixed bed reactor filled with metallic packing. Chem Eng Sci 122:78–85
24. Lim YA, Babu P, Kumar R, Linga P (2013) Morphology of carbon dioxide-hydrogen-cyclopentane hydrates with or without sodium dodecyl sulfate. Cryst Growth Des 13:2047–2059
25. Yagasaki T, Matsumoto M, Andoh Y, Okazaki S, Tanaka H (2014) Effect of bubble formation on the dissociation of methane hydrate in water: A molecular dynamics study. J Phys Chem B 118:1900–1906
26. Zhang JS, Lee S, Lee JW (2007) Kinetics of methane hydrate formation from sds solution, pp 6353–6359
27. Zhang J, Lee JW (2009) Effect of sodium dodecyl sulfate on the supercooling point of ice and clathrate hydrates. Energy Fuels 23:3045–3047
28. Zhong Y, Rogers RE (2000) Surfactant effects on gas hydrate formation. Chem Eng Sci 55:4175–4187
29. World THE, Surface OF The world of surface science *, pp 14–24
30. Fuhrhop JH, Koning J (1994) *Membranes and molecular assemblies*, Monographs in supramolecular chemistry, The Royal Society of Chemistry, ISBN 978-0-85186-732-8
31. He Y, Sun M-T, Chen C, Zhang G-D, Chao K, Lin Y, Wang F (2019) Surfactant-based promotion to gas hydrate formation for energy storage. J Mater Chem A 7:21634–21661
32. Karaaslan U, Parlaktuna M (2002) Promotion effect of polymers and surfactants on hydrate formation rate. Energy Fuels 16:1413–1416
33. Karaaslan U, Parlaktuna M (2000) Surfactants as hydrate promoters? Energy Fuels 14:1103–1107
34. Okutani K, Kuwabara Y, Mori YH (2008) Surfactant effects on hydrate formation in an unstirred gas / liquid system : an experimental study using methane and sodium alkyl sulfates 63:183–194
35. Kumar A, Sakpal T, Linga P, Kumar R (2013) Influence of contact medium and surfactants on carbon dioxide clathrate hydrate kinetics. Fuel 105:664–671
36. Veluswamy HP, Ang WJ, Zhao D, Linga P (2015) In fl uence of cationic and non-ionic surfactants on the kinetics of mixed hydrogen / tetrahydrofuran hydrates. Chem Eng Sci 132:186–199
37. Rogers RE, Kothapalli C, Lee MS, Woolsey JR (2008) Catalysis of Gas Hydrates by Biosurfactants in Seawater-Saturated Sand/Clay. Can J Chem Eng 81:973–980
38. Arora A, Cameotra SS, Kumar R, Balomajumder C, Singh AK, Santhakumari B, Kumar P, Laik S (2016) Biosurfactant as a promoter of methane hydrate formation: thermodynamic and kinetic studies. Sci Rep 6:1–13
39. Rogers R, Zhang G, Dearman J, Woods C (2007) Investigations into surfactant/gas hydrate relationship. J Pet Sci Eng 56:82–88
40. Barakat Y, Fortney LN, Schechter RS, Wade WH, Yiv SH, Graciaa A (1983) Criteria for structuring surfactants to maximize solubilization of oil and water. II. Alkyl benzene sodium sulfonates. J Colloid Interface Sci 92:561–574
41. Kunieda H, Shinoda K (1982) Correlation between critical solution phenomena and ultralow interfacial tensions in a surfactant/water/oil system. Chem Soc Japan 55:1777–1781
42. Daimaru T, Yamasaki A, Yanagisawa Y (2007) Effect of surfactant carbon chain length on hydrate formation kinetics. J Pet Sci Eng 56:89–96
43. Dicharry C, Diaz J, Torré JP, Ricaurte M (2016) Influence of the carbon chain length of a sulfate-based surfactant on the formation of CO_2, CH_4 and CO_2–CH_4 gas hydrates. Chem Eng Sci 152:736–745
44. Roy S, Mehra A, Bhowmick D (1997) Prediction of solubility of nonpola—gases in micellar solutions of ionic surfactants. J Colloid Interface Sci 196:53–61

45. Luo H, Sun CY, Peng BZ, Chen GJ (2006) Solubility of ethylene in aqueous solution of sodium dodecyl sulfate at ambient temperature and near the hydrate formation region. J Colloid Interface Sci 298:952–956
46. Peng BZ, Chen GJ, Luo H, Sun CY (2006) Solubility measurement of methane in aqueous solution of sodium dodecyl sulfate at ambient temperature and near hydrate conditions. J Colloid Interface Sci 304:558–561
47. Pandey JS, Daas YJ, Solms N (2019) Von insights into kinetics of methane hydrate formation in the presence of surfactants
48. Di Profio P, Arca S, Germani R, Savelli G (2005) Surfactant promoting effects on clathrate hydrate formation: Are micelles really involved? Chem Eng Sci 60:4141–4145
49. Zhang JS, Lee S, Lee JW (2007) Does SDS micellize under methane hydrate-forming conditions below the normal Krafft point? J Colloid Interface Sci 315:313–318
50. Albertí M, Costantini A, Laganá A, Pirani F (2012) Are micelles needed to form methane hydrates in sodium dodecyl sulfate solutions? J Phys Chem B 116:4220–4227
51. Kobayashi I, Ito Y, Mori YH (2001) Microscopic observations of clathrate-hydrate films formed at liquid/liquid interfaces. I. Morphology of hydrate films. Chem Eng Sci 56:4331–4338
52. Watanabe K, Imai S, Mori YH (2005) Surfactant effects on hydrate formation in an unstirred gas/liquid system: An experimental study using HFC-32 and sodium dodecyl sulfate. Chem Eng Sci 60:4846–4857
53. Yoslim J, Linga P, Englezos P (2010) Enhanced growth of methane—propane clathrate hydrate crystals with sodium dodecyl sulfate, sodium tetradecyl sulfate, and sodium hexadecyl sulfate surfactants. J Cryst Growth 313:68–80
54. Karanjkar PU, Lee JW, Morris JF (2012) Surfactant effects on hydrate crystallization at the water-oil interface: hollow-conical crystals. Cryst Growth Des 12:3817–3824
55. Tajima H, Kiyono F, Yamasaki A (2010) Direct observation of the effect of sodium dodecyl sulfate (SDS) on the gas hydrate formation process in a static mixer. Energy Fuels 24:432–438
56. Mitarai M, Kishimoto M, Suh D, Ohmura R (2015) Surfactant effects on the crystal growth of clathrate hydrate at the interface of water and hydrophobic-guest liquid. Cryst Growth Des 15:812–821
57. Hayama H, Mitarai M, Mori H, Verrett J, Servio P, Ohmura R (2016) Surfactant Effects on Crystal Growth Dynamics and Crystal Morphology of Methane Hydrate Formed at Gas/Liquid Interface. Cryst Growth Des 16:6084–6088
58. Hayama H, Mitarai M, Mori H, Ohmura R (2016) Methane hydrate crystal growth at the gas/liquid interface in the presence of sodium dodecyl sulfate. Procedia Eng 148:339–345
59. Wang F, Wang L, Wang C, Guo G, Liu G, Luo S, Guo R (2015) Direction controlled methane hydrate growth. Cryst Growth Des 15:5112–5117
60. Wang F, Jia ZZ, Luo SJ, Fu SF, Wang L, Shi XS, Wang CS, Guo RB (2015) Effects of different anionic surfactants on methane hydrate formation. Chem Eng Sci 137:896–903
61. Botimer JD, Dunn-Rankin D, Taborek P (2016) Evidence for immobile transitional state of water in methane clathrate hydrates grown from surfactant solutions. Chem Eng Sci 142:89–96
62. Molokitina NS, Nesterov AN, Podenko LS, Reshetnikov AM (2019) Carbon dioxide hydrate formation with SDS: Further insights into mechanism of gas hydrate growth in the presence of surfactant. Fuel 235:1400–1411
63. Ando N, Kuwabara Y, Mori YH (2012) Surfactant effects on hydrate formation in an unstirred gas/liquid system: an experimental study using methane and micelle-forming surfactants. Chem Eng Sci 73:79–85
64. Wang F, Liu GQ, Meng HL, Guo G, Luo SJ, Guo RB (2016) Improved methane hydrate formation and dissociation with nanosphere-based fixed surfactants as promoters. ACS Sustain Chem Eng 4:2107–2113
65. Daniel-David D, Guerton F, Dicharry C, Torré JP, Broseta D (2015) Hydrate growth at the interface between water and pure or mixed CO2/CH4 gases: Influence of pressure, temperature, gas composition and water-soluble surfactants. Chem Eng Sci 132:118–127

66. Zhang JS, Lo C, Somasundaran P, Lu S, Couzis A, Lee JW (2008) Adsorption of sodium dodecyl sulfate at THF hydrate/liquid interface. J Phys Chem C 112:12381–12385
67. Zhang JS, Lo C, Somasundaran P, Lee JW (2010) Competitive adsorption between SDS and carbonate on tetrahydrofuran hydrates. J Colloid Interface Sci 341:286–288
68. Lo C, Zhang JS, Couzis A, Somasundaran P, Lee JW (2010) Adsorption of cationic and anionic surfactants on cyclopentane hydrates. J Phys Chem C 114:13385–13389
69. Schramm LL, Stasiuk EN, Marangoni DG (2003) Surfactants and their applications. Annu Rep Prog Chem Sect C 99:3–48
70. Scamehorn JF, Schechter RS, Wade WH (1982) Adsorption of surfactants on mineral oxide surfaces from aqueous solutions. I: Isomerically pure anionic surfactants. J Colloid Interface Sci 85:463–478
71. Somasundaran P, Fuerstenau DW (1966) Mechanisms of alkyl sulfonate adsorption at the alumina-water interface. J Phys Chem 70:90–96
72. Zerpa LE, Salager JL, Koh CA, Sloan ED, Sum AK (2011) Surface chemistry and gas hydrates in flow assurance. Ind Eng Chem Res 50:188–197
73. Song JH, Couzis A, Lee JW (2010) Investigation of macroscopic interfacial dynamics between clathrate hydrates and surfactant solutions. Langmuir 26:18119–18124
74. Torré JP, Dicharry C, Ricaurte M, Daniel-David D, Broseta D (2011) CO2 capture by hydrate formation in quiescent conditions: in search of efficient kinetic additives. Energy Procedia 4:621–628
75. Torre J, Ricaurte M, Dicharry C, Broseta D (2012) CO2 enclathration in the presence of water-soluble hydrate promoters: Hydrate phase equilibria and kinetic studies in quiescent conditions. Chem Eng Sci 82:1–13
76. Li X, Chen C, Chen Y, Li Y, Li H (2015) Kinetics of methane clathrate hydrate formation in water-in-oil emulsion. Energy Fuels 29:2277–2288
77. Dalmazzone D, Hamed N, Dalmazzone C (2009) DSC measurements and modelling of the kinetics of methane hydrate formation in water-in-oil emulsion T / K 64:2020–2026
78. Mu L, Li S, Ma QL, Zhang K, Sun CY, Chen GJ, Liu B, Yang LY (2014) Experimental and modeling investigation of kinetics of methane gas hydrate formation in water-in-oil emulsion. Fluid Phase Equilib 362:28–34
79. Xiang CS, Peng BZ, Liu H, Sun CY, Chen GJ, Sun BJ (2013) Hydrate formation/dissociation in (Natural Gas + Water + Diesel Oil) emulsion systems. Energies 6:1009–1022
80. Lv X, Shi B, Zhou S, Peng H, Lei Y, Yu P (2018) Study on the growth rate of natural gas hydrate in water-in-oil emulsion system using a high-pressure flow loop. RSC Adv 8:36484–36492
81. Ding K, Zhong DL, Lu YY, Le Wang J (2015) Enhanced precombustion capture of carbon dioxide by gas hydrate formation in Water-In-Oil emulsions. Energy Fuels 29:2971–2978
82. A water droplet size distribution dependent modeling of hydrate formation in water/oil emulsion
83. Yegya Raman AK, Venkataramani D, Bhagwat S, Martin T, Clark PE, Aichele CP (2016) Emulsion stability of surfactant and solid stabilized water-in-oil emulsions after hydrate formation and dissociation. Colloids Surf A Physicochem Eng Asp 506:607–621
84. Nashed O, Partoon B, Lal B, Sabil KM, Shariff AM (2018) Review the impact of nanoparticles on the thermodynamics and kinetics of gas hydrate formation. J Nat Gas Sci Eng 55:452–465
85. Said S, Govindaraj V, Herri JM, Ouabbas Y, Khodja M, Belloum M, Sangwai JS, Nagarajan R (2016) A study on the influence of nanofluids on gas hydrate formation kinetics and their potential: application to the CO_2 capture process. J Nat Gas Sci Eng 32:95–108
86. Arjang S, Manteghian M, Mohammadi A (2013) Effect of synthesized silver nanoparticles in promoting methane hydrate formation at 4.7 MPa and 5.7MPa. Chem Eng Res Des 91:1050–1054
87. Rahmati-Abkenar M, Manteghian M, Pahlavanzadeh H (2017) Experimental and theoretical investigation of methane hydrate induction time in the presence of triangular silver nanoparticles. Chem Eng Res Des 120:325–332
88. Rahmati-Abkenar M, Manteghian M, Pahlavanzadeh H (2017) Nucleation of ethane hydrate in water containing silver nanoparticles. Mater Des 126:190–196

89. Kakati H, Mandal A, Laik S (2016) Promoting effect of Al2O3/ZnO-based nanofluids stabilized by SDS surfactant on CH4+C2H6+C3H8 hydrate formation. J Ind Eng Chem 35:357–368
90. Aliabadi M, Rasoolzadeh A, Esmaeilzadeh F, Alamdari AM (2015) Experimental study of using CuO nanoparticles as a methane hydrate promoter. J Nat Gas Sci Eng 27:1518–1522
91. Najibi H, Mirzaee Shayegan M, Heidary H (2015) Experimental investigation of methane hydrate formation in the presence of copper oxide nanoparticles and SDS. J Nat Gas Sci Eng 23:315–323
92. Kumar A, Kushwaha OS, Rangsunvigit P, Linga P, Kumar R (2016) Effect of additives on formation and decomposition kinetics of methane clathrate hydrates: application in energy storage and transportation. Can J Chem Eng 94:2160–2167
93. Veluswamy HP, Kumar S, Kumar R, Rangsunvigit P, Linga P (2016) Enhanced clathrate hydrate formation kinetics at near ambient temperatures and moderate pressures: application to natural gas storage. Fuel 182:907–919
94. Mech D, Gupta P, Sangwai JS (2016) Kinetics of methane hydrate formation in an aqueous solution of thermodynamic promoters (THF and TBAB) with and without kinetic promoter (SDS). J Nat Gas Sci Eng 35:1519–1534
95. Kakati H, Mandal A, Laik S (2016) Effect of SDS/THF on thermodynamic and kinetic properties of formation of hydrate from a mixture of gases (CH4+C2H6+C3H8) for storing gas as hydrate. J Energy Chem 25:409–417
96. Pan Z, Liu Z, Zhang Z, Shang L, Ma S (2018) Effect of silica sand size and saturation on methane hydrate formation in the presence of SDS. J Nat Gas Sci Eng 56:266–280
97. Øtergaard KK, Anderson R, Llamedo M, Tohidi B (2002) Hydrate phase equilibria in porous media: Effect of pore size and salinity. Terra Nov 14:307–312
98. Moroi Y, Matuura R (1988) Thermodynamics of solubilization into surfactant micelles: effect of hydrophobicity of both solubilizate and surfactant molecules. J Colloid Interface Sci 125:456–462
99. Mandal A, Laik S (2008) Effect of the promoter on gas hydrate formation and dissociation. Energy Fuels 22:2527–2532
100. Nashed O, Partoon B, Lal B, Sabil KM, Shariff AM (2019) Investigation of functionalized carbon nanotubes' performance on carbon dioxide hydrate formation. Energy 174:602–610
101. Choi JW, Chung JT, Kang YT (2014) CO2 hydrate formation at atmospheric pressure using high efficiency absorbent and surfactants. Energy 78:869–876
102. Yang M, Song Y, Jiang L, Zhao Y, Ruan X, Zhang Y, Wang S (2014) Hydrate-based technology for CO2 capture from fossil fuel power plants. Appl Energy 116:26–40
103. Liu Z, Pan Z, Zhang Z, Liu P, Shang L, Li B (2018) Effect of porous media and sodium dodecyl sulphate complex system on methane hydrate formation. Energy Fuels 32:5736–5749
104. Pandey JS, Daas YJ, von Solms N (2020) Methane hydrate formation, storage and dissociation behavior in unconsolidated sediments in the presence of environment-friendly promoters
105. Bui T, Phan A, Monteiro D, Lan Q, Ceglio M, Acosta E, Krishnamurthy P, Striolo A (2017) Evidence of structure-performance relation for surfactants used as antiagglomerants for hydrate management. Langmuir 33:2263–2274
106. Sicard F, Bui T, Monteiro D, Lan Q, Ceglio M, Burress C, Striolo A (2018) Emergent properties of antiagglomerant films control methane transport: implications for hydrate management. Langmuir 34:9701–9710
107. Bui T, Sicard F, Monteiro D, Lan Q, Ceglio M, Burress C, Striolo A (2018) Antiagglomerants affect gas hydrate growth. J Phys Chem Lett 9:3491–3496
108. Phan A, Bui T, Acosta E, Krishnamurthy P, Striolo A (2016) Molecular mechanisms responsible for hydrate anti-agglomerant performance. Phys Chem Chem Phys 18:24859–24871
109. Tokiwa Y, Sakamoto H, Takiue T, Aratono M, Matsubara H (2015) Effect of alkane chain length and counterion on the freezing transition of cationic surfactant adsorbed film at alkane mixture—water interfaces. J Phys Chem B 119:6235–6241
110. Lei Q, Bain CD (2004) Surfactant-induced surface freezing at the alkane-water interface. Phys Rev Lett 92:2–5

111. Huo Z, Freer E, Lamar M, Sannigrahi B, Knauss DM, Sloan ED (2001) Hydrate plug prevention by anti-agglomeration. Chem Eng Sci 56:4979–4991
112. York JD, Firoozabadi A (2009) Effect of brine on hydrate antiagglomeration. Energy Fuels 23:2937–2946
113. York JD, Firoozabadi A (2008) Alcohol cosurfactants in hydrate antiagglomeration. J Phys Chem B 112:10455–10465
114. Aman ZM, Haber A, Ling NNA, Thornton A, Johns ML, May EF (2015) Effect of brine salinity on the stability of hydrate-in-oil dispersions and water-in-oil emulsions. Energy Fuels 29:7948–7955
115. Striolo A, Phan A, Walsh MR (2019) Molecular properties of interfaces relevant for clathrate hydrate agglomeration. Curr Opin Chem Eng 25:57–66
116. Stern LA, Circone S, Kirby SH, Durham WB (2001) Anomalous preservation of pure methane hydrate at 1 atm. J Phys Chem B 105:1756–1762
117. Kang HJ, Yang Y, Ki MS, Shin MS, Choi J, Cha JH, Lee D (2016) A concept study for cost effective NGH mid-stream supply chain establishing strategies. Ocean Eng 113:162–173
118. Rehder G, Eckl R, Elfgen M, Falenty A, Hamann R, Kähler N, Kuhs WF, Osterkamp H, Windmeier C (2012) Methane hydrate pellet transport using the self-preservation effect: a techno-economic analysis. Energies 5:2499–2523
119. Zhang G, Rogers RE (2008) Ultra-stability of gas hydrates at 1 atm and 268.2 K. Chem Eng Sci 63:2066–2074
120. Subramani A, Jacangelo JG (2015) Emerging desalination technologies for water treatment: a critical review. Water Res 75:164–187
121. Kang KC, Linga P, Park KN, Choi SJ, Lee JD (2014) Seawater desalination by gas hydrate process and removal characteristics of dissolved ions (Na+, K+, Mg2+, Ca2+, B3+, Cl−, SO42−). Desalination 353:84–90
122. Park KN, Hong SY, Lee JW, Kang KC, Lee YC, Ha MG, Lee JD (2011) A new apparatus for seawater desalination by gas hydrate process and removal characteristics of dissolved minerals (Na+, Mg2+, Ca2+, K+, B3+). Desalination 274 91–96
123. He T, Nair SK, Babu P, Linga P, Karimi IA (2018) A novel conceptual design of hydrate based desalination (HyDesal) process by utilizing LNG cold energy. Appl Energy 222:13–24
124. Li F, Chen Z, Dong H, Shi C, Wang B, Yang L, Ling Z (2018) Promotion effect of graphite on cyclopentane hydrate based desalination. Desalination 445:197–203
125. Hong S, Moon S, Lee Y, Lee S, Park Y (2019) Investigation of thermodynamic and kinetic effects of cyclopentane derivatives on CO2 hydrates for potential application to seawater desalination. Chem Eng J 363:99–106
126. McAuliffe C (1966) Solubility in water of paraffin, cycloparaffin, olefin, acetylene, cycloolefin, and aromatic hydrocarbons. J Phys Chem 70:1267–1275
127. Herslund PJ, Thomsen K, Abildskov J, Von Solms N (2014) Modelling of cyclopentane promoted gas hydrate systems for carbon dioxide capture processes. Fluid Phase Equilib 375:89–103
128. Herri JM, Bouchemoua A, Kwaterski M, Brântuas P, Galfré A, Bouillot B, Douzet J, Ouabbas Y, Cameirao A (2014) Amélioration de la sélectivité du captage du CO_2 dans les semi-clathrates hydrates en utilisant les ammoniums quaternaires comme promoteurs thermodynamiques. Oil Gas Sci Technol 69:947–968
129. Ho-Van S, Bouillot B, Douzet J, Babakhani SM, Herri JM (2019) Cyclopentane hydrates-a candidate for desalination? J Environ Chem Eng 7:103359
130. Erfani A, Varaminian F (2016) Kinetic promotion of non-ionic surfactants on cyclopentane hydrate formation. J Mol Liq 221:963–971
131. Brown EP, Koh CA (2016) Micromechanical measurements of the effect of surfactants on cyclopentane hydrate shell properties. Phys Chem Chem Phys 18:594–600
132. Henry D, Gilles B, Jean-Philippe T, Christophe, D, Philippe G (2019) Evaluation of the performance of a new biodegradable AA-LDHI in cyclopentane hydrate and CH4/C3H8 gas hydrate systems. SPE Middle East Oil Gas Show Conf. MEOS, Proc, 2019-March, pp 1–13

133. Linga P, Kumar R, Englezos P (2007) Gas hydrate formation from hydrogen/carbon dioxide and nitrogen/carbon dioxide gas mixtures. Chem Eng Sci 62:4268–4276
134. Kumar R, Linga P, Englezos P (2006) Pre post combustion capture of carbon dioxide via hydrate formation. IEEE EIC Clim Chang Technol Conf EICCCC 2006:1–7
135. Xu CG, Li X (2014) Sen Research progress of hydrate-based CO_2 separation and capture from gas mixtures. RSC Adv 4:18301–18316
136. Ho LC, Babu P, Kumar R, Linga P (2013) HBGS (hydrate based gas separation) process for carbon dioxide capture employing an unstirred reactor with cyclopentane. Energy 63:252–259
137. da Lirio CFS, Pessoa FLP, Uller AMC (2013) Storage capacity of carbon dioxide hydrates in the presence of sodium dodecyl sulfate (SDS) and tetrahydrofuran (THF). Chem Eng Sci 96:118–123
138. Jiang, L. Le, Li AR, Xu JF, Liu YJ (2016) Effects of SDS and SDBS on CO2Hydrate formation, induction time, storage capacity and stability at 274.15 K and 5.0 MPa. Chem Select 1:6111–6114
139. Yang M, Liu W, Song Y, Ruan X, Wang X, Zhao J, Jiang L, Li Q (2013) Effects of additive mixture (THF/SDS) on the thermodynamic and kinetic properties of CO2/H2 hydrate in porous media. Ind Eng Chem Res 52:4911–4918
140. Song Y, Wang X, Yang M, Jiang L, Liu Y, Dou B, Zhao J, Wang S (2013) Study of selected factors affecting hydrate-based carbon dioxide separation from simulated fuel gas in porous media. Energy Fuels 27:3341–3348
141. Pandey JS, von Solms N (2019) Hydrate stability and methane recovery from gas hydrate through CH_4–CO_2 replacement in different mass transfer scenarios. Energies 12:2309
142. Frankcombe TJ, Kroes GJ (2007) Molecular dynamics simulations of type-sII hydrogen clathrate hydrate close to equilibrium conditions. J Phys Chem C 111:13044–13052
143. Struzhkin VV, Militzer B, Mao WL, Mao HK, Hemley RJ (2007) Hydrogen storage in molecular clathrates. Chem Rev 107:4133–4151
144. Ozaki M, Tomura S, Ohmura R, Mori YH (2014) Comparative study of large-scale hydrogen storage technologies: Is hydrate-based storage at advantage over existing technologies? Int J Hydrogen Energy 39:3327–3341
145. Trueba AT, Radović IR, Zevenbergen JF, Kroon MC, Peters CJ (2012) Kinetics measurements and in situ Raman spectroscopy of formation of hydrogen-tetrabutylammonium bromide semi-hydrates. Int J Hydrogen Energy 37:5790–5797
146. Veluswamy HP, Chin WI, Linga P (2014) Clathrate hydrates for hydrogen storage: The impact of tetrahydrofuran, tetra-n-butylammonium bromide and cyclopentane as promoters on the macroscopic kinetics. Int J Hydrogen Energy 39:16234–16243
147. Du J, Wang L, Liang D, Li D (2012) Phase equilibria and dissociation enthalpies of hydrogen semi-clathrate hydrate with tetrabutyl ammonium nitrate. J Chem Eng Data 57:603–609
148. Florusse LJ, Peters CJ, Schoonman J, Hester KC, Koh CA, Dec SF, Marsh KN, Sloan ED (2004) Stable low-pressure hydrogen clusters stored in a binary clathrate hydrate. Science (80-.) 306:469–471
149. Veluswamy HP, Linga P (2013) Macroscopic kinetics of hydrate formation of mixed hydrates of hydrogen/tetrahydrofuran for hydrogen storage. Int J Hydrogen Energy 38:4587–4596
150. Alavi S, Ripmeester JA (2007) Hydrogen-gas migration through clathrate hydrate cages. Angew. Chemie—Int. Ed. 46:6102–6105
151. Okuchi T, Moudrakovski IL, Ripmeester JA (2007) Efficient storage of hydrogen fuel into leaky cages of clathrate hydrate. Appl Phys Lett 91:2005–2008
152. Kumar R, Klug DD, Ratcliffe CI, Tulk CA, Ripmeester JA (2013) Low-pressure synthesis and characterization of hydrogen-filled ice Ic. Angew Chemie—Int Ed 52:1531–1534
153. Lu H, Wang J, Liu C, Ratcliffe CI, Becker U, Kumar R, Ripmeester J (2012) Multiple H 2 occupancy of cages of clathrate hydrate under mild conditions. J Am Chem Soc 134:9160–9162
154. Grim RG, Kerkar PB, Sloan ED, Koh CA, Sum AK (2012) Rapid hydrogen hydrate growth from non-stoichiometric tuning mixtures during liquid nitrogen quenching. J Chem Phys 136

155. Veluswamy HP, Chen JY, Linga P (2015) Surfactant effect on the kinetics of mixed hydrogen/propane hydrate formation for hydrogen storage as clathrates. Chem Eng Sci 126:488–499
156. Di Profio P, Arca S, Rossi F, Filipponi M (2009) Comparison of hydrogen hydrates with existing hydrogen storage technologies: energetic and economic evaluations. Int J Hydrogen Energy 34:9173–9180
157. Veluswamy HP, Hong QW, Linga P (2016) Morphology study of methane hydrate formation and dissociation in the presence of amino acid
158. Carter BO, Wang W, Adams DJ, Cooper AI (2010) Gas storage in "Dry Water" and "Dry Gel" clathrates. Langmuir 26:3186–3193
159. Liu Y, Chen B, Chen Y, Zhang S, Guo W, Cai Y, Tan B, Wang W (2015) Methane storage in a hydrated form as promoted by leucines for possible application to natural gas transportation and storage. Energy Technol 3:815–819
160. Scott MJ, Jones MN (2000) The biodegradation of surfactants in the environment. Biochim Biophys Acta—Biomembr 1508:235–251
161. Campbell JM (1992) Gas conditioning and processing, vol 2. The equipment modules, ISBN 0-9703449-0-2
162. Sloan ED, Koh CA (2007) Clathrate hydrates of natural gases, 3rd edn, ISBN 9781420008494
163. Carroll J (2014) Natural gas hydrates—a guide for engineers, 3rd edn, ISBN 978-0-12-800074-8
164. Anderson BJ, Tester JW, Borghi GP, Trout BL (2005) Properties of inhibitors of methane hydrate formation via molecular dynamics simulations. J Am Chem Soc 127:17852–17862
165. Xiao C, Adidharma H (2009) Dual function inhibitors for methane hydrate. Chem Eng Sci 64:1522–1527
166. Kelland MA, Moi N, Howarth M (2013) Breakthrough in synergists for kinetic hydrate inhibitor polymers, hexaalkylguanidinium salts: Tetrahydrofuran hydrate crystal growth inhibition and synergism with polyvinylcaprolactam. Energy Fuels
167. Sa JH, Kwak GH, Lee BR, Park DH, Han K, Lee KH (2013) Hydrophobic amino acids as a new class of kinetic inhibitors for gas hydrate formation. Sci Rep 3:1–7
168. Lee D, Go W, Seo Y (2019) Experimental and computational investigation of methane hydrate inhibition in the presence of amino acids and ionic liquids. Energy
169. Klamt A (1995) Conductor-like screening model for real solvents: a new approach to the quantitative calculation of solvation phenomena. J Phys Chem
170. Eckert F, Klamt A (2002) Fast solvent screening via quantum chemistry: COSMO-RS Approach. AIChE J
171. Klamt A, Eckert F, Arlt W (2010) COSMO-RS: an Alternative to simulation for calculating thermodynamic properties of liquid mixtures. Annu Rev Chem Biomol Eng
172. Klamt A, Schüürmann G (1993) COSMO: a new approach to dielectric screening in solvents with explicit expressions for the screening energy and its gradient. J Chem Soc Perkin Trans 2
173. Dickens GR, Quinby-Hunt MS (1997) Methane hydrate stability in pore water: A simple theoretical approach for geophysical applications. J Geophys Res B Solid Earth 102:773–783
174. Kirkwood JG (1934) Theory of solutions of molecules containing widely separated charges with special application to zwitterions. J Chem Phys
175. Nashed O, Dadebayev D, Khan MS, Bavoh CB, Lal B, Shariff AM (2018) Experimental and modelling studies on thermodynamic methane hydrate inhibition in the presence of ionic liquids. J Mol Liq
176. Pieroen AP (1955) Gas hydrates—approximate relations between heat of formation, composition and equilibrium temperature lowering by "inhibitors". Recl des Trav Chim des Pays-Bas 74:995–1002
177. Partoon B, Wong NMS, Sabil KM, Nasrifar K, Ahmad MR (2013) A study on thermodynamics effect of [EMIM]-Cl and [OH-C2MIM]-Cl on methane hydrate equilibrium line. Fluid Phase Equilib

178. Javanmardi J, Moshfeghian M, Maddox RN (1998) Simple method for predicting gas-hydrate-forming conditions in aqueous mixed-electrolyte solutions. Energy Fuels 12:219–222
179. Javanmardi J, Moshfeghian M, Maddox RN (2001) An accurate model for prediction of gas hydrate formation conditions in mixtures of aqueous electrolyte solutions and alcohol. Can J Chem Eng 79:367–373
180. Bavoh CB, Partoon B, Lal B, Gonfa G, Foo Khor S, Sharif AM (2017) Inhibition effect of amino acids on carbon dioxide hydrate. Chem Eng Sci 171:331–339
181. Xiao C, Wibisono N, Adidharma H (2010) Dialkylimidazolium halide ionic liquids as dual function inhibitors for methane hydrate. Chem Eng Sci 65:3080–3087
182. Atilhan M, Pala N, Aparicio S (2014) A quantum chemistry study of natural gas hydrates. J Mol Model 20:1–15
183. Tariq M, Atilhan M, Khraisheh M, Othman E, Castier M, García G, Aparicio S, Tohidi B (2016) Experimental and DFT approach on the determination of natural gas hydrate equilibrium with the use of excess N_2 and choline chloride ionic liquid as an inhibitor. Energy Fuels 30:2821–2832
184. Choudhary N, Das S, Roy S, Kumar R (2016) Effect of polyvinylpyrrolidone at methane hydrate-liquid water interface. Application in flow assurance and natural gas hydrate exploitation. Fuel 186:613–622
185. Mohamed NA, Tariq M, Atilhan M, Khraisheh M, Rooney D, Garcia G, Aparicio S (2017) Investigation of the performance of biocompatible gas hydrate inhibitors via combined experimental and DFT methods. J Chem Thermodyn 111:7–19
186. Bellucci MA, Walsh MR, Trout BL (2018) Molecular dynamics analysis of anti-agglomerant surface adsorption in natural gas hydrates 122
187. Fang B, Ning F, Hu S, Guo D, Ou W, Wang C, Wen J, Sun J, Liu Z, Koh CA (2020) The effect of surfactants on hydrate particle agglomeration in liquid hydrocarbon continuous systems: a molecular dynamics simulation study. RSC Adv 10:31027–31038
188. Bhattacharjee G, Choudhary N, Barmecha V, Kushwaha OS, Pande NK, Chugh P, Roy S, Kumar R (2019) Methane recovery from marine gas hydrates: a bench scale study in presence of low dosage benign additives. Appl Energy 253:113566
189. Sammalkorpi M, Karttunen M, Haataja M (2007) Structural properties of ionic detergent aggregates: a large-scale molecular dynamics study of sodium dodecyl sulfate. J Phys Chem B 111:11722–11733
190. Chun BJ, Choi JII, Jang SS (2015) Molecular dynamics simulation study of sodium dodecyl sulfate micelle: Water penetration and sodium dodecyl sulfate dissociation. Colloids Surfaces A Physicochem Eng Asp 474:36–43
191. Kitabata M, Fujimoto K, Yoshii N, Okazaki S (2016) A molecular dynamics study of local pressures and interfacial tensions of SDS micelles and dodecane droplets in water. J Chem Phys 144:224701
192. Bresme F, Faraudo J (2004) Computer simulation studies of newton black films. Langmuir 20:5127–5137
193. Hande VR, Chakrabarty S (2016) Exploration of the presence of bulk-like water in AOT reverse micelles and water-in-oil nanodroplets: the role of charged interfaces, confinement size and properties of water. Phys Chem Chem Phys 18:21767–21779
194. Volkov NA, Tuzov NV, Shchekin AK (2016) Molecular dynamics study of salt influence on transport and structural properties of SDS micellar solutions. Fluid Phase Equilib 424:114–121
195. Poghosyan AH, Arsenyan LH, Gharabekyan HH, Falkenhagen S, Koetz J, Shahinyan AA (2011) Molecular dynamics simulations of inverse sodium dodecyl sulfate (SDS) micelles in a mixed toluene/pentanol solvent in the absence and presence of poly(diallyldimethylammonium chloride) (PDADMAC). J Colloid Interface Sci 358:175–181
196. Poghosyan AH, Arsenyan LH, Shahinyan AA, Koetz J (2016) Polyethyleneimine loaded inverse SDS micelle in pentanol/toluene media. Colloids Surf A Physicochem Eng Asp 506:402–408

197. Fujimoto K, Yoshii N, Okazaki S (2012) Free energy profiles for penetration of methane and water molecules into spherical sodium dodecyl sulfate micelles obtained using the thermodynamic integration method combined with molecular dynamics calculations. J Chem Phys 136:014511
198. Fujimoto K, Yoshii N, Okazaki S (2012) Molecular dynamics study of free energy of transfer of alcohol and amine from water phase to the micelle by thermodynamic integration method. J Chem Phys 137:094902
199. Brodskaya EN (2012) Computer simulations of micellar systems. Colloid J 74:154–171
200. Kinning DJ, Winey KI, Thomas EL (1988) Structural transitions from spherical to nonspherical micelles in blends of poly(styrene-butadiene) diblock copolymer and polystyrene homopolymers. Macromolecules 21:3502–3506
201. Choudhary N, Hande VR, Roy S, Chakrabarty S, Kumar R (2018) Effect of sodium dodecyl sulfate surfactant on methane hydrate formation: a molecular dynamics study. J Phys Chem B 122:6536–6542
202. Khurana M, Yin Z, Linga P (2017) A review of clathrate hydrate nucleation. ACS Sustain Chem Eng 5:11176–11203
203. Walsh MR, Koh CA, Sloan DE, Sum AK, Wu DT (2009) Microsecond simulations of spontaneous methane hydrate nucleation and growth. Science (80-.) 326:1095–1098
204. Michalis VK, Tsimpanogiannis IN, Stubos AK, Economou IG (2016) Direct phase coexistence molecular dynamics study of the phase equilibria of the ternary methane-carbon dioxide-water hydrate system. Phys Chem Chem Phys 18:23538–23548
205. Kumar A, Veluswamy HP, Linga P, Kumar R (2019) Molecular level investigations and stability analysis of mixed methane-tetrahydrofuran hydrates: implications to energy storage. Fuel 236:1505–1511
206. Carver TJ, Drew MGB, Rodger PM (1995) Inhibition of crystal growth in methane hydrate. J Chem Soc, Faraday Trans 91:3449–3460
207. Mehrabian H, Bellucci MA, Walsh MR, Trout BL (2018) Effect of Salt on Antiagglomerant Surface Adsorption in Natural Gas Hydrates. J Phys Chem C 122:12839–12849
208. Mehrabian H, Walsh MR, Trout BL (2019) In Silico Analysis of the Effect of Alkyl Tail Length on Antiagglomerant Adsorption to Natural Gas Hydrates in Brine. J Phys Chem C 123:17239–17248
209. Jiménez-Ángeles F, Firoozabadi A (2018) Hydrophobic hydration and the effect of NaCl salt in the adsorption of hydrocarbons and surfactants on clathrate hydrates. ACS Cent Sci 4:820–831
210. Naullage PM, Bertolazzo AA, Molinero V (2019) How do surfactants control the agglomeration of clathrate hydrates?. ACS Cent, Sci
211. Bavoh CB, Lal B, Nashed O, Khan MS, Keong LK, Bustam MA (2016) COSMO-RS: An ionic liquid prescreening tool for gas hydrate mitigation. *Chinese J. Chem. Eng*
212. Klamt A, Eckert F (2000) COSMO-RS: a novel and efficient method for the a priori prediction of thermophysical data of liquids. Fluid Phase Equilib 172:43–72
213. Klamt A, Eckert F, Hornig M, Beck ME, Brger T (2002) Prediction of aqueous solubility of drugs and pesticides with COSMO-RS. J Comput Chem 23:275–281
214. Klamt A (2012) Solvent-screening and co-crystal screening for drug development with COSMO-RS. J Cheminform 4:1–2
215. Paduszyński K (2017) An overview of the performance of the COSMO-RS approach in predicting the activity coefficients of molecular solutes in ionic liquids and derived properties at infinite dilution. Phys Chem Chem Phys 19:11835–11850
216. Xia L, Wang J, Liu S, Li Z, Pan H (2019) Prediction of CO_2 solubility in ionic liquids based on multi-model fusion method. Processes 7:258

Demulsification

Surfactants as Integral Components of Chemical Demulsifiers

Abubakar A. Umar, Nobert I. Nnakenyi, Muhammad K. Abba, and I. H. Roy-Omeni

Abstract Emulsions are thermodynamically unstable systems, since they will separate to reduce the interfacial area between the oil phase and the water phase, as a function of time. As a metastable system, surfactant molecules, amphiphilic polymers or solid particles must be present before a stable emulsion system is formed. These components of an emulsion system are called emulsifiers. The relative balance of the hydrophilic and lipophilic properties of these emulsifiers is known to be the most important parameter dictating the emulsion type, whether an oil-in-water (o/w) or water-in-oil (w/o) emulsion. Irrespective of the emulsion type formed, demulsification is a costly exercise in the oil and gas industry. This chapter describes the fundamental role played by surface active agents (surfactants) as integral components of a chemical demulsifier.

Keywords Emulsions · Demulsification · Demulsifiers · Surfactants · Oil/water interface

1 Introduction

An exhaustive literature searches on emulsions during this study led to the understanding that there are certain universal rules that form the basic theory of emulsions formation and breaking. Firstly, crude oil emulsions are comprised essentially of immiscible liquids. Separation should be the normal susceptibility of these liquids in the presence of density difference between the immiscible liquids. Secondly, the rate of gravitational settling rate depends on the surface tension of the dispersed droplets. When these internal phase droplets are large, their surface tensions as a function of mass is smaller than when they are small. Therefore, any technique that would increase the sizes of the dispersed droplets (via coalescence mostly)

A. A. Umar (✉) · N. I. Nnakenyi · I. H. Roy-Omeni
Nigeria Petroleum Development Company (NPDC), Benin City, Nigeria
e-mail: abubakar_g02934@utp.edu.my

M. K. Abba
The University of Salford, Manchester M4 5WT, UK

T. Solling et al. (eds.), *Surfactants in Upstream E&P*, Petroleum Engineering,
https://doi.org/10.1007/978-3-030-70026-3_16

will enhance water separation rate and subsequent demulsification. Thirdly, environmental influence on emulsion stability is also key. An emulsion that is stable in a given environment might separate instantly when a different environment is introduced to it. Lastly, a stable emulsion would form only in the presence of emulsifiers. Any technique that would alter, eliminate or neutralize the emulsifiers would lead to a successful separation of the water from oil.

2 The Emulsification Process—Surfactant Action

The formation of crude oil emulsion involves the commingling of crude oil and water (brine) when there is adequate mixing or agitation and in the presence of a surface-active agent (surfactant) that serves as an emulsifier.

How rigorous the two phases are being mixed as well as the amount or concentration of emulsifiers present is very critical in the formation of emulsions [1, 2]. There exist various mixing sources within reach during crude oil production. These sources provide what is called the amount of shear. These sources include; the flow of fluid through reservoir rock, fluid flow through tubing, bottom-hole perforations, flowlines, production headers (pipes connecting the wellhead to the manifold), chokes or orifices, several surface equipment as well as during the release of gas bubbles when there is change is fluid phase [3, 4]. Although a universal theory that has been hypothesized for all emulsions types and behaviors does not exist, several hypotheses have been suggested to explain variations in the processes of emulsions formation and stabilization [5]. Various researchers [6–15] have studied and identified several factors responsible for emulsions formation and stabilization, with variations in the significance of roles played by each factor.

The roles of surfactants in emulsion stabilization has been well established. Two major mechanisms have been identified, via which surfactants stabilize emulsions. These are: (a) by adsorption at the liquid–liquid interface, and (b) by altering the stability of solids at the interface.

2.1 Surfactants Adsorption at Liquid–Liquid Interface

The adsorption of surfactants at liquid interfaces can influence emulsion stability by lowering interfacial tension, increasing surface elasticity, increasing electric double layer repulsion (ionic surfactants), and possibly increasing surface viscosity [16]. Also, surfactant nature can control the arrangement of the phases in an emulsion, that is, which phase will form the dispersed versus continuous phase. As discussed earlier, several experimental predictive methods based on anticipated surfactant positioning at the interface exist [16–18]. These include the Bancroft's rule, the oriented wedge theory, the hydrophile–lipophile balance (HLB), and the volume balance value [17]. Among all the methods, the HLB has been the most widely used. The HLB

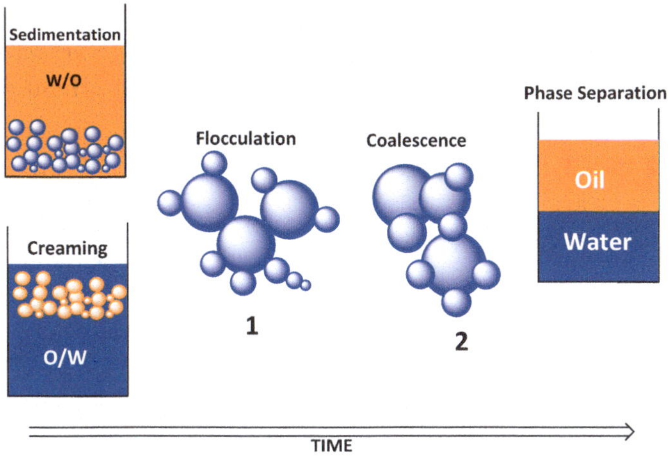

Fig. 1 Schematic representation of the various breakdown processes in emulsions. Modified from Alvarado et al. [22]

dimensionless scale ranges from 0 to 20 for non-ionic surfactants; a low HLB (<9) refers to a lipophilic surfactant (oil soluble) and a high HLB (>11) to a hydrophilic (water soluble) surfactant. Most ionic surfactants have HLB values greater than 20. Water-in-oil (w/o) surfactants show HLB values in the range 3–8 while oil-in-water (o/w) emulsifiers possess HLB values of about 8–18 [18–20]. Additionally, surfactant exchanges between the interface and the bulk can drastically lower interfacial visco-elasticities [21]. Figure 2 is a schematic surfactants types used to stabilize food emulsions (Fig. 1).

3 The Demulsification Process

Knowledge of the nature of emulsions and demulsification is important in resolving any emulsion [24, 25]. Nearly three-fourths of all crude oil produced must be treated. The percentage of crude needing treatment increases as fields become more mature and more water is produced. Demulsification most often is attained through the use of surfactants, heat and electric treaters. It is thought that over 75% of all oilfield emulsions are treated with chemicals usually containing alkylene oxide condensates. Although acids and bases are known to affect emulsion stability; they rarely are used in the oil field except in batch treatment when the stability of the emulsion is very severe [26, 27]. Separating produced water from crude oil is a problem that is as old as the oil and gas industry itself. In other words, this menace has existed since the advent of the oil industry. In the beginning of the Industry, the problem was managed by settling the free water from oil in open tanks or pits. The middle phase between clean water and clean oil, also known as "sludge," was disposed of normally by

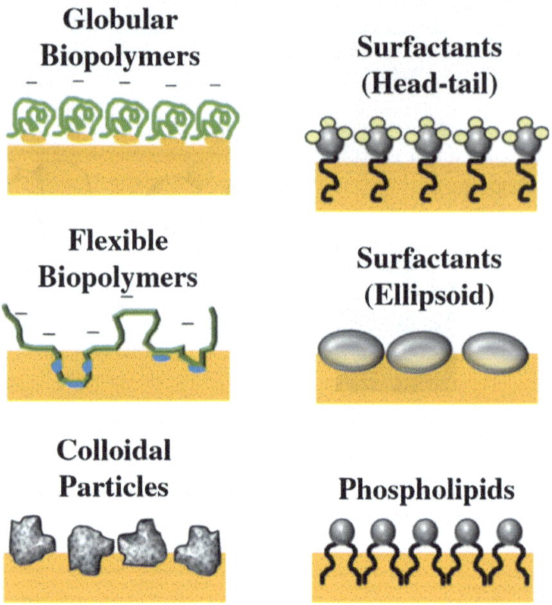

Fig. 2 Schematics of surfactants types used to stabilize food emulsions [23]

burning. It was not until the turn of the century that study and experience showed that, the "sludge" is an emulsion of crude oil and water and that large amounts of commercial oil can be recovered from this emulsion [28]. Several researchers have highlighted the necessity of demulsification for decades [29].

Kim [30] highlights the necessity of breaking emulsions in many practical applications such as the petroleum industry, coating, painting, and wastewater treatment in environmental technology. This is due to the occurrence of unwanted emulsions at many points within the process plant. These emulsions if not well treated can endanger the entire process or lead to added cost of operation.

Gafonava and Yarranton [31] in their work, pointed out the relevance of demulsification as the use of steam and caustic injection or combustion processes for in-situ recovery of heavy crude oils is complicated by the production of viscous emulsions of oil, water and clay. The demulsification of crude oil emulsions forms an integral part of crude oil production system. Even though demulsification of water-in-oil emulsions are conducted using either of four methods; chemical, mechanical, electrical and thermal, other methods such as pH adjustment, filtration, and membrane separation can also be used.

4 Demulsification Techniques

Although different types of techniques all aimed at; destroying the effects of the film surrounding the water droplets, bringing coalescence and providing an opportunity for undisturbed settling of water droplets through the oil- have been proposed and used at several applications [32], chemical demulsification is the most widely used and offers the advantage of completely preventing emulsions when injected at early stage of the treatment, and breaking emulsions at significantly lower temperatures, thus saving fuel and reducing crude oil volume and gravity losses [33, 34]. As a cost-effective and convenient method, chemical demulsification has been employed in the oil and gas industry to break up w/o emulsions. Chemical demulsifiers are amphiphilic compounds, which can destabilize emulsions by changing the interfacial film properties, such as interfacial tension, mechanical strength, elasticity and thickness of interfacial regions to promote coalescence, or through flocculation of water droplets [35]. Of all the methods of demulsification, the chemical method is the most widely used technique and it involves the use of chemical demulsifiers to accelerate the emulsion breaking process [30]. Figure 1 presents the detailed breakdown processes in an emulsion.

In emulsion treatment, the emulsion with large dispersed water percentage is less stable and easier to treat than one with a relatively small percentage of water [36]. This is largely because the presence of large dispersed water phase gives more opportunity for the droplets to join, collide, and subsequent coalescence into larger drops. This opportunity does not present itself for the few isolated drops in an emulsion with low water percentage. In an analogous way, a fresh emulsion is easier to treat than an aged emulsion. Among other reasons, this is because a prolonged ageing of emulsions in a tank permits the settling out of some of the water droplets (by gravity) leaving fewer droplets of water than if the emulsion is a fresh one. Again, less opportunity for the water droplets to collide and coalesce [30]. The breakdown of an emulsion process is majorly classified into five. These are: creaming and sedimentation, flocculation, Ostwald Ripening (Disproportionation), coalescence and phase inversion [9, 37]. Brief explanations with regards each of the demulsification step are provided in subsequent subsections.

4.1 Creaming and Sedimentation

This process is due to external forces; usually gravitational or centrifugal. When such forces surpass the thermal motion of the droplets (Brownian motion), a concentration gradient builds up in the system with the bigger droplets moving faster to the top (if their density is lower than that of the medium) or to the bottom (if their density is larger than that of the medium) of the container. In the limiting cases, the droplets may form a close-packed (random or ordered) array at the top or bottom of the system (as in points 1 and 2 in Fig. 1) with the remainder of the volume occupied by the continuous liquid phase [9, 37].

4.2 Flocculation

This breakdown process refers to the aggregation of the droplets into larger units (without any change in primary droplet size). This is due to the van der Waals attraction that is universal with all dispersed systems. Flocculation occurs when there is not enough repulsion to keep the droplets apart to distances where the Vander-Waals attraction is weak. Flocculation may be strong or weak, depending on the magnitude of the attractive energy involved [9, 37].

4.3 Ostwald Ripening (Disproportionation)

This process is consequent of the finite solubility of the liquid phases. Liquids that are referred to as being immiscible often have mutual solubilities that are not negligible. With emulsions, which are usually poly-disperse, the smaller droplets will have larger solubility when compared with the larger ones (due to curvature effects). With time, the smaller droplets disappear, and their molecules diffuse to the bulk and become deposited on the larger droplets. With time, the droplet size distribution shifts to larger values [9, 37].

4.4 Coalescence

This refers to the process of thinning and disruption of the liquid film between the droplets with the result of fusion of two or more droplets to form larger ones. The limiting case for coalescence is the complete separation of the emulsion into two distinct liquid phases. The driving force for coalescence is the surface or film fluctuations which results in close approach of the droplets whereby the van der Waals forces is strong thus preventing their separation [9, 37].

4.5 Phase Separation

This refers to the process whereby there will be an exchange between the disperse phase and the medium. For example, an o/w emulsion may with time or change of conditions invert to a w/o emulsion. In many cases, phase inversion passes through a transition state whereby multiple emulsions are produced [9, 37].

5 The Chemical Demulsification Technique

Chemical demulsifiers used in treating emulsions are similar to emulsifiers (they are also surface-active agents) [34]. The pioneer of the chemical method of treating emulsions was William S. Barnickel, a pharmaceutical chemist who, later became interested in this method of approach to treating oil fields in mid-continent [1]. The first chemical reagents used were of the inorganic type. This discovery was then followed by the discovery of the ability of soaps and soap variations to act as demulsifiers.

Monson [38] highlighted stages of demulsifier formulations prior to the knowledge of polyether condensates synthesis. That era witnessed the direct use of chemicals such as Turkey red oil, sulphuric acid, sulphated castor oil, mahogany soaps, polyamines and polyhydric alcohols as demulsifiers. The technology of alkylene oxide condensation then followed in the 1940s and opened a new phase of formulating demulsifiers with condensation products of ethylene, propylene and butylene oxides.

Post 1945, companies involved in surfactant technology had access to the whole branch of chemistry. This made possible the large-scale production of condensed polymer via ethylene and propylene oxides [39]. This led to the wide appearance of a new class of nonionic detergents in the oilfield markets. The condensation products of ethylene oxide were found to be water soluble, and the high reactivity of the Oxirane ring made it valuable in several chemical reactions. It was also found that the propylene oxide gave poly condensation products their tendency to be oil soluble.

The search for effective demulsifiers has continued since then, and the oil and gas industry and service providers have been tirelessly working towards solving this production menace [12, 40–46]. However, due to the complexity of an emulsion, and the variations in the nature and type of emulsifiers responsible for each emulsion type, a universal demulsifier that can solve all emulsion problems is far from being a reality. In fact, such attempt could amount to a work in futility.

Chemicals employed as demulsifiers may be ordinary surface-active agents (surfactants). These surfactants can be cationic or anionic. Cationic surfactants such as quaternary amines $(NR_1R_2R_3R4)^+$, where R can be any alkyl or aryl group have been used greatly in demulsifier formulation. Anionic surfactants such as sodium dodecybenzenesul-fonates ($RPhSO_3Na$), petroleum sulfonates ($RSO_3^-M^+$) and sodium di-iso-octylsulfonosuccinates [$ROOCC(CH_2COOR)H\ SO_3\ Na^1$, trade name Aerosol OT].

Apart from the ionic surfactants (cationic and anionic), nonionic surfactants have been used as well in the preparation of chemical demulsifiers. Surfactants such as fatty alcohol ethers [$CH_3(CH_2)_{10}CH_2O(C_2H_4O)_nH$], fatty esters [$(CH_3(CH_2)_{10}COO(C_2H_4O)_nH$], alkyl phenol ethers [$R\text{-}Ph\text{-}O\text{-}(C_2H_4O)_nH$], polyoxypropylene glycol ethers, and fatty amides. In the same vein, zwitterionic surfactants that are pH dependent, such as alkylbetaine derivatives [$RCH_2COO^-N^+(CH_3)_2$] have been used as demulsifiers too. There are situations where a more specific demulsifier is needed, and simple copolymers of ethylene oxide (EO) and propylene oxide (PO) may be used singly or as a blend in the presence of a surfactant to produce a good chemical demulsifier [47].

There has been an intricate classification of these surfactants in the literature (as shown in Fig. 2). There are several bases on which surfactants are classified, majorly; a classification that is based on usage and another classification based on their dissociation in water. Although they have largely been classified based on their usage, this classification is not as important as their classification based on their dissociation in water [48, 49]. In the Oil and Gas industry, for example, there is an increased deployment of surfactants during drilling fluids formulation. When formulating oil-based drilling fluid, for instance, surfactants are employed to serve as emulsifiers and/or wetting agents. However, the formulation of water-based drilling muds requires applications of diverse techniques that may involve oil in water (o/w) emulsification, prevention of wellbore instabilities using shell swelling inhibitors, and so on [13].

In this section, we will discuss surfactants classifications based on their dissociation in water.

5.1 Anionic Surfactants

Anionic surfactants are those surface-active agents that dissociate in water into an amphiphilic anion (such as SO_4^{2-}, COO^-, and SO_3^-) and a corresponding cation that is either an alkaline metal (Na+, K+) or quaternary ammonium. Anionic surfactants are the most extensively utilized surfactants in several industries largely due to its temperature stability and tolerance. They include the sulfonates of alkylbenzene (most common detergents), soaps (fatty acid), agents used in foam production (like lauryl sulfate), agents used to wet surfaces (wetting agents like di-alkyl sulfosuccinate), dispersants (like lignosulfonates), and so on [13]. It is believed that more than half of the world surfactants produced are Anionic surfactants [49]. This moiety bears a negative charge, as can be seen in a soap: $C_{17}H_{35}COO^-\ Na^+$, sodium stearate [50–52].

5.2 Cationic Surfactants

The second class of surfactants is the cationic surfactants. These surfactants are dissociated in water into an amphiphilic cation and an anion, most usually of the halide group. In general, cationic surfactants are more expensive than anionic surfactants, due to the high-pressure hydrogenation reaction involved during their synthesis process. Thus, cationic surfactants are exclusively used when cheaper substitutes are not available [48, 49, 51]. These moieties are positively charged, as can be seen in quaternary ammonium salts: $(C_{18}H_{37})_2\ N^+(CH_3)_2Cl^-$, dimethyl dioctadecyl ammonium chloride [8, 49, 53].

5.3 Zwitterionic Surfactants

There are circumstances where a surfactant molecule demonstrates both anionic and cationic dissociations. This class of surfactants is termed zwitterionic surfactants or amphoteric. Several natural substances exhibit this behavior; substances such as amino-acids and phospholipids, and other synthetic products like betaines or sulfobetaines [16, 49]. For these surfactants to solubilize, there must be positive and negative charge in the molecule, as can be seen in $C_{12}H_{35}N^{+}(CH_3)_2CH_2COO^{-}$, Ɓ-N-alkyl amino propionic acid [37, 50, 51].

5.4 Nonionic Surfactants

There is a class of surfactants that do not ionize in aqueous solutions and, thus, are called nonionic surfactants. This is because the hydrophilic group consists of a non-dissociated functional group. Exemplars of such functional groups include alcohol, phenol, ether, ester, or amide. By heavy usage, these surfactants come second with around 45% industrial production [16, 49]. The solubility of these surfactants is realized only when polyoxyethylene oxide groups solvate in water. Example of these surfactants are $C_9H_{19}C_6H_4(OCH_2CH_2)_9OH$ and nonyl phenol ethoxylate [37, 50, 51].

6 Surfactants as Demulsifiers

Due to their functionalities and several surface activities reported in many literatures [16, 49, 51], surfactants have been proven to be the integral parts of a chemical demulsifier. Chemicals (demulsifiers) are traditionally employed as interfacial tension (IFT) reducers. The effectiveness of such chemicals is enhanced by time, rate/strength of mixing, and temperature. Sufficient mixing and adequate time are required to achieve close contact of the chemical with the dispersed phase (either oil, in an oil-in-water emulsion, or water, in a water-in-oil emulsion). A specified minimum temperature is needed to guarantee the accomplishment of the functions of the chemical demulsifier. The reduction in viscosity as well as the effectiveness of chemical depends on fulfilling of a defined minimum temperature. The increase in chemical effectiveness may be a result of the decrease in viscosity of the oil phase [54]. One of the fundamental and principal reasons for the universal deployment of surfactants is their extraordinary ability to modify the properties of surfaces and interfaces, and thus, have an impact on technical processes and outcomes. The applicability of surfactants in the industry is quite distinct and covers a lot of areas. Surfactants are employed to advantage in the manufacturing industry, pharmaceuticals, agrochemicals, food processing, fuels, and adhesives, etc. Due to their surface activities, they

Table 1 Some examples of surfactant applications in the petroleum industry [50]

Gas/liquid systems	Producing oil well and well-head foams Oil flotation process froth Distillation and fractionation tower foams Fuel oil and jet fuel tank (truck) foams Foam drilling fluid Foam fracturing fluid Foam acidizing fluid Blocking and diverting foams Gas-mobility control foams
Liquid/liquid systems	Emulsion drilling fluids Enhanced oil recovery in situ emulsions Oil sand flotation process slurry Oil sand flotation process froths Well-head emulsions Heavy oil pipeline emulsion Fuel oil emulsions Asphalt emulsion Oil spill emulsions Tanker bilge emulsions
Liquid/solid systems	Reservoir wettability modifiers Reservoir fines stabilizers Tank/vessel sludge dispersants Drilling mud dispersants

have received a wide spectrum of applications in the oil and gas industry as well. Table 1 shows some of the applications of surfactants in the oil and gas industry.

The initial action in a well-organized emulsion breaking (demulsification) procedure is the characterization of the emulsion, whether it is an oil-in-water (O/W), water-in-oil (W/O) or multiple emulsion, the amount and type of immiscible phases, the existence of a protecting interfacial film round the dispersed droplets and the reactivity of the emulsifiers. The demulsification, therefore, involves two steps. Firstly, agglomeration or coagulation of the dispersed droplets must take place. Secondly, these agglomerated droplets must coalesce. Phase separation is only possible after two steps have occurred. It is worthy of note that, either of these two steps can determine the rate of demulsification of an emulsion. An emulsion can occasionally be broken by adjusting the temperature or by applying mechanical shear. More regularly, chemicals (demulsifiers) are applied to neutralize the impact of the emulsifier, accompanied by mechanical methods to conclude the phase separation [37, 55].

For this purpose, a wide range of chemical demulsifiers is available to effect this separation. These demulsifiers are classified based on their chemical structures, as discussed below.

6.1 Demulsifier Classifications and Selection

Generally, the chemicals employed in developing demulsifiers are classified according to their chemical structure, how they are applied, and the type of oil used in formulating them. Two major groups of chemicals are used: Non-ionic demulsifiers and Ionic demulsifiers. Although, classifying demulsifiers is not as easy as stated here, several w/o demulsifiers are polymeric nonionic chemicals, common with a complex comb or branched structures, having molecular weights (Mw) in the range of 2000–50,000 [56, 57]. Anionic and cationic polymers are used depending on the type of emulsifier that stabilized the emulsion, or as wetting agents. The most conventional classification of water-in-oil demulsifiers is summarized below.

- Polyalkoxylate block copolymers and ester derivatives
- Alkylphenol–aldehyde resin alkoxylates
- Polyalkoxylates of polyols or glycidyl ethers
- Polyamine polyalkoxylates and related cationic polymers (mainly for oil-in-water resolution)
- Polyurethanes (carbamates) and polyalkoxylate derivatives
- Hyperbranched polymers
- Vinyl polymers
- Polysilicones (also used as demulsifier boosters).

It is a fact that there is no universal demulsifier that can work for all emulsion problems. As the crude properties change over the life of a field, or from one reservoir to another, the efficiency of the demulsifier also change. Normally, at the beginning of a well's water production, the emulsions produced are tough to break. As the field matures and the amount of produced-water increases, the emulsion stability may change, so also the emulsifiers may change. Thus, it is conventional to examine demulsifier performance every 2–3 years [54]. Therefore, it is safe to say that demulsifiers are purpose-tailored, which means they are formulated with different primary active agents and precursors to serve the particular purpose they are designed for, depending on the type of emulsion (continuous phase) as well as the nature of emulsifiers. Some of the primary active agents in commercial demulsifiers include ethoxylated (either cross-linked or not) propylene oxide (PO)—ethylene oxide (EO) polymers or alkylphenol resins. The demulsifiers are formulated to give distinct properties, that includes including hydrophile-lipophile balance (HLB), solubility, the diffusion rate into the interface, and its efficacy at destabilizing the interface. Demulsifiers are normally added to the continuous phase, inside which they must then spread to the interface and interrupt the stabilizing film at the interface. The demulsifier should ordinarily be added adequately far, upstream to allow this process to take place, and so that droplet coalescence occurs before the emulsion get to the separating vessel [16, 53, 58].

There have been several studies with regards to how the structure of a demulsifier is related to its performance. For example, Zhang et al. [58] studied the demulsification of dendritic copolymers with the aim of establishing a correlation between their molecular structure and crude oil emulsions demulsifications. In a similar study, Wang et al. [59] studied the demulsification efficiency of six dendrimers with the same polyamidoamine (PAMAM) primary structure but distinct terminals. The dendrimers were investigated by surface tension estimation at the air–water interface. The results of the study showed that the dendrimers with proper structures have great potential for breaking crude oil emulsions.

Further in 2010, Wang et al. [60] further synthesized another set of six surfactants made of dendritic polyether using varying proportions of propylene oxide (PO) and ethylene oxide (EO) utilizing phenol-amine resin as a precursor. A study of surface tension and demulsification efficiency of these polyether surfactants shows that the six dendritic polyethers work like conventional surfactants. The findings from this study show that the structure of the polyethers can affect the demulsification efficiency, of which the principal determinant is the molecular block. Fink [61] expounded some of the common precursor chemicals that are used in the formulation of demulsifiers. Some of them are stated below, as outlined by Fink [61].

6.1.1 Polyalkylene Oxides

Polyalkylene oxides are precursor substances with the following general structure:

$$HO\text{-}(CH_2\text{-}CHR\text{-}O)_X\text{-}H$$

The most essential precursors are the oxides of polyethylene polypropylene oxide, and polybutylene. These oxides are also called polyalkylene glycol (PAG), although this name is only suitable strictly for derivatives of 1,2-diols.

6.1.2 Polyamines

Polyamines are normally open-chain compounds that have primary, secondary, or tertiary amino groups. On the other hand, polyimines can be used.

7 Requirements of a Demulsifier

A triumphant breaking of emulsion requires an effective technique. In this work, a chemical demulsification technique is being considered. Both the processes involved in the emulsification and demulsification of emulsions are complex. Nevertheless, the process of demulsification is by no means the reverse of emulsification [62, 63]. This is particularly the case in the petroleum industry. An efficient demulsification of water-in-crude oil emulsions requires that a clear understanding of the characteristics of the emulsion in question should be achieved. Also, of paramount importance is the nature of the film at the oil/water interface, hence the causes of emulsion stability. Consequently, choosing a demulsification protocol requires that the key factors responsible for the emulsion stability be identified, identify those properties of these stabilizers and how they can be modified in order to achieve destabilization. Finally, a source of sufficient energy that will promote coalescence must be identified and introduced into the system as well as finding the most favourable conditions that allow phase separation [47].

Angle [64] made a comprehensive review in one of his works, from which it would appear that the fundamental requirements on demulsifiers are the abilities to possess one or more of the following properties:

(1) strong affinity to the oil/water interface with the capability to destabilize the protecting film that forms around the dispersed droplet and/or to alter the contact angle of the solids which may be a component of the interfacial film;
(2) ability to flocculate the droplets;
(3) ability to increase coalescence by creating pathways for water's natural attraction to water; and
(4) improvement of film drainage and thinning of the inter-droplet lamella by producing changes to the interfacial rheology such as lowered interfacial viscosity and improved compressibility.

7.1 Major Types of Demulsifier Chemicals

In the Oil industry, the most employed polymer in demulsification process is a surfactant that possesses both hydrophilic and hydrophobic groups [65]. The polymeric surfactant when added to the crude oil emulsion place itself in the interface between the water and oil molecules. While the hydrophilic groups orient themselves towards the water, the hydrophobic ends orient themselves towards the oil [66].

Nowadays, the best polymeric surfactants used world over, are derivatives of alkoxylated materials [65, 67].

Since these polymeric surfactants are alkoxylated, they are considered nonionic polymers. However, there are times when mixtures of nonionic, cationic, or anionic materials are used together. This depends on the characteristics of the crude oil. Etoxylated nonionic surfactants are effective multipurpose and versatile substances. Commercial products are obtained by reaction of ethylene oxide with a hydroprobe having an active hydrogen group (e.g. fatty acids, alkylphenols or fatty alcohols) in the presence of suitable catalysts. In Appendix C, a brief history of chemical demulsifiers used in breaking w/o emulsions since the work of Barnickel [68] is presented.

A lot of work has been done by different researchers from academia as well as the oil industry on the characteristics of the major types of demulsifiers that are employed in breaking w/o emulsions. Reported in Table 2 are some of the characterization of the major types of demulsifier chemicals. The actions of a chemical demulsifier are stated later in this section, and no single chemical can provide all the four required

Table 2 Class of water-in-oil demulsifiers

Demulsifier class	Functions/comments
Polyalkoxylate block copolymers	The hydroxyl groups on the edges of these polymers can be ethoxylated with EO to form EO/PO/EO block copolymers, which are linear demulsifiers. These copolymers are moderately poor demulsifiers by themselves [69, 70]
Alkylphenol–aldehyde resin alkoxylates	This class has been in existence for many decades, because they are easy to manufacture and have good performance. Contains phenol groups, and as such are not environmentally friendly, also considered as endocrine disrupters in marine species [61, 71]
Polyesteramines and related polymers	They are extremely surface active, can resolve emulsions at low dosage, produce sharp o/w interface [72, 73]
Polyurethanes and polyalkoxylate derivatives	A well-known class that contains carbamate functional groups. Prepared by condensing polyisocyanates and polyglycols with terminal hydroxyl groups. They perform well at moderate concentrations, but they delay coalescence at higher concentrations. Also, environmentally not friendly [74, 75]
Hyperbranched polymers	They include hyperbranches of polyesteramides, polyurethanes, polyurea, polyamides, polyethers, polyesters and so on [85]. Some can be reasonably biodegradable while some are not and can function as demulsifiers for both o/w and w/o emulsions [76–78]
Polysilicones (also as demulsifier boosters)	Some can be very good demulsifiers (e.g. block copolymers of polyoxyalkylene-polysiolxane), while others (like dimethyl siloxane) can function as demulsifier boosters [79, 80]
Oxyalkylated phenols	They are good wetting agents, used in blends, reduce interface pad build-up and are fair to poor demulsifiers [80, 81]

actions. Thus, commercial demulsifiers are exclusive mixtures of these compounds (i.e., surfactants and additives) [34, 62].

However, it is an acceptable practice to blend the selected mixtures (30–60%) of proprietary, solid chemicals with a suitable solvent such as heavy aromatic naphtha (HAN) or isopropyl alcohol (IPA) to obtain a liquid that pours at the lowest expected temperature. W/O demulsifiers are always very insoluble in water and highly soluble in oil so that they can diffuse rapidly through the continuous crude oil phase and reach the droplet interfaces [82–84]. Typically, these oil soluble demulsifiers are formulated in organic solvents alone such as toluene, xylene, tetrahydrofuran, dioxane, lower alcohols and light gasoline fractions having boiling limits from 50 to 200 °C.

Sometimes, they are formulated in co-solvents comprising organic solvents and water where in the organic solvents are usually C_3–C_{10} alkanols, ethylene diamine, diethyl triamine or ethanolamines including diethanol-amine [40].

However, using organic solvents in a demulsifier formulation has several disadvantages like increased cost, flammability, and toxicity. Thus, a demulsifier formulation, which does not include organic solvents would represent an improvement in the art of demulsification [65].

7.2 Functions of an Effective Demulsifier

An effective chemical demulsifier performs the following actions [34, 85]:

1. It must have a strong attraction to the water–oil interface. The demulsifier must displace and/or neutralize the action and/or position of the emulsifying agents already on the droplet surface.
2. Flocculation action: The demulsifier must neutralize any repulsive electrical charges between the dispersed water droplets and so allow the droplets to touch each other.
3. Coalescence action: The demulsifier must permit small droplets to combine and form large drops; large enough to settle. This requires that the film surrounding and stabilizing the droplets is being ruptured.
4. Wetting ability (solids wetting): Demulsifiers must prevent fines at the droplet interface from physically blocking coalescence. Clays, drilling muds, and iron sulfides fines can be water wet so that they leave the film interface and migrate into the water droplet. Asphaltenes and waxes can be dissolved or oil wet to disperse them into the continuous crude phase.

7.3 Mechanism of Chemical Demulsification

Theories of how demulsifiers perform are incomplete [34]. Their incompleteness is due to the failure of the theories to explain the extreme specificity of the various types of chemicals on any crude-water mixture. The wide variety of reagents needed for effective handling of different emulsions suggests that the mechanisms of performance of the chemical reagents on the film surrounding the water droplets are complex and cannot be adequately explained using a single theory [86]. However, two valid generalizations concerning the mechanism of the chemical demulsifiers hold. First, effective demulsifiers have high molecular weights that are comparable to those of the natural surfactants. Secondly, if used as emulsifying agents, they tend to cause emulsions opposite in type (i.e., o/w) to the produced w/o emulsions [48, 49].

Yang et al. [79] in their study found out that the demulsification process can occur via three mechanisms. (1) They can absorb at the w/o interface, displace indigenous surfactant into bulk phases and make interfacial film sufficiently weak and promote coalescence. (2) They also can act as flocculating agents, joining together small droplets, which then coalesce and grow. (3) Alternatively, or simultaneously, they can act as wetting agents and wet fine solids present in the crude oil, hence reduce their ability to stabilize emulsions.

However, other traditional theories as to why demulsifiers work also exist [26]. According to these theories, the demulsifier 'neutralizes' the emulsifying agent; in other words, breaking a w/o emulsion requires a chemical that would normally produce a reverse or o/w emulsion. Another explanation is that the demulsifying chemical makes the film surrounding the water droplet very rigid. When the oil drop expands on being heated, the film is ruptured. Alternatively, if the chemical makes the film to diminish, then heat is not required to burst the film [34, 42].

Nuraini et al. [81] in their study made a careful selection of four groups of demulsifiers; i.e. amine, natural, polyhydric, and alcohol demulsifier groups aiding in breaking of stable emulsions. They found out that amine demulsifier groups demonstrated the highest efficiency in breaking the emulsion, when compared to polyhydric, alcohol, and natural groups. They also found out that a demulsifier efficiency depends on two-factors; solubility of demulsifier (either in water or oil) and molecular weight of demulsifier. Figure 3 presents a chemical demulsification mechanism.

Krawczyk et al. [87] in a study of chemical demulsification of petroleum emulsions investigated the factors affecting the coalescence and interfacial behavior of w/o emulsions in the presence of oil soluble demulsifiers. They attributed the performance of a chemical demulsifier to the rupture of the thin film separating droplets,

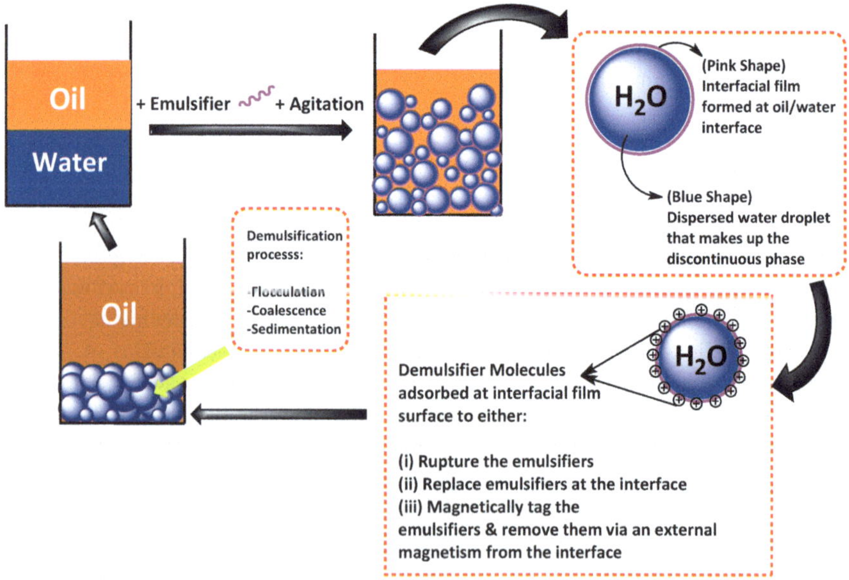

Fig. 3 Chemical demulsification mechanism

which affects the rate of coalescence of the droplets, and that the higher the rate of adsorption, the more effective the demulsifier is.

Kim [30] reported a novel procedure for measuring rheological properties of a water/oil/water film, characteristics of which actively determine whether demulsification occurs or not, and the mechanism of demulsification. Determinations of dynamic interfacial tension and activities with a standard drop-volume apparatus support the water/oil/water film results. Although the measurements of static interfacial tension could not correlate with the demulsifier performance data, the data, nonetheless, provide valuable insight into physicochemical mechanisms. The study shows that the rapidly dispersing, low molecular weight components of demulsifiers break emulsions by maintaining a low dynamic interfacial tension during film drainage.

In another study, Kang et al. [88] investigated the influence of demulsifier on interfacial film between oil and water and used that to explain the mechanism of demulsification. The study established that the mechanism exhibited by these demulsifiers is partial replacement of the emulsifiers, which led to the interfacial elasticity decrease.

The demulsification mechanism reported by Erica et al. [89] in a study of demulsification mechanism of asphaltene-stabilized w/o emulsions by a polymeric EO-PO demulsifier described gradual infiltration of the demulsifier into the asphaltene film. The demulsifier was more surface-active than asphaltenes and thus exhibited an effective competition with asphaltenes to occupy the interface, penetrating asphaltene films and softening them under shear and compression. The softness of asphaltenes films was found to increase with increasing demulsifier leading to faster kinetics of asphaltene film penetration at higher demulsifier dosages.

7.4 *Formulation of a Chemical Demulsifier*

Chemical demulsifiers are surfactants that can destabilize emulsions. For a demulsifier to perform effectively, it must counteract the emulsifying agent stabilizing the emulsion, and also promote aggregation and coalescence of the dispersed phase into large droplets that can settle out of the continuous phase [9, 47, 90]. Since chemical demulsifiers are surfactants; meaning they are surface active, they usually possess a reasonable degree of toxicity (EC50 < 10 mg/l) [57]. This is mostly applicable to cationic surfactants that are used in many production chemical products.

When formulating a demulsifier, certain properties must be put in mind. The demulsifier should have a strong attraction to the interface (good surfactant properties) and should be able to migrate rapidly through the continuous phase (in other words, should be oil-soluble) to reach the droplet interface [37, 50]. After the demulsifier concentrate at the oil/water interface, it counteracts the emulsifying agent and promotes flocculation of the dispersed phase. In the flocculated system, the emulsifier film is still continuous, thus it is a requirement that the demulsifier must neutralize the emulsifier and facilitate the rupture of the droplet interfacial film, leading to coalescence.

Addition of ethylene oxide and/or propylene oxide to formaldehyde resins and to diamines and higher functional amines (triamines, tetramines etc.) yields a variety of modified polymers that perform well at relatively low concentrations [50].

The low molecular weight demulsifiers can be transformed into high-molecular-weight products by reactions with diacids, diepoxides, di-isocynates, and aldehydes. This gives room for tailoring demulsifier chemistry to accommodate various oil gravities and surfactant properties, and to adjust surface activity and the rate at which demulsifiers move to the interface [37].

Fatty amine derivatives as surfactants have been studied for their biodegradability. Toxic surfactants would kill bacteria, which can affect the biodegradation results. (Biodegradation tests are usually carried out at 2–40 mg/l but toxicity tests are at µg/l). Generally, the longer the alkyl tail, the more toxic is the surfactant. Below a single tail of 8–10 carbon atoms (depending on the hydrophilic-lipophilic balance), the toxicity decreases significantly [57, 91].

8 Comparison Between Two Different Demulsifiers with Different Interfacial Properties (Interfacial Activity)

In order to describe how the surface activity of a demulsifier affects its performance, two different chemical demulsifiers were tested, after their interfacial tensions between oil and water were determined. An optical analyzer TURBISCAN was used to determine the performance of the demulsifiers and the result is shown below.

As discussed in Sect. 5, for a chemical to qualify as an effective chemical demulsifier, it must possess certain properties, one of which is its ability to exhibit surface activity. In a study of demulsification of water-in-crude oil emulsions, [92] found out that surfactants effective as demulsifiers reduced surface tension of water by more than 25 dynes-cm^{-1}. By exhibiting interfacial activity, it means that when the compound is added to a liquid at low concentration, it should be able to adsorb on the surface or interface of the system and reduce the surface or interfacial excess free energy, and a surface activity is achieved when the number of carbon atoms in the hydrophobic tail are higher than 8 [93]. There have been several efforts to establish a correlation between the performance of a demulsifier and other demulsifier properties like their molecular structures, hydrophilic-lipophilic balance (HLB), of the surfactants used as demulsifiers, interfacial tension, and so on. This work does not claim to achieve such, and only makes use of an established criterion of the ability of the demulsifier to reduce the surface tension of water by 25–30 dynes-cm^{-1} [93] as a good demulsifier.

The equilibrium interfacial tension measurements as well as the TURBISCAN procedure were carried out using the method described in our recent publication [29]. Two different commercial demulsifiers, DPG 144 and DTG 123 were used in the study. The result showed that DPG 144 reduced the water surface tension by 27 dynes cm^{-1} (27×10^{-4} mN/m) while DTG 123 reduced by surface tension of water by barely 10–11 dynes cm^{-1}. To further verify the performance of these two demulsifiers, the TURBISCAN test was run and the results are shown below.

Figure 4 shows the destabilization profile of a water-in-crude emulsion in the presence of DTG 123. The presence of wavy pattern at the bottom of the sample suggests that this part of the sample is inhomogeneous; and it is represented by 0–18 mm of the sample bottle. Beyond that point is an additional behavioral feature reported as the second wall effect as explained by Akther et al. 2007 [94], from 18 to 32 mm of the bottle length. The top of the sample is however characterized by "clarification", where oil migrate from the bottom of the sample to the top, causing a fall in backscattering. It is believed that the phase separation caused a line pattern (striations to form on the wall of the cell). As there is no significant increase or decrease at the bottom of the sample, the water phase is mixed with a bit of oil, thus would require longer time to separate. This suggests that the demulsifier is not very effective.

Figures 5 shows the backscattering profiles of a water-in-crude oil. The same concentration of DTG 123 used in the first instance is used here for DPG 144. It can be seen from Fig. 5 that the emulsion ceases to be stable and a lot of water settles at the bottom of the bottle. From 0 mm to around 26 mm of the sample bottle witnesses settling of water (defined by increase in signal—sedimentation). Beyond this point, there is a drop in backscattering (from 26 to 40 mm of the sample bottle). This phase is the clarification zone, expressed by drop in signal due to the presence of oil phase absorbing the light. The ability of DPG 144 demulsifier to break more water than DTG 123 suggests that it is a better demulsifier.

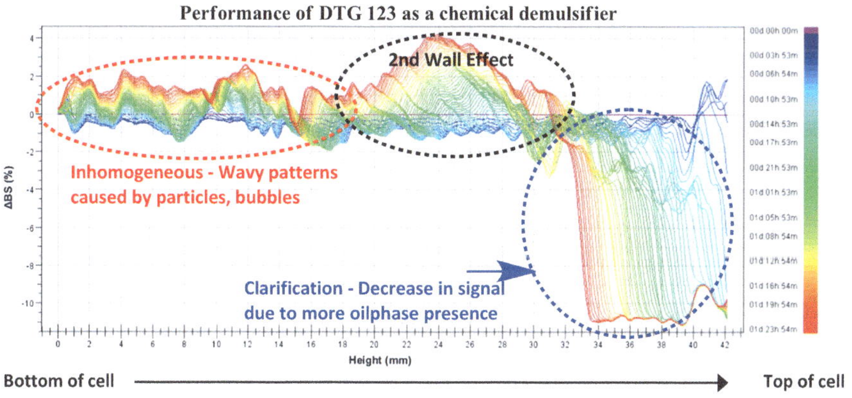

Fig. 4 Delta Backscattering profile for a stable water-in-crude emulsion evaluated by the Turbiscan Lab® AGS. Data are reported as a function of time (0–48 h) and sample height of the emulsion sample (0–42 mm)

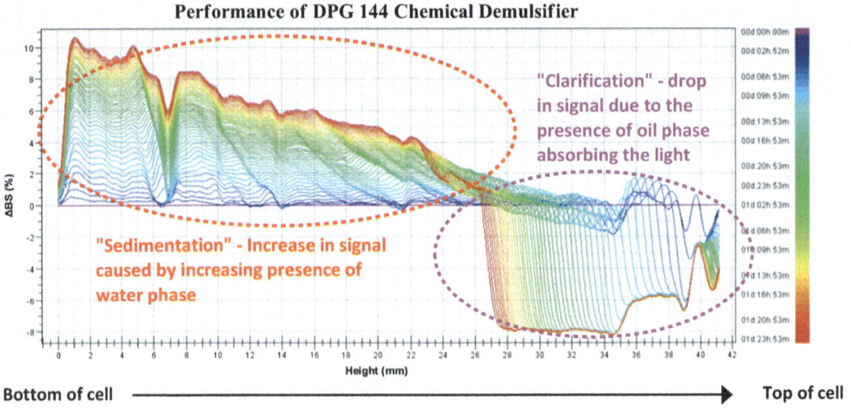

Fig. 5 Delta Backscattering profile for a stable water-in-crude emulsion evaluated by the Turbiscan Lab® AGS. Data are reported as a function of time (0–48 h) and sample height of the emulsion sample (0–42 mm)

9 Conclusion

As a conclusion, studies have shown that surfactants exhibit surface activity, which means their molecules will adsorb preferentially at the interfaces of air and water, oil and water, and solid/water. Surfactants are amphiphilic or amphipathic, and thus they can dissolve in nonpolar medium with the other part in the polar medium. Surfactant molecules form oriented monolayers at interfaces and show surface activity. Owing to their characteristic surface activity and tendency for molecular self-assembly, they can control and modify both the physical and chemical properties of the phase boundary between different phases of liquids encountered in almost all chemical reactions. This ability makes them the integral components in a demulsifier formulation. Although lowering in interfacial tension and adsorption of demulsifier at the crude oil/water interface is not the final and most important condition that determines a demulsifier efficiency, it has been shown in this study that the surface activity of a surfactant is a good condition for its efficiency as a demulsifier.

References

1. Kokal SL (2005) Crude oil emulsions: a state-of-the-art review. SPE Prod Fac 20(01):5–13
2. Kokal SL, Al Ghamdi A (2006) Oil/water separation experience from a large oil field. SPE Product Oper 21(03):365–371
3. Langevin D et al (2004) Crude oil emulsion properties and their application to heavy oil transportation. Oil Gas Sci Technol 59(5):511–521
4. Fingas M, Fieldhouse B (2009) Studies on crude oil and petroleum product emulsions: water resolution and rheology. Colloids Surf A 333(1):67–81

5. Fingas M, Fieldhouse B (2014) Water-in-oil emulsions: formation and prediction. Spill Sci Technol 225
6. Becher P (1988) Basic theory, measurement and applications. In: Encyclopedia of emulsion technology, vol 3. Dekker, New York
7. Sheu EY, Mullins OC (1998) Structures and dynamics of asphaltenes. Plenum Press, pp 337–423
8. Schramm LL (1992) Petroleum emulsions. ACS Publication
9. Sjoblom J (2005) Emulsions and emulsion stability: surfactant science series/61-Chp 14, vol 132. CRC Press
10. Clayton W (1923) The theory of emulsions and emulsification. J. & A. Churchill
11. Chevalier Y, Bolzinger M-A (2013) Emulsions stabilized with solid nanoparticles: pickering emulsions. Colloids Surf Physicochem Eng Aspects 439:23–34
12. Abdel-Raouf M (2012) Crude oil emulsions: composition stability and characterization. BoD–Books on Demand
13. Umar AA et al (2018) A review of petroleum emulsions and recent progress on water-in-crude oil emulsions stabilized by natural surfactants and solids. J Petrol Sci Eng 165:673–690
14. Umar A, Saaid I, Sulaimon A (2017) The roles of polar compounds in the stability and flow behavior of water-in-oil emulsions. ICIPEG 2016. Springer, pp 643–653
15. Urdahl O, Sjöblom J (1995) Water-in-crude oil emulsions from the Norwegian continental shelf. A stabilization and destabilization study. J Dispersion Sci Technol 16(7):557–574
16. Schramm LL (2000) Surfactants: fundamentals and applications in the petroleum industry. In: Surfactants in solutions. Cambridge University Press
17. Dingcong W (2002) A study of identifying the emulsion type of surfactant: volume balance value. J Colloid Interface Sci 247(2):389–396
18. Pasquali RC, Sacco N, Bregni C (2009) The studies on hydrophilic–lipophilic balance (HLB): sixty years after William C. Griffin's pioneer work (1949–2009). Lat Am J Pharm 28(2):313–317
19. Zafeiri I et al (2017) Emulsions co-stabilised by edible pickering particles and surfactants: the effect of HLB value. Colloid Interface Sci Commun 17:5–9
20. Surfluh J (1937) Prevention of oil-field emulsions. In: Drilling and production practice. American Petroleum Institute
21. Powell KC, Damitz R, Chauhan A (2017) Relating emulsion stability to interfacial properties for pharmaceutical emulsions stabilized by Pluronic F68 surfactant. Int J Pharm 521(1):8–18
22. Alvarado V, Wang X, Moradi M (2011) Stability proxies for water-in-oil emulsions and implications in aqueous-based enhanced oil recovery. Energies 4(7):1058–1086
23. McClements DJ, Jafari SM (2017) Improving emulsion formation, stability and performance using mixed emulsifiers: a review. Adv Colloid Interface Sci
24. Kokal S, Wingrove M (2000) Emulsion separation index: from laboratory to field case studies. In: SPE annual technical conference and exhibition. Society of Petroleum Engineers
25. Kokal S (2002) Crude oil emulsions: a state-of-the-art review. In: SPE annual technical conference and exhibition. Society of Petroleum Engineers
26. Jones T, Neustadter E, Whittingham K (1978) Water-in-crude oil emulsion stability and emulsion destabilization by chemical demulsifiers. J Can Petrol Technol 17(02)
27. Strassner J (1968) Effect of pH on interfacial films and stability of crude oil-water emulsions. J Petrol Technol 20(03):303–312
28. Moroney JR, Berg MD (1999) An integrated model of oil production. Energy J 20(1)
29. Umar AA et al (2020) Magnetic polyester bis-MPA dendron nanohybrid demulsifier can effectively break water-in-crude oil emulsions. J Market Res 9(6):13411–13424
30. Kim Y-H, Wasan D, Breen P (1995) A study of dynamic interfacial mechanisms for demulsification of water-in-oil emulsions. J Colloids Surf A Physicochem Eng Aspects 95(2–3):235–247
31. Gafonova OV, Yarranton HW (2001) The stabilization of water-in-hydrocarbon emulsions by asphaltenes and resins. J Colloid Interface Sci 241(2):469–478
32. Myers D (1990) Surfaces, interfaces and colloids. Wiley-Vch, New York etc.

33. Thompson D, Taylor A, Graham D (1985) Emulsification and demulsification related to crude oil production. Colloids Surf 15:175–189
34. Manning FS, Thompson RE (1991) Oilfield processing of petroleum: natural gas, vol 1. Pennwell books
35. Feng X et al (2009) Mechanistic study on demulsification of water-in-diluted bitumen emulsions by ethylcellulose. Langmuir 26(5):3050–3057
36. Meyer P (1964) Chemelectric treating a new phase in the electrical dehydration of oil emulsions. In: SPE production automation symposium. Society of Petroleum Engineers, Kansas
37. Schramm LL (2014) Emulsions, foams, suspensions, and aerosols: microscience and applications. Wiley
38. Monson T (1969) Chemical resolution of emulsions. In: Surface operations in petroleum production, II. Elsevier
39. Becker J (1997) Crude oil waxes, emulsions, and asphaltenes. Pennwell Books
40. Merchant Jr P, Lacy SM (1988) Water based demulsifier formulation and process for its use in dewatering and desalting crude hydrocarbon oils. Google Patents
41. Kim YH, Wasan DT (1996) Effect of demulsifier partitioning on the destabilization of water-in-oil emulsions. Ind Eng Chem Res 35(4):1141–1149
42. Abdel-Azim AAA , Zaki NN, Maysour NES (1998) Polyoxyalkylenated amines for breaking water-in-oil emulsions: effect of structural variations on the demulsification efficiency. Polym Adv Technol 9(2):159–166
43. Dalmazzone C, Noïk C (2001) Development of new" green" demulsifiers for oil production. In: SPE international symposium on oilfield chemistry. Society of Petroleum Engineers
44. Dalmazzone C, Noik C, Komunjer L (2005) Mechanism of crude-oil/water interface destabilization by silicone demulsifiers. SPE J 10(01):44–53
45. Darling DF (1982) Recent advances in the destabilization of dairy emulsions. J Dairy Res 49(4):695–712
46. Mat H et al (2005) Study on demulsifier formulation for treating Malaysian crude oil emulsion
47. Sjoblom J (2001) Encyclopedic handbook of emulsion technology. CRC Press
48. Salager J-L (1990) The fundamental basis for the action of a chemical dehydrant. Influence of the physical and chemical formulation on the stability of an emulsion. Int Chem Eng 30(1):103–116
49. Salager J-L (2002) Surfactants types and uses. FIRP booklet, (E300A)
50. Schramm LL, Stasiuk EN, Marangoni DG (2003) 2 Surfactants and their applications. Annu Rep Sect "C" (Phys Chem) 99:3–48
51. Migahed M, Al-Sabagh A (2009) Beneficial role of surfactants as corrosion inhibitors in petroleum industry: a review article. Chem Eng Commun 196(9):1054–1075
52. Fingas M (2014) Handbook of oil spill science and technology, vol 1. Wiley
53. Shehzad F, Hussein IA, Kamal MS, Ahmad W, Sultan AS, Nasser MS (2018) Polymeric surfactants and emerging alternatives used in the demulsification of produced water: a review. Polym Rev 58(1):63–101
54. Maurice S, Ken A (2008) Emulsions and oil treating equipment. Gulf Professional Publishing, Hardbound
55. Li Y et al (2017) A versatile and efficient approach to separate both surfactant-stabilized water-in-oil and oil-in-water emulsions. Sep Purif Technol 176:1–7
56. Bhardwaj A, Hartland S (1994) Kinetics of coalescence of water droplets in water-in-crude oil emulsions. J Dispers Sci Technol 15(2):133–146
57. Kelland MA (2014) Production chemicals for the oil and gas industry. CRC press
58. Zhang Z, Xu G, Wang F, Dong S, Chen Y (2005) Demulsification by amphiphilic dendrimer copolymers. J Colloid Interface Sci 282(1–4)
59. Wang J, Li C-Q, Li J, Yang J-Z (2007) Demulsification of crude oil emulsion using polyamidoamine dendrimers. Sep Sci Technol 42(9):2111–2120
60. Wang J, Hu F-L, Li C-Q, Li J, Yang Y (2010) Synthesis of dendritic polyether surfactants for demulsification. Sep Purif Technol 73(3):349–354

61. Fink J (2015) Petroleum engineer's guide to oil field chemicals and fluids Gulf Professional Publishing, Chp 23, p 705
62. Matijevic E, Good RJ (2012) Surface and colloid science, vol 12. Springer Science & Business Media
63. Sjöblom J (2012) Emulsions: a fundamental and practical approach, vol 363. Springer Science & Business Media
64. Angle CW (2001) Chemical demulsification of stable crude oil and bitumen emulsions in petroleum recovery—a review. Marcel Dekker, New York, pp 541–594
65. Hanapi B et al (2006) Study on demulsifier formulation for treating Malaysian crude oil emulsion. Internet website. http://eprints.utm.my/2768/1/74004.pdf
66. Fink J (2003) Oil field chemicals. Elsevier
67. Al-Sabagh AM et al (2013) Synthesis and evaluation of new demulsifiers incorporating linear alkyl benzene moiety for treating water-in-oil emulsion. J Dispers Sci Technol 34(7):996–1007
68. Barnickel WS (1922) Apparatus for treating natural emulsions of petroleum. Google Patents
69. Mansur CR et al (2004) PLURONIC× TETRONIC polyols: study of their properties and performance in the destabilization of emulsions formed in the petroleum industry. J Colloid Interface Sci 271(1):232–240
70. Alvarez F et al (2011) Dissipative particle dynamics (DPD) study of crude oil– water emulsions in the presence of a functionalized co-polymer. Energy Fuels 25(2):562–567
71. Beyer J et al (2001) Ecotoxicology of oilfield chemicals: the relevance of evaluating low-dose and long-term impact on fish and invertebrates in marine recipients. In: SPE international symposium on oilfield chemistry. Society of Petroleum Engineers
72. Al-Sabagh AM, Kandile NG, Noor El-Din MR (2011) Functions of demulsifiers in the petroleum industry. Sep Sci Technol 46(7):1144–1163
73. Ali N et al (2014) Synthesis of P (MMA-co-AA-co-DVB) Fe 3 O 4/magnetic composite nanoparticles. In: Proceedings of 2014 11th international Bhurban conference on applied sciences & technology (IBCAST). IEEE, Islamabad, Pakistan
74. Reeve PFD (2002) Method of inhibiting the formation of oil and water emulsions. Google Patents
75. Peña AA, Hirasaki GJ, Miller CA (2005) Chemically induced destabilization of water-in-crude oil emulsions. Ind Eng Chem Res 44(5):1139–1149
76. Feng X et al (2008) Biodegradable polymer for demulsification of water-in-bitumen emulsions. Energy Fuels 23(1):451–456
77. Killat GR, Conklin JR (1984) Use of quaternized polyamidoamines as demulsifiers. Google Patents
78. Conklin JR, Wilson LR (1984) Use of heterocyclic ammonium polyamidoamines as demulsifiers. Google Patent
79. Yang M, Stewart A, Davies G (1996) Interactions between chemical additives and their effects on emulsion separation. In: SPE annual technical conference and exhibition. Society of Petroleum Engineers
80. Pensini E et al (2014) Demulsification mechanism of asphaltene-stabilized water-in-oil emulsions by a polymeric ethylene oxide–propylene oxide demulsifier. J Energy Fuels 28(11):6760–6771
81. Nurainia M, Abdurahmanab H, Kholijaha A (2011) Effect of chemical breaking agents on water-in-crude oil emulsion system. Int J 2(4)
82. Madaan V et al (2014) Emulsion technology and recent trends in emulsion applications. Int Res J Pharm 5(7):533–542
83. McMahon AJ (1992) Interfacial aspects of water-in-crude oil emulsion stability. Emulsions—a fundamental and practical approach. Springer, pp 135–156
84. Manousakis M, Avranas A (2013) Dynamic surface tension studies of mixtures of hydroxypropylmethylcellulose with the double chain cationic surfactants didodecyldimethylammonium bromide and ditetradecyldimethylammonium bromide. J Colloid Interface Sci 402:237–245

85. Fink J (2015) Petroleum engineer's guide to oil field chemicals and fluids. Gulf Professional Publishing. GPP
86. Müller FH, Weiss A (2007) Emulsions, vol 63. Steinkopff
87. Krawczyk MA, Wasan DT, Shetty CJI (1991) Chemical demulsification of petroleum emulsions using oil-soluble demulsifiers. J Indus Eng Chem Res 30(2):367–375
88. Kang W et al (2006) Influence of demulsifier on interfacial film between oil and water. J Colloids Surf A Physicochem Eng Aspects 272(1–2):27–31
89. Pensini E et al (2014) Demulsification mechanism of asphaltene-stabilized water-in-oil emulsions by a polymeric ethylene oxide–propylene oxide demulsifier. Energy Fuels 28(11):6760–6771
90. Hunter TN et al (2008) The role of particles in stabilising foams and emulsions. Adv Coll Interface Sci 137(2):57–81
91. Kilpatrick PK (2012) Water-in-crude oil emulsion stabilization: review and unanswered questions. Energy Fuels 26(7):4017–4026
92. Bhardwaj A, Hartland S (1993) Study of demulsification of water-in-crude oil emulsion. J Dispers Sci Technol 14(5):541–557
93. Farn RJ (2008) Chemistry and technology of surfactants. Wiley
94. Akther S, Hwang J-Y, Lee H-M (2007) Effects of the water quality on the dispersion properties of bentonites used for drilling fluid. J Mineral Soc Korea 20(1):21–33

Conclusion

This book provides a compiled detailed account of surfactants in the upstream oil industry, and potential applications in drilling, cEOR, conformance control, gas injection, stimulation, corrosion inhibition, hydrate inhibition, and demulsification are defined precisely. Water-based drilling fluids are widely accepted in drilling operations due to superior rheology and filtration properties, enhanced shale inhibition characteristics, and environmentally friendly nature. The selection of surfactants for inhibition of swelling and hydration mainly depends on reservoir properties, types of swelling clay content in the reservoir formation, and the interaction of the surfactant molecules with the formation rock. Usually, all kinds of surfactants inhibit shale swelling; however, the use of cationic gemini surfactants in the formulation of water-based drilling fluids resulted in improved shale inhibition properties. Surfactant flooding in EOR is usually challenged by the surfactant adsorption and harsh reservoir conditions. Field trials of biosurfactant floods have led to increased recovery from brown fields in the last two decades. However, more research is required to select optimal microbial strains, to produce new and more efficient biosurfactants and to develop methodologies for recycling of biosurfactants. In addition, low salinity water flooding with conventional surfactant flooding can be an effective EOR solution, and laboratory results have shown significant incremental oil recovery (32% of OOIP). The presence of divalent ions and pH of the low salinity surfactant solution lower the interfacial tension and change the wettability of the reservoir rock from oil-wet to water-wet regime. Conformance problems occur when the injected fluids pass through the high permeability zone and make channels through reservoir rocks, decreasing the sweep efficiency. The bentonite clay modification using ammonium-based cationic surfactants have shown satisfactory results and exhibited potential applicability as a conformance control system. In foam displacement for better oil recovery, foamability and foam stability are considered significant concerns. CO_2-philic surfactants are used to overcome foam stability, as well as they also handle CO_2 mobility issues. The headgroup of CO_2-philic surfactants interact with water, and the tail has an affinity for CO_2 to attain a certain degree of stability for the foam. Hydraulic fracturing is becoming an increasingly sought topic of creating an extensive fracture network in low permeability formations and achieving economic

T. Solling et al. (eds.), *Surfactants in Upstream E&P*, Petroleum Engineering,
https://doi.org/10.1007/978-3-030-70026-3

production results. However, groundwater can enter into the oil-producing zones during this process, which can have adverse repercussions on both production and environmental aspects. The introduction of surfactant in fracturing operation help in mitigating stability, flow back, and water-blockage issues. Surfactants are useful clean-up additives and reduce the number of residues or precipitates remaining within the reservoir formation post-application. Hydrochloric acid (HCl) is widely used in large industrial applications such as cleaning, pickling, descaling, and etching of mild steel. Therefore, corrosion inhibitors are essential for saving machinery, particularly those made mostly of mild steel. Different types of surfactants have been used as a corrosion inhibitors; gemini surfactants are new and have gained significant interest in corrosion inhibition in recent years. Quaternary ammonium-based surfactants have exhibited high inhibition efficiency in acidic media up to 97% for carbon steel corrosion. Gas hydrates are the ice-like structures that arise by an interaction between gas molecules and water to result in blockages in oil and gas pipelines. Non-ionic surfactants such as Surfynol 465 and ethoxylated fatty alcohols demonstrated inhibition of hydrate growth. The production of water during crude oil and gas production is also a challenging task in the oilfield industry that impacts the operation cost because water production enhances with time. Non-ionic surfactants, as a demulsifier, proved to have excellent efficiency in breaking water in oil emulsion.

The use of surfactants for specific oilfield applications is preceded by careful screening; this includes surfactant type (cationic, anionic, non-ionic or zwitterionic), chemical structure with specific functionalities, and development of the structure-property relationship. Further research should expand the reservoir properties, clay types, interaction of surfactant molecules with the formation rock, salinity, and temperature, which will help design new surfactant-based materials. Hopefully the wide variety will provide the reader with proper inspiration to guide the design of increasingly more advanced surfactant molecules.

The manufacturer's authorised representative in the EU is Springer Nature Customer Service Centre GmbH, Europaplatz 3, 69115 Heidelberg, Germany. If you have any concerns regarding our products, please contact ProductSafety@springernature.com

Printed and bound by CPI Group (UK) Ltd, Croydon, CR0 4YY
30/09/2026
02225695-0017